WE WANT TO HEAR FROM YOU!!

By sharing your opinions about this book, you will help us ensure that you are getting the most value for your textbook dollars. After you've used the book for awhile, please fill out this form fold, tape and drop it in the mail.

Course Title:_____ Text Title & Author: _____

1. Are you a major in this subject? ❏ yes ❏ no ❏ undecided
   Were you required to purchase this text? ❏ yes ❏ no

2. Did you purchase this book: ❏ for yourself? ❏ for yourself and at least one other student?
   Was a copy available when you needed one? ❏ yes ❏ no

3. Was a study guide available for purchase? ❏ yes ❏ no ❏ don't know
   If yes, did you purchase it? ❏ yes ❏ no
   Might you purchase it in the future? ❏ yes ❏ no

4. Were any other supplements to the text available (for example, software, a workbook, etc.)?
   ❏ yes ❏ no   If yes, what? _____
   Did you purchase any other supplement? ❏ yes ❏ no

5. How far along in the course are you? ❏ only starting ❏ less than midway
   ❏ more than midway ❏ completed

6. How much have you used this text? ❏ only skimmed it ❏ read/studied a few chapters
   ❏ read/studied most chapters ❏ read/studied entire text

7. Have you read the introductory material (such as the preface)? ❏ yes ❏ no
   Do you feel you know how to effectively use this book? ❏ yes ❏ no

8. Even if you've only skimmed the text, please rate your perception of it in terms of the following:

a) Value as a reference	❏ highly valuable	❏ somewhat valuable	❏ not valuable
b) Readability	❏ consistently clear	❏ sometimes clear	❏ generally unclear
c) Illustrations/photos	❏ very effective	❏ somewhat effective	❏ ineffective
d) Design/use of color	❏ very effective	❏ somewhat effective	❏ ineffective
e) Study help in the text	❏ very effective	❏ somewhat effective	❏ ineffective
f) Level	❏ too difficult	❏ appropriate	❏ too easy/not challenging
g) Problems	❏ too difficult	❏ appropriate	❏ too easy/not challenging
h) OVERALL PERCEPTION:	❏ better than average	❏ average	❏ less than average

9. Do you find the examples in the text relevant to you? ❏ yes ❏ no
   Note any that you find particularly relevant _____

10. By looking at the text, do you think it treats the subject as interestingly as possible?
    ❏ yes ❏ no ❏ hard to tell

11. What do you like most about this book?_____
    What *don't* you like about this book? _____

12. At the end of the semester, what do you intend to do with this text?
    ❏ keep for future reference ❏ sell back to bookstore or other students ❏ unsure

THANK YOU FOR YOUR HELP!

WILEY

The Sciences

An Integrated Approach

James Trefil

Robert M. Hazen

George Mason University

John Wiley & Sons, Inc.

New York • Chichester • Brisbane • Toronto • Singapore

ACQUISITIONS EDITOR Kaye Pace
DEVELOPMENTAL EDITOR Barbara Heaney
MARKETING MANAGER Catherine Faduska
SENIOR PRODUCTION EDITOR Nancy Prinz
DESIGN A Good Thing, Inc.
MANUFACTURING MANAGER Susan Stetzer
PHOTO EDITOR Lisa Passmore
ILLUSTRATION COORDINATER Edward Starr
COVER PHOTOGRAPH Pat O'Hara
This book was set in Stone Serif by Bi-Comp, Inc. and printed and
bound by Von Hoffmann Press, Inc.

Recognizing the importance of preserving what has been written, it is a
policy of John Wiley & Sons, Inc. to have books of enduring value published
in the United States printed on acid-free paper, and we exert our best
efforts to that end.

John Wiley & Sons, Inc., places great value on the environment and is actively involved in efforts to
preserve it. Currently, paper of high enough quality to reproduce full-color art effectively contains a
maximum of 10% recovered and recycled post-consumer fiber. Wherever possible, Wiley uses paper
containing the maximum amount of recycled fibers. In addition, the paper in this book was manufactured by a mill whose forest management programs include sustained yield harvesting of its timber-
lands. Sustained yield harvesting principles ensure that the number of trees cut each year does not
exceed the amount of new growth.

TOTAL 10% RECYCLED PAPER
ALL POST-CONSUMER WASTE

Library of Congress Cataloging in Publication Data:
Trefil, James S., 1938—
    The sciences : an integrated approach / James Trefil and Robert M.
Hazen.
        p.   cm.
    Includes bibliographical references (p.   ).
    ISBN 0-471-58931-4 (pbk. : alk. paper)
    1. Science. 2. Science—Study and teaching(Higher)—United
States. I. Hazen, Robert M., 1948–   . II. Title.
Q161.2.T74 1995
500—dc20   94-36672
                                                                                   CIP

Printed in the United States of America

10 9 8 7 6 5 4 3 2 1

Preface

Pick up a newspaper any day of the week and you will find a dozen articles that relate to science and technology. There are always stories on the weather, energy, the environment, medical advances—the list goes on and on. Is the average American college graduate prepared to understand the scientific component of these issues? In most cases the answer is no.

A week prior to graduation at a major U.S. university, 25 seniors, selected at random, were asked a simple question: "What is the difference between an atom and a molecule?" Only one-third of the students could answer the question correctly. Even allowing for the festive mood of the graduates (all of whom had a full year of laboratory science), this result does not instill confidence in our ability to produce students who are in command of rudimentary facts about their physical universe. There can be little doubt that we are faced with a generation of students who complete their education without learning even the most basic concepts about science. They lack the critical knowledge to make informed decisions regarding environmental issues, resource management, and medical advances.

Science Education Today

We have written this book to address two problems that pervade the organization and presentation of science at U.S. colleges and universities. First, many introductory science courses are geared toward science majors. Specialization is vital for these students—typically a small percent of all college students—who must learn an appropriate vocabulary and develop skills in experimental method and mathematical manipulations to solve problems. Unfortunately, few of these science majors ever gain a broad overview of the sciences, nor do they understand how their chosen specialty fits into the larger scheme of science and society. A physics major may never learn about the modern revolution in genetics and the biology major is unlikely to appreciate the nature and importance of semiconductors in our technological society.

Specialization, however, is not well suited for the majority of students—the nonscience majors, for whom experimental technique and mathematical rigor often divorce science from its familiar day-to-day context. Introductory science courses designed for science majors fail to foster scientific understanding among the nonscience majors. Without a broad context, many students neither understand the distinctive process of science, nor do they retain the abstract content of what they have been taught. Ultimately, this needlessly narrow approach to science education alienates most nonmajors, who graduate with the perception that science is difficult, boring, and irrelevant to their everyday interests.

The second problem with most introductory science courses at the college level, even among those science courses specifically designed for nonscientists, is that they rarely integrate physics, astronomy, chemistry, Earth science, and biology. Such departmentally based courses cannot produce graduates who are broadly literate in science. The students who take introductory geology learn nothing about lasers or nuclear reactions, while those who take "Physics for Poets" remain uninformed about the underlying causes of earthquakes and volcanoes. And neither physics nor geology classes touch on such vital modern fields as genetics, environmental chemistry, space exploration, or materials science. Therefore, students must take courses in at least four departments to gain a basic overview of the sciences.

Perhaps most disturbing, few students—science majors or nonmajors—ever learn how the often arbitrary divisions of specialized knowledge fit into the overall sweep of the sciences. In short, traditional science curricula of most colleges and universities fail to provide the basic science education that is necessary to understand the many scientific and technological issues facing our society.

This first edition of *The Sciences: An Integrated Approach* is the culmination of a collective effort of dozens of faculty and thousands of students who used a preliminary edition during the 1993–1994 academic year. The content and scope of this book was shaped by their numerous constructive comments and suggestions—an integrated effort that is reflected throughout the text.

The Need for a New Science Education

The central goal of science education must be to give every student the ability to place important public issues such as the environment, energy, and medical advances in a scientific context. Students should understand the scientific process, be familiar with the role of experiments in probing nature, and recognize the importance of mathematics in describing its behavior. They should be able to read and appreciate popular accounts of major discoveries in physics, astronomy, chemistry, geology, and biology, as well as advances in medicine, information technology, and new materials. And, most important, they should understand that a few universal laws describe the beavhior of our physical surroundings—laws that operate every day, in every action of our lives. Achieving this kind of scientific proficiency requires a curriculum quite different from the traditional, departmentally based requirements for majors.

Most societal issues concerning science and technology draw on a broad range of knowledge. For example, to understand the debate over nuclear waste disposal, one needs to know how nuclei decay to produce radiation (physics), how radioactive atoms interact with their environment (chemistry), how radioactive elements from waste can enter the biosphere (Earth science), and how the radiation will affect living things (biology). These scientific aspects must be weighed with other societal issues—economics, energy demand, perceptions of risk, and demographics, for example. Other public issues, such as global warming, space research, alternative energy sources, and AIDS prevention also depend on a spectrum of scientific concepts, as well as other social concerns.

Many recent studies recognize the urgent need for reform in science education and advocate an interdisciplinary approach. For example, reports by the American Association for the Advancement of Science (*Science for All Americans—Project 2061,* Washington, 1989; *The Liberal Art of Science,* Washington, 1990), by the National Research Council (*National Science Education Standards,* Washington, 1994), and by the White House (*Science in the National Interest,* Washington, 1994), which all document this need. However, appropriate interdisciplinary courses are not yet widely taught nor have any appropriate textbooks or other support materials been available. This text, based on a course "Great Ideas in Science," which has been developed over the past seven years at George Mason University, is an attempt to fill those gaps. Our approach recognizes that science forms a seamless web of knowledge about the universe. Our integrated course encompasses physics, chemistry, astronomy, Earth sciences, and biology, and emphasizes general principles and their application to real-world situations rather than esoteric detail.

Many ways exist to achieve this synthesis, but any general treatment should take advantage of the fact that virtually everything in science is based on a few simple overarching principles. Newton's laws, the atomic model, natural selection, and the genetic code would make every scientist's list of these "great ideas." By returning to general science courses for all students, we have the means to achieve our goal—to produce college graduates who appreciate that scientific understanding is one of the crowning achievements of the human mind, that the universe is a place of magnificent order, and that science provides the most powerful means to discover knowledge that can help us understand and shape our world.

The Organization of This Book

This text adopts a distinctive and innovative approach to science education, based on the principle that general science courses are a key to a balanced and effective college-level science education for nonmajors and a broadening experience for science majors. We organize the text around a series of 25 scientific concepts. The most basic principle, the starting point of all science, is the idea that the universe can be studied by observation and experiment (Chapter 1). A surprising number of students—even science majors—have no clear idea how this central concept sets science apart from religion, philosophy, and the arts as a way to understand our place in the cosmos.

Once students understand the nature of science and its practice, they can appreciate some of the basic principles shared by all sciences: Newton's laws governing force and motion (Chapter 2); the laws of thermodynamics that govern energy and entropy (Chapters 3 and 4); the equivalence of electricity and magnetism (Chapters 5 and 6); and the atomic structure of all matter (Chapters 7–10). These concepts apply to everyday life, explaining, for example, the compelling reasons for wearing seat belts, the circulation of the blood, the dynamics of a pot of soup, the regulation of public airwaves, and the rationale for dieting. In one form or another, all of these ideas appear in virtually every elementary science textbook, but often in abstract form. As educators we must strive to make them part of every student's day-to-day experience.

Having established these general principles, we go on to examine specific natural systems such as atoms, the Earth, or living things. The realm of the nucleus (Chapter 11) and subatomic particles (Chapter 12), for example, must follow the basic rules governing all matter and energy. An optional chapter on the theory of relativity (Chapter 13) examines the consequences of a universe in which all observers discern the same laws of nature.

Plate tectonics and the cycles of rocks, water, and the atmosphere unify the Earth sciences (Chapters 14–16). The laws of thermodynamics, which decree that no feature on the Earth's surface is permanent, can be used to explain geologic time, gradualism, and the causes of earthquakes and volcanoes. The fact that matter is composed of atoms tells us that individual atoms in the Earth system— for example, in a grain of sand, a gold ring, or a student's most recent breath— have been recycling for billions of years.

In sections on astronomy and cosmology (Chapters 17 and 18) students learn that stars and planets form and move as predicted by Newton's laws, that stars eventually burn up according to the laws of thermodynamics, that nuclear reactions fuel stars by the conversion of mass into energy, and that stars produce light as a consequence of electromagnetism.

Living things (Chapters 19–25) are arguably the most complex systems that scientists attempt to understand. We identify seven basic principles that apply to all living systems: all living things obey the laws of chemistry and physics; all living things incorporate a few simple molecular building blocks; all living things are made of cells; all living things use the same genetic code; all living things evolved by natural selection; living things use many strategies to maintain and reproduce life; and interdependent collections of living things (ecosystems) recycle matter while energy flows through them.

The text has been designed so that four chapters—quantum mechanics (8), particle physics (12), relativity (13), and cosmology (18)—may be skipped without loss of continuity.

Special Features

In an effort to aid student learning and underscore the interconnections among the sciences, we have attempted to relate scientific principles to each student's everyday life. To this end, we have incorporated several distinctive features throughout the book:

Great Ideas.   Each chapter begins with a statement of a great unifying idea or theme in science, so that students immediately grasp the chief concept of that chapter. These statements are not intended to be recited or memorized, but rather to provide a framework for placing everyday experience in a broad context.

Random Walks.   Each chapter also begins with "A Random Walk," in which we tie the chapter's main theme to common experiences, such as eating, driving a car, or suntanning. These "Random Walks" grew out of our idea of the perfect class: during every class period, we would meet outdoors and walk until we saw something that would illustrate that day's topic.

The Human Body.   To help show the interdisciplinary nature of the many concepts we introduce, we have included sections on human physiology and medical advances in most chapters. These sections, for example, relate the anatomy of the eye to electromagnetic radiation, the operation of the central nervous system to electrical conductivity, and the processes of aging to the second law of thermodynamics. In each case we underscore how scientific principles relate immediately to our physical well-being.

Technology.   The application of scientific ideas to commerce, industry, and other modern technological concerns is perhaps the most immediate way students encounter science. In most chapters we include examples of these technologies, such as petroleum refining, microwave ovens, and nuclear medicine.

Science in the Making.   These historical episodes trace the process of scientific discovery as well as portray the lives of central figures in science. In these episodes we have tried to illustrate the scientific method, examine the interplay of science and society, and reveal the role of serendipity in scientific discovery.

Thinking More about.   Each chapter ends with a section that addresses a social or philosophical issue tied to science, such as federal funding of science, nuclear waste disposal, the human genome project, and priorities in medical research.

Mathematical Equations and Worked Examples.   Unlike the content of many science texts, formulae and mathematical derivations play a subsidiary role in our treatment. We rely much more on real-world experience and everyday vocabulary. We believe, however, that every student should understand the role of mathematics in science. Therefore, we have included a few key equations and appropriate worked examples in many chapters. Whenever an equation is introduced, it is presented in three steps: first as an English sentence, then as a word equation, and finally in its traditional symbolic form. In this way, students can focus on the meaning rather than on the abstraction of the mathematics. We also include an appendix on English and SI units.

Science by the Numbers.   We also think that students should understand the importance of simple mathematical calculations in making estimates and determining orders of magnitude. Thus we have incorporated many nontraditional calculations of this kind—for example—how much solid waste is generated in the United States, how long it would take to erode a mountain, and how many people were required to build Stonehenge.

Key Words.   We believe that most science texts suffer from too much complex vocabulary, and we have avoided any unnecessary jargon. Nevertheless, the scientifically literate student must be familiar with many words and concepts that appear regularly in newspaper articles or other material for general readers. In each chapter a number of these words appear in **boldface** type, and they are summarized at the end of the chapters. For example, in Chapter 11 on nuclear physics, key words include proton, neutron, isotope, radioactivity, half life, radiometric dating, fission, fusion, and nuclear reactor—all terms likely to appear in the newspaper.

Many other scientific terms are important, although more specialized; we have highlighted these terms in *italics*. We strongly recommend that students be expected to know the meaning and context of key words but not be expected to memorize these italicized words. We encourage all adopters of this text to provide their own lists of key words and other terms—both ones we have omitted and ones they would eliminate from our list.

Questions.   We feature four levels of end-of-chapter questions. "Review Questions" test the most important factual information covered in the text. "Discussion Questions" are also based on material in the text, but they examine student comprehension, and they explore applications and analysis of the scientific concepts. "Problems" are quantitative questions that require students to use mathematical operations. Finally, "Investigations" require additional research outside the classroom. Each instructor should decide which levels of questions are most appropriate for his or her students. We welcome suggestions for additional questions, which will be added to the next edition of this text.

Illustrations.   Students come to any science class with years of experience dealing with the physical universe. Everyday life provides an invaluable science laboratory—the physics of sports, the chemistry of cooking, and the biology of being alive. This book has thus been extensively illustrated with familiar color images in an effort to amplify the key ideas and principles. All diagrams and graphs have been designed for maximum clarity and impact.

Supplements

The supplements that accompany the first edition of *The Sciences: An Integrated Approach* assist both the instructor and the student.

The *Instructor's Manual* was prepared by Gail Steehler and Karen Adkisson, both of Roanoke College. The manual contains teaching suggestions, lecture notes, answers to problems in text, practice questions and problems, a list of supplemental readings and films, as well as ideas for beginning and ending lectures. In addition, "Connecting Back" and "Connecting Ahead" sections give instructors suggestions on how to connect the material from different chapters.

The *Test Bank,* prepared by Cynthia Roubie of George Mason University, contains approximately 1250 multiple choice, true/false, and short-answer conceptual questions, as well as essay questions for each chapter.

The *Laboratory Manual,* prepared by Victor Stanionis and the Iona College Science Faculty, contains laboratory experiments that have been written and class-tested at Iona College as well as a few that have been contributed from instructors at other schools.

Acknowledgments

The development of this text has benefited immensely from the help and advice of numerous people. Students in our "Great Ideas in Science" course at George Mason University have played a central role in designing this text. Approximately 1100 students, the majority of whom were nonscience majors, have enrolled in the course over the past seven years. They represent a diverse cross section of American students: more than half were women, while many minority, foreign-born, and adult learners were enrolled. Their candid assessments of course content and objectives as well as their constructive suggestions for improvements have helped to shape our text.

We are also grateful to members of the Core Science Course Committee at George Mason University, including Richard Diecchio (Earth Systems Science), Don Kelso and Harold Morowitz (Biology), Minos Kafatos and Jean Toth-Allen (Physics), and Suzanne Slaydon (Chemistry), who helped to design many aspects of this treatment.

We thank the many teachers across the country who are developing integrated science courses. Their letters to us and responses to our publisher's survey inspired us to write this text.

We especially thank those professors who used and class-tested the preliminary edition, sharing with us the responses of their students and their own analyses. Their classroom experience helped us shape the first edition.

Lauretta Bushar, Beaver College

Tim Champion, Johnson C. Smith University

Ben Chastain, Samford University

Marvin Goldberg, Syracuse University

James Grant and Michael Held, St. Peter's College

Jim Holler, University of Kentucky

Patricia Hughey, Lansing Community College

Joseph Ledbetter and Rick Saparano, Contra Costa College

Leigh Mazany, Dalhousie University

Donald Miller, University of Michigan at Dearborn

Lynn Narasimhan, DePaul University

Ervin Poduska, Kirkwood Community College

Gail Steehler, Roanoke College

Jim Yoder, Hesston College

We have also benefited from detailed chapter reviews by experts in physics, astronomy, chemistry, Earth science, and biology: Paul Fishbane, University of Virginia; John Graham, Carnegie Institution of Washington; Gerry Karp, formerly of the University of Florida, Gainesville; Hallie M. Krider, University of Connecticut; Larry Kodosky, Oakland Community College; Harold Morowitz, George Mason University; Bjorn Mysen, Geophysical Laboratory; Selwyn Sacks, Carnegie Institution of Washington; and John S. Thompson, Texas A&M University at Kingsville.

We gratefully acknowledge the significant contributions of the many reviewers who commented on earlier drafts of this edition. Their insights and teaching experience have been invaluable to us:

Debra J. Barnes, Contra Costa College

Doug Bingham, West Texas State University

Larry Blair, Berea College

Virginia R. Bryan, Southern Illinois University

Joe C. Burnell, University of Indianapolis

W. Barkley Butler, Indiana University of Pennsylvania

Ben B. Chastain, Samford University

LuAnne Clark, Lansing Community College

John Cobley, University of San Francisco

Stan Cohn, DePaul University

Marjorie Collier, St. Peter's College

Rod Cranson, Lansing Community College

Phillip D. Creighton, Salisbury State University

Whitman Cross II, Southern Museum of Flight

John Cruzan, Geneva College

E. Alan Dean, University of Texas at El Paso

J.-Cl. De Bremaeker, Rice University

Normand A. Dion, Franklin Pierce College

William Faissler, Northeastern University

John Freeman, Rice University

William Fyfe, University of Western Ontario

Biswa Ghosh, Hudson County College

J. Howard Hargis, Auburn University

Dennis Hibbert, North Seattle Community College

Dave Hickey, Lansing Community College

Patricia M. Hughey, Lansing Community College

Roger Koeppe, Oklahoma State University

Diona Koerner, Marymount College

Charles J. Kunert, Concordia College

Kathleen H. Lavoie, University of Michigan-Flint

Joseph E. Ledbetter, Contra Costa College

Robert W. Lind, University of Wisconsin-Plattesville

Sam LittlePage, University of Findlay

Becky Lovato, Lansing Community College

Bruce MacLaren, Eastern Kentucky University

Donald Miller, University of Michigan at Dearborn

Michael J. Neilson, University of Alabama

Joseph Ruchlin, Lehman College of CUNY

Frederick D. Shannon, Houghton College

Howard J. Stein, Grand Valley State University

Herbert H. Stewart, Florida Atlantic University

Barbara E. Warkentine, Lehman College of CUNY

Jim Yoder, Hesston College

Iris Knell at George Mason University assisted in manuscript preparation, offered many suggestions for editorial improvements, and provided an atmosphere of cheerful efficiency that helped us to meet the tight production schedule. Judith Peatross cheerfully and efficiently compiled the glossary for us under a tight deadline.

Finally, we applaud the many people at John Wiley who proposed this book and helped to shape every aspect of its planning and production. Our editor Kaye Pace, with grace and persuasion, encouraged us to pursue this project, while she mustered the support of her Wiley colleagues for this unique effort. Barbara Heaney worked closely with us to craft a unique approach to teaching science; her ideas and dedication are reflected in every page. Anne Boynton-Trigg obtained and analyzed the dozens of external reviews that guided so much of this first edition; her perceptive synopses allowed us to address many of the ideas and concerns of science faculty. Erica Liu, Eric Stano, Brent Peich, and Jennifer Brady provided editorial assistance throughout the developmental process, and guided the production of supplements to the text.

Nancy Prinz supervised the accelerated production schedule, handling the countless technical details with efficiency and skill. Pedro Noa designed the text in a way that illustrates the integration of the sciences, and Lucille Buonocore managed the overall production process. Lisa Passmore displayed great resourcefulness in obtaining the numerous photos for this edition, while Edward Starr coordinated the development of other illustrations. Catherine Faduska and Anna Saar supervised the marketing of *The Sciences* with creativity and good humor.

To all the staff at John Wiley we owe a great debt for their enthusiastic support, constant encouragement, and sincere dedication to science education reform.

Brief Contents

Contents

Chapter 11   The Nucleus of the Atom

Chapter 12   The Ultimate Structure of Matter

Chapter 13   Albert Einstein and the Theory of Relativity

Chapter 24 Strategies of Life

Great Idea   Living things use many different strategies to deal with the problems of acquiring matter and energy, adapting to the local environment, eliminating wastes, and reproducing.

Chapter 25 Ecosystems

Great Idea   Ecosystems, interdependent collections of living things, recycle matter while energy flows through them.

Why Study Science

Pick up your local newspaper any morning of the week and glance at the headlines. On a typical day you'll see articles about the weather, environmental concerns, and long-range planning by one of your local utility companies. There might be news about a new treatment for AIDS, an earthquake in California, or government spending for cancer research. Inside the paper will be stories about the job market, foreign trade, and the latest records in sports. The editorial pages might feature comments on genetically engineered tomatoes at your grocery store, plans for a new incinerator for solid waste, or perhaps a trial involving DNA fingerprinting. What do all of these stories have in common? They all affect your life in one way or another, and they all depend to a significant degree on science.

We live in a world of matter and energy, forces and motions. Everything we experience in our lives—every action we take and every sensation we feel—takes place in an ordered universe with regular and predictable phenomena. You have learned to survive in this universe, so many of these scientific ideas are second nature to you. When you drive a car, cook a meal, or play a pickup game of basketball, you instinctively take advantage of a few simple physical laws. As you eat, sleep, work, or play you experience the world as a biological system and must come to terms with the natural laws governing all living things.

Science gives us a powerful tool to understand how our world works and how we interact with our physical surroundings. Science not only incorporates basic ideas and theories about how our universe behaves, but it also provides a framework for learning more and tackling new questions and concerns that come our way. Science represents our best hope for predicting and coping with natural disasters, curing diseases, and discovering new materials and new technologies with which to shape our world. Science also provides an unparalleled view of the magnificent order and symmetry of the universe and its workings—from the unseen world of the atomic nucleus, to the inconceivable vastness of space.

So why should you study science? Chances are you are not going to be a professional scientist—only about one in every hundred Americans is. You probably won't ever have to predict the weather or cure a disease, much less calculate the orbit of a planet or the acidity of a solution. So why not just leave science to the experts? The answer is simple: By learning about the central ideas and methods of science, you will be in a much better position to make informed decisions about issues that affect your work, family, and other aspects of your daily life.

The chances are that your job will depend on advances in science and technology. New technologies are a driving force in economics, business, and even many aspects of law: Semiconductor advances, agricultural methods, and information processing have altered our world. Biological research and drug development plays a crucial role in the medical professions: Genetic diseases, AIDS vaccines, and nutritional information appear in the news every day. New materials and manufacturing techniques drive our industries and present constant challenges to workers in sales and advertising: Plastics, superconductors, and composite materials have changed the way we shape our environment. Even professional athletes must constantly evaluate and use new and improved gear and rely on improved medical treatments and therapies. By studying science you will not only be better able to incorporate these advances into your professional life, but you will also better understand the process by which the advances were made.

Science is no less central to your daily life. As a consumer, you are besieged by new products and processes, not to mention a bewildering variety of warnings about health and safety. As a taxpayer you must vote on issues that directly affect your community—energy taxes, recycling proposals, government spending on research, and more. As a living being, you must make informed decisions about diet and lifestyle. And as a parent, you will have to nurture and guide your children through an ever more complex world. A firm grasp of the principles and methods of science will help you make life's important decisions in a more informed way.

By studying science you become empowered to understand and deal with our changing world, both on the job and at home. But wait, there's more. As an extra bonus, you will be poised to share in the excitement of the scientific discoveries that, week-by-week, transform our understanding of the universe and our place in it.

Science opens up astonishing, unimagined worlds—bizarre life forms in the deep oceans, exploding stars in deep space, and aspects of the history of life and our world more wondrous than any fiction. Dinosaurs, black holes, superconductors, mass extinctions, space travel, and much more await the informed science reader.

With this book as an introduction, you too can share in the greatest ongoing adventure of the human race: the adventure of science.

A Request from the Authors

You are part of an experiment in science education. Colleges have traditionally required students to take one or two semesters of departmentally based science courses—introductory physics, chemistry, or biology, for example. This integrated science course takes a different approach. *The Sciences: An Integrated Approach* targets nonscience majors who take science as part of their general education. Our main objectives are to explore central principles of physical and life sciences, to examine the role these principles play in everyday events, and to investigate the interactions of science and society. Our approach emphasizes the interdisciplinary scientific nature of human health, technology, environmental concerns, and other issues. It has been designed to relate scientific principles to your everyday experience and to help you cope with these important issues.

You can help us to achieve these objectives by suggesting improvements for subsequent editions of this book. What aspects of the book did you like? What could we improve? Were there sections that seemed confusing or too complicated? Were the examples relevant and the problems helpful? Please take some time to let us know, by writing to James Trefil and Robert Hazen, Robinson Professors, George Mason University, Fairfax, VA 22030-4444. We are eager to incorporate your comments and suggestions in subsequent editions, and thank you in advance for sharing in this development.

The Sciences

Science: A Way of Knowing

Science is a way of asking and answering questions about the physical universe.

A Random Walk
Making Choices

Our lives are filled with choices. What shall I wear? What should I eat? Where shall I go and how should I get there? Every day we have to make dozens of decisions, each based in part on the knowledge that actions in a physical world have predictable consequences.

When you pull into a gas station, for example, you have to decide what sort of gasoline to put into your car. Over a period of time you may try several different types, observing how your car responds to each. You may keep a record of mileage and cost; perhaps even notes on whether you were driving in the city or on high-speed roads. In the end, you may conclude that a particular brand and grade suits your car best, and buy that one in the future.

This simple example illustrates the way we normally learn about the universe. First, we look at the world to see what is there and learn how it works. Then we generalize, making rules that seem to fit what we see. And finally, we apply those rules to situations we've never encountered, fully expecting them to work.

At the level of choosing a brand of gasoline, there seems to be nothing earth-shattering about this. But the same basic procedure can be applied in a more formal and quantitative way when we want to understand the workings of a distant star or a living cell. In these cases, the enterprise is called science, and the people who do it for a living are called scientists.

The Scientific Method

Science is a way of asking and answering questions about the physical universe. It is not simply a set of facts or a catalog of answers, but rather a way of conducting an ongoing dialogue with our physical surroundings. Like any human activity, science is enormously varied and rich in subtleties. Nevertheless, there are a few basic steps that, taken together, can be said to comprise the **scientific method.**

Observation

If our goal is to learn about the world, the first thing we have to do is look around us and see what's there. This statement seems obvious to us in our technological age, yet throughout much of history learned men and women rejected the idea that you can understand the world simply by observing it.

A Platonic philosopher living during the Golden Age of Athens would have argued that one cannot deduce the true nature of the universe by trusting to the senses. The senses lie, he would have said. Only the use of reason and the insights of the human mind can lead us to true understanding. In his famous book *The Republic,* Plato compared human beings to people living in a cave, watching shadows on a wall but unable to see the objects causing the shadows. In just the same way, he argued, observing the physical world will never put us in contact with reality, but will doom us to a lifetime of wrestling with shadows. Only with the "eye of the mind" can we break free from illusion and arrive at the truth.

In the Middle Ages in Europe, a similar frame of mind was to be found, but with a devout trust in received wisdom replacing the use of human reason as the ultimate tool in the search for truth. There is a story (probably apocryphal) about a debate in an Oxford college on the question of how many teeth a horse has. One learned scholar got up and quoted the Greek scientist Aristotle on the subject, and another quoted the theologian St. Augustine to put forward a different answer. Finally, a young monk at the back of the hall got up and noted that since there was a horse outside, they could settle the question by looking in its mouth. At this point, the manuscript states, the assembled scholars "fell upon him, smote him hip and thigh, and cast him from the company of educated men."

As both of these examples illustrate, there are valid ways to attack the problem of learning about the physical world without actually making observations and measurements. They are not, however, the methods of science, nor did they produce the kinds of advanced technologies and knowledge we associate with modern societies. These other attempts to understand our place in the cosmos were, however, perfectly self-consistent and were pursued by people every bit as intelligent as we are. In the next chapter we will see how human beings gradually came to understand that observation has an important role to play in learning about the universe.

In the remainder of this book, we will differentiate between **observations,** in which we observe nature without manipulating it, and **experiments,** in which we manipulate some aspect of nature and observe the outcome. An astronomer, for example, observes distant stars without changing them, while a chemist experiments by mixing materials together and seeing what happens.

Identifying Patterns and Regularities

When we observe a particular phenomenon over and over again, we begin to get a sense of how nature behaves. We start to recognize patterns in nature. Eventually, we will generalize our experience into a synthesis that summarizes what we have learned about the way the world works. We may, for example, notice that whenever we drop something, it falls. This statement would represent a summary of the results of many observations.

It often happens that at this stage scientists summarize the results of their observations in mathematical form, particularly if they have been making quantitative measurements. In the case of a falling object, for example, they might be measuring the time it takes an object to fall a certain distance, rather than just noticing that the object falls. The first step would be to collect data in the form of a table (see Table 1–1).

Table 1–1 • Measurements of Falling Objects

Time of Fall (seconds)	Distance of Fall (meters)
1	5
2	20
3	45
4	80
5	125

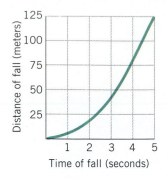

Figure 1–1

Measurements of a falling object can be presented visually in the form of a graph. Time of fall (on the horizontal axis) is plotted versus distance of fall (on the vertical axis).

These data could also be presented in the form of a graph, in which distance is plotted against time. (See Figure 1–1.)

After preparing tables and graphs of their data, scientists would notice that the longer something falls, the farther it travels. Furthermore, the distance isn't simply proportional to the time of fall. If one object falls for twice as long as another, it will travel four times as far; if it falls three times longer, it will travel nine times as far; and so on. This statement can be summarized in three ways (a format that we'll use throughout this book):

▶ **In words:**

The distance traveled is proportional to the square of the time of travel;

▶ **In the form of a word equation:**

distance = constant × (time)^2;

▶ **In symbolic form:**

$d = k \times t^2$.

(The constant, k, has to be determined from the measurements. We'll return to the subject of constants in the next chapter.)

Mathematics is a concise language that allows scientists to communicate their results and to make very precise predictions, but anything that can be said in an equation can also be said (albeit in a less concise way) in a plain English sentence. When you encounter equations in your science courses, you should always ask "What English sentence does this equation represent?" This routine will keep the mathematics from obscuring the simple ideas that lie behind most equations.

Not every scientific idea has to be stated this precisely, though. A scientist studying the gradual encroachment of a forest into an abandoned field, for example, might notice that certain plants seem to follow each other—weeds, followed by scrub trees, followed by pines, followed by hardwoods, for example. We can test the statement that a succession of plant types will be observed everywhere in a particular climate zone, so it is a part of scientific inquiry.

Hypothesis and Theory

Once we have summarized experimental and observational results, we can form an **hypothesis**—a tentative guess—about how the world works. In the case of our everyday experience with falling objects, this hypothesis could be very simple. We could say "When I drop something, it falls." In other cases, the formation of the hypothesis may be more complicated, and the hypothesis may be stated in the form of mathematical equations. When confronted with a new phenomenon, scientists often weigh several different hypotheses at once, much as a detective in a murder mystery may consider several different suspects.

The word **theory** refers to a description of the world that covers a relatively large number of phenomena and has met many observational and experimental tests. After observing hundreds of dropped objects, for example, we could state a theory such as "In the absence of wind resistance, all objects fall a distance proportional to the square of the time of the fall." Just as a detective announces a theory at the conclusion of a murder mystery, so, too, scientists reach a conclusion based on their observations of nature.

Prediction and Testing

In science, every hypothesis and theory must be tested by using them to make **predictions** about how a particular system will behave, then observing nature to see if the system behaves as predicted. For example, if we hypothesize that all objects fall when they are dropped, then that idea can be tested by dropping all sorts of objects. Each drop constitutes a test of our prediction, and the more successful tests, the more confidence we have that the hypothesis is correct.

As long as we restrict our tests to solids or liquids on the Earth's surface, the hypothesis is consistently confirmed. Test a helium-filled balloon, however, and we discover a clear exception to the rule. The balloon falls up. The original hypothesis, which worked so well for most objects, fails for certain gases. And more tests would show there are other limitations. If you were an astronaut in a space shuttle, every time you held something out and let it go it would just float in space. Evidently our hypothesis is invalid in space, as well.

This example illustrates an important aspect about the testing of hypotheses and theories in science. Tests do not necessarily prove or disprove a theory; instead, they often serve to define the range of situations under which the theory is valid. We may, for example, observe that nature behaves in a certain way only at high temperatures or only at low ones; only at low velocities or only at high ones. Such limitations indicate that the original theory doesn't cover enough ground and has to be replaced by something more general.

When a theory or group of related theories has been tested extensively and seems to apply everywhere in the universe—when we have had enough experience with it to have a lot of confidence that it is true—we generally elevate the theory to a new status. We call it a **law of nature**—an overarching statement of how the universe works. In the example of

$$D = k \times T^2$$

Equations allow us to describe with precision the behavior of objects in our physical world. One such equation predicts the behavior of falling objects.

falling objects, we shall see that the theory that "objects fall when dropped" has been replaced by Isaac Newton's more sophisticated and general laws of motion, and the law of universal gravitation. These laws of nature describe and predict the motion of dropped objects both on the Earth and in space and are, therefore, a more successful set of statements than the original hypothesis. We will discuss them in more detail in the next chapter.

We will encounter many such laws in this book, all backed by millions of observations and measurements. It is important, however, to remember where these laws come from. They are not written on tablets of stone, nor are they simply good ideas that someone once had. They arise from repeated and rigorous observation and testing. They represent our best understanding of how nature works.

Remember—we never stop questioning the validity of our hypotheses, theories, or laws of nature. Scientists constantly think up new, more rigorous experiments to test the limits of our theories. In fact, one of the central tenets of science is:

> **Every law of nature is subject to change,
> based on new observations.**

The Scientific Method in Operation

The elements of observation, hypothesis formation, prediction, and testing together comprise the scientific method. In practice, you can think of the method as working as shown in Figure 1–2. It is a never-ending cycle in which observations lead to hypotheses, which lead to more observations.

If observations confirm a hypothesis, then more tests may be devised. If the hypothesis fails, then the new observations are used to revise it, after which the revised hypothesis is tested again. Scientists continue this process until the limits of existing equipment are reached, in which case researchers often try to develop better instruments to do even more tests.

Figure 1–2

The scientific method can be illustrated as an endless cycle of collecting observations (data), identifying patterns and regularities in the data, forming hypotheses, making predictions, and collecting more observations.

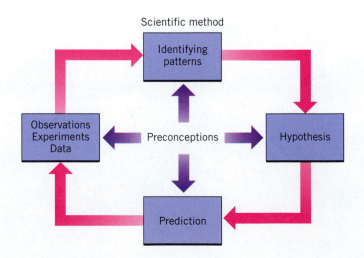

If and when it appears that there's just no point to going further, the hypothesis may be elevated to a theory, or even a law of nature.

Several important points should be made about the scientific method:

1. Scientists are not required to observe nature with an "open mind," with no preconceptions about what they are going to find. Although some scientists do try to operate this way, most experiments and observations are designed and undertaken with a specific hypothesis in mind, and most researchers have a preconception about whether that hypothesis is right or wrong. Perhaps the most important point about the scientific method is that scientists have to believe the results of their experiments and observations, whether they fit preconceived notions or not. Science does not demand that we have no ideas when we enter the cycle in Figure 1–2—only that we be ready to change those ideas if the evidence forces us.

2. There is no "right" place to enter the cycle. Scientists can (and have) started their work by making extensive observations, but they can also start with a theory and test it. It makes no difference where you enter the cycle—eventually the scientific process takes you all the way around.

3. Observations and experiments must be reported in such a way that anyone with the proper equipment can verify the results. Scientific results, in other words, must be **reproducible**.

4. The cycle is continuous; it has no end. Science does not provide final answers, nor is it a search for ultimate truth. Instead, it is a way of producing successively more detailed and exact descriptions of wider and wider areas of the physical world—descriptions that allow us to predict more of the behavior of that world with higher and higher levels of confidence.

5. Finally, we should stress that while the orderly cycle shown in Figure 1–2 provides a useful framework to help us think about science, it shouldn't be thought of as a kind of rigid cookbook-style set of steps to follow. Because science is done by human beings, it involves occasional bursts of intuition, sudden leaps, a joyful breaking of the rules, and all the other characteristics we associate with other human activities.

Science in the Making

Dimitri Mendeleev and the Periodic Table

The discoveries of previously unrecognized patterns in nature provide scientists with some of their most exhilarating moments. Dimitri Mendeleev (1834–1907), a popular chemistry professor at the Technological Institute of St. Petersburg in Russia, experienced such a breakthrough in 1869 as he was tabulating data for a new chemistry textbook.

The mid-nineteenth century was a time of great excitement in chemistry. Almost every year saw the discovery of one or two new chemical elements, and new apparatus and processes were greatly ex-

Dimitri Mendeleev recognized regular patterns in the properties of known chemical elements, and thereby devised the first periodic table of elements.

panding the repertoire of laboratory and industrial chemists. In such a stimulating field, it was no easy job to keep up to date with all the developments and summarize them in a textbook. In an effort to consolidate the current state of knowledge about the most basic chemical building blocks, Mendeleev listed various properties of the 63 known chemical elements (substances that could not be divided by chemical means). He arranged his list in order of increasing atomic weight, and then noted the distinctive chemical behavior of each element.

Examining his list, Mendeleev realized an extraordinary thing: elements with similar chemical properties appeared at regular, or *periodic,* intervals. One group of elements, including lithium, sodium, potassium, and rubidium (he called them group-one elements), formed compounds with chlorine in a one-to-one ratio. Immediately following the group-one elements in the list were beryllium, magnesium, calcium, and barium—group-two elements that form compounds with chlorine in a one-to-two ratio, and so on.

As other similar patterns emerged from his list, Mendeleev realized that the elements could be arranged in the form of a table (Figure 1–3). Not only did this so-called periodic table highlight previously unrecognized relationships among the elements, but it also revealed obvious gaps—places where as yet undiscovered elements must lie.

The power of Mendeleev's periodic table of the elements was underscored when several new elements, with atomic weights and chemical properties just as he had predicted, were discovered in the following years.

The discovery of the periodic table ranks as one of the great achievements of science. It was so important, in fact, that Mendeleev's students carried a large poster of it behind his coffin in his funeral procession. ■

Figure 1–3

The first published version of Dimitri Mendeleev's periodic table of the elements revealed regular patterns in the chemical behavior of known elements, as well as obvious gaps where as-yet undiscovered elements must lie.

TABLE I Distribution of the Elements in Groups and Series

Group	I.	II.	III.	IV.	V.	VI.	VII.	VIII.
Series 1	H	—	—	—	—	—	—	
" 2	Li	Be	B	C	N	O	F	
" 3	Na	Mg	Al	Si	P	S	Cl	
" 4	K	Ca	Sc	Ti	V	Cr	Mn	Fe . Co . Ni . Cu
" 5	(Cu)	Zn	Ga	Ge	As	Se	Br	
" 6	Rb	Sr	Y	Zr	Nb	Mo	—	Ru . Rh . Pd . Ag
" 7	(Ag)	Cd	In	Sn	Sb	Te	I	
" 8	Cs	Ba	La	Ce	Di?	—	—	— . —
" 9	—	—	—	—	—	—	—	
" 10	—	—	Yb	—	Ta	W	—	Os . Ir . Pt . Au
" 11	(Au)	Hg	Tl	Pb	Bi	—	—	
" 12	—	—	—	Th	—	U	—	
Higher oxides	R_2O	R_2O_2	R_2O_3	R_2O_4	R_2O_5	R_2O_6	R_2O_7	RO_4
	—	RO	—	RO_2	—	RO_3	—	
Hydrogen compounds	—	—	—	RH_4	RH_3	RH_2	RH	

The Human Body

William Harvey and the Circulation of the Blood

You have probably been taught that blood circulates in your body, but stop and think for a moment. How do we know? One of the great puzzles faced by scientists who studied the human body was deducing the role played by the blood. The English physician William Harvey (1578–1657) gave us our current picture of the circulation, in which blood is pumped from the heart to all parts of the body through arteries, and returned to the heart through veins. His experiments reveal the scientific method at work.

Before Harvey's work several competing hypotheses had been proposed. Some scientists had taught that blood didn't move at all, but simply pulsed in response to pumping of the heart. Others taught that the arteries and veins constituted different systems, with blood in the veins flowing from the liver to the various parts of the body, where it was absorbed and its nutrients taken in. Harvey, on the other hand, adopted the hypothesis that blood circulates through a connected system of arteries and veins. When confronted with such conflicting hypotheses, a scientist must devise experiments that test the distinctive predictions of each competing idea.

To establish the circulation of the blood, Harvey did a number of things. First, he performed careful dissections of animals to trace out the veins and arteries. Second, he undertook studies of live animals, often killing them so that he could observe the veins and arteries as the heart stopped beating. Then, as now, animals were sometimes sacrificed to advance medical science (see Investigation 7). Finally, Harvey performed a series of classical experiments to establish that blood in the veins did indeed flow back to the heart, rather than simply being absorbed in tissue like a stream of water in the desert. One of those ex-

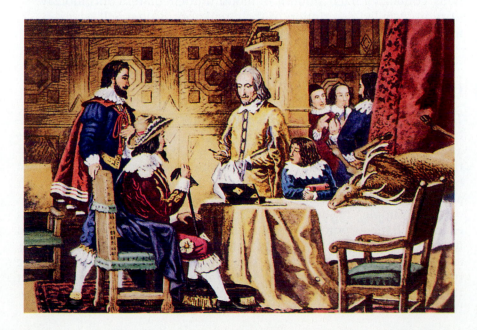

William Harvey conducted a series of experiments to demonstrate conclusively that blood circulates through veins and arteries by the pumping action of the heart.

Figure 1–4
One of William Harvey's famous experiments on the circulation of the blood tested the hypothesis that blood flows from veins to the heart. Harvey first applied a tourniquet to a subject's arm and had the subject squeeze something to raise the veins (*a*). Pressing down on the vein caused it to gradually subside (*b*), indicating that the blood was indeed flowing back to the heart.

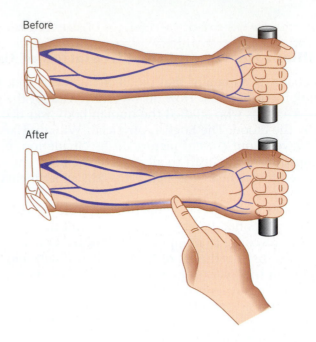

Before

After

periments is shown in Figure 1–4. A tourniquet was applied to a subject's arm, and he was asked to squeeze something so that the veins filled with blood and "popped." (You have probably done the same thing when having blood drawn in a doctor's office). Harvey would then press down on the vein and note that it would subside (indicating that the blood was leaving it) on the side toward the heart. This result, of course, is just the opposite of what would occur if blood were flowing from the liver to the extremities. Based on this experiment, and many others like it, Harvey eventually came to see that the body contains a fixed amount of blood that circulates continuously. (A more detailed description of the circulatory system will be given in Chapter 24). ■

■ Other Ways of Knowing

The central idea of science revolves around the notion that we can discover laws that describe how nature works by reproducible observations and measurements. *Every* idea in science is subject to test at any time—there are no ultimate truths that lie beyond question. If an idea cannot be tested in a manner that yields reproducible results, then it simply isn't a part of science.

For example, a scientist can ask whether a particular painting was executed in the seventeenth century. He or she could use various chemical tests to find the age of the paint, study the canvas, X-ray the painting, and so on. The statement about the age of the painting may be wrong (the painting, for example, may turn out to be a modern forgery), but the statement itself can be tested. It is, therefore, an acceptable scientific hypothesis.

But there are questions that the methods of science cannot answer. No physical or chemical test will tell us whether or not the painting is beautiful, or tell us how we are to respond to it. These questions are simply outside the realm of science.

In fact, the methods of science are not the only way to answer questions that matter in our lives. Science provides us with a way of tackling questions about the physical world—how it works and how we can shape it to our needs. But many questions—some would say the most important questions—lie beyond the scope of science and the scientific methods. Some of these questions are deeply philosophical: What is the meaning of life? Why does the world hold so much suffering? Is there a God? Other important questions are more personal: What career should I choose? Whom should I marry? Should I have children? We cannot answer these questions by the cycle of observation, hypothesis, and testing. For answers, we turn instead to religion, philosophy, and the arts.

A symphony, a poem, and a painting are not, in the end, objects to be studied scientifically. This is not a criticism—they are not supposed to be. These art forms address different human needs than science does, and they use different methods. The same can be said about religious faith. Strictly speaking, there should be no conflict between science and religion, because they deal with different aspects of life. Conflicts arise only when zealots on either side try to push their methods into areas where they aren't applicable.

Pseudoscience

Many kinds of inquiry—creationism, extrasensory perception (ESP), unidentified flying objects (UFOs), astrology, crystal power, reincarnation, or many of the other myriad notions you see in the tabloids at supermarket checkout counters—fail the elementary test that defines sciences. None of these subjects, collectively labeled **pseudoscience**, can be tested in the sense that we are using the term. There is no reproducible test you can imagine that will convince the believers in these notions that their ideas are wrong. Yet, as we have seen, the central property of scientific ideas is that they are testable and could be wrong, at least in principle. Pseudoscience lies outside the domain of science, in the realm of belief or dogma.

Science by the Numbers

Astrology

Astrology is a very old system of beliefs that most modern scientists would call a pseudoscience. The central belief of astrology is that the positions of objects in the sky at a given time (a person's birth, for example) determine a person's future. Astrology was part of a complex set of omen systems developed by the Babylonians, and it was practiced by many famous astronomers well into modern times.

If you were in a spaceship above the Earth's atmosphere, you could see the Sun and the stars at the same time. As the Earth traveled around the Sun, you would see the stars in back of the Sun change. The band of background stars through which the Sun appears to move is called the zodiac. The stars of the zodiac are customarily divided into twelve constellations, which are called "signs" or "houses." At any time, the Sun, the Moon, and the planets all appear in one of these constellations, and a diagram showing these positions is called a horoscope. The constellation in which the Sun appeared at the time of your birth is your "Sun sign," or, simply, your "sign."

Astrologers have a complex (and far from unified) system in which each combination of heavenly bodies and signs is believed to signify particular things. The Sun, for example, is thought to indicate the outgoing, expressive aspects of one's character, the Moon the inner-directed ones, and so on.

Scientists reject astrology for two reasons. First, there is no known way that planets and stars could exert a significant influence on a child at birth. It is true, as we shall learn in Chapter 2, that they exert a gravitational force on the infant, but the gravitational force exerted by the delivering physician (who is smaller but closer) is much greater than that exerted by any celestial object.

But more importantly, scientists reject astrology because it just doesn't work. Over the millennia, there has been no evidence at all that the stars can predict the future.

You can test the ideas of astrology for yourself, if you like. Try this: Have a member of the class take the horoscopes from yesterday's newspaper and type them on a sheet of paper without indicating which horoscope goes with which sign. Then ask members of your class to indicate the horoscope that best matches the day they actually had. Have them write their birthday (or sign) on the paper as well.

If people just picked horoscopes at random, you would expect about 1 person in 12 to pick the horoscope corresponding to his or her sign. Are the results of your survey any better than that? What does this tell you about the predictive power of astrology? ■

The Organization of Science

Scientists investigate all sorts of systems, from the tiniest elementary particles to living cells to the human body to forests to the Earth to stars to the entire cosmos. In all of this vast sweep, the same scientific method can be applied. Men and women have been carrying out this task for hundreds of years, and by now we have a pretty good understanding of how many parts of the system work. And, in the process, we have found a kind of overall organization to scientific knowledge.

The organization of science is analogous to a large spiderweb (Figure 1–5). Around the periphery of the web are all the phenomena examined by scientists, from atoms to fish to comets. Moving toward the web's center we find the cross-linking hypotheses that scientists have developed to explain how these phenomena work. The farther in we move, the more general these hypotheses become and the more they explain. And

Figure 1–5
The interconnected web of
scientific knowledge.

at the very center of the web, connecting all the parts and holding the entire structure together, we find a small number of very general principles that have attained the rank of laws of nature.

No matter where you start on the web, no matter what part of nature you are investigating, you will eventually come to one of the ideas at the central core. Everything that happens in the universe happens because one or more of these laws is operating.

The hierarchical organization of scientific knowledge provides an ideal way to approach the study of nature. At the center of any scientific question are a few laws of nature—we often call them "great ideas of science." We will begin by looking at seven of these laws that are accepted and used by all scientists, no matter what their field of research. These overarching principles of science recur over and over again as we study different parts of the world. You will find that many of these ideas and their consequences seem quite simple—perhaps even obvious—because you are intimately familiar with the physical world in which these laws of nature operate all the time.

After an introduction to these general principles, we look at how the scientific method is applied in specific areas of nature. We examine the nature of materials and the atoms that make them, for example, and look at the chemical reactions that form them. We explore the planet on which we live and discover how mountains, oceans, rivers, and plains are formed and evolve over time. We move beyond Earth to consider the stars and galaxies and ask how they came to be. We look at the diversity of living things on our planet, try to understand how it came to be the way it is, and describe how living things are connected to each other.

By the time you have finished this journey, you will have touched on many of the great truths about the physical universe that scientists have deduced over the centuries. You will discover how the different parts of our universe operate and how all the parts fit together, and you will know that there are still great unanswered questions that drive scientists today. You will understand some of the great scientific and technological challenges that face our society and, more importantly, you will know enough about how the world works to deal with many of the new problems that will arise in the future.

The Organization of Scientists

Science is done by scientists. Hollywood films have created many stereotypes of scientists—the absent-minded professor with frizzy white hair, the mad scientist intent on destroying the world, the bespectacled nerd in a white lab coat. Popular fiction seems to portray the scientist alternately as an eccentric genius with solutions to every human crisis, or an arrogant, self-centered villain who pursues forbidden knowledge. Most people think of scientists as white and male.

Our own experience with our colleagues does not support these stereotypes. There are all kinds of scientists—women and men of every race, religion, and national heritage. There are scholarly types and party animals, honest men and women and people you wouldn't trust as far as you could throw them. Scientists, in short, come in pretty much the same spectrum of personality types as accountants, butchers, and automobile mechanics. And as far as appearances go, all we can say is that neither of the authors, despite a combined total of 50 years in research careers, has *ever* worn a white lab coat.

Scientific Specialties

When modern science first started in the seventeenth century, it was possible for a person to know almost all there was to know about the physical world and the "three kingdoms" of animals, vegetables, and minerals. In the seventeenth century, Isaac Newton could do forefront research in astronomy, in the physics of moving objects, in the behavior of light, and in mathematics. Today, there is so much more knowledge and so much more complexity that no one could possibly be at the frontier in such a wide variety of fields. Scientists today must specialize. They must choose a field—biology, chemistry, physics, and so on—and study it at great length. Even within these disciplines, there are different subspecialties.

In biology, for example, a student may elect to study behavior of ants in a colony, growth stages of a mushroom, energy-producing chemical reactions in a cell, or the body's defense mechanisms against a virus. The amount of information and expertise required to get to the frontier in any of these fields is so large that most students have to ignore almost everything else to learn their specialty.

Within subspecialties, there are different ways to approach problems. Some scientists, called theorists, spend their time imagining universes that might exist. Other scientists, the experimentalists, observe and experiment in order to determine which of the possible universes we actually live in. Both kinds of scientists need to work together to make progress.

Basic Research, Applied Research, and Technology

There are many ways to study the physical universe, and many reasons for doing so. Many scientists are simply interested in finding out how the world works—in knowledge for its own sake. They are engaged in **basic research** and may be found studying anything from the behavior of sea creatures to distant stars. Although discoveries made by these researchers may have profound effects on society (see the discussion of the discovery of the electric generator in Chapter 3, for example), that is not the primary goal of the scientist engaged in basic research.

Many other scientists approach their work with specific practical goals in mind. They wish to develop **technology,** which is the application of the results of science to specific commercial or industrial goals. These scientists are said to be doing **applied research,** and their ideas are often translated into practical systems by large scale **research and development (R & D)** projects.

Government laboratories, private industry, and colleges and universities all support both basic and applied research, but most large scale R & D (as well as most applied research) is done in government and industry labs (see Table 1–2).

Table 1–2 • Some Important Research Laboratories in the United States

Facility	Type	Location
Argonne National Laboratory	Govt/Univ	Near Chicago, IL
AT&T Bell Laboratories	Industrial	Murray Hill, NJ
Brookhaven National Laboratory	Government	Long Island, NY
Dupont R & D Center	Industrial	Wilmington, DE
Fermi National Accelerator Lab	Govt/Univ	Near Chicago, IL
IBM Watson Research Laboratory	Industrial	Yorktown Hts, NY
Keck Telescope	University	Mauna Kea, HI
Los Alamos National Laboratory	Government	Los Alamos, NM
National Institutes of Health	Government	Bethesda, MD
Oak Ridge National Laboratory	Government	Oak Ridge, TN
Stanford Linear Accelerator	Govt/Univ	Stanford, CA
Texas Center for Superconductivity	University	Houston, TX
United States Geological Survey	Government	Reston, VA
Upjohn Pharmaceuticals Laboratory	Industrial	Kalamazoo, MI
Woods Hole Oceanographic Institution	University	Woods Hole, MA

Technology

Buckeyballs: A Technology of the Future?

An extraordinary discovery, announced in 1990, reveals the close relationships among pure scientific research, applied research, and technology. For centuries the element carbon was known in only two basic forms—the soft black mineral graphite, and the hard transparent gemstone diamond. But in 1985, chemists at Sussex University in England and Rice University in Texas found evidence for a totally new form of carbon, in which 60 carbon atoms bond together in a ball-shaped molecule (Figure 1–6). The distinctive linkage of the atoms, much like the geodesic domes of architect-inventor Buckminster Fuller, led scientists to dub the new material buckminsterfullerene, or ''buckeyballs'' for

Figure 1–6

Buckeyballs are soccer-ball-like molecules of 60 carbon atoms (red spheres) that form crystals in which these round molecules stack together like oranges at the grocery store.

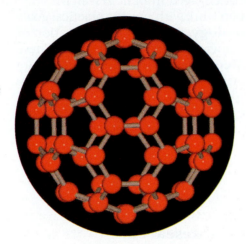

short. Buckeyballs, though completely unexpected, at first excited little attention outside a small research community, for the material had no known uses and it could be produced only in minute quantities. Nevertheless, the lure of the new form of an important chemical element kept several research groups busy studying the stuff. With no obvious practical applications, these early buckeyball studies were examples of basic research.

A major advance came in May of 1990, when a team of German chemists discovered a way to produce and isolate large quantities of buckeyball crystals in a simple and inexpensive device. With the possibility of commercial scale production, an explosion of applied buckeyball research followed. Thousands of scientists, including teams at most major industrial and government laboratories, jumped on the buckeyball bandwagon, and hundreds of scientific articles documented an astonishing range of chemical and physical properties for the new carbon.

Among the preliminary findings: Buckeyballs and closely related materials may contribute to a new generation of versatile electronic materials, powerful lightweight magnets, atom-sized ball bearings, and super-strong building materials. With such extraordinary prospects on the horizon, buckeyball investigations may soon become the domain of engineers developing new technologies—new kinds of batteries for automobiles, carbon-based girders for skyscrapers, unparalleled lubricants, and other products as yet undreamed.

Within the next few years the first buckeyball products may appear at your hardware store, as engineers take the results of applied scientific research and use them to design large-scale production facilities. When the discovery is big enough, the transition from basic research to new technologies may be rapid indeed! ■

Becoming a Scientist

How or why do people become scientists? Here are some reasons from a few well-known scientists, in their own words.

Carl Djerassi, biochemist and inventor of "the pill," as quoted in his autobiography, *The Pill, Pygmy Chimps, and Degas' Horse* (Basic Books, 1992): "In my mind, and most likely also in that of my mother and father,... the tacit assumption was that I would eventually follow in my physician-parents' steps, but no seed of inspiration had yet been sown. Given my fairly logical frame of mind and argumentative nature, I might have turned to law; when I reflect on the ancient-history books I devoured in Europe, and the pleasure I derived in subsequent years from exploring...sites of early civilizations, I can imagine a career in archaeology. As it happened, the person who sowed and sprinkled one of the first chemical seeds was Nathan Washton, the inspiring freshman chemistry teacher at Newark Junior College, where I started with the standard premedical course requirements of chemistry and biology."

Richard Feynman, Nobel Prize–winning physicist, as quoted in his autobiography, *Surely You're Joking, Mr. Feynman!* (Norton, 1985): "When I was a kid I had a "lab." It wasn't a laboratory in the sense that I would measure, or do important experiments. Instead, I would play: I'd make a motor, I'd make a gadget that would go off when something passed a photocell, I'd play around with selenium; I was piddling around all the time."

Judith Howard, crystallographer and chemist, Durham University, England: "My father had lots of old books, on physics mostly, and I always liked science at school, but it was at university that it became a serious affection and more especially so when I went to Oxford to work with Dorothy Hodgkin, the Nobel prizewinner."

Ho-Kwang (Dave) Mao, geophysicist and member of the National Academy of Science, as related in *The New Alchemists* by Robert Hazen (Random House, 1993): "Mao, the son of a Chinese nationalist general, was born in Shanghai in 1941 but grew up in Taiwan after his family fled the mainland with the defeated Nationalists.... All students enrolling in Taiwan's university system listed their choices for school and subject major and then took a difficult six-part entrance examination. The highest scorers received their first selections. Mao placed in the top 5 percent, but by the time his score came up, his top choice—the physics department at Taiwanese National University—was filled. In fact, he had to settle for the university's geology department, his eighth pick. One point higher on the six-hundred-point exam and Mao would have studied food production as an agricultural chemist; one point lower and he would now be a vet."

Lewis Thomas, physician and medical researcher, as quoted in his book, *The Youngest Science* (Viking, 1983): "My father took me along on house calls whenever I was around the house, all through my childhood. He liked company, and I liked watching him and listening to him.... I'm quite sure my father always hoped I would want to become a doctor, and that must have been part of the reason for taking me along on visits. But the general drift of his conversation was to make clear to me, early on, the aspect of medicine that troubled him most all through his professional life; there were so many people needing help, and so little he could do for any of them."

Communication Among Scientists

Sometimes it's easier to do your homework with other students than by yourself, and the same is true of the work that scientists do. It is very hard to work in isolation, and scientists often seek out other people with whom to converse and collaborate. The popular stereotype of the lonely genius changing the course of history just doesn't describe the world of the working scientist. The next time you walk down the hall of a science department at your university, you will probably see faculty and students deep in conversation, talking and scribbling on blackboards. This is the simplest type of scientific communication—direct contact between colleagues.

Scientific meetings provide a more formal and structured forum for communication. Every week of the year, at conference retreats and convention centers across the country, groups of scientists gather to trade ideas. You may notice that science stories in your newspapers often originate in the largest of these meetings, where thousands of scientists converge at one time and a cadre of science reporters with their own special briefing room is poised to publicize exciting results. Scientists often hold off announcing important discoveries until they can make a splash at such a well-attended meeting and press conference.

One thing you would be sure to notice at these meetings is that they are truly international. At almost any scientific gathering you will find people from many different countries. There is no difference between an electron in Japan, Brazil, or the United States, and this universality is reflected in the composition of the scientific community.

Finally, scientists communicate with each other in writing. In addition to rapid communications such as letters, fax, and electronic mail, almost all scientific fields have specialized journals to publish the results of research. The system works like this: When a group of scientists have finished a piece of research and want to communicate their results, they write a concise paper describing exactly what they've done, giving the technical details of their method so that others can reproduce the data,

An extraordinary range of scientific periodicals keep professionals, as well as nonscientists, informed about the latest important research.

and stating their results and conclusions. Many thousands of specialty journals cover all aspects of science and technology, but if the results are especially newsworthy, scientists often submit their work to the editors of one of a few high-profile periodicals such as *Science* or *Nature,* which are read weekly by hundreds of thousands of scientists. More specialized research will find its way to journals with more limited readership. A physicist, for example might submit his or her paper to a journal called *Physical Review,* a chemist to *Journal of the American Chemical Society,* an earth scientist to *American Mineralogist,* and a medical researcher to *The New England Journal of Medicine* or *Cell.*

The journal editor sends the submitted manuscript to one or more knowledgeable scientists who act as referees. These reviewers, whose identities are not revealed to the authors, read the paper carefully, checking for mistakes, misstatements, or shoddy procedures. If they tell the editor that the work passes muster, it will probably be published. Often, such a recommendation is accompanied by a list of necessary modifications and corrections. This system, called **peer review,** is one of the cornerstones of modern science.

The presence of peer review and a clear protocol for entering new results into the scientific literature explains why scientists get so upset when one of their colleagues tries to bypass the system and announce results at a press conference. Such work has not been subject to the thorough review process, and no one can be sure that it meets established standards. When the results turn out to be unreproducible, overstated, or just plain wrong, as they were in the case of cold fusion (see Chapter 11), it damages the credibility of the entire scientific community.

Scientific Societies

Professional organizations form an important part of the life of a scientist. Ever since 1660, the founding year of the Royal Society of London for the Promotion of Natural Knowledge, scientists have recognized the need for organizations devoted to the promotion of their craft. Scientists in

Table 1–3 • A Few Major Scientific Awards

Name of Prize	Subject Area
Arthur Day Medal	Earth Science
Field Prize	Mathematics
Japan Prize	A subject not covered by Nobel Prizes
Nobel Prize	Chemistry, Medicine, Physics
Priestley Award	Chemistry

America established many scientific societies of their own, beginning in 1769 with the American Philosophical Society, "held at Philadelphia, for promoting useful knowledge." Benjamin Franklin and Thomas Jefferson were both active members.

Many American scientists now join the **American Association for the Advancement of Science**, universally referred to as the AAAS ("triple-A S"). This Washington-based organization represents all branches of the physical, biological, and social sciences. With approximately 133,000 members, the AAAS is among the largest scientific societies in the world. It holds a national meeting and several regional meetings every year, publishes the weekly periodical *Science*, and is a strong force in establishing science policy and promoting science education.

The **National Academy of Sciences**, based in Washington, DC, serves a very different function. Founded by Abraham Lincoln and Congress during the Civil War to help the government with scientific and technical problems, the National Academy provides professional advice for the government on policy issues, ranging from environmental risks and natural resource management to education and funding for science research. Membership in the Academy is limited to just 60 new members per year—approximately 2,000 scientists in all—who are elected in semiannual ballots by the members. Election to fellowship in the National Academy is a mark of high distinction among American scientists. Scientists have also created hundreds of prizes and awards to recognize excellence in specific fields of research (Table 1–3).

Table 1–4 • Major Scientific Societies in the United States

Name of Society	Membership
American Chemical Society	144,000
American Geophysical Union	30,000
American Institute of Biological Sciences	90,000
American Institute of Physics	90,000
Geological Society of America	17,000

In addition to general science organizations, there are hundreds of societies devoted to specific fields of study. A few of the largest are summarized in Table 1–4.

Funding for Science ◼

An overwhelming proportion of funding for American scientific research comes from various agencies of the federal government (Table 1–5). The **National Science Foundation,** with an annual budget of about $2.5 billion, supports research and education in all areas of science. Other agencies, including the Department of Energy, the Department of Defense, the Environmental Protection Agency, and the National Aeronautic and Space Administration, fund research and science education in their own particular areas of interest, while Congress may appropriate additional money for special projects. Funding for basic and applied research in medicine and biology also comes from the **National Institutes of Health**, headquartered in Bethesda, Maryland.

An individual scientist seeking funding for research will usually submit a grant proposal—an outline of the planned research together with a statement about why the work is important—to the appropriate federal agency. That agency asks panels of independent scientists to rank proposals in order of importance, and funds as many as it can. Depending on the field, a proposal has anywhere from a 10% to 40% chance of being successful. This money from federal grants buys experimental equipment and computer time, pays the salaries of researchers, and supports advanced graduate students. Without this support, science in the United States would all but come to a halt.

Table 1–5 ● Your Tax Dollars: 1994 Federal Science Funding

Agency or Department	Funding (in millions of $)
Department of Agriculture	$ 1,313.0
Department of Energy	
General science and research	1,615.1
Environmental restoration; nuclear waste management	5,181.9
Environmental Protection Agency	353.6
NASA	
Research and development; space station	9,475.3
Space flight and construction of facilities	5,419.2
National Institutes of Health	10,942.1
National Institutes of Standards and Technology	518.7
National Oceanographic and Atmospheric Administration	1,927.8
National Science Foundation	3,027.8

The funding of science by the federal government is one place where the opinions and ideas of the citizen, through his or her elected representatives, have a direct effect on the development of science. As of 1993, the total research and development budget of the United States government exceeded 18 billion dollars each year.

As you might expect, there are many debates among scientists and politicians about how this research money should be spent. One constant point of contention, for example, concerns the question of basic versus applied research. How much money should we put into the latter, which can be expected to show a quick payoff, as opposed to the former, which may not have a payoff for years (if it has one at all)?

Another continuing debate concerns the division of money between "big" and "little" science. Most scientists work in small groups, concentrating on specialized research. This method of operation is called "small science," and many of the important discoveries we'll be talking about in later chapters came from this tradition. The synthesis of high-temperature superconductors (Chapter 10), the solution of the DNA double helix structure (Chapter 23), and the discovery of evidence for dinosaur extinction (Chapter 24) are recent examples of major advances made by small research teams.

On the other hand, sometimes science is done at huge, centralized (and very expensive) installations. Today, there are proposals to build a $35 billion space station and conduct a $5 billion Human Genome Project (see Chapter 22). (A partially completed $10 billion machine called the Superconducting Supercollider [see Chapter 12] was canceled by Congress in 1993.) These projects epitomize "big science." Many of the important discoveries we'll discuss later were made at these sorts of establishments.

The conflict, of course, arises when we ask how many small science projects are to be canceled to pay for the expensive instruments needed to conduct big science. This sort of conflict is not easy to resolve—in fact, the two authors of this book often find themselves in disagreement on specific policy issues related to big versus little science.

THINKING MORE ABOUT RESEARCH

Funding Priorities

Sometimes questions of research funding get caught up in questions of public policy. For example, there have been 800,000 cases of AIDS diagnosed in the United States since 1980, primarily among male homosexuals and intravenous drug users. Over that same period, some 350,000 Americans have died of cancer *every year*. Yet every year since 1990, the budget for AIDS research at the National Institutes of Health exceeded that for cancer.

Critics of this policy argue that research money should be spent on those diseases that affect the greatest number of people, and that the federal policy has been distorted by a vocal minority. Supporters argue that AIDS, an incurable and invariably fatal disease, is a *potential* threat to many more people than cancer, and point to the tens of millions of heterosexual men and women who have died of the disease in Africa as a portent of what could happen if a cure or vaccine for the disease is not found soon.

As so often happens, there is no "scientific" solution to this problem. What do you think the proper course for the government ought to be? Should we spend more to combat a disease that is already killing many people or one that is relatively minor now, but could potentially be even more deadly? What nonscientific arguments have to be brought to bear in making decisions like these?

Summary

Science is a way of learning about our physical universe. The *scientific method* relies on making *reproducible observations* and *experiments*, which may suggest general trends and *hypotheses*, or *theories*. Hypotheses, in turn, lead to *predictions* that can be tested with more observations and experiments. Successful hypotheses may, after extensive testing, be elevated to the status of *laws of nature*, but are always subject to further test. Science and the scientific method differ from other ways of knowing, including religion, philosophy, and the arts, and differ from *pseudosciences*.

Science is organized around a hierarchy of fundamental principles. Overarching concepts about forces, motion, matter, and energy apply to all scientific disciplines. Additional great ideas relate to specific systems—molecules, cells, planets, or stars. This body of scientific knowledge forms a seamless web, in which every detail fits into a larger, integrated picture of our universe.

Scientists engage in *basic research,* whose goal is solely the acquisition of knowledge, *applied research,* and *research and development* (R & D) aimed at specific problems. *Technology* is developed by this process.

Scientific societies such as the *American Association for the Advancement of Science* and the *National Academy of Sciences* facilitate the scientific process by organizing professional meetings, publishing research results, and fostering science education. Scientific results are communicated in *peer-reviewed* publications. The federal government plays the important role of funding most scientific research and advanced science education in the United States. Much basic research is funded by the *National Science Foundation* and the *National Institutes of Health*.

Key Terms About Science

scientific method	reproducible	American Association for the Advancement of Science (AAAS)
observation	pseudoscience	National Academy of Sciences
experiment	basic research	National Science Foundation
hypothesis	technology	National Institutes of Health
theory	applied research	
prediction	research and development	
law of nature	peer review	

Review Questions

1. Describe the steps in the scientific method.
2. What distinguishes a hypothesis from a theory?
3. What is the difference between an experiment and an observation?
4. What are ways of knowing that are not science?
5. What is the difference between basic and applied research?
6. What is peer review?
7. Name three important scientific organizations.
8. What is the National Science Foundation?
9. Who pays for most scientific research in the United States?
10. Give examples of big and little science.

Discussion Questions

1. Why are studies of unidentified flying objects (UFOs) considered to be pseudoscience by most scientists?

2. Which of the following statements could be tested scientifically?
   a. Most of the energy coming from the Sun is in the form of visible light.
   b. Unicorns exist.
   c. Shelley wrote beautiful poetry.
   d. The Earth was created over four billion years ago.
   e. Diamond is harder than steel.
   f. Diamond is more beautiful than ruby.

3. How does the work of Harvey illustrate the scientific method? The work of Mendeleev?

4. Scientists are currently investigating whether certain microscopic organisms can clean up toxic wastes. How might you set up an experiment to determine that you had found such an organism?

5. Categorize the following examples as basic research or applied research.
   a. The discovery of a new galaxy.
   b. The development of a better method to fabricate rubber tires.
   c. The breeding of a new variety of disease-resistant chicken.
   d. A study of the diet of parrots in a tropical rain forest.
   e. The identification of a new chemical element.
   f. The improvement of a method to extract the element gold from stream gravels.

6. The claim is sometimes made that the cycle of the scientific method produces closer and closer approximations to "reality." Is this a scientific statement? Why or why not?

7. Following World War II, in the 1950s, many movies portrayed evil physicists and rocket scientists, who seemed ready to destroy the world. In more recent movies biologists are more often the villains. What discoveries and events may have influenced this change?

8. A recent television commercial claimed that an antacid consumed "47 times its own weight in excess stomach acid." How might you test this statement in the laboratory? As a consumer, what additional questions would you ask before deciding to buy this product? Are all of these questions subject to the scientific method?

Problems

1. Susan has kept careful records of driving speed versus fuel efficiency. She has noted that traveling 10 miles per hour (mph) she averages 22 miles per gallon of gasoline (mpg). Similarly, she gets 26 mpg at 20 mph, 29 mpg at 30 mph, 31 mpg at 40 mph, 32 mpg at 50 mph, and 29 mpg at 60 mph. Describe and illustrate some of the ways you might present these data. What additional data would you like to obtain to improve your description?

2. Measure the height and weight of 10 friends and present these data both in a table and graphically. What trends do you observe? Why might physicians find such a table useful?

Investigations

1. What is the closest major government research laboratory to your school? The closest industrial laboratory?

2. Describe a program of scientific research carried out by a member of your school's faculty. How is the scientific method employed in this research?

3. Identify a current piece of legislation relating to science or technology (perhaps an environmental or energy bill). How did your representatives in Congress vote on this issue?

4. Look at a recent newspaper article about science funding. Are the projects described big or little science?

5. Find a science story in a newspaper or popular magazine. Did it originate at a scientific meeting? Which one?

6. How were scientists depicted in the novel and film of *Jurassic Park* by Michael Crichton? Were you convinced by these portrayals? Why?

7. Was Harvey justified in his use of animals in his studies of the circulatory system? What limits should scientists accept in research using animals? What national organizations are involved in this debate?

8. Design an experiment to test the relative ability of three different brands of paper towel to absorb water. What data would you need to collect?

9. Malaria, the deadliest infectious disease in the world, kills more than 2,000,000 people (mostly children in poor countries) every year. The an-

nual malaria research budget in the United States is less than a million dollars—a minuscule fraction of spending on cancer, heart disease, and AIDS. Should the United States devote more research funds to this disease, which does not occur in North America? Why or why not?

Additional Reading

Alvarez, Luis. *Alvarez: Adventures of a Physicist.* New York: Basic Books, 1987.

Gross, Paul and Norman Leavitt. *Higher Superstition.* Baltimore: Johns Hopkins University Press, 1994.

Harre, Rom. *Great Scientific Experiments,* New York: Oxford University Press, 1981.

Holton, Gerald. *Thematic Origins of Scientific Thought: Kepler to Einstein.* Cambridge, MA: Harvard University Press, 1988.

Kevles, Daniel. *The Physicists.* Cambridge, MA: Harvard University Press, 1971.

Kuhn, Thomas. *The Structure of Scientific Revolutions.* University of Chicago Press, 1962.

The Ordered Universe

One set of laws—Newton's laws of motion and gravity—predict the behavior of objects on Earth and in space.

A Random Walk
Cause and Effect

When you wake up in the morning, you expect the Sun to be shining. Sometime in the evening, you expect it to go down. When you turn the key in your car, you expect it to start. When you flip a light switch, you expect a light to go on.

Our world is filled with ordinary events like these, events that we take so much for granted that we scarcely notice them. Yet they set the background for the way we think about the world. We believe in cause-and-effect because it is so much a part of our lives.

The regular passage of seasons, with the shortening and lengthening of days and gradual changes in temperature, provides a template for our lives. We plant and harvest crops, purchase wardrobes, and even plan vacations around this predictable cycle, in the sure knowledge that we must adapt and prepare for nature's cycles. Indeed, the predictability of our physical world has become the central principle of science—an idea so important that science could never have developed had it not been true.

The Night Sky

Among the most predictable things in the universe are the lights we see in the sky at night—the stars and planets. Modern men and women, living in large metropolitan areas, are no longer very conscious of the richness of the night sky's shifting patterns. But think about the last time you were out in the country on a clear moonless night, far from the lights of town. There, the stars seem very close, very real. Before the nineteenth century's development of artificial lighting, human beings often experienced jet black skies filled with brilliant pinpoint stars.

The sky changes—it's never quite the same from one night to the next. Living with this display all the time, our ancestors noticed regularities in the arrangement and movements of stars and planets, and they wove these almost lifelike patterns into their religion and mythology. They learned that when the Sun rose in a certain place, it was time to plant crops because spring was on its way. They learned that there were certain times of the month when a full Moon would illuminate the ground, allowing them to continue harvesting and hunting after sunset. To these people, knowing the behavior of the sky was not an intellectual game or an educational frill—it was an essential part of their lives. It is no wonder, then, that astronomy, the study of the heavens, was one of the first sciences to develop.

By relying on their observations and records of the regular motions of the stars and planets, ancient observers of the sky were perhaps the first humans to accept the most basic tenet of science:

> **Physical events are predictable and quantifiable.**

Without the predictability of physical events, as we saw in Chapter 1, the scientific method could not proceed.

Stonehenge

No symbol of humankind's early preoccupation with astronomy is more dramatic than Stonehenge, the great prehistoric stone monument on Salisbury Plain in southern England. The structure consists of a large circular bank of earth surrounding a ring of single upright stones, which in turn encircle a horseshoe-shaped structure of five giant stone archways. Each arch is constructed from three massive blocks—two vertical supports several meters tall capped by a great stone lintel. The open end of the horseshoe aligns with an avenue that leads northeast to another large stone, called the "heel stone" (see Figure 2–1).

Stonehenge was built in spurts over a long period of time, starting in about 2800 B.C. Despite various legends assigning it to the Druids, Julius Caesar, the magician Merlin (who was supposed to have levitated the stones from Ireland), or other mysterious unknown races, archaeologists have shown that it was built by several groups of people, none of

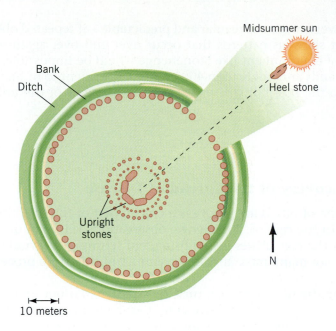

Bank

Ditch

Midsummer sun

Heel stone

Upright stones

N

10 meters

Figure 2–1
Stonehenge is built so that someone standing at the center will see the Sun rise over the heel stone on midsummer's morning.

whom had a written language and some of whom even lacked metal tools. Why would these people expend such a great effort to erect one of the world's great monuments?

Stonehenge, like many similar structures scattered around the world, was built to mark the passage of time. It was a calendar based on the movement of objects in the sky. The most famous astronomical function of Stonehenge was to measure the seasons. In an agricultural society, after all, you have to know when it's time to plant the crops, and you can't always tell by looking at the weather. At Stonehenge, this job was done by sighting through the stones. On midsummer's morning, for example, someone standing in the center of the monument will see the Sun rising directly over the heel stone.

The building of a structure like Stonehenge required accumulation of a great deal of knowledge about the sky, knowledge that could only have been gained through many years of observation. Without a written language, people would have had to pass complex information about the movements of the Sun, the Moon, and the planets from one generation to the next. How else could they have aligned their stones so perfectly that modern-day Druids in England could still greet the midsummer sunrise over the heel stone?

But as impressive as Stonehenge the monument is, Stonehenge the symbol of universal regularity and predictability is even more impressive.

If the universe were not regular and predictable—if repeated observation could not show us patterns that occur over and over again—the very concept of a monument like Stonehenge would be impossible. And yet, there it stands after more than 4000 years, a testament to human ingenuity and to the possibility of predicting the behavior of the universe we live in.

Science in the Making

The Discovery of the Spread of Disease

The process of observing nature is a crucial part of the scientific method. In the case of Stonehenge, observations led to an understanding about the regularities in the rising and setting of heavenly bodies, but there are many more modern examples of this same process at work.

During the nineteenth century, for example, Europe experienced an epidemic of cholera, a severe and often fatal intestinal disease. No one knew the cause of the disease (the discovery of the germ theory of disease was still decades in the future), but doctors could record their observations about the illness. A list of cholera outbreaks was chronicled in 1854 by John Snow (1823–1858), a distinguished London physician who pioneered the use of anesthetics and assisted at the birth of Queen Victoria's children. He noticed that in London the diseases seemed to be centered around certain public squares. At that time, most people in London got their drinking water from fountains and pumps in those squares. The Broad Street pump, located in a place

Figure 2–2

John Snow plotted the number of cases of cholera versus the date for residents in the vicinity of the Broad Street pump. The number of cases declined in early September because most residents fled the area, but few new cases occurred after the pump handle was removed on September 8.

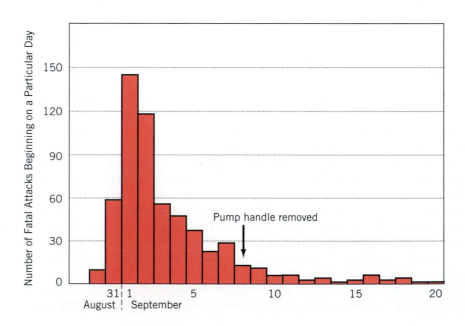

called Golden Square, seemed to be the focus of a particularly large number of cases. Upon investigation, Snow found that the square was surrounded by a large number of homes where human waste was dumped into backyard pits. He argued that these findings suggested the disease was somehow related to contamination of the water supply (Figure 2–2).

Driven by Snow's findings, the city of London (and soon all major population centers) began to require that human waste be carried away from dwellings in sewers. Thus Snow's discovery of a regularity in nature (in this case between disease and polluted water) was the foundation upon which modern sanitation and public health systems are based.

Just as the builders of Stonehenge had no idea of the structure of the solar system, or why the heavens behave as they do, Snow had no idea of why keeping human waste out of the drinking water supply should eliminate a disease such as cholera. It wasn't until the early 1890s, in fact, that the German scientist Robert Och first suggested that the disease was caused by a particular bacterium, *Vibrio cholerae,* that is carried in human waste. ■

Science by the Numbers

Ancient Astronauts

Confronted by a monument such as Stonehenge, with its precise orientation and epic proportions, some people refuse to accept the notion that it could have been built by the ingenuity and hard work of ancient peoples. Instead, they evoke some outside intervention—frequently in the form of visitors from other planets—whose handiwork survives in the monument today. Many ancient monuments, including the pyramids of Egypt, the Mayan temples of Central America, and the giant statues of Easter Island, have been ascribed to these mysterious aliens.

Such conjecture is unconvincing unless you first show that building the monument was beyond the capabilities of the indigenous people. Suppose, for example, that Columbus had found a building like the World Trade Center when he landed in America. Both the ability to produce the materials (steel, glass, and plastic, for example) and the ability to construct a building more than 100 stories tall were beyond the abilities of Native Americans at that time. A reasonable case could have been made for the intervention of ancient astronauts or some other advanced intelligence.

Is Stonehenge a similar case? The materials—local stone—were certainly available to anyone who wanted to use them. Working and shaping stone was also a skill, albeit a laborious one, that was available to early civilizations. The key question, then, is whether people without steel tools or wheeled vehicles could have moved the stones from the quarry to the construction site.

The largest stone, about 10 meters (more than 30 feet) in length, weighs about 50 metric tons (50,000 kilograms, or about 100,000 pounds) and had to be moved overland some 30 kilometers (20 miles) from quarries to the north. Could this massive block have been moved by primitive people, equipped only with wood and ropes?

While Stonehenge was being built, it snowed frequently in southern England. This means that the stones could have been hauled on sleds. A single person can easily haul 100 kilograms on a sled (think of pulling a couple of your friends). How many people would it take to haul a 50,000 kilogram stone?

50,000 kilograms/100 kilograms pulled by each person = 500 people

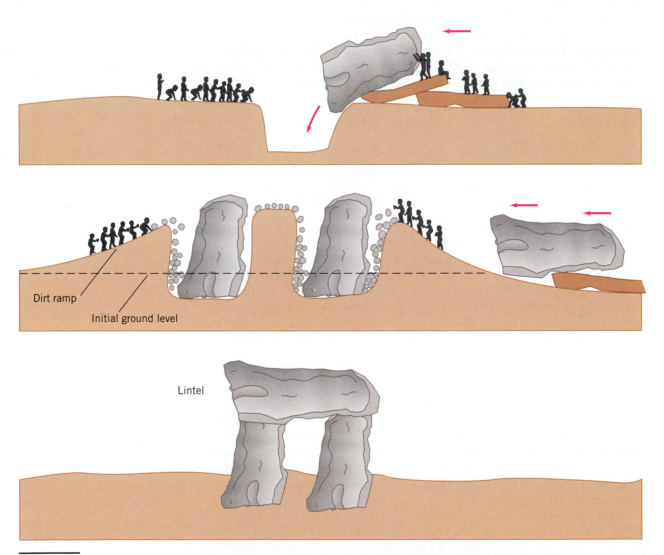

Figure 2–3

Perhaps the most puzzling aspect of the construction of Stonehenge is the raising of the giant lintel stones. Three steps in the process were probably (*a*) to dig a pit for each of the upright stones; (*b*) to pile dirt into a long sloping ramp up to the level of the two uprights so that the lintel stones could be rolled into place; and (*c*) cart away the dirt, thus leaving the stone archway.

Organizing this many people for the job would have been a major social achievement, of course, but there's nothing physically impossible about it.

Scientists cannot absolutely disprove the possibility that Stonehenge was constructed by some strange, forgotten technology. But why evoke such alien intervention when the concerted actions of a dedicated, hard-working human society would have sufficed? All of us are fascinated and awed by the mysterious and unknown, and an ancient structure like Stonehenge, standing stark and bold on the Salisbury Plain, certainly evokes these feelings.

Yet, when confronted with phenomena in a physical world, we should accept the simplest explanation as the most likely. This procedure is called *Occam's razor,* after William of Occam, a fourteenth-century English philosopher who argued that "postulates must not be multiplied without necessity"—that is, given a choice, the simplest solution to a problem is most likely to be right (Figure 2–3). Scientists thus reject the notion of ancient astronauts building Stonehenge, and they relegate such speculation to the realm of pseudoscience. ■

The Birth of Modern Astronomy

When you look up at the night sky you see a dazzling array of objects. Thousands of visible stars fill the heavens and appear to move each night in stately circular arcs centered on the north pole star. The relative positions of these stars never seem to change, and closely spaced groups of stars called constellations have been given names like the Big Dipper and Leo the Lion. Moving across this fixed starry background are the Earth's Moon, with its regular succession of phases, and half-a-dozen planets that wander through the zodiac. You might also see swift streaking meteors or long-tailed comets—transient objects that grace the night sky from time to time.

What causes these objects to move, and what do those motions tell us about the universe in which we live?

The Historical Background: Ptolemy and Copernicus

Since before recorded history people have observed the characteristic motions of objects in the sky and tried to explain them. Most societies created legends and myths tied to these movements, and some (the Babylonians, for example) had long records of sophisticated astronomical observations. It was the Greeks, however, who devised the first astronomical explanations that incorporated elements of modern science.

Ancient scholars believed in the perfection of the heavens. The universe must be finite and spherical, they said, and all motions in the heavens must be represented by uniform motion of heavenly objects on spheres. Given these constraints, perhaps the greatest challenge in modeling the heavens is the occasional *retrograde,* or backward, motion of planets. Most of the time a planet like Mars appears to move from east to west across the fixed background of stars. But every so often the planet's

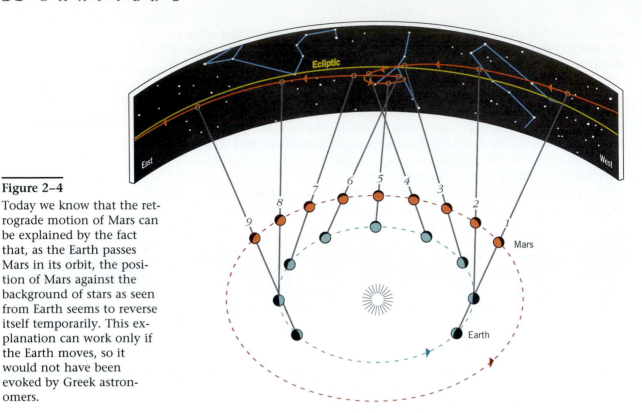

Figure 2–4

Today we know that the retrograde motion of Mars can be explained by the fact that, as the Earth passes Mars in its orbit, the position of Mars against the background of stars as seen from Earth seems to reverse itself temporarily. This explanation can work only if the Earth moves, so it would not have been evoked by Greek astronomers.

motion is retrograde: that is, for a few weeks it appears to slow, stop, and reverse direction with respect to the fixed backdrop of stars. What combination of circular motions could cause such behavior (see Figure 2–4)?

Claudius Ptolemy, an Egyptian-born Greek astronomer and geographer who lived in Alexandria in the second century A.D., proposed the first plausible explanation for complex celestial motions. Working with the accumulated observations of earlier Babylonian and Greek astronomers, he put together a singularly successful model—a theory, to use the modern term—about how the heavens had to be arranged to produce the display we see every night. In the Ptolemaic description of the universe, the Earth sat unmoved at the center. Around it, on a concentric series of rotating spheres, moved the stars and planets. The model was carefully crafted to take account of observations. The planets, for example, were attached to small spheres rolling inside of the larger spheres so that their uneven retrograde motion across the sky could be understood.

This system remained the best explanation of the universe for almost 1500 years. It successfully predicted planetary motions, eclipses, and a host of other heavenly phenomena, and was one of the longest-lived scientific theories ever devised.

During the first decades of the sixteenth century, however, a Polish cleric by the name of Nicolas Copernicus (1473–1543) devised a competing hypothesis that was to herald the end of Ptolemy's crystal spheres. His ideas were published in 1543 under the title *On the Revolutions of the Spheres*. Copernicus retained the notions of a spherical universe with circular orbits, and even kept the ideas of spheres rolling within a sphere, but he asked a simple and extraordinary question: "Is it possible to construct a model of the heavens whose predictions are as accurate as Ptolemy's, but in which the Sun, rather than the Earth, is at the center?" We do not know how Copernicus, a busy man of affairs in medieval Poland, conceived this question, nor do we know why he devoted his spare time for most of his adult life to answering it. We do know, however, that in 1543, for the first time in over a millennium, there was a serious alternative to the Ptolemaic system (see Figure 2–5).

Observations: Tycho Brahe and Johannes Kepler

With the publication of the Copernican theory astronomers were confronted by two competing models of the universe. The Ptolemaic and Copernican systems differed in a fundamental way—a way that had far-reaching implications about the place of humanity in the universe. They both described possible universes, but in one the Earth, and by implication humankind, was no longer at the center. The astronomers' task: decide which model best describes the universe we actually live in.

To resolve the question astronomers had to compare the predictions of the two competing hypotheses—predictions of the exact time a planet or the Moon would appear in an exact location in the heavens, for example—to the observations of what was actually seen in the sky. When they performed these observations, a fundamental problem became apparent. Although the two models made different predictions about various things—the position of a planet at midnight, for example, or the time of moonrise—the differences were too small to be measured with equipment available at the time. The telescope had not yet been invented,

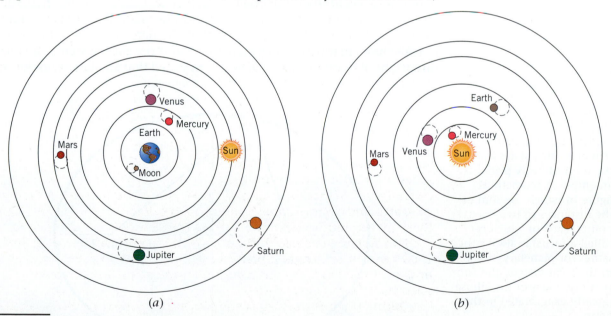

(a)   (b)

Figure 2–5

The Ptolemaic (a) and Copernican (b) systems both assumed that all orbits are circular. The fundamental difference is that Copernicus placed the Sun at the center.

Johannes Kepler
(1571–1630)

and astronomers had to record planetary positions by depending entirely on naked eye measurements with awkward instruments. Until the accuracy of measurement was improved, the question of whether or not the Earth was at the center of the universe couldn't be decided.

Some scientists thrive on experimental challenges, and they revel in devising new tricks for making measurements better than anyone else before. The Danish nobleman Tycho Brahe (1546–1601) was such a scientist. Though abducted in infancy by his uncle, Tycho was raised in comfort and given the best possible education. His scientific reputation was firmly established at the age of 25, when he observed and described a new star in the sky (in fact, a supernova—see Chapter 17). Within the next five years, the Danish king had given him the island of Hveen off the coast of Denmark and funds to build a royal observatory there.

Tycho built his career on the design and use of vastly improved observational instruments. He determined the position of each star or planet with a "quadrant," a large sloping device something like a gunsight, recording each position as two angles. If you were to do this today, you might, for example, measure the angle up from the horizon, and the angle around from due north. He constructed his sighting device of carefully selected materials, and he learned to correct his measurements for the inevitable thermal contraction of brass and iron components that occurred during the cold Danish nights. Over a period of 25 years, he accumulated precise data on the positions of the planets with these instruments, compiling the most accurate record of planetary positions ever assembled.

When Tycho died in 1601, his data passed into the hands of his assistant, Johannes Kepler (1571–1630), a German mathematician who had joined Tycho two years before. Kepler was skilled in mathematics, and he was able to analyze Tycho Brahe's decades of planetary data in new ways. In the end, Kepler found that the data could be summarized in three basic mathematical statements about the solar system. The most important of these (shown in Figure 2–6) stated that all planets, including

Figure 2–6

Kepler's first law, shown schematically, states that the orbit of every planet is an ellipse, a geometrical figure in which the sum of the distances to two fixed points (each of which is called a focus) is always the same. For planetary orbits in the solar system, the Sun is at one focus of the ellipse (greatly exaggerated in this figure).

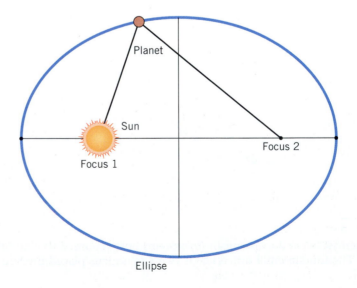

the Earth, orbit the Sun in elliptical, not circular, paths. In this picture, the spheres-within-spheres are gone, because ellipses fully account for the observed planetary motions. Not only do Kepler's laws give a better description of what is observed in the sky, but they present a simpler picture of the solar system as well.

Previous astronomers had assumed that planetary orbits in the heavens must be ideal, and thus circular, and many believed on theological grounds that the Earth had to be the center of a spherical universe. In science, such assumptions of ideality may guide thinking, but they must be replaced when observations prove them wrong.

The work of Brahe and Kepler firmly established that the Earth is not at the center of the universe, that planetary orbits are not circular, and that the answer to the contest between the Ptolemaic and Copernican universes is "neither of the above." This research also illustrates a recurrent point about scientific progress. The ability to answer scientific questions—even questions dealing with the most fundamental aspects of human existence—often depends on the kinds of instruments scientists have at their disposal, and the ability of scientists to apply advanced mathematical reasoning to their data.

At the end of this historical episode astronomers had Kepler's laws that describe *how* the planets in the solar system moved, but they had no idea of *why* planets behaved the way they did. The answer to that question was to come from an unexpected source.

The Birth of Mechanics

Mechanics is an old word for the branch of science that deals with the motions of material objects. A rock rolling down a hill, a ball thrown into the air, and a sailboat skimming over the waves are all fit subjects for this science. Since ancient times, philosophers had speculated on why things move the way they do, but it wasn't until the seventeenth century that our modern understanding of the subject began to emerge. For military rulers concerned with the behavior of cannonballs and other projectiles, mechanics was becoming a science of more than abstract interest.

Galileo Galilei

The Italian physicist and philosopher Galileo Galilei (1564–1642) was, in many ways, a forerunner of the modern scientist. A professor of mathematics at the University of Padua, he quickly became an advisor to the powerful court of the Medici at Florence as well as a consultant at the Arsenal of Venice, the most advanced naval construction center in the world. He invented many practical devices, such as the first thermometer, the pendulum clock, and the proportional compass that draftsmen still use today. Galileo is also famous as the first to observe the heavens with a telescope, which he built after hearing of the instrument from others (Figure 2–7). His astronomical writings, which supported the Sun-centered Copernican model of the universe, led to his trial, conviction, and eventual imprisonment by the Inquisition.

Figure 2–7

Telescopes used by Galileo Galilei in his astronomical studies.

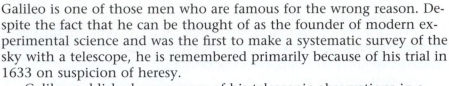

Science in the Making

The Heresy Trial of Galileo

Galileo is one of those men who are famous for the wrong reason. Despite the fact that he can be thought of as the founder of modern experimental science and was the first to make a systematic survey of the sky with a telescope, he is remembered primarily because of his trial in 1633 on suspicion of heresy.

Galileo published a summary of his telescopic observations in a book called *The Starry Messenger.* This book, clearly written in Italian, the language of the people, rather than Latin, the language of scholars, made Copernican ideas available to the educated public. There were complaints that these ideas violated Catholic Church doctrine, and in 1616 Galileo was called before the College of Cardinals. What happened at this meeting is not clear—the Catholic Church later claimed that he had been warned not to discuss Copernican ideas unless he treated them, as Copernicus had, as hypotheses, while Galileo claimed he had not been warned.

In any case, the situation remained in this unsettled state until 1632, when Galileo published a book called *A Dialogue Concerning Two World Systems,* which was a long defense of the Copernican system. This publication led to the famous trial, in which Galileo purged himself of suspicion of heresy by denying that he held the views in his book. He was already an old man by this time, and he spent his last few years under virtual house arrest in his villa near Florence.

The legend of the trial of Galileo, in which an earnest seeker after truth is crushed by a rigid hierarchy, bears little resemblance to the historical events. The Catholic Church had not banned Copernican ideas. Copernicus, after all, was a savvy Church politician who knew how to get his ideas across without ruffling feathers. Indeed, there were seminars on the system given at the Vatican in the years before Galileo. Furthermore, Galileo's arguments in favor of the system were pretty unconvincing—much of the *Dialogues,* for example, is taken up by a completely incorrect discussion of the tides. Finally, his confrontational tactics—putting the Pope's favorite arguments into the mouth of a foolish character in the book, for example—brought a predictable reaction that earlier, more reasonable approaches had not. As always happens, the simple myth associated with a historical event dissolves into something much more complex when we look at the event itself.

A footnote: In 1992, the Catholic Church reopened the case of Galileo and, in effect, issued a retroactive "not guilty." The grounds for the reversal were that the original judges had not separated questions of faith from questions of scientific fact. ■

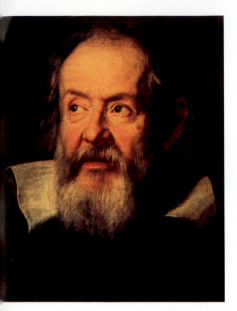

Galileo Galilei (1564–1642)

The Father of Experimental Science

From the point of view of scientists, Galileo's greatest achievement was his work on experimental technique. You can see why by considering his studies on the behavior of objects thrown or dropped on the surface of the Earth. Greek philosophers, using pure reason, had taught that heavier objects must fall faster than light ones. In a series of classic experiments, Galileo showed that this was not the case—that at the surface of the Earth all objects speed up at the same rate as they fall downward. Ironically, Galileo probably never performed the one experiment for which he is most famous—dropping two different weights from the Leaning Tower of Pisa to see which would land first.

To describe falling objects, it is necessary to make precise measurements of two variables: distance and time. Distance measurements posed little problem for Galileo and his contemporaries, but they did not have timepieces precise enough to measure the brief times it took objects to fall straight down. While previous workers had simply observed the behavior of objects around them as best they could, Galileo constructed a special apparatus designed purely to measure how things fall (see Figure 2–8). He slowed down the time of fall by rolling large balls down an inclined plane crafted of brass and hard wood, and measured the time it took the ball to travel different distances. By increasing the angle of elevation of the plane, he could make the ball move faster and faster until, at ninety degrees, the ball would fall freely.

To understand Galileo's results, you have to understand the distinction between three terms: speed, velocity, and acceleration.

▶ **In words:**

 Speed is the distance an object travels divided by the time that it takes to travel that distance.

▶ **In equation form:**

 speed (in m/s) = distance (in m)/time (in s),

 where m and s stand for the units meters and seconds, respectively.

▶ **In symbols:**

 $$s = \frac{d}{t}$$

Galileo's apparatusi inclined plane

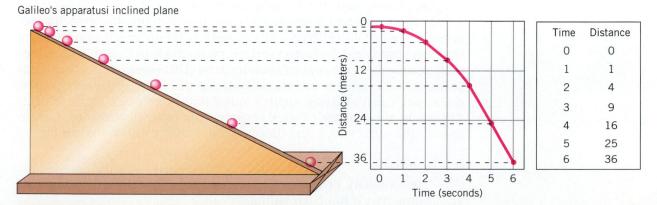

Time	Distance
0	0
1	1
2	4
3	9
4	16
5	25
6	36

Figure 2–8
Galileo's falling-ball apparatus with a table of measurements and a graph of distance versus time.

Velocity has the same numerical value as speed, but velocity always includes information on the direction of travel. The speed of a car might be 40 miles per hour, for example, while the velocity is 40 miles per hour due west. Velocity and speed are measured in units such as meters per second, feet per second, or miles per hour.

Acceleration measures the rate of change of velocity, and, like velocity, acceleration has a direction. You're accelerating if you're speeding up, slowing down, or changing direction.

▶ **In words:**

Acceleration is the amount of change in velocity divided by the time it takes the change to occur.

▶ **In equation form:**

$$\text{acceleration (in m/s}^2) = \frac{\text{final velocity} - \text{initial velocity (in m/s)}}{\text{time (in s)}}$$

▶ **In symbols:**

$$a = \frac{(v_f - v_i)}{t}$$

When velocity changes, it may be by a certain number of meters per second in each second. Consequently, the units of acceleration are (meter/second)/second (described as "meters per second per second," and usually abbreviated m/s^2), where the first "meters per second" refers to the velocity, and the last "per second" to the time it takes to change.

Whenever an object changes speed or direction, it accelerates. When you step on the gas pedal in your car, for example, the car accelerates forward. When you slam on the brakes, the car accelerates backward (in everyday English we call this motion *deceleration*). When you go around a curve in your car, even if the car's speed stays exactly the same, the car is still being accelerated because the direction of the velocity is changing. The most thrilling amusement park rides combine these different kinds of acceleration—speeding up, slowing down, and changing direction in wiggles, bumps, tight turns, and rapid spins.

Galileo wanted to discover the rules that describe the velocity and acceleration of falling objects. His experiments convinced him that any object dropped near the Earth's surface, no matter how heavy or light, falls with exactly the same constant acceleration. This constant acceleration is so important in science that we assign it a special letter, *g*, the acceleration due to the Earth's gravity. Galileo found that every object at the Earth's surface, in the absence of wind resistance or other disturbance, falls in exactly the same way. Every dropped object has a velocity given by:

▶ **In words:**

The velocity of a falling object is proportional to the length of time that it has been falling.

▶ **In equation form:**

velocity (in m/s) = the constant *g* (in m/s^2) × time (in s),

▶ **In symbols:**

$$v = g \times t$$

The velocity, of course, is always directed downward. This equation tells us that an object that falls two seconds achieves a velocity twice that of an object that falls only one second (Figure 2–8). In fact, a penny can kill you if it's dropped from high enough.

The constant g is nothing more than a numerical value for a specific acceleration—the very important acceleration that we (and all other objects) experience at the Earth's surface. (Note that the Moon and other planets will have their own very different surface accelerations; g applies *only* to the Earth's surface.) The value of g can be determined by measuring the actual fall rate of objects in a laboratory (Figure 2–9). Galileo found that g is given by

$$g = 9.8 \text{ m/s}^2 = 32 \text{ feet/s}^2.$$

This equation tells us that in the first second a falling object accelerates from a stationary position to a velocity of 9.8 m/s, (about 22 miles per hour) straight down. After two seconds the velocity doubles to 19.6 m/s, after three seconds it triples to 29.4 m/s, and so on.

With Galileo's work, scientists began to isolate and observe the motion of material objects in nature and to summarize their results into mathematical relationships. As to why bodies should behave this way, however, they had no suggestions. And there was certainly little reason to believe that the measurements of falling objects at the Earth's surface had anything at all to do with motions of planets and stars in the heavens.

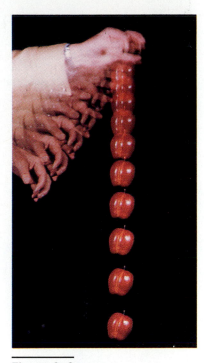

Figure 2–9
The accelerated motion of a falling apple is captured by a multiple-exposure photograph. In each successive time interval the apple falls farther.

The Human Body

Effects of Extreme Acceleration

You experience accelerations every day of your life. Just lying in bed you feel an acceleration equal to g, due to the Earth's gravitational pull. When you travel in a car or plane, ride an elevator, and especially when you enjoy amusement park rides, your body is subjected to additional accelerations, though rarely exceeding $2g$. But jet pilots and astronauts experience accelerations many times that caused by the Earth's gravitational pull during takeoffs, sharp turns, and emergency ejections. What happens to the human body under extreme acceleration, and how can equipment be designed to reduce the risk of injury? In the early days of rocket flights and high-speed-jet design, government scientists had to know.

Controlled laboratory accelerations were most often produced by rocket sleds (Figure 2–10), which may reach accelerations exceeding $10g$ for short bursts, and centrifuges—rotating metal arms—that can sustain accelerations of several g. Researchers quickly discovered that muscles and bones behave as an effectively rigid framework. Sudden extreme acceleration, such as that experienced in a car crash, may cause

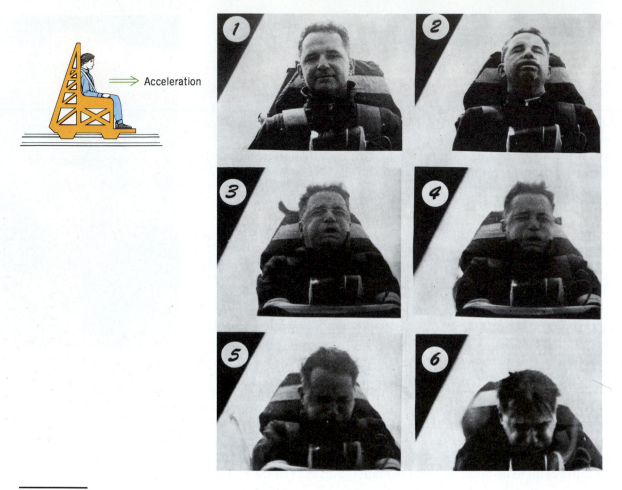

Acceleration

Figure 2–10
Colonel John Stapp experienced extreme acceleration in rocket sled experiments. The severe contortion of soft facial tissues was recorded by a movie camera.

damage, but these parts can withstand the more gradual changes in acceleration associated with flight.

The body's fluids, on the other hand, shift and flow under sustained acceleration. Blood may thus concentrate in some places and drain from others. As a result, a pilot in a sitting position who sustains an acceleration of 5 or 6g for more than a few seconds will usually experience a blackout, followed by unconsciousness. Greater accelerations can be tolerated in the prone position adopted by the first astronauts, who had to endure sustained 8g conditions during takeoffs.

One of the authors (J.T.) once rode in a centrifuge and experienced an 8g acceleration. The machine itself was a gray, egg-shaped capsule located at the end of a long steel arm. When in operation, the arm moved in a horizontal circle. Funny things happen at 8g. For example, the skin in your face is pulled down, so that it's hard to keep your mouth open to breathe. The added weight feels like a very heavy person sitting on your chest, and it is hard to keep your eyes focused.

There is, however, one advantage to having had this particular experience. Now, whenever he encounters the question "What is the most you have ever weighed?" on a medical form, the author can write "1600 pounds." ■

Isaac Newton and the Universal Laws of Motion

The English scientist Isaac Newton (1642–1727), arguably the most brilliant scientist who ever lived, synthesized the work of Galileo and others into a statement of the basic principles that govern the motion of everything in the universe, from stars and planets to clouds, cannonballs, and the muscles in your body. We call these results **Newton's laws of motion.** They sound so simple and obvious that it's hard to realize that they represent the results of centuries of experiment and observation, and even harder to appreciate what an extraordinary effect they had on the development of science.

By all accounts Isaac Newton had a troubled childhood, and he is said to have displayed a decidedly neurotic personality. He was separated from his mother as an infant and raised by his grandmother on a family farm in Lincolnshire, England. The young Newton was interested in mechanical devices, and eventually enrolled as a student at Cambridge University.

For most of the 1665–66 school year the University was closed due to the Great Plague that devastated much of Europe. Isaac Newton spent the time at the Lincolnshire farm, reading and thinking about the physical world. There he began thinking through his extraordinary discoveries in the nature of motion, as well as pivotal advances in optics and mathematics.

Three basic laws form the centerpiece of Newton's description of motions.

Isaac Newton (1642–1727)

The First Law

A moving object will continue moving in a straight line at a constant speed, and a stationary object will remain at rest, unless acted on by an unbalanced force.

Newton's first law seems so obvious that it scarcely needs to be stated—if you leave an object alone, it won't change its state of motion. In order to change it, you have to push it or pull on it, thus applying a force. Yet virtually all scientists from the Greeks to Copernicus would have argued that the first law is wrong. They believed that since the circle is the most perfect geometrical shape, objects will move in circles unless something interferes. They believed that their heavenly spheres would keep turning without any outside force acting (indeed, they had to believe this or face the question of why the spheres didn't just run down).

Figure 2–11
Newton's first law states
that a force is required to
set objects into motion.

Figure 2–11
Newton's first law states that a force is required to set objects into motion.

Newton, basing his arguments on observations and the work of his predecessors, turned this notion around. An object left to itself will move in a straight line, and if you want to get it to move in a circle, you have to apply a force (Figure 2–11). You know this is true—if you swing something around your head, it will move in a circle only as long as you hold on to it. Let go, and off it goes in a straight line.

This simple observation led Newton to recognize two different kinds of motion. An object is in **uniform motion** if it travels in a straight line at constant speed. *All other motions are called acceleration.* Accelerations can involve changes of speed, changes of direction, or both.

Newton's first law tells us that when we see an acceleration, something must have acted to produce that change. We define a **force** as something that produces a change in the state of motion of an object. In fact, we will use the first law of motion extensively in this book to tell us how to recognize when a force, particularly a new kind of force, is acting.

The tendency of an object to remain in uniform motion is called *inertia*. A body at rest tends to stay at rest because of its inertia, while a moving body tends to keep moving because of its inertia. We often use this idea in everyday speech; for example, we may talk about the inertia in a company or government organization that is resistant to change.

The Second Law

The acceleration produced on a body by a force is proportional to the magnitude of the force and inversely proportional to the mass of the object.

If Newton's first law of motion tells you when a force is acting, then the second law of motion tells you what the force does when it acts. This law conforms to our everyday experience—it is easier to move a bicycle than a car, easier to lift a child than an adult, easier to push a ballerina than a defensive tackle.

Newton's second law is most useful when expressed as an equation.

▶ **In words:**

> The greater the force, the greater the acceleration; but the more massive the object being acted on by a given force, the smaller the acceleration.

▶ **In equation form:**

> force = mass (in kg) × acceleration (in m/s^2)

▶ **In symbols:**

> $F = m \times a$

This equation, well known to generations of physics majors, tells us that if we know the forces acting on a system of known mass, we can predict its future motion—its acceleration. The equation conforms to our intuition that an object's acceleration is a balance between two factors: force and mass.

On the one hand, a force causes the acceleration. The greater the force, the greater the acceleration. The harder you throw a ball, the faster (and therefore farther) it goes. Mass measures the amount of matter in any object. The greater the object's mass—the more "stuff" you have to accelerate—the less effect a given force is going to have. A given force will accelerate a golf ball more than a bowling ball, for example. Newton's second law of motion thus defines the balance between force and mass in producing an acceleration.

Newton's first law defines the concept of force as something that causes a mass to accelerate, but the second law goes much further. It tells us the exact magnitude of the force necessary to cause a given mass to achieve a given acceleration. Because force equals mass times acceleration, the units of force must be the same as mass times acceleration. Mass is measured in kilograms (kg) and acceleration in meters per second per second (m/s^2), so the unit of force is the "kilogram-meter-per-second-squared" (kg-m/s^2), which is called the *newton*. The symbol for the newton is N.

EXAMPLE 2–1: From Zero to Ten in Less than a Second

What is the force needed to accelerate a 100-kilogram sprinter from rest to a speed of 10 meters per second (a very fast run) in a half second?

▶ **Reasoning:** We must first find the acceleration, and then use Newton's second law to find the force

▶ **Solution:** The first step is to find the acceleration:

$$\text{acceleration (in m/s}^2) = \frac{\text{final velocity} - \text{initial velocity (in m/s)}}{\text{time (in s)}}$$

$$= \frac{10 \text{ m/s} - 0 \text{ m/s}}{0.5 \text{ s}}$$

$$= \frac{10 \text{ m/s}}{0.5 \text{ s}}$$

$$= 20 \text{ m/s}^2$$

What force is needed to produce this acceleration? From Newton's second law:

$$\text{force (in N)} = \text{mass (in kg)} \times \text{acceleration (in m/s}^2)$$

$$= 100 \text{ kg} \times 20 \text{ m/s}^2$$

$$= 2{,}000 \text{ newtons} \blacktriangle$$

The second law of motion does not imply that every time a force acts, motion must result. A book placed on a table still feels the force of gravity, and you can push against a wall without moving it. In these situations, the atoms in the table or the wall shift around and exert their own force that balances the one that acts on them. It is only the net, or unbalanced, force that actually gives rise to acceleration.

The Third Law

For every action there is an equal and opposite reaction.

Newton's third law of motion tells us that whenever a force is applied to an object, that object simultaneously exerts an equal and opposite force (Figure 2–12). When you push on the wall, for example, it instantaneously pushes back on you—you can feel the force on the palm of your hand. In fact, the force the wall exerts on you is equal in magnitude (but opposite in direction) to the force you exert on it.

The third law of motion is perhaps the least intuitive of the three. We tend to think of our world in terms of causes and effects, in which big or fast objects exert forces on smaller, slower ones: a car slams into a tree, a batter drives the ball into deep left field, a boxer punches his opponent's eye. But in terms of Newton's third law it is equally valid to think of these events the "other way around." The tree stops the car's motion, the baseball alters the swing of the bat, and the opponent's eye blocks the thrust of the boxer's glove, thus exerting a force and changing the direction and speed of the punch.

Forces *always* act simultaneously in pairs. You can convince yourself of this fact by thinking about any of your day's myriad activities. As you sit in a chair reading this book your weight exerts a force on the chair, but the chair exerts an equal and opposite force (called a contact force) on you, preventing you from falling to the floor. The book feels heavy in your hand as it presses down, but your hands hold the book up, exerting an equal and opposite force. You may feel a slight draft from an open window or fan, but as the air exerts that gentle force on you, your skin just as surely exerts an equal and opposite force on the air, causing it to change its path.

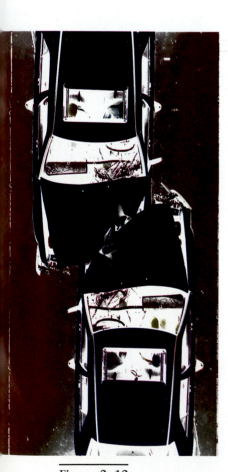

Figure 2–12

Two automobiles in a collision exert equal and opposite forces on each other.

The space shuttle *Discovery* rises from its launch pad at Cape Canaveral, Florida. As hot gases accelerate violently out the rocket's engine, the shuttle experiences an equal and opposite acceleration that lifts it into orbit.

Newton's Laws at Work

Every motion in your life—indeed, every motion in the universe—involves the constant interplay of all three of Newton's laws. The laws of motion never occur in isolation, but rather are interlocking aspects of every object's behavior. The interdependence of Newton's three laws of motion can be envisioned by a simple example. Imagine a boy standing on roller skates holding a stack of baseballs. He throws the balls, one by one. Each time he throws a baseball, the first law tells us that he has to exert a force so that the ball accelerates. The third law tells us that the baseball will exert an equal and opposite force on the boy. This force acting on the boy will, according to the second law, cause him to recoil backward.

And while the example of the boy and the baseballs may seem a bit contrived, it exactly illustrates the principle by which fish swim and rockets fly. As the fish moves its tail, it applies a force against the water. The water, in turn, pushes back on the fish and propels it forward. In the rocket motor, forces are exerted on hot gases, accelerating them out the tail end. By the argument just given above, this means that an equal and opposite force must be exerted on the rocket, propelling it forward. Every rocket, from simple fireworks to a space shuttle, works this way.

Isaac Newton's three laws of motion form a comprehensive description of all possible motions, as well as the forces that lead to them. In and of themselves, however, Newton's laws do not say anything about the nature of those forces. In fact, much of the progress of science during and after Newton's time has been associated with the discovery and elucidation of the forces of nature.

The Universal Force of Gravity

Gravity is the most obvious force in our daily lives. It holds you down in your chair and keeps you from floating off into space. It guarantees that when we drop things they fall. The effects of what we call gravity were known to the ancients, and its quantitative properties were studied by Galileo and many of his contemporaries, but Isaac Newton revealed its universality.

Figure 2–13

An apple falling, a ball being thrown, a space shuttle orbiting the Earth, and the orbiting Moon, all display the influence of the force of gravity.

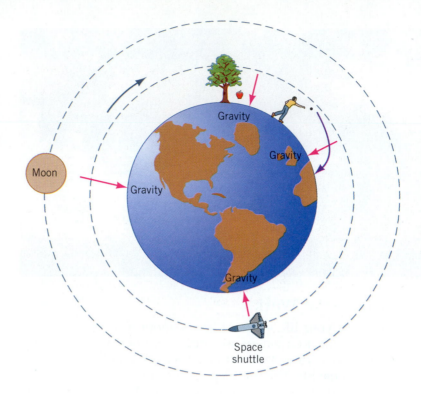

Everyone knew that gravity pulled things toward the Earth, but until Newton most people assumed that the phenomenon was local, something that operated only near the planet's surface. People believed that farther out, in the realm of the stars and planets, different rules applied. They would say that "terrestrial gravity" operated on the Earth and "celestial gravity" operated in the heavens, but the two had little to do with each other. Isaac Newton discovered that these two seemingly different kinds of gravity were, in fact, one and the same. In modern language, in a remarkable coming together of seemingly disparate elements, he unified earthly and heavenly gravity.

By Newton's account, he experienced his great insight in an apple orchard. He saw an apple fall and, at the same time, saw the Moon in the sky behind it. He wondered whether the gravity that caused the apple to move downward could extend far outward to the Moon, supplying the force that kept it in its orbit.

Look at the problem this way: If the Moon goes around the Earth, then it isn't moving in a straight line. From the first law of motion it follows that a force must be acting on it. Newton hypothesized that this was the same force that made the apple fall; the familiar force of gravity (see Figure 2–13).

Eventually, he realized that the orbits of all the planets could be understood if gravity was not restricted to the surface of the Earth but was a force found throughout the universe. He formulated this insight (an insight that has been overwhelmingly confirmed by observations) in what is called **Newton's law of universal gravitation.**

▶ **In words:**

Between any two objects in the universe there is an attractive force (gravity) that is proportional to the masses of the objects and inversely proportional to the square of the distance between them.

▶ **In equation form:**

$$\text{force (in newtons)} = G \times \frac{\text{mass}_1 \text{ (in kg)} \times \text{mass}_2 \text{ (in kg)}}{[\text{distance (in m)}]^2},$$

▶ **In symbols:**

$$F = G \times \frac{m_1 \times m_2}{d^2},$$

G is a number known as the *gravitational constant* (see below).

In everyday words, this law tells us that the more massive two objects are, the greater the force between them will be; the farther apart they are, the less the force will be.

The Gravitational Constant, *G*

When we say that *A* is *directly proportional* to *B*, we mean that if *A* increases, *B* must increase by the same proportion. If *A* doubles then *B* must double as well. We can state this idea in mathematical form by writing:

$$A = k \times B,$$

where *k* is a number known as the constant of proportionality between *A* and *B*. This equation tells us that if we know the constant *k* and either *A* or *B*, we can calculate the exact value of the other. The constant of proportionality in a relationship, then, is a useful thing to know.

The gravitational constant, *G*, is a constant of direct proportionality; it expresses the exact numerical relation between the masses of two objects and their separation, on the one hand, and the force between them on the other:

$$F = G \times \frac{m_1 \times m_2}{d^2}.$$

Unlike *g*, however, which applies only to the Earth's surface, *G* is a universal constant, that applies to any two masses anywhere in the universe.

Henry Cavendish, a scientist at Oxford University in England, first measured *G* in 1798 by using the experimental apparatus shown in Figure 2–14. Cavendish suspended a dumbbell made of two small lead balls by a stiff wire, and he fixed two larger lead spheres near the suspended balls. The gravitational attraction between the hanging lead balls and the fixed spheres caused the wire to twist slightly. By measuring the amount of twisting force, or *torque*, on the wire, Cavendish could calculate the gravitational force on the dumbbells. This force, *F*, together with a knowledge

Figure 2–14
The Cavendish balance measures the universal gravitational constant G. This experimental device balances the gravitation attractive force between the suspended balls and fixed spheres, against the force exerted by a twisted wire.

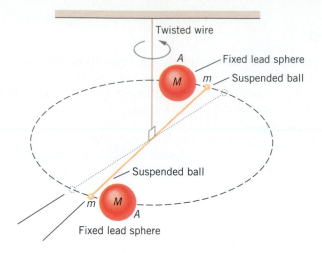

of the masses of the dumbbells (m_1 in the equation) and the heavy spheres (m_2), as well as their final separation (d), gave him the numerical value of everything in Newton's law of universal gravitation except G, which he then calculated using simple arithmetic.

In metric units, the value of G is 6.67×10^{-11} m^3/sec^2-kg, or 6.67×10^{-11} N-m^2/kg^2 (recall that N is the symbol for a newton, the unit of force). This constant appears to be universal, holding true everywhere in our universe. Modern versions of the Cavendish balance have succeeded in measuring G to an accuracy of five decimal places.

Weight and Gravity

The law of universal gravitation says that there is a force between *any* two objects in the universe: two dancers, a Pacific crab and an Atlantic crab, this book and you; all exert forces on each other. Forces between such small objects are usually insignficant, but gravitational forces between large objects shape the universe. The formation of the solar system (see Chapter 14) and the stars of the Milky Way (Chapter 17) were controlled by these gravitational forces, while the forces between the Moon and the Earth's oceans cause the tides.

The gravitational attraction between you and the Earth would pull you down if you weren't standing on the ground. As it is, the ground exerts an equal and opposite force to cancel gravity, a force you can feel in the soles of your feet. If you were standing on a scale, the gravitational pull of the Earth would pull you down until a spring or other mechanism in the scale exerted this opposing force. In this case, the size of that counterbalancing force registers on a display and you call it your weight.

Weight, in fact, is just the force of gravity on an object located at a particular point. Weight depends on where you are—on the surface of the Earth you weigh one thing, on the surface of the Moon another, and in the depths of interstellar space you would weigh next to nothing. You even weigh a little less on a high mountaintop than you do at sea level, because you are farther from the Earth's center. Weight contrasts with your **mass** (the amount of matter), which stays the same no matter where you go.

EXAMPLE 2–2: Weighty Matters

A cantaloupe has a mass of 0.5 kilogram. What does it weigh?

▶ **Reasoning:** To answer this question, we have to calculate the force of gravity exerted on the cantaloupe at the Earth's surface.

▶ **Solution:** The relation between mass and force is:

force (in N) = mass (in kg) × acceleration (in m/s^2), or, in the case of the Earth's surface,

weight = mass × g
    = 0.5 kg × 9.8 m/sec^2
    = 4.9 kg m/sec^2
    = 4.9 N (newtons)

This value is the weight of the cantaloupe. Note that "kilogram" is not a unit of weight, despite popular usage to the contrary (see Appendix A). ▲

Although the followers of Newton tended to think about the gravitational force in terms of its effect on astronomy, the downward pull of gravity has played an important role in the development of life on our planet. A plant like seaweed, for example, that lives immersed in water doesn't have to have a rigid support structure. Forces exerted by the surrounding water partially nullify the force of gravity. This is why you feel lighter when you float in a swimming pool.

Life on Earth originally developed in water, and it wasn't until about 400 million years ago that living things first emerged on land. Unlike seaweed, land-dwelling plants do need the ability to support their weight against gravity. You can, for example, think of a tree trunk as a device for supporting the weight of the tree so that the leaves can intercept sunlight high above the ground. The human skeleton is, of course, another example of an adaptation to life on land—another way of overcoming the challenge of gravity.

EXAMPLE 2–3: Gravitational Forces and the Value of g

What is the gravitational force between a 75-kilogram person and the Earth?

▶ **Reasoning:** We can solve this problem in two different ways. First, try substitution into the universal equation for gravitational force:

$$\text{force (in newtons)} = G \times \frac{\text{mass}_1 \text{ (in kg)} \times \text{mass}_2 \text{ (in kg)}}{\text{distance (in m)}^2},$$

In this equation the constant G is 6.67×10^{-11} N-m^2/kg^2, the first mass (the person) is 75 kilograms, and the second mass (Earth) is approximately 6.02×10^{24} kilograms. The distance between the person and the Earth is approximately the radius of the Earth, about 6,400 km (6.40×10^6 m).

▶ **Solution:** Substituting these values into the gravity equation,

$$\text{Force} = (6.67 \times 10^{-11} \, \text{N-m}^2/\text{kg}^2) \times \frac{75 \, \text{kg} \times 6.02 \times 10^{24} \, \text{kg}}{6.40 \times 10^6 \, \text{m}^2}$$

$$= (6.67 \times 10^{-11} \, \text{N-m}^2/\text{kg}^2) \times \frac{4.52 \times 10^{26} \, \text{kg}^2}{4.10 \times 10^{13} \, \text{m}^2}$$

$$= \frac{3.015 \times 10^{16} \, \text{N-m}^2}{4.10 \times 10^{13} \, \text{m}^2}$$

$$= 735 \, \text{newtons}$$

▶ **Reasoning:** An easier approach to the solution is to remember Newton's second law, which states

$$\text{Force (in N)} = \text{mass (in kg)} \times \text{acceleration (in m/s}^2),$$

At the Earth's surface, the acceleration is g, and Newton's second law tells us that the force needed to produce that acceleration is

$$\text{force} = \text{mass} \times g,$$

where g is 9.8 m/s^2, the acceleration due to Earth's gravity.

▶ **Solution:** Substituting into this equation,

$$\text{force} = 75 \, \text{kg} \times 9.8 \, \text{m/s}^2$$

$$= 735 \, \text{kg-m/s}^2$$

$$= 735 \, \text{newtons} \; \blacktriangle$$

The two different approaches to calculating the gravitational force in Example 2–3 reveal a close relationship between the universal constant G, and the Earth's gravitational acceleration g. Compare the two equations for gravitational force used above:

$$\text{force} = G \times \frac{\text{mass} \times M_E}{(R_E)^2},$$

and,

$$\text{force} = \text{mass} \times g.$$

Equating the right sides of these two equations,

$$\text{mass} \times g = G \times \frac{\text{mass} \times M_E}{(R_E)^2},$$

and dividing both sides by mass,

$$g = \frac{G \times M_E}{R_E^2}.$$

But we have already measured the values of G, M_E, and R_E:

$$g = (6.67 \times 10^{-11} \, \text{N-m}^2/\text{kg}^2) \times \frac{6.02 \times 10^{24} \, \text{kg}}{(6.40 \times 10^2 \text{m})^2}$$

$$= \frac{4.015 \times 10^{14} \text{ N-m}^2/\text{kg}}{4.10 \times 10^{13} \text{ m}^2}$$
$$= 9.8 \text{ N-kg}$$
$$= 9.8 \text{ m/s}^2.$$

Thus the value of the Earth's gravitational acceleration, g, can be calculated from Newton's universal equation for gravity.

This result is extremely important. For Galileo, g was a number to be measured, but whose value he could not predict. For Newton, on the other hand, g was a number that could be calculated purely from the size and mass of the Earth. Because we understand where g comes from, we can now predict the appropriate value of gravitational acceleration not only for the Earth, but for any body in the universe, provided we know its mass and radius.

EXAMPLE 2–4: Weight on the Moon

The mass of the Moon is $M_M = 7.18 \times 10^{22}$ kg, and its radius R_M is 1738 km. If your mass is 75 kg, what would you weigh on the Moon?

▶ **Reasoning:** Once again we have to calculate the force exerted on an object at the surface of an astronomical body. This time both the mass and the radius of the body are different from that of the Earth, although G is the same.

▶ **Solution:** From the equation that defines weight, we have,

$$\text{weight} = G \times \frac{\text{first mass} \times \text{second mass}}{\text{distance}^2}$$
$$= G \times \frac{75 \text{ kg} \times M_M}{R_M{}^2}$$
$$= (6.67 \times 10^{-11} \text{ N-m}^2/\text{kg}^2) \times \frac{75 \text{ kg} \times 7.18 \times 10^{22} \text{ kg}}{(1.738 \times 10^6 \text{ m})^2}$$
$$= (6.67 \times 10^{-11} \text{ N-m}^2/\text{kg}^2 \times \frac{5.39 \times 10^{24} \text{ kg}^2}{3.02 \times 10^{12} \text{ m}^2}$$
$$= \frac{(3.60 \times 10^{14} \text{ N-m}^2}{3.02 \times 10^{12} \text{ m}^2}$$
$$= 119 \text{ newtons}$$

This weight is about one-sixth of the weight the same object would have on Earth, *even though its mass is the same in both places.* ▲

Newton bequeathed a picture of the universe that is beautiful and ordered. The planets orbit the Sun in stately paths, forever trying to move off in straight lines, forever prevented from doing so by the inward tug of gravity. The same laws that operate in the cosmos operate on Earth, and these laws were discovered by the application of the scientific method. To a Newtonian observer, the universe was like a clock. It had been wound up by God and was ticking along according to His laws. Newton and his followers were persuaded that in carrying out their work, they were discovering what was in the mind of God when He created the universe.

Of all celestial phenomena, none seemed more portentous and magical than comets, yet even these chance wanderers were subject to Newton's laws. In 1682 British astronomer Edmond Halley (1656–1742) used Newtonian logic to compute the orbit of the comet that bears his name, and he predicted its return in 1758. The "recovery" of Halley's Comet on Christmas Eve of that year was celebrated around the world as a triumph for the Newtonian system.

 ## Technology

Working with Microgravity

Gravity is a force that acts on every object in the universe, but the force of gravity is not the same everywhere. Until recently, all scientific observations and experiments had to be performed under the full gravitational pull at the Earth's surface. With the advent of extended space missions, however, scientists now have the opportunity to study phenomena in *microgravity,* a nearly weightless environment. The National Aeronautics and Space Administration (NASA) has selected microgravity research as a key opportunity for developing new technologies in space.

Of special importance are microgravity experiments on proteins—large molecules, often with many thousands of atoms, that control chemical processes in all living things (see Chapter 21). Thousands of different kinds of proteins, each with a different chemical job, have been identified, but only a few of these complex molecules have been fully described. Biologists study proteins by trying to arrange the molecules into the regular pattern of a crystal, and then probing the crystals with X-rays to determine their atomic structures. This information is critical for learning how proteins work in healthy organisms, and how they may be defective in a diseased organism. Diabetes, for example, is a disease in humans and other mammals in which a vital protein named insulin, used for digesting sugar, is not properly manufactured by the body. Knowledge of insulin's structure is therefore a key part of understanding diabetes.

A major difficulty in protein research is that many of these molecules do not form good crystals in Earth-based laboratories. Evidently, the force of gravity pulls on the molecules and inhibits them from adopting a regular pattern. A 14-day space shuttle mission in June of 1992 demonstrated that much better-quality protein crystals can be grown in the microgravity of space than on Earth (Figure 2–15). According to NASA, these crystals could play a key role in "establishing the structural foundations of molecular biology and biochemistry and have important applications to the fields of medicine, drug design, and agriculture." ■

Figure 2–15

Protein crystals grown on the Space Shuttle (*top*) have a different shape and better internal order than those grown in Earth-based laboratories (*bottom*).

THINKING MORE ABOUT ORDER

Predictability

The Newtonian universe seemed regular and predictable in the extreme. Indeed, from the point of view of the Newtonians, if you knew the present state of a system and the forces acting on it, the laws of motion would allow you to predict its entire future. This notion was taken to the extreme by the French mathematician Pierre Simon Laplace (1749–1827), who proposed the notion of the *Divine Calculator*. His argument (in modern language) was this: If we knew the position and velocity of every atom in the universe and we had infinite computational power, then we could predict the position and velocity of every atom in the universe for all future times. There is no distinction in this argument between an atom in a rock and an atom in your hand. Thus Laplace would say that all of your movements are completely determined by the laws of physics to the end of time. You cannot choose your future. What is to be was determined from the very beginning.

This argument raises many interesting questions, and you might want to try it out in your next late-night bull session. However, it has been rendered moot by two modern developments in science. One of these, the Heisenberg uncertainty principle (see Chapter 8), tells us that at the level of the atom it is impossible to know simultaneously and exactly both the position and velocity of any particle. Thus you can never get the information the Divine Calculator needs to begin working.

More recently, scientists working with computer models have discovered that there are many systems in nature that can be described in simple Newtonian terms but whose futures are, to all intents and purposes, unpredictable. These situations are called "chaotic systems," and their field of study is called *chaos*.

White water in a mountain stream is a familiar chaotic system to think about. If you put two chips of wood down on the upstream side of the rapids, those chips (and the water on which they ride) will be widely separated by the time they get to the end. This is true no matter how small you make the chips, or how close together they are at the beginning. If you know the exact initial position of a chip and every aspect of the waterway in such a system with complete mathematical precision, you can in principle predict where it will come out downstream. But if there is the slightest error in your initial description, no matter how small, the actual position of the chip and your prediction will differ, often wildly. Every measurement in the real world has some error associated with it, so it is never possible to determine the exact position of the chip at the start of its trip. For all practical purposes you cannot predict where it will come out *even if you know all the forces acting on it*.

The existence of chaos, then, tells us that there are some systems in nature where the philosophical conclusions drawn from the Newtonian vision of the universe simply don't apply. Of course, most situations with which we are familiar are not chaotic. The fact that we can guide a spacecraft to Jupiter or predict the orbits of comets tells us that most of our world is comfortably predictable. But systems such as the turbulent flow of fluids, the dripping of a faucet, and the beating of a human heart may be chaotic in some circumstances.

Many scientists also believe that the flow of the Earth's atmosphere and the long-term development of weather is chaotic, and it may turn out that very complex things such as ecosystems (see Chapter 25) behave this way as well. If this is true, what implications might there be when governments have to deal with issues such as global warming (see Chapter 25) and the preservation of endangered species? How confident do you have to be that something is going to happen before you start taking steps to avoid it?

Summary

Since before recorded history people have observed regularities in the heavens, and have built monuments such as Stonehenge to help order their lives. Models, such as the Earth-centered system of Ptolemy and the Sun-centered system of Copernicus, attempted to explain these regular motions of stars and planets. Astronomers such as Tycho Brahe made ever-more-precise measurements of star and planet positions. These data led mathematician Johannes Kepler to propose his laws of planetary motion, which, among other things, state that planets orbit the Sun in elliptical orbits, not circular orbits as had been previously assumed.

Meanwhile, Galileo Galilei and other scientists investigated the science of *mechanics*—the way things move near the Earth's surface. These workers recognized two fundamental different kinds of motion: *uniform motion* in a constant *speed* and direction (*velocity*), and *acceleration,* which entails a change in either speed or direction of travel. Galileo devised experiments to study falling objects, and he discovered that all things fall the same way, at the constant rate of acceleration of 9.8 meters per second per second.

Isaac Newton combined the work of Kepler, Galileo, and others in his sweeping *laws of motion* and the *universal law of gravitation*. Newton realized that nothing accelerates without a *force* acting, and that the amount of acceleration is proportional to the force applied, but inversely proportional to the *mass*. He also pointed out that forces always act in pairs. This understanding of forces and motions led Newton to describe the most obvious force in our daily lives—*gravity*. He demonstrated that the same force that pulls a falling apple to the Earth causes the Moon to curve around the Earth in its elliptical orbit. Indeed, the force of gravity operates everywhere, with pairs of forces between every pair of masses in the universe.

Key Terms About Motion and Forces

mechanics	Newton's laws of motion	Newton's law of universal gravitation
speed	uniform motion	weight
velocity	force	mass
acceleration	gravity	

Key Equations (and their units)

speed (m/s) = distance (m)/time (s).

acceleration

$$(m/s^2) = \frac{\text{final velocity} - \text{initial velocity (m/s)}}{\text{time (s)}}$$

velocity of a falling object (m/s) = g (m/s^2) × Time (s)

force (N) = Mass (kg) × Acceleration (m/s^2)

$$\text{force (N)} = G \times \frac{\text{first mass (kg)} \times \text{second mass (kg)}}{\text{distance (m)}^2},$$

force (N) = mass (kg) × g = weight

Constants

g = 9.8 m/s^2 = acceleration due to gravity on Earth

G = 6.67 × 10^{-11} N-m^2/kg^2 = universal gravitational constant

Review Questions

1. How did Stonehenge allow ancient people to make predictions?

2. Why do scientists argue that Stonehenge was not built by ancient astronauts?

3. What are some of the objects you see in the night sky?

4. Describe the main features of the Ptolemaic and Copernican systems of the universe.

5. What did Tycho Brahe try to do to resolve the question of the structure of the universe?

6. What was Kepler's role in interpreting Tycho Brahe's data?

7. What is mechanics?

8. How did Galileo study falling objects?

9. Define speed, velocity, and acceleration.

10. What are Isaac Newton's three laws of motion?

11. According to Newton, what are the two kinds of motion in the universe?

12. Why is gravity called a *universal* force?

13. What is the difference between the constants g and G?

13. What similarity did Newton see between the Moon and an apple?

14. What is the difference between weight and mass?

15. How did Edmond Halley support Newton's theories?

16. What is a chaotic system?

Discussion Questions

1. Which of the following is in uniform motion, and which is in accelerated motion?
   a. a car heading north at 35 mph
   b. a car going around a curve at 50 mph
   c. a dolphin leaping out of the water
   d. an airplane cruising at 30,000 feet at 300 mph

2. Which, if any, of the following objects does not exert a gravitational force on you?
   a. this book
   b. the Sun
   c. the nearest star
   d. a distant galaxy

3. What pairs of forces act in the following situations?
   a. A pitcher throws a fast ball.
   b. An apple falls to the ground.
   c. The Moon orbits the Earth.
   d. A pencil rests on your desk.

4. In what sense is the Newtonian universe simpler than Ptolemy's? Suppose observations had shown that the two did equally well at explaining the data. Construct an argument to say that Newton's universe should still be preferred.

5. Why don't the planets just fly off into space? What keeps them in their orbits?

6. How did Henry Cavendish's experiment fit into the scientific method?

Problems

1. What do you weigh in pounds? In newtons?

2. If a race car completes a 2-mile oval track in 58 seconds, what is its average speed? Did the car accelerate during the 58 seconds?

3. If your car goes from 0 to 60 miles per hour in 6 seconds, what is your acceleration?

4. What is your mass in kilograms?

5. Estimate the largest force that you can exert without mechanical aids. How would you exert this force?

6. What would you weigh on a planet four times as massive as the Earth with a radius twice that of the Earth's value?

7. How much force must be applied to accelerate a 20-kg satellite to counter the Earth's gravitational acceleration of 9.8 m/s^2?

8. In Chapter 1 we discussed astrology. Calculate the gravitational force on a newborn infant exerted by a star the size of the Sun one light-year (9.5×10^{15} m) away. Compare it to the gravitational force exerted by a 100-kg physician 0.1 m away.

Investigations

1. Read the Bertolt Brecht play *Galileo,* which dramatizes Galileo Galilei's heresy trial. Discuss the dilemma faced by scientists whose discoveries offend conventional ideas. What areas of scientific research does today's society find offensive or immoral? Why?

2. The concept of predestination plays an important role in some kinds of theology. What is it? How does it relate to Laplace? To chaos?

3. What other kinds of models of the universe did old civilizations develop? Look up those of the Mayans, the Chinese, and the Indians of the American Southwest.

4. Find out how Galileo came to the idea of the pendulum clock. What did he actually observe that led him to this development?

5. Drop a wadded-up sheet of paper and a flat one side by side. Which reaches the ground first? Why? What would happen if this experiment were done in a vacuum?

Additional Reading

Christiansen, Gale E. *In the Presence of the Creator: Isaac Newton and His Times.* New York: Macmillan/Free Press, 1984.

Cohen, I. Bernard. *The Birth of the New Physics.* New York: W.W. Norton, 1985.

Drake, Stillman. *Galileo: Pioneer Scientist.* Toronto: University of Toronto Press, 1990.

Finnochiaro, Maurice A. *The Galileo Affair: A Documentary History.* Berkeley: University of California Press, 1989.

Gleick, James. *Chaos.* New York: Viking, 1987.

Hawking, Gerald S. *Stonehenge Decoded.* New York: Doubleday, 1965.

Energy

The many different kinds of energy are interchangeable, and the total amount of energy in an isolated system is conserved.

A Random Walk
The Great Circle of Energy

Hundreds of millions of years ago, energy was generated in the flaming core of the Sun. For thousands of years, that energy percolated outward to the Sun's surface, then, in a mere eight minutes, made the trip through empty space to the Earth in the form of sunlight. Unlike other bits of energy that were reflected back into space by clouds or simply served to warm the Earth's soil, this particular energy was absorbed by algae floating on the warm ocean surface.

Through a process known as photosynthesis, the light energy was transformed into energy holding together the complex molecules in that living organism. Eventually the algae, along with countless others, died and sank to the bottom of the ocean. Over the long eons, more sediment buried the organic material deeper and deeper. Under the influence of pressure and heat, it was eventually transformed into petroleum.

Then, a short while ago, engineers brought that petroleum with its stored energy up out of the ground.

At a refinery, the large molecules were broken down into gasoline, and the gasoline was shipped to your town. A few days ago you put it into the tank of your car.

The last time you drove you burned that gasoline, freeing the stored energy and using it to move your car. When you parked the car, the hot engine slowly cooled, and that bit of heat energy, after having been delayed for a few hundred million years, was radiated out into space to continue its voyage away from the solar system. As you read these words, the energy you freed yesterday has long since left the solar system, and is out in the depths of interstellar space.

All the energy you use every day, like the energy in our story, is in transit. But no matter how complicated the story, no matter how many different forms that energy takes, there remains one central fact about it: the total amount of energy doesn't change. Energy is neither created nor destroyed, but just changes form. This fact, as we shall see, is one of the great laws of nature that govern our universe.

Scientifically Speaking

Whenever you ride in a car, climb a flight of stairs, or just take a breath, you use energy. At this moment trillions of cells in your body are hard at work turning yesterday's food into chemical energy that will keep you alive today. Energy in the atmosphere generates sweeping winds and powerful storms, while the ocean's energy drives mighty currents and incessant tides. Meanwhile, deep within the Earth, energy in the form of heat is moving the continent on which you are standing.

Energy is all around you—in the ever-shifting atmosphere and restless seas, in simple bacteria and mighty redwood trees, in brilliant sunlight and shimmering moonlight. Energy affects everything in the physical world, and the laws that govern its behavior are among the most important concepts in science.

In any situation where energy is expended you will find one thing in common. If you look at the event closely enough you will find that, in accord with Newton's laws of motion (Chapter 2), a force is being exerted on an object to make it move. When your car burns gasoline, the fuel's energy ultimately turns the wheels of your car, which then exert a force on the road; the road exerts an equal and opposite force on the car, pushing it forward. When you climb the stairs, your muscles exert a force that lifts you upward against gravity. When you breathe, your lungs exert a force as they pull air in and push air out. Even in your body's cells, a force is exerted on molecules in chemical reactions. Energy, then, is intimately connected with the application of a force.

In everyday conversation we speak of someone having lots of energy, of a song sounding energetic, or an athlete being energized. In science, the term "energy" has a precise definition that is somewhat different from the ordinary meaning. To see what scientists mean when they talk about energy, we must first introduce the familiar concept of work.

Work

In the lexicon of physics, we say that **work** is done whenever a force is exerted over a distance. When you picked up this book, for example, your muscles applied a force equal to the weight of the book over a distance of a foot or so. You did work.

This definition of work differs considerably from everyday usage. From a physicist's point of view, if you accidently drive into a tree and smash your fender, work has been done, because a force deformed the car's metal a measurable distance. On the other hand, a physicist would say that you haven't done any work if you spend an hour in a futile effort to move a large boulder, no matter how tired you get. Even though you have exerted a considerable force, the distance over which you exerted it is negligible.

Physicists provide an exact mathematical definition to their notion of work.

▶ **In words:**

Work is equal to the force that is exerted times the distance over which it is exerted.

▶ **In equation form:**

work (in joules) = force (in newtons) × distance (in meters),

where a joule is the unit of work, as defined below.

▶ **In symbols:**

$$W = F \times d$$

Work is done when a force is exerted over a distance.

In practical terms, even a small force can do a lot of work if it is exerted over a long distance.

As you might expect from the equation above, the units of work are equal to a force times a distance. In the metric system of units, where force is measured in *newtons*, work is measured in "newton-meters." This unit is given the special name "joule," after the English scientist James Prescott Joule (1818–1889), one of the first people to understand the properties of energy. One **joule** is defined as the amount of work done when you exert a force of one newton through a distance of one meter:

1 joule of work = 1 newton of force × 1 meter of distance.

In the English system of units (see Appendix A), where force is measured in pounds, work is measured in a unit called the *foot-pound* (usually abbreviated ft-lb).

EXAMPLE 3–1: Working Against Gravity

How much work do you do when you lift a 10-kilogram suitcase one meter off the ground?

▶ **Reasoning:** We must first calculate the force exerted by a 10-kilogram mass before we can determine work. From the previous chapter, we know that to lift a 10-kilogram mass against the acceleration of gravity (9.8 meters per second^2) requires a force given by:

force = mass × g
     = 10 kg × 9.8 m/sec^2
     = 98 newtons.

▶ **Solution:** Then, from the equation for work,

work = force × distance
    = 98 newtons × 1 meter
    = 98 joules ▲

Athletes must release their energy quickly and with great power to succeed in events such as the javelin throw.

Energy

Energy is defined as the ability to do work. If a system is capable of exerting a force over distance, then it possesses energy. The amount of a system's energy, which can be recorded in joules or foot-pounds (the same units used for work) is a measure of how much work the system might do. When you run out of energy, you simply can't do any work.

Power

Power provides a measure of both the amount of work that's done (or, equivalently, the amount of energy expended), and the time it takes to do it. In order to complete a physical task quickly, you must generate more power than if that same amount of work is done more slowly. If you run up a flight of stairs, your muscles need to generate more power than they would if you walked up that same flight. A power hitter in baseball swings the bat faster, converting the energy in his muscles more quickly than most other players.

Scientists define **power** as the rate at which work is done, or the rate at which energy is expended.

▶ **In words:**

> power is the amount of work done, divided by the time it takes to do it.

▶ **In equation form:**

$$\text{power (in watts)} = \frac{\text{work (in joules)}}{\text{time (in seconds)}},$$

where the watt is the unit of power, as defined below.

▶ **In symbols:**

$$P = W/t$$

If you do more work, or do a task in a shorter time, you use more power. Also, because the work performed equals the energy expended:

$$\text{power (in watts)} = \frac{\text{energy (in joules)}}{\text{time (in seconds)}},$$

In the metric system, power is measured in watts, after James Watt (1736–1819), the Scottish inventor who developed the modern steam engine that powered the Industrial Revolution. The **watt**, a unit of measurement that you probably encounter every day, is defined as the expenditure of 1 joule of energy in 1 second:

$$1 \text{ watt of power} = \frac{(1 \text{ joule of energy})}{(1 \text{ second of time})}.$$

When you change a light bulb, for example, you look at the rating of the new bulb to see whether it's 60, 75, or 100 watts—a measure of the rate of energy that the light bulb consumes when it is operating. Almost any hand tool or appliance in your home will be labeled with a power rating in watts. The unit of 1000 watts (corresponding to an expenditure of 1000 joules per second) is called a **kilowatt**, a commonly used measurement of electrical power. The English system, on the other hand, uses the more colorful *horsepower*, which is defined as 550 foot-pounds per second.

The equation defining power as energy divided by time, introduced above, may be rewritten as follows:

$$\text{energy (in joules)} = \text{power (in watts)} \times \text{time (in seconds)}$$

This equation is very important, because it allows you (and the electric company) to calculate how much energy you consume (and how much you have to pay for). Note from this equation that, while the joule is the

Table 3–1 ● **Important Terms**

Quantity	Definition	Units
Force	mass × acceleration	newtons
Work	force × distance	joules
Energy	ability to do work	joules
Energy	power × time	joules
Power	work/time = energy/time	watts

standard scientific unit for energy, energy can also be measured in units of power × time, such as the familiar kilowatt-hour (often abbreviated kwh) that appears on your electric bill.

Table 3–1 summarizes the important terms we've used for force, work, energy, and power.

Science in the Making

James Watt and the Horsepower

The horsepower, a unit of power with a colorful history, was devised by James Watt so that he could sell his steam engines. Watt knew that the main use of his engines would be in mines, where owners traditionally used horses to drive pumps that removed water. The easiest way to promote his new engines was to tell the mining engineers how many horses each engine would replace. Consequently, he did a series of experiments to determine how much energy a horse could generate over a given amount of time. He found that, on average, a healthy horse can do 550 ft-lbs of work every second over an average working day. Watt defined this unit to be the horsepower, and rated his engines accordingly. We still use the unit (the engines of virtually all cars and trucks are rated in horsepower), although we seldom build engines to replace horses these days. ■

EXAMPLE 3–2: Paying the Piper

A typical CD system uses 250 watts of electrical power. If you play your system for three hours in an evening, how much energy do you use? If energy costs eight cents a kilowatt-hour how much do you owe the electrical company?

▶ **Reasoning and Solution:** The total amount of energy you use will be given by:

energy = power × time
= 250 watts × 3 hours
= 750 watt-hours

Since 750 watts equals 0.75 kilowatt,

energy = 0.75 kilowatt-hour.

The cost will be:

8 cents per kilowatt-hour × 0.75 kilowatt-hour = 6 cents. ▲

Types of Energy ▮

Because energy—the ability to do work—appears in such a wide variety of physical systems, there are many different kinds of energy. The identification of these varieties posed a great challenge to science in the nineteenth century. The division of energy into different types is somewhat arbitrary, but it will prove convenient to recognize two very broad categories. **Kinetic energy** is energy associated with moving objects, while stored or **potential energy** is energy waiting to be released.

Kinetic Energy

Think about a cannonball flying through the air. When the iron ball hits a wooden target it exerts a force on the fibers in the wood, splintering and pushing them apart and creating a hole. Work has to be done to make that hole—fibers have to be moved aside, which means that a force must be exerted over the distance they move. When the cannonball hits the piece of wood, it does work, and a cannonball in flight clearly has the *ability* to do work because of its motion. This energy of motion is what we have called kinetic energy.

You can find countless examples of kinetic energy in nature. A fish moving through water, a bird flying, and a predator catching its prey all have kinetic energy. So do a speeding car, a flying frisbee, a falling leaf, and a running child.

Our intuition tells us that two factors govern the amount of an object's kinetic energy. First, heavier objects have more energy than light ones: a bowling ball traveling 10 meters per second (a very fast sprint) carries a lot more kinetic energy than a golf ball traveling at the same velocity. In fact, kinetic energy is proportional to mass—double the mass, double the kinetic energy.

Second, the faster something is moving, the greater the force it is capable of exerting. A high-speed collision causes much more damage than a fender bender in a parking lot. It turns out that an object's kinetic energy increases as the square of its velocity. A car moving 40 miles per hour has four times as much kinetic energy as one moving 20 mph, while at 60 mph a car carries nine times as much kinetic energy as at 20 mph. Thus a modest increase in speed can cause a large increase in kinetic energy.

These ideas are combined in the equation for kinetic energy:

▶ **In words:**

Kinetic energy equals the mass of the moving object times the square of that object's velocity, multiplied by the constant ½.

▶ **In equation form:**

kinetic energy (in joules) = ½ × [mass (in kg)] × [velocity (in m/s)]^2

▶ **In symbols:**

$E = \frac{1}{2} mv^2$

EXAMPLE 3–3: Bowling Balls and Baseballs

What is the kinetic energy of a 4-kilogram (about 8-pound) bowling ball rolling down a bowling lane at 10 meters per second (about 22 miles per hour)? Compare this energy to that of a 250-gram (about half-a-pound) baseball traveling 50 meters per second (almost 110 miles per hour). Which object would hurt more if it hit you (i.e., which object has the greater kinetic energy)?

▶ **Reasoning:** We have to substitute numbers into the equation for kinetic energy.

▶ **Solution:** For the 4-kg bowling ball traveling at 10 m/s:

$$\begin{aligned}\text{energy (in joules)} &= \frac{1}{2} \times \text{mass (in kg)} \times [\text{velocity (in m/s)}]^2 \\ &= \frac{1}{2} \times 4 \text{ kg} \times (10 \text{ m/s})^2 \\ &= \frac{1}{2} \times 4 \text{ kg} \times (100 \text{ m}^2/\text{s}^2) \\ &= 200 \text{ kg-m}^2/\text{s}^2 \\ &= 200 \text{ joules}\end{aligned}$$

For the 250-g baseball traveling at 50 m/s:

$$\begin{aligned}\text{energy (in joules)} &= \frac{1}{2} \times \text{mass (in kg)} \times [\text{velocity (in m/s)}]^2 \\ &= \frac{1}{2} \times 250 \text{ g} \times (50 \text{ m/sec})^2\end{aligned}$$

A gram is a thousandth of a kilogram, so 250 g = 0.25 kg:

$$\begin{aligned}&= \frac{1}{2} \times 0.25 \text{ kg} \times (2500 \text{ m}^2/\text{s}^2) \\ &= 312.5 \text{ kg-m}^2/\text{s}^2 \\ &= 312.5 \text{ joules}\end{aligned}$$

Even though the bowling ball is much more massive than a baseball, a hard hit baseball carries more kinetic energy than a typical bowling ball because of its high velocity. ▲

Potential Energy

Almost every mountain range in the country has a "balancing rock"—a boulder precariously perched on top of a hill so that it looks as if a little push would send it tumbling down the slope. If the balancing rock were to fall it would clearly acquire kinetic energy, and it would do "work" on anything it smashed. The balancing rock has the ability to do work even though it's not doing work right now, and it's not necessarily going to be doing work any time in the near future. The boulder possesses energy just by virtue of having been lifted up.

This kind of energy—energy that could result in the exertion of a force over distance but is not doing so now—is called potential energy. In the case of the balancing rock, it is called *gravitational potential energy,* because it is the force of gravity that would cause the rock to move and be capable of exerting its own force.

An object that has been lifted above the surface of the Earth possess an amount of gravitational potential energy exactly equal to the total

The influential nineteenth-century American painter Thomas Cole produced a series of five large works called "The Course of Empire." A human civilization in the foreground is seen to pass from infancy to opulence to destruction, while in the background of each canvas a distinctive large boulder sits precariously perched on a cliff edge. The great potential energy of the boulder is obvious to the viewer; Cole's point was that the most transient of natural phenomena outlasts the most permanent works of humans. (Collection of The New-York Historical Society.)

amount of work you would have to do to lift it from the ground to its present position.

▶ **In words:**

> The gravitational potential energy of any object equals its weight (the force of gravity exerted by the object) times its height above the ground.

▶ **In equation form:**

> Gravitational Potential Energy (in joules)
> = mass (in kg) × g (in m/s^2) × height (in m),

where g is the acceleration due to gravity at the Earth's surface (see Chapter 2).

▶ **In symbols:**

> $E = mgh$

In Example 3–1 we saw that it requires 98 joules of energy to lift a 10-kilogram suitcase 1 meter into the air. This is the amount of work that would be done if the suitcase were allowed to fall, and it is the amount of gravitational potential energy stored in the elevated suitcase.

We encounter many other kinds of potential energy besides the gravitational kind in our daily lives. *Chemical potential energy* is stored in the gasoline that moves your car, the batteries that power your radio, and the food you eat. All animals depend on the chemical potential energy of food, and all living things rely on molecules that store chemical energy for future use. In each of these situations energy is stored in the chemical bonds between atoms (see Chapter 9).

Wall outlets at your home and work provide a means to tap into *electrical potential energy,* waiting to turn a fan or drive your vacuum cleaner. A tightly coiled spring, a flexed muscle, and a stretched rubber band contain *elastic potential energy,* while a refrigerator magnet carries *magnetic potential energy*. In every case, energy is stored, ready to do work (Figure 3–1).

Heat, or Thermal Energy

Two centuries ago scientists understood the behavior of kinetic and potential energy, but the nature of heat was far more elusive. We now know that atoms and molecules—the minute particles that make up all matter—move around and vibrate, and therefore these particles possess kinetic energy. The tiny forces that they exert will be experienced only by other atoms and molecules, but that small scale doesn't make the force any less real. If molecules in a material begin to move more rapidly, they have more kinetic energy and are capable of exerting greater forces on each other in collisions. If you touch an object whose molecules are moving fast, the collisions of those molecules with molecules in your hand will exert greater force, and you will perceive the object to be hot.

(a)

(b)

(c)

Figure 3–1
Potential energy comes in many forms. (*a*) A rock ready to fall possesses gravitational potential energy. (*b*) A lighted fuse contains chemical potential energy. (*c*) A drawn bow has elastic potential energy. All are examples of potential energy about to be converted into kinetic energy.

What we normally call *heat,* therefore, is simply **thermal energy**—the kinetic energy of atoms and molecules.

Science in the Making

Discovering the Nature of Heat

What is heat? How would you apply the scientific method to determine its origins? That was the problem facing scientists 200 years ago.

In many respects, they realized, heat behaves like a fluid. It flows from place to place, and it seems to spread out evenly like water that has been spilled on the floor. Some objects soak up heat faster than others, and many materials seem to swell up when heated, just like waterlogged wood. Thus, in 1800, after years of observations and experiments, many physicists mistakenly accepted the theory that heat is an invisible fluid—they called it "caloric," the Latin word for heat. According to the caloric theory of heat, the best fuels like coal are saturated with caloric, while ice is virtually devoid of the substance.

The caloric theory of heat eventually failed as new observations failed to match the theory's predictions. In particular, the practical experience of machinists just did not bear out the idea of heat as a fluid. If heat is a fluid, then each object must contain a fixed quantity of that substance. But Benjamin Thompson (1753–1814), an American who spent some time as a cannon maker, discovered that the heat generated in cannon boring had nothing at all to do with the quantity of brass to be drilled. Sharp tools, he found, cut brass quickly with minimum heat generation, while dull tools made slow progress and produced prodigious amounts of heat.

Thompson proposed an alternative hypothesis. He suggested that heat is nothing more than a consequence of the mechanical energy of friction, not some theoretical, invisible fluid. He proved his point by immersing an entire cannon-boring machine in water, turning it on, and watching the heat generated turn the water to steam. British chemist and popular science lecturer Sir Humphry Davy (1778–1829) further dramatized Thompson's point when he generated heat by rubbing two pieces of ice together on a cold London day.

The work of Thompson, Davy, and others inspired English researcher James Prescott Joule to devise a special experiment to test the predictions of the rival theories. As shown in Figure 3–2, Joule's apparatus employed a weight that was lifted up and attached to a rope. The rope turned a paddle wheel immersed in a tub of water. The weight had gravitational potential energy and, as it fell, that energy was converted into kinetic energy of the rotating paddle. The paddle wheel's kinetic energy, in turn, was transferred to kinetic energy of water molecules. As Joule suspected, the water heated up. Heat, he declared, is just another form of energy. ■

Figure 3–2

Joule's experiment demonstrated that heat is another form of energy by showing that the kinetic energy of a paddle wheel is transferred to heat energy of the agitated water.

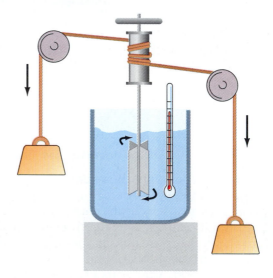

The Human Body

Energy in the Blood

A different approach to the problem of understanding the nature of heat was taken in the early 1840s by a young German physician named Julius Mayer (1814–1878). Soon after passing the state medical examinations, he traveled as a ship's doctor on a Dutch merchant vessel to what we now call Indonesia. In those days, one of the common methods of treating many diseases was a procedure known as bleeding, in which blood was drained from a patient's veins (presumably to let out the bad influences that were causing the disease). Mayer noticed that the blood in the veins of sailors just arrived from Europe was much redder than that of long-time residents in the tropics.

Mayer reasoned that this observation meant that the high temperatures of the tropics allowed the body to operate at a lower metabolism (the body's rate of energy production) and still maintain the same body temperature. In other words, there was an equivalence between heat (as produced by the body) and the kinds of energy stored in food (what we would call chemical energy today). Upon his return to Europe, Mayer did further studies to support his conclusion that heat was a form of energy, but it's clear that his main inspiration came from watching blood flowing from a patient's arm in his ship's sick bay in a sweltering tropical harbor.

In modern language, we would say that Mayer's insight was related to the fact that the body takes in chemical potential energy stored in food molecules, then uses chemical reactions (see Chapter 20) to convert that energy into heat (to maintain temperature) and other forms of energy (e.g., kinetic energy in muscles). The oxygen that we breathe in plays a crucial role in this process—in effect, it is needed to "burn" energy-rich molecules. If less energy is needed to maintain temperature, less oxygen is needed for the body's chemistry, and more oxygen will be found in the blood in the veins—a situation that makes blood appear redder, as Mayer observed. ■

Wave Energy

Anyone who has watched surf battering a seashore knows that waves carry energy. In the case of water waves, the type of energy involved is obvious. Large amounts of water are in rapid motion, and therefore possess kinetic energy. It is this energy that we see released when waves hit the shore.

Other kinds of waves possess energy, as well. For example, when a **sound wave** is generated, molecules in the air are set in motion and the energy of the sound wave is associated with the kinetic energy of those molecules. In Chapter 6 we will meet another important kind of wave—waves associated with electromagnetic radiation, such as the radiant energy (light) that streams from the Sun. This sort of wave stores its energy in changing electrical and magnetic fields.

Mass as Energy

The discovery that certain atoms such as uranium spontaneously release energy as they disintegrate—the phenomenon of *radioactivity*—led to the realization in the early twentieth century that mass is also a form of energy. This principle is the focus of Chapter 11, but the main idea is summarized in Albert Einstein's most famous equation:

▶ **In words:**

Every object at rest contains potential energy equivalent to the product of its mass times a constant, which is the speed of light squared.

▶ **As an equation:**

energy (in joules) = mass (in kg) × [speed of light (in m/s)]^2

▶ **In symbols:**

$E = mc^2$,

where c is the symbol for the speed of light, a constant equal to 300,000,000 meters per second (3×10^8 m/s).

This equation, which has achieved the rank of a cultural icon, tells us that it is possible to create energy from mass and to use energy to create mass. (Note: This equation does *not* mean the mass has to be traveling at the speed of light; the mass is assumed to be at rest.) Furthermore, because the speed of light is so great, the amount of energy stored in even a tiny amount of mass is enormous.

EXAMPLE 3–4: Lots of Potential

According to Einstein's equation, how much potential energy is contained in the mass of a 1-gram grape sitting on your desk?

▶ **Reasoning and Solution:** Substitute the mass, 1 gram, into Einstein's famous equation. Remember that 1 gram is a thousandth of a kilogram, and the speed of light is a constant, 3×10^8 m/s.

energy (in joules) = mass (in kg) × [speed of light (in m/s)]^2
$$= 0.001 \text{ kg} \times (3 \times 10^8 \text{ m/s})^2$$
$$= 0.001 \times 9 \times (10^{16} \text{ kg-m}^2/\text{s}^2)$$
$$= 9 \times 10^{13} \text{ joules}$$

The energy contained in 1 gram of mass is prodigious: almost 10 trillion joules or 25 million kilowatt-hours. The average American family uses about 1000 kilowatt-hours of electricity per month, so a single grape (if we had the means to convert its mass entirely to electrical energy) could satisfy your home's energy needs for the next 10,000 years! ▲

In practical terms, Einstein's equation pointed the way for using mass to generate electricity in nuclear power plants. This technology means a few pounds of nuclear fuel is enough to power an entire city.

The Interchangeability of Energy

You know from everyday experience that energy can be changed from one form to another (Table 3–2). Plants absorb light streaming from the Sun and convert that radiant energy into the stored chemical energy of cells and plant tissues. You eat those plants and convert that chemical energy into the kinetic energy of your muscles—energy of motion that in turn can be converted into gravitational potential energy when you climb a flight of stairs, elastic potential energy when you stretch a rubber band, or heat when you rub your hands together.

You observe energy changing form thousands of times every day. When you strike a match, the kinetic energy of the match head is transformed into heat as it rubs against the striking surface, and this heat triggers a chemical reaction that produces a flame. The energy in the flame comes from stored chemical energy in the match head. Similarly, hydroelectric power plants convert the gravitational potential energy of dammed-up water into electrical energy, which in turn can be transformed to heat or light energy in your home. When you drive your car chemical energy stored in gasoline converts into kinetic energy of the moving car, and eventually into waste heat.

The lesson from these examples is clear:

> **The many different forms of energy are interchangeable.**

Energy in one form can be converted into others.

Bungee jumping provides a dramatic illustration of this rule (Figure 3–3). Bungee jumpers climb to a high bridge or the top of a construction crane, where elastic cords are attached to their ankles. Then they launch themselves into space and fall toward the ground until the cords stretch, slow them down, and stop their fall.

From an energy point of view, the bungee jumper uses the chemical potential energy generated from food to walk up to the launching platform. The work that had been done against gravity provides the jumper with gravitational potential energy. During the long descent, that gravitational potential energy diminishes, while the jumper's kinetic energy simultaneously increases. As the cords begin to stretch, the jumper slows down and kinetic energy gradually is converted to stored elastic potential

A pole vaulter uses the energy of food to power his muscles. Kinetic energy of running is converted into elastic potential energy of the bent pole. That elastic potential energy converts to kinetic energy, giving the skilled vaulter sufficient gravitational potential energy to clear the bar.

Table 3–2 ● Some Kinds of Energy

Potential Energy	Kinetic Energy	Other
Gravitational	Moving objects	Mass
Chemical	Heat	
Elastic	Sound and other waves	
Electromagnetic		

Figure 3–3
Bungee jumping provides a dramatic example of energy changing from one form to another. Can you identify points at which the jumper has maximum kinetic energy? Maximum gravitational potential energy? Maximum elastic potential energy? Does the jumper ever have a combination of all three? What other kinds of energy are also involved?

energy in the cords. Eventually, the gravitational potential energy that the jumper had at the beginning is completely transferred to the stretched elastic cords, which then rebound, converting some of that stored elastic energy back into kinetic energy and gravitational potential energy. All the time, some of the energy is also converted to heat—heat in the stressed cord, frictional heat on the jumper's ankles, and heat in the air as it is pushed aside.

Every form of energy on our list can be converted to every other form of energy: This is one of the most fundamental properties of the universe in which we live.

The First Law of Thermodynamics: Energy Is Conserved

Scientists always look for constants in their efforts to describe a changing universe. Is the total number of atoms or electrons constant in the universe? Is the total amount of electrical charge fixed? Any statement that says that a quantity in nature does not change—that it is conserved—is called a **conservation law.**

Before describing the conservation law that relates to energy, we must first introduce the idea of a **system.** You can think of a system as an imaginary box into which you put some matter and some energy that you'd like to study. Scientists might want to study systems containing a pan of water, a forest, or the entire planet Earth. Doctors examine your nervous *system,* astronomers explore the solar *system,* and biologists observe a variety of eco*systems*. In each case, our investigation of nature is simplified by focusing on one small part of the universe.

If the object under study can exchange matter and energy with its surroundings—for example, a pan full of water that is heated on a stove and gradually evaporates, then we have an *open system.* An open system is like an open box where you can take things out and put things back in. If matter and energy do not freely exchange with their surroundings, as in a tightly shut box, then the system is said to be *closed* or *isolated.* Earth and its primary source of energy, the Sun, together make a system that may be thought of for most purposes as closed, because there are no significant amounts of matter or energy being added from outside sources.

The most important conservation law in the sciences is the conservation of energy. This law is also called the **first law of thermodynamics**. (*Thermodynamics*—literally the study of the movement of heat—is a term used for the science of heat, energy, and work.) The law can be stated as follows:

> **In an isolated system the total amount of energy, including heat energy, is conserved**

This law tells us that, although the kind of energy in a given system can change, the total amount cannot. For example, when that bungee jumper hurls herself into space, the total amount of gravitational energy she had at the beginning of the fall is still present in her and her surroundings at any time during the fall. When she's moving, some of the gravitational potential energy is converted into kinetic energy, some into elastic potential energy, and some into heat. At each point during the fall, however, the sum of kinetic, elastic potential, gravitational potential, and heat energies has to be the same as the gravitational energy at the beginning.

Energy is something like an economy with an absolutely fixed amount of money. You can earn it, store it in a bank or under your pillow, and spend it here and there when you want to. But the total amount of money doesn't change just because it passes through your hands. Likewise, in any physical situation you can shuffle energy from one place to another. You could take it out of the account labeled "kinetic" and put into the account labeled "potential," you could spread it around into accounts labeled "chemical potential," "elastic potential," "heat" and so on, but the first law of thermodynamics tells us that you can never have more energy than you started with.

 ## Science by the Numbers

Diet and Calories

The first law of thermodynamics has a great deal to say about the American obsession with weight and diet. Human beings take in energy with their food, energy we usually measure in calories. (Note that the calorie we talk about in foods is defined as the amount of energy needed to raise the temperature of a kilogram of water 1 degree C, a unit we will later call a kilocalorie.) When a certain amount of energy is taken in, the first law says that only one of two things can happen to it: it can be converted into work and waste heat, or it can be stored. If we take in more energy than we expend, the excess is stored in fat. If, on the other hand, we take in less than we expend, energy must be removed from storage to meet the deficit, and the amount of body fat decreases.

Here are a couple of rough rules you can use to calculate calories in your diet:

1. Under most circumstances, normal body maintenance uses up about 15 calories per day for each pound of body weight.
2. You have to take in about 3500 calories to gain a pound of fat.

Suppose you weigh 150 pounds. Then, to keep your weight constant, you have to take in

150 pounds × 15 calories/pound = 2250 calories per day.

If you wanted to lose one pound (3500 calories) a week (7 days), you would have to reduce your daily calorie intake by:

3500 calories/7 days = 500 calories per day

Another way of saying this is that you would have to reduce your calorie intake to 1750 calories—the equivalent of skipping dessert every day.

Alternatively, the first law says you can increase your energy use through exercise. Roughly speaking, to burn off 500 calories you'd have to run 5 miles, bike 15 miles, or swim for an hour.

It's a whole lot easier to refrain from eating than to burn off the weight by exercise. In fact, most researchers now say that the main benefit of exercise in weight control has to do with its ability to help people control their appetites. ■

Science in the Making

Lord Kelvin and the Age of the Earth

The first law of thermodynamics provided physicists with a powerful tool for describing and analyzing their universe. Every isolated system, the law tells us, has a fixed amount of energy. Naturally, one of the first systems that scientists considered was the Earth and Sun.

British physicist William Thomson (1824–1907), knighted as Lord Kelvin, asked a simple question: How much energy could be stored in the Earth? And, given the present rate at which energy radiates out into space, how old might it be? Though simple, these questions had profound implications for philosophers and theologians who had their own ideas about the Earth's relative antiquity. Some Biblical scholars believed that the Earth could be no more than a few thousand years old. Most geologists, on the other hand, saw evidence in layered rocks to suggest an Earth at least hundreds of millions of years old, while biologists also required vast amounts of time to account for the gradual evolution of life on Earth. Who was correct?

Kelvin believed, as did most of his contemporaries, that the Earth had formed from a contracting cloud of interstellar dust (see Chapter 14). He thought that the Earth had started as a hot body because impacts of large objects on it early in its history must have converted huge amounts of gravitational potential energy into heat. He used new developments in mathematics to calculate how long it would take for the hot Earth to cool to its present temperature. He assumed that there were no sources of energy inside the Earth, and found that the age of the Earth had to be less than about 100 million years. He soundly rejected the geologists' and biologists' claims of an older Earth as being incompatible with the laws of physics.

Seldom have scientists come to such a bitter impasse. Two competing theories about the age of the Earth, each supported by seemingly sound observations, were at odds. The calculations of the physicist

William Thomson, Lord Kelvin (1824–1907)

seemed unassailable, yet the observations of biologists and geologists in the field were equally meticulous. What could possibly resolve the dilemma? Had the scientific method failed? The solution to the dilemma came from a totally unexpected source when scientists discovered in the 1890s that rocks hold a previously unknown source of energy—radioactivity (see Chapter 11), in which heat is generated by the conversion of mass. Lord Kelvin's rigorous age calculations were in error only because he and his contemporaries were unaware of this critical component of the Earth's energy budget. The Earth, we now know, gains approximately half of its internal heat from the energy of radioactive decay. Revised calculations suggest an Earth several billions of years old, in conformity with geological and biological observations. ■

The Great Chain of Energy

Think about the epic chain of events that led up to the last time you turned on a light bulb. Hundreds of millions of years ago, energy was generated by the conversion of mass in the core of the Sun—the process called nuclear fusion (Chapter 11). Through atomic-scale collisions and other processes, this energy eventually worked its way to the surface of the Sun. From there, light energy streamed outward into space in the form of electromagnetic waves—ordinary sunlight. Most of that energy left the solar system and was radiated into outer space, but a portion reached the Earth. Some of that radiant energy was reflected by clouds or particles in the atmosphere, so that it, too, was lost in the vastness of space, but a tiny fraction of the original energy from the Sun came down to the surface of the Earth where it was absorbed.

Some of that absorbed energy was taken in by plants and used to drive the chemical reactions of life. Those biological processes created molecules that formed the fiber of trees and shrubs in a tropical forest and stored the sunlight as chemical potential energy. As time went by, the plants and shrubs died, were buried under the ooze of the tropical swamp, and began a long stay within the Earth. The organic matter in the plants was gradually transformed by the Earth's interior pressure and temperatures into coal. Through all this process the original energy contained in the sunlight remained, passing through an energy chain from living plant material to buried organic matter to coal.

Then, a few months ago, miners brought that coal from its underground reservoir to the surface of the Earth. The coal was shipped to a power plant and burned to make electricty. When you turned on the light, you tapped into that energy and used it to help you see. Most of the electrical energy was used to produce the heat of the glowing light filament, but a small fraction of the energy was converted directly into light. That light energy was itself converted into heat when it was absorbed by objects around you.

But that's not the end of the story. When you turned on the light, the energy that was stored in the sunlight was eventually entirely converted into heat. Some of this heat raised the temperature of the light bulb and its socket, some warmed the walls of your room slightly, some of it went into motion of the air that was heated when it absorbed light. Wherever it went, however, the result is the same. Because you turned on your light, the Earth is a little bit warmer than it would have been had you not done so. The net effect of this warmth is that the Earth will radiate a bit more heat into space.

From the energy point of view, then, the whole story about the creation of coal and your use of it can be thought of as nothing more than a modest delay in the radiation of energy from the Sun into the depths of space. The energy that does stop on the Earth may be stored for millions of years (short times on an astronomical scale), but eventually it, too, winds up being radiated back into space. Energy flows through our planet, and we use it while it's passing through.

The flow of energy never ceases. There is no point at which we can say "Here is where energy is created," or "Here is where it is destroyed." Energy flows through the Earth and every other system in the universe, constantly changing form, constantly shifting from one type to another, but always remaining the same in total quantity.

Technology

A Better Way to Burn

Many engineers and scientists work to provide new ways of supplying energy sources for the future. Some of this work, as we shall see in Chapter 10, concentrates on finding new ways to tap things such as solar energy. But some of the work is devoted to finding new and less polluting ways of using familiar sources, such as coal.

The fluidized bed combustor is an example of that sort of technology. As shown in Figure 3–4, this device is a high-tech "furnace" in which coal is first ground into a powder, mixed with a much larger amount of an inert material such as sand or limestone, and then supported on a blast of air as it burns. In this situation, the fine particles float around and act more like a liquid than a collection of solids. The floating, burning assembly is called a "fluidized bed."

This way of burning coal has many advantages over the standard boiler. For one thing, since most of the material in the bed isn't actually burning, once the bed is heated up, the coal is burned much more efficiently than in a normal boiler. In addition, the pipes carrying the water to be heated can be put right down in the bed, so that a lot more of the heat actually gets to the water.

In addition to being more efficient, the fluidized bed combustor is also less polluting than a normal boiler. For example, some of the sulfur impurities in the coal combine with limestone in the bed to form a solid slag, which doesn't go out the smokestack. Furthermore, the increased efficiency of burning means that the combustor can operate at a lower temperature than a normal boiler, so that fewer nitrogen molecules in the air will be combined with oxygen as the coal burns. Thus this method of burning coal contributes much less to acid rain (see Chapter 25) than standard methods.

This modern coal-burning furnace is one example of a large number of advanced technologies that scientists and engineers are trying to

Figure 3–4

Some day coal-burning furnaces may employ this efficient, less-polluting "fluidized bed" design. Air jets suspend coal particles, which burn efficiently and facilitate efficient heat transfer from the burning fuel. This technology allows more energy to be extracted from the coal and reduces pollution from sulfur impurities in the coal.

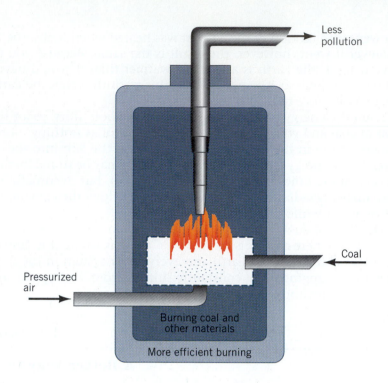

Less pollution

Coal

Pressurized air

Burning coal and other materials

More efficient burning

develop in their search for nonpolluting energy sources. It's not spectacular, and you won't see it in headlines. Nevertheless, it's one of those "nickel-and-dime" projects that could very likely make our lives just a little bit better in the future. ■

THINKING MORE ABOUT ENERGY

Fossil Fuels

The first law of thermodynamics contains two central ideas: (1) the energy around us constantly shifts from one form to another, and (2) the total amount of energy is unchanging. At first glance this natural law seems like good news—it seems to imply that we can use the same energy over and over again and never run out. But not all kinds of energy are equal.

All life is rich in the element carbon, which plays a key role in virtually all the chemicals that make up our cells. Life uses the Sun's energy, directly through photosynthesis or indirectly through food, to form these carbon-based substances, which are rich in chemical potential energy. When living things die, they may collect in layers at the bottoms of ponds, lakes, or oceans. Over time, as the layers become buried, the Earth's temperature and pressure may alter the chemicals of life into deposits of fossil fuels.

Fossil fuels, carbon-rich deposits of ancient life that burn with a hot flame, have been the most important energy source during the past century and a half. *Coal*, *oil (petroleum)*, and *natural gas*, the most common fossil fuels, are consumed in prodigious quantities around the world (Figure 3–5). They now account for approximately 90% of all energy consumed by industrial nations. In the United States alone approximately 1 billion tons of coal and 2.5 billion barrels of petroleum are used every year.

Geologists estimate that it takes tens of millions of years of gradual burial under layers of sediments, combined with the transforming

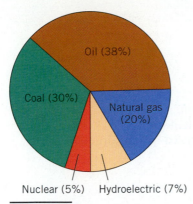

Figure 3–5

Sources of energy for industrial notation. Note that most of our energy comes from fossil fuels.

effects of temperature and pressure, to form a coal seam or petroleum deposit. Coal forms from layer upon layer of plants that thrived in vast ancient swamps, while petroleum represents primarily the organic matter once contained in plankton, microscopic organisms that float near the ocean's surface. While these natural processes continue today, the rate of coal and petroleum formation in the Earth is only a small fraction of the fossil fuels being consumed. For this reason, fossil fuels are classified as *nonrenewable resources*.

One consequence of this situation is clear. Humans cannot continue to rely on fossil fuels forever. Reserves of high-grade crude oil and the cleanest-burning varieties of coal may last less than 100 more years. Less-efficient forms of fossil fuels, including lower grades of coal and *oil shales* in which petroleum is dispersed through solid rock, could be depleted within a few centuries. All the energy now locked up in those valuable energy reserves will still exist, but in the form of unusable heat radiating far into space.

Given the irreversible consequences of burning up our fossil fuel reserves, what steps should we take to promote energy conservation? Should energy be taxed at a higher rate? Should we assume that new energy sources will become available as they are needed?

Summary

Work, measured in *joules* (or foot-pounds), is defined as a force applied over a distance. You do work every time you move an object. Every action of our lives requires *energy* (also measured in joules), which is the ability to do work. *Power,* measured in *watts* or *kilowatts,* indicates the rate at which energy is expended.

Energy comes in several varieties. *Kinetic energy* is the energy associated with moving objects such as cars or cannonballs. *Potential energy,* on the other hand, is stored energy, ready for use—the chemical energy of coal, the elastic energy of a coiled spring, the gravitational energy of dammed-up water, or the electrical energy in your wall socket. *Thermal energy* or *heat* is the form of kinetic energy associated with vibrating atoms and molecules. Energy can also take the form of waves, such as *sound waves* or light waves. And early in the twentieth century it was discovered that mass is also a form of energy. All around us energy constantly shifts from one form to another, and all of these kinds of energy are interchangeable.

The most fundamental idea about energy, expressed in the *first law of thermodynamics*, is that it is *conserved*—the total amount of energy in an isolated *system* never changes. Energy can shift back and forth between the different kinds, but the sum of all energy is constant.

Key Terms About Energy

work (measured in joules)	watt and kilowatt	sound wave
joule	kinetic energy	conservation law
energy (measured in joules)	potential energy	system
power (measured in watts or kilowatts)	thermal energy (heat)	first law of thermodynamics

Key Equations

work (in joules) = force (in newtons) × distance (in meters)

power (in watts) = $\dfrac{\text{energy (in joules)}}{\text{time (in seconds)}}$,

energy (in joules) = power (in watts) × time (in seconds)

kinetic energy (in joules) = ½ × [mass (in kg)] × [velocity (in m/s)]²

gravitational potential energy (in joules) = g × mass (in kg) × height (in m)

energy associated with mass at rest (in joules) = mass (in kg) × [speed of light (in m/s)]²

Constant

$c = 3 \times 10^8$ m/s = speed of light

Review Questions

1. What is the scientific definition of work? How does it differ from ordinary English usage?

2. Is the kilowatt-hour a unit of energy or power? How about the kilowatt?

3. What is the difference between the watt and the horsepower?

4. What is the difference between the joule and the kilowatt-hour? Who uses which unit?

5. What is the difference between energy and power?

6. List some different kinds of energy. Explain how they differ from each other.

7. Find something in your classroom or dorm room that possesses gravitational potential energy.

8. What is the relation between the perception that something is hot and the motion of the molecules in it?

9. Describe the historical process by which the notion of heat as a form of energy was developed.

10. Can waves carry energy? If so, give an example.

11. What does it mean to say that different forms of energy are interchangeable?

12. What does it mean to say that energy is conserved?

13. How did the discovery that mass is a form of energy resolve the debate over the age of the Earth?

14. Explain what it means to say "Energy flows through the Earth."

Discussion

1. How does the discovery of heat as a form of energy illustrate the scientific method?

2. You use energy to heat your home. What ultimately happens to the energy that you pay for in your heating bill?

3. Think about your activities today. Pick one of them and identify the chain of energy that led to it. Where will the energy in that chain eventually wind up?

4. What happens on a hot summer day when the energy demand on your electrical-generating plant exceeds its energy output?

5. What kinds of energy are present in the following systems:
   a. water behind a dam
   b. a swinging pendulum
   c. an apple on an apple tree
   d. a uranium atom deep in the Earth

6. Identify four sources of energy around us that are constantly being renewed. What sources of energy do we use that are not constantly renewed?

7. Describe how you might convert the elastic potential energy of a coiled spring into heat energy. How might you convert heat energy into gravitational potential energy?

8. Plants and animals are still dying and falling to the ocean bottom today. Why, then, do we not classify fossil fuels as renewable resources?

9. Some people say that you lose more calories by eating celery than you gain. How could that be?

Problems

1. How much work against gravity do you do when you climb a flight of stairs 3 meters high? Compare this work to that done by a 60-watt light

bulb in an hour. How many flights of stairs would you have to climb to equal the work of the light bulb?

2. Would you rather be hit by a 1-kilogram mass traveling 10 meters per second, or a 2-kilogram mass traveling 5 meters per second?

3. Compared to a car moving at 10 miles per hour, how much kinetic energy does that same car have when it moves at 20 miles per hour? 30? 60? What do these numbers suggest to you about the difficulty of stopping a car as its speed increases?

4. According to Einstein's famous equation, $E = mc^2$, how much energy is contained in a pound of feathers? A pound of lead? (Note: You will first need to convert pounds into kilograms.)

5. A small air compressor operates on a 1.5-horse-power electric motor for 8 hours a day. How much energy is consumed by the motor daily? If electricity costs 10 cents a kilowatt-hour, how much does it cost to run the compressor each day. (Note: 1 horsepower equals about 750 watts.)

6. Joules and kilowatt-hours are both units of energy. How many joules are equal to one kilowatt-hour?

Investigations

1. Look at your most recent electric bill and find the cost of one kilowatt-hour in your area. Then,
   a. Look at the back of your CD player or an appliance and find the power rating in watts. How much does it cost for you to operate the device for one hour?
   b. If you leave a 100-watt light bulb on all the time, how much will you pay in a year of electric bills?
   c. If you had to pay $10.00 for a high-efficiency bulb that provided the same light as the 100-watt bulb with only 10 watts of power, how much would you save per year of electric bills, assuming you use the light five hours per day? Would it be worth your while to buy the energy-efficient bulb if the ordinary bulb cost $1.00?

2. What kind of fuel is used at your local power plant? What are the implications of the first law of thermodynamics regarding our use of fossil fuels? Our use of solar energy?

3. Keep track of the calorie content of what you eat for a few days. If you kept eating at this rate, what would your weight eventually become?

4. Check your household's electric bills for the past year and calculate your total electric consumption for the year.
   a. How many 10-kilogram weights would you have to raise 100 meters to produce a gravitational potential energy equal to this consumption.
   b. How much mass is equal to this consumption ($E = mc^2$).

5. Investigate the history of the controversy between Lord Kelvin and his contemporaries regarding the age of the Earth. When did the debate begin? How long did it last? What kinds of evidence did biologists, geologists, and physicists use to support their differing calculations of the Earth's age?

6. In this chapter we introduced several energy units: the joule, the foot-pound, the kilowatt-hour, and the calorie. There are other energy units as well, including the BTU, the erg, the electron-volt, and many more. Find a table of energy units and their conversion factors. Arrange these units in order of smallest to largest. Why are there so many different energy units? Who uses which ones and why?

7. At one point the Clinton–Gore administration advocated the imposition of a BTU tax on energy. What is this tax? Where does the name "BTU tax" come from? How might it be implemented? What are the pros and cons? If new revenues were essential to prevent the budget deficit from growing, how would you vote on legislation to impose a BTU tax? Why?

Additional Readings

Burchfield, Joe D. *Lord Kelvin and the Age of the Earth.* Chicago: University of Chicago Press, 1990.

Cardwell, D.S.L. *From Watt to Clausius: The Rise of Thermodynamics in the Early Industrial Age.* Ithaca NY: Cornell University Press, 1971.

Fowler, John M. *Energy and the Environment* (second edition). New York, McGraw-Hill, 1984.

Heat and the Second Law of Thermodynamics

Energy always goes from a more useful to a less useful form.

A Random Walk
The Cafeteria

Next time you're in your local cafeteria think about an obvious but extraordinary fact: Nature seems to have a direction. Many events in your daily life progress only one way in time.

As you go through the food line, think about how each course is prepared. You can peel a piece of fruit, but never put it back together. It's easy to scramble eggs, but impossible to unscramble them. You can cook vegetables, but there's no way to uncook them. And popcorn, once popped, can't be unpopped. But why should this be? Nothing in Newton's laws of motion or the law of gravity suggests that events work only in one way. Nothing that we have learned about energy explains the directionality of nature.

At the cafeteria you've probably noticed that foods and drinks that are very hot or very cold display a similar kind of direction. A glass of ice water gradually gets warmer, while a plate of hot food cools—heat spreads out uniformly. Ice cream gradually melts, while hot fudge sauce cools and hardens. These everyday events are so familiar that we take them for granted, yet underlying the popping of popcorn and the cooling of a hot drink is one of nature's most subtle and fascinating laws—the second law of thermodynamics.

Nature's Direction

Think about the dozens of directional events that happen every day. A drop of perfume quickly pervades an entire room with scent, but you'd be hard pressed to collect all those perfume molecules into a single drop again. Your dorm room seems to get messy in the course of the week all by itself, but it takes time and effort to clean it up. And you constantly experience the inevitable, irreversible process of aging.

The first law of thermodynamics—conservation of energy—in no way forbids events to progress in the "wrong" direction. In fact, it would take exactly as much energy to unscramble an egg, if we could do it, as it takes to scramble it. The energy of a room with perfume molecules dispersed throughout is the same as the energy of the room with those molecules tightly bottled. And the energy that went into strewing things about your room is exactly the same energy it takes to put everything back again. Yet there seems to be some natural tendency for things to become less orderly with time.

This directionality in nature can be traced, ultimately, to the behavior of atoms and molecules in materials. If you hold an object in your hand, for example, its atoms are moving at more or less the same average velocity. If you introduce one more atom into this collection—an atom that is moving much faster than any of its neighbors—you will have a situation in which many atoms move slowly and one atom moves fast.

Over the course of time, the fast atom will collide again and again with the others. In each collision it will probably lose some of its energy, much as a fast-moving billiard ball slows down after it collides with a couple of other billiard balls. If you wait long enough, the fast-moving atom will share its energy with all the other atoms. Consequently, every atom in the collection will be moving slightly faster than those in the original collection, and there will be no single fast atom.

If you watch this collection of atoms for a long period of time, it is extremely unlikely that the collisions will ever arrange themselves in such a way that one atom moves very fast in a collection of very slow ones. In the language of the physicists, the original *state* with only one fast atom is highly improbable. Over the course of time, any unlikely initial state will evolve into a more probable state—a situation like the one in which all the atoms have approximately the same energy.

The tendency of all systems to evolve from improbable to more probable states accounts for the directionality that we see in the universe around us. There is no reason from the point of view of energy alone that improbable situations can't occur. Fifteen slow-moving billiard balls have enough energy to produce one fast-moving ball. The fact that this situation doesn't occur in nature is an important clue as to how things work at the atomic level (Figure 4–1).

Nineteenth-century scientists discovered the underlying reasons for nature's directionality by studying heat, the motion of atoms and molecules. Before dealing with the details of these discoveries, as summarized in the second law of thermodynamics, let's consider some properties of heat.

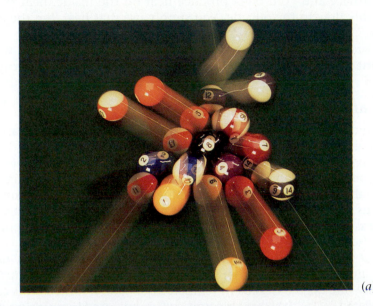

(*a*)

Figure 4–1

Many events work in only one direction. (*a*) As a pool game begins the cue ball contains all the kinetic energy, but this energy soon becomes distributed evenly through all the balls. (*b*) A building collapses, transforming the highly ordered structure into a disordered pile of rubble. (*c*) A flower, once cut, begins to wilt and decay.

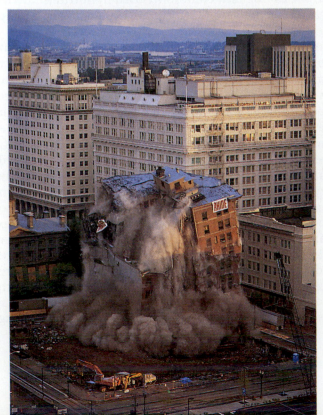

(*b*)

(*c*)

Coming to Terms with Heat

Atoms never sit still. They are always moving, and in the process they distribute their kinetic energy—what we called thermal energy, or heat. If you have ever tried to warm a house during a cold winter day, you have practical experience of this fact. If you turn off the furnace, the heat in the house gradually leaks away to the outside, and the house begins to get cold. The only way you can keep things warm is to keep adding more heat. Similarly, our bodies constantly produce energy to maintain our core body temperature close to 98.6 degrees Fahrenheit (37°C). Both your furnace and your body produce heat on the inside, heat that will inevitably flow to the outside. You simply use that heat "on the fly," as it were.

In order to understand the nature of heat and its movement, we need to define three closely related terms: heat, temperature, and heat capacity.

Heat and Temperature

In everyday conversation we often use the words temperature and heat interchangeably, but to scientists the words have somewhat different meanings. Heat is a form of energy that flows from a hot object to a colder object. Any object that is hot stores and can transfer a fixed and measurable amount of this kind of energy. A gallon of boiling water contains more heat energy than a pint of boiling water. Heat, therefore, is a measure of the quantity of atomic kinetic energy contained in every object (see Chapter 3).

Heat is often measured in *calories,* a common unit of energy defined as the amount of heat required to raise one gram of room-temperature water by one degree Celsius in temperature. (Don't confuse this calorie, abbreviated cal, with the calorie unit commonly used in nutrition discussions. The dietary calorie, abbreviated Cal, equals 1000 calories. See Appendix A.)

Temperature, on the other hand, is a relative term and reflects how vigorously atoms in a substance are moving and colliding. Two objects are defined to be at the same **temperature** if no heat energy flows spontaneously from one to the other. A gallon of boiling water, therefore, is at the same temperature as a pint of boiling water. The difference in temperature between two objects is one of the factors that determines how quickly heat energy will be transferred between those two objects: the larger the temperature difference, the more rapidly heat energy will be transferred. You may have seen this phenomenon at work in the summertime when violent winds, often accompanied by a line of thunderstorms, precede a cold front. The strong winds are a consequence of rapid heat transfer between two air masses at very different temperatures.

Temperature scales provide a convenient way to compare the temperatures of two objects. Many different temperature scales have been proposed; all scales and temperature units are arbitrary, but every scale requires two easily reproduced temperatures for calibration. The freezing and boiling points of pure water are commonly used standards today in the Fahrenheit scale (32 and 212 degrees for freezing and boiling,

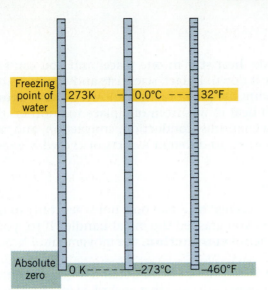

respectively) and the Celsius scale (where 0 and 100 degrees correspond to freezing and boiling water, respectively). The Kelvin temperature scale uses the same degree as the Celsius scale, but it defines 0 Kelvins as **absolute zero,** which is the coldest attainable temperature—the temperature at which it is impossible to extract any heat energy at all from atoms or molecules. The temperature of absolute zero is approximately −273°C, or −460°F. It turns out, therefore, that freezing and boiling occur at about 273 and 373 Kelvins, respectively (Figure 4–2).

Heat Capacity

Heat capacity is a measure of the ability of a material to absorb heat energy, and is defined as the quantity of heat required to raise the temperature of one gram of that material by 1°C. Water displays the largest heat capacity of any common substance; by definition, one calorie is required to raise the temperature of a gram of water by 1°C. By contrast, you know that metals heat up quickly in a fire, so a small amount of heat energy can cause a significant increase in the metal's temperature.

   Think about the last time you boiled water in a copper-bottomed pot. It doesn't take long to raise the temperature of a copper pot to above the boiling point of water because copper, like most other metals, can't hold much heat energy. In fact, one calorie will raise the temperature of a gram of copper by about 10°C. But water is a different matter; it must absorb 10 times more energy per gram than copper to raise its temperature. Thus, even at the highest stove setting, it can take several minutes to boil a pot of water. This ability of water to store heat energy plays a critical role in the Earth's climate, which is moderated by the relatively steady temperatures of the oceans.

Heat Transfer

You can't make heat stay in one place, and you can't confine it—you can only slow it down. In fact, scientists and engineers have spent many decades studying the phenomenon known as **heat transfer**—the process by which heat moves from one place to another. Heat transfers by three basic mechanisms: conduction, convection, and radiation, each of which is important to different aspects of everyday experience.

Conduction

Have you ever reached for a pan on a hot stove, only to have your fingers scorched when you grasped the metal handle? If so, you have had first-hand experience of **conduction,** the movement of heat by atomic collisions.

As shown in Figure 4–3, conduction works through the action of individual atoms or molecules that are linked together by chemical bonds. If a piece of metal is heated at one end, the atoms at that end begin to move faster. When they vibrate and collide with atoms farther away from the heat source, they are likely to transfer energy to those atoms, so that the farther molecules will begin to move faster as well. A chain of collisions occurs, with atoms progressively farther and farther away from the heat source moving faster and faster as time goes by.

Looking at this phenomenon from the outside, it appears that heat somehow flows like a liquid from the heat source through the metal. If you consider the atomic level, however, you realize that there's nothing particularly mysterious about this process. Conduction of heat results from collisions between vibrating atoms or groups of atoms, called molecules. When a fast-moving object collides with a slow-moving one, the fast object will usually slow down and the slow object will speed up.

Figure 4–3

Steelworkers test a molten sample from the furnace. Why does the worker at the left use thick insulated gloves to hold the cool end of the rod? Heat transfers from the hot end to the cool end by the process of conduction.

Conduction of heat is responsible for a large part of the heating bills in homes and office buildings. The process works like this: The air inside a house in winter is kept warmer than the air outside, so that the molecules of the air inside are moving faster than the molecules in the air outside. When the molecules inside collide with materials in the wall (a window pane, for example), they impart some of their heat energy to the molecules in the wall. At that point, conduction takes over and the heat is transferred to the outside of the wall. There the heat energy is transferred to the great outdoors by convection and radiation, processes that we will describe in a moment. The key point, however, is that you can think about every part of your house as being a kind of conduit carrying heat from the interior to the outside.

One way of slowing down the flow of heat out of a house is to add insulation to the walls or to use special kinds of glass for the windows. These processes work because materials differ in their **thermal conductivity**—their ability to transfer heat energy from one molecule to the next by conduction. Have you ever noticed that a piece of wood at room temperature feels "normal," while a piece of metal at the same temperature feels cold to the touch? The wood and metal are at exactly the same temperature, but the metal feels cold because it is a good *heat conductor*—it moves heat rapidly away from your skin, which is generally warmer than air temperature. The wood, on the other hand, is a good *heat insulator*—it impedes the flow of heat and so it feels comparatively warm. You wouldn't want to live in a house made entirely of metal, and we usually use relatively good insulators such as wood or masonry as primary building materials. The insulation in your home is designed to have especially low thermal conductivity, so that heat transfer is slowed down (but never completely stopped). Thus, when you use special insulated window panes or put certain kinds of insulation in your walls, you make it more difficult for heat to flow outside, and thereby allow yourself to use the heat longer before it ultimately leaks away.

Convection

Let's look carefully at a pot of boiling water on the stove. On the surface of the water you will see a rolling, churning motion as the water moves and mixes. If you put your hand above the water, you feel heat. Heat has been transferred from the water at the bottom of the pot to the top by **convection,** the transfer of heat by the bulk motion of the water itself, as shown in Figure 4–4.

Water cools

Cooler water sinks   Hot water rises   Cooler water sinks

Figure 4–4

Convection. Heat is transferred by the bulk motion of the water.

Water near the bottom of the pan expands as it is heated by the flames. Therefore, it weighs less per unit volume than the colder water immediately above it. A situation like this, with colder, denser water above and warmer, less dense water below, is unstable. The denser fluid tends to descend and displace the less dense fluid, which in turn begins to rise. Consequently, the warm water from the bottom rises to the top, while the cool water from the top sinks to the bottom (Figure 4–4). In convection, masses of water move in bulk and carry the fast-moving molecules with them. Heat is transferred by the actual physical motion of these masses of water.

Convection is a continuous, cyclic process as long as heat is added to the water. As cool water from the top of a pot arrives at the bottom it begins to be heated by the burner. In the same way, as the hot water gets to the top, its heat is sent off into the air. The water on the top cools and contracts, while the water on the bottom heats and expands. The original situation is repeated continuously with the less dense fluid on the bottom always rising and the more dense fluid on the top always sinking. This transfer of fluids results in a kind of a rolling motion, which you see when you look at the surface of boiling water. Each of these regions of rising and sinking water is called a **convection cell**.

The areas of clear water, which seem to be bubbling up, are the places where warm water is rising. The places where bubbles and scum tend to collect—the places that look rather stagnant—are where the cool water is sinking, leaving behind whatever passengers (such as minerals) it happened to be carrying at the surface. Heat is carried from the burner through the convection of the water, and is eventually transferred to the atmosphere.

Convection is a very efficient way of transferring heat. If you watch water in a pot on the stove through the entire heating process, you will notice that for a while the surface of the water simply gets hotter and hotter. During this period, the heat transfers to the surface by ordinary conduction through the water molecules, not by bulk motion of the water. Soon, however, the water contains too much heat to be transferred by conduction alone, and the water starts to move—first in little bubbles here and there, then in fully defined convection cells. Convection sets in, in other words, when there is too much heat to be transferred by conduction alone.

Convection is a very common process in nature. You can find all sorts of examples, from the small-scale circulation of cold water in a glass of ice tea, to air rising above a radiator or toaster, to large-scale motions of the earth's atmosphere. You may even have seen convection cells in operation in large urban areas. When you're in the parking lot of a large shopping mall on a hot summer day, you can probably see the air shimmer. What you are seeing is air, heated by the hot asphalt, rising upward. Some place farther away, perhaps out in the countryside, cooler air is falling. The shopping center with all its concrete is called a "heat island" and is the source of rising air. It is the hot part of a convection cell, while the rest of the convection cell is the downward flowing air elsewhere.

You may also have noticed that the temperature in big cities is usually a few degrees warmer than in outlying suburbs. Cities help create their own weather because they are heat islands where convection cells develop. Rainfall is typically higher in cities than in the surrounding atmosphere precisely because the cities set up convection cells that draw in cool moist air from surrounding areas.

Technology

Home Insulation

Today's home builders take heat convection and conduction very seriously. An energy efficient dwelling has to hold onto its heat in the winter and remain cool in the summer. A variety of high-tech materials provide effective solutions to this insulation problem.

Fiberglass, the most widely used insulation, is made of loosely intertwined strands of glass. It works by minimizing the opportunities for conduction and convection of heat out of your home (see Figure 4–5). Solid glass is a rather poor heat insulator, but it takes a long time for heat to move along a thin, twisted glass fiber, and even longer for heat to transfer across the occasional contact points between pairs of crossed fibers. A clothlike mat of fiberglass, furthermore, disrupts air flow and prevents heat transfer by convection. A thick, continuous layer of fiberglass in your walls and ceiling thus acts as an ideal barrier to the flow of heat.

If our houses were constructed with solid walls, then fiberglass would serve all our needs, but windows pose a special problem. Have you ever sat near an old window on a cold winter day? Old-style single-pane windows conduct heat rapidly, contributing to convection cells (cold drafts) in old houses. But how do we let light in without letting heat out? One solution is double-pane windows with sealed, airtight spaces between the panes that greatly restrict heat conduction. In addition, builders employ a variety of caulking and foam insulation to seal any possible leaks around windows and doors. ■

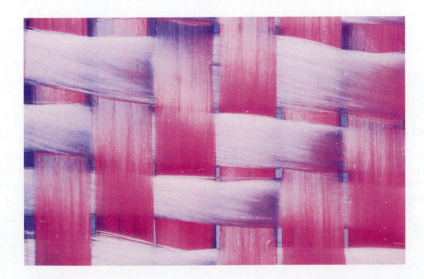

Figure 4–5
The interweaving of fibers in an ordinary piece of fiberglass, magnified in this photomicrograph, reduce heat transfer by convection and conduction.

Feathers provide birds with a natural insulation.

Animal Insulation: Fur and Feathers

Houses aren't the only place where insulation can be seen in our world. Two kinds of animals—birds and mammals—maintain a constant body temperature despite the temperature of their surroundings, and both have evolved methods to control the flow of heat into and out of their bodies. Part of these strategies involve the use of insulating materials—furs, feathers, and fat—that serve to slow down that flow. Since most of the time an animal's body is warmer than the environment, the most common situation is one in which the insulation works to keep heat in.

Whales, walruses, and seals are example of animals that have thick layers of fat to insulate them from the cold arctic waters in which they swim. Fat is a poor conductor of heat, and plays much the same role in their bodies as the fiberglass insulation in your attic.

Feathers are another kind of insulation—in fact, most biologists believe that feathers evolved first as a kind of insulation to help birds maintain their body temperature, and only later were adapted for flight. Feathers are made of light, hollow tubes connected to each other by an array of small interlocking spikes. They have some insulating properties themselves, but their main effect comes from the fact that they trap air next to the body, and, as we pointed out above, stationary air is a rather good insulator.

Birds often react to extreme cold by contracting muscles in their skin so that the feathers fluff out. This has the effect of increasing the thickness (and hence the insulating power) of the layer of trapped air. (Incidentally, birds need insulation more than we do because their normal body temperature is 41°C or 106°F.)

Hair (or fur) is actually made up of dead cells similar to those in the outer layer of the skin. Like feathers, hair serves as an insulator in its own right and traps a layer of air near the body. In some animals (for example, polar bears) the insulating power of the hair is increased because each hair contains tiny bubbles of trapped air. The reflection of light from these bubbles makes polar bear fur appear white—the strands of hair are actually translucent.

Hair grows from follicles in the skin, and small muscles allow animals to make their hair stand up to increase its insulating power. Human beings, who evolved in a warm climate, have lost much of their body hair as well as the ability to make most of it stand up. There is a reminder of our mammal nature, however, in the phenomenon of "goose bumps," which is the attempt by muscles in the skin to make the nonexistent hair stand up.

Radiation

Everyone has had the experience of coming in on a cold day and finding a fire in the fireplace or an electric heater glowing red hot. The normal reaction is to walk up to the source of heat, hold out your hands, and feel the warmth moving into your skin.

How did the heat energy get from the fire to your hands? It couldn't have done so by conduction—it's too hard to carry heat through the air that way. It couldn't have been convection either, because you don't feel a hot breeze. The air in the room is almost stationary.

What you experience is the third kind of heat transfer—**radiation,** or the transfer of heat by electromagnetic radiation, which is a form of wave energy that we will discuss in Chapter 6. A fire or an electric heater radiates heat in the form of *infrared radiation*. This radiation travels like light from the source of heat to your hand, where it is absorbed and converted into kinetic energy of molecules. You perceive heat because of the energy that the infrared radiation carries to your hand (Figure 4–6).

Every object in the universe radiates energy. Under normal circumstances, as an object gives off radiation to its surroundings it also receives radiation from those surroundings. Thus a kind of equilibrium is set up, and there is no net loss of energy because the object is at the same temperature as its surroundings. If, however, the object is at a higher temperature than its surroundings, it will radiate more energy than it receives. Your body, for example, constantly radiates energy into its cooler surroundings, energy that can be detected easily at night with infrared goggles. You will continue to radiate this energy as long as your body processes the food that keeps you alive (see Chapter 21).

Radiation, such as infrared, is the only kind of energy that can travel through the emptiness of space. Conduction requires atoms or molecules that can vibrate and collide with each other. Convection requires atoms or molecules of liquid or gas in bulk, so that they can move. But radiation doesn't require anything in the environment to facilitate it; radiation can even travel through a vacuum. The energy that falls on Earth in the form of sunlight—almost all of the energy that sustains life on Earth—travels through 93 million miles of intervening empty space in the form of radiation.

In the real world all three types of heat transfer—conduction, convection, and radiation—occur all the time. Any one can occur by itself, in

Figure 4–6
A fire transfers most of its energy by infrared radiation.

combination with another, or all three can occur simultaneously. At this moment, heat is being generated in your body. It can travel through tissues by conduction and is radiated from the surface of your skin. Think about the heat generated in a pot of boiling water. Heat moves through the metal bottom and sides of the pot by conduction, through the boiling water by convection, and from the sides and surface of the water by radiation. In fact, everywhere in the natural world heat is constantly being transferred by these three mechanisms.

The Human Body

Temperature Regulation

Like other mammals, human beings maintain a constant body temperature, regardless of the temperature in the environment. The laws of heat transfer apply to the human body, of course, but the way the body actually maintains its temperature is quite complex.

If the body temperature starts to rise, blood vessels near the surface of the skin dilate, so that the blood can carry more heat to the surface, where it can be radiated away. This is why you often appear flushed after being in the sun for a while. In addition, you start to sweat. The purpose of sweating is to put water on your skin, then use body heat to evaporate that water. In effect, the water molecules carry the heat away.

When the temperature falls, the blood vessels near the surface contract, lowering the ability of the blood to carry heat to the surface. In extreme cases, this situation can lead to frostbite, in which cells die because they are denied oxygen and other substances normally carried to them by the blood. Cold also causes the metabolism of the body to increase, generating more heat. Shivering, for example, is a response in which the heat generated by involuntary contractions of the muscles is used to counterbalance the falling temperature.

Incidentally, you may recall that you often start shivering at the onset of a fever. This seemingly paradoxical response occurs because, in response to the disease, the body's internal "thermostat" is adjusted to a higher than normal setting. You shiver and feel cold because, when this happens, normal body temperature is below the new setting of the thermostat and more heat is needed to raise the body's temperature. ■

The Second Law of Thermodynamics

Throughout the universe the behavior of energy is regular and predictable. According to the first law of thermodynamics the total amount of energy is constant, though energy may change from one form to another over and over again. Energy in the form of heat can flow from one place to another by conduction, convection, and radiation. But, as we pointed out earlier, everyday experience tells us that there is more to the behavior of energy: there is a direction to energy's flow. Hot things tend to cool off, cold things tend to warm up, and an egg, once broken, can never be reassembled. These common-sense ideas are the domain of the second law of thermodynamics, which is one of the most fascinating and powerful ideas in science.

The **second law of thermodynamics** places restrictions on the ways heat and other forms of energy can be transferred and used to do work. We will explore three different statements of this law:

1. Heat will not flow spontaneously from a cold to a hot body.
2. You cannot construct an engine that does nothing but convert heat to useful work.
3. Every isolated system becomes more disordered with time.

Although these three statements appear very different, they are actually logically equivalent—given any one statement, you can derive any of the others as a consequence. Given the statement that heat flows from hot to cold objects, for example, a physicist could produce a set of mathematical steps that would show that no engine can convert heat to work with 100% efficiency. In this sense, the three statements of the law all say the same thing.

1. Heat Will Not Flow Spontaneously from a Cold to a Hot Body

The first statement of the second law of thermodynamics relates to the relative temperature of two objects. If you take an ice-cream cone outside on a hot summer afternoon, it will melt. In the language of energy, heat will flow from the warm atmosphere to the cold ice-cream cone and cause its temperature to rise above the melting point. By the same token, if you take a cup of hot chocolate outside on a cold day, it will cool as heat energy flows from the cup into the surroundings. These experiences are so much a part of our everyday life that we scarcely notice them.

From the point of view of energy alone, there is no reason why things should work this way. Energy would be conserved if heat stayed put, or even if heat flowed from an ice-cream cone into the warm atmosphere, making the ice-cream cone colder than it was at the beginning. Our everyday experience (and many experimental confirmations) convince us that our universe does not work this way. In our universe heat flows in only one direction—from hot to cold. It does not go the other way spontaneously. This everyday observation may seem trivial, but in this statement is hidden all the mystery of those changes that make the future different from the past—the directionality in the universe.

At the molecular level the explanation of this version of the second law is easy to see. If two objects collide, and one of them is moving faster than the other, chances are that the slower object will be speeded up and the faster object slowed down by the collision. It's unlikely that events will go the other way. Thus, as we saw in the discussion of heat conduction, faster-moving molecules tend to share their energy with slower-moving ones. On the macroscopic scale, this process is seen as heat flowing from warm regions to cold ones by conduction.

For the second law to be violated, the molecules in a substance would have to conspire so that collisions would cause slower-moving molecules to slow down even more, giving up their energy to faster molecules so they could go even faster. Our experience tells us that this doesn't happen, and the second law takes this experience and makes it into a general law of nature.

The second law does *not* state that it is impossible for heat to flow from a cold to a hot body. Indeed, you know that in every refrigerator that's exactly what happens. When the refrigerator is operating, heat energy is removed from the colder inside to the warmer outside—a fact that you can verify by putting your hand under the refrigerator and feeling the warm air coming off it. The second law merely says that this action cannot take place *spontaneously,* of its own accord. If you wish to cool something in this way, you must supply energy. In fact, an alternative statement of the second law of thermodynamics could be: A refrigerator won't work unless it's plugged in.

The second law doesn't tell you that you can't make ice cubes, only that you can't make ice cubes without expending energy. Paying the electric bill, of course, is another piece of our everyday experience.

2. You Cannot Construct an Engine that Does Nothing but Convert Heat to Useful Work

The second statement of the second law of thermodynamics places a severe restriction on the way we can use energy. At first glance, this statement about heat and work seems to have very little to do with the idea that heat never flows from cold to hot. Yet the two statements are logically equivalent—given the one, the other must follow.

Energy is defined as the ability to do work. This second statement of the second law tells us that whenever energy is transformed from heat to another type—from heat to an electrical current, for example—some of that heat must be dumped into the environment and is unavailable to do work. The energy is neither lost nor destroyed, but it can't be used to make electricity to play your radio or gasoline to drive your car.

Scientists and engineers have defined a special term, efficiency, to quantify the loss of useful energy. **Efficiency** is the amount of work you get from an engine, divided by the amount of energy you put into it.

In Chapter 3 we discovered that heat and other forms of energy are interchangeable and the total amount of energy is conserved. From the point of view of the first law of thermodynamics, there is no reason why energy in the form of heat could not be converted to electrical energy with 100% efficiency. But the second law of thermodynamics tells us that such a process is not possible. The flow of energy, like time, has a direction. Another way of stating this law is to say that energy always goes from a more useful to a less useful form.

Your car engine provides a familiar example of this everyday rule of nature. In the engine an exploding mixture of gasoline and air creates a very high-temperature, high-pressure gas that pushes down on a piston. The motion of this piston is converted into rotational motion of a series of machine parts that eventually turn the car's wheels. In the real world, some of the energy is lost because of friction; but, in fact, the second law of thermodynamics would restrict our use of the energy even if friction did not exist, and even if every machine in the world was perfectly designed.

Look closely at the various stages of an engine's operation (Figure 4–7). There seems to be no obvious reason why energy in the form of heat in the exploding air-gas mixture could not be converted with 100% efficiency into energy of motion of the engine's piston, which is translated into the motion of the car. But you can't just think about the downward motion of a piston—what engineers call the *power stroke*—in the operation of your car's engine. If that was all the engine did, then the engine in every car could turn over only once. The problem is that once you have the piston pushed all the way down—once you have extracted all the useful work you're going to extract from the air-gas mixture—you have to return the piston to the top of the cylinder so that the cycle can be repeated (in actuality, the pistons in modern cars go up, down, and up again before they get back to the point where they can return energy to the system). In other words, in order to reset the engine to its original position so that more useful work can be done, some energy has to be expended to lift the piston back up.

Ignore for a moment the fact that a real engine is more complicated than the one we are discussing. Suppose that all you had to do was to lift the piston up after you had gotten the work from it. The cylinder is full of air, and consequently when you lift the piston up the air will be compressed and heated. In order to return the engine to the precise state it was in before the explosion, the heat from this compressed air has to be taken away. In practice, it is dumped into the atmosphere.

Figure 4–7

The cycle of an engine's piston. (*a*) The beginning of the intake stroke; (*b*) The middle of the intake stroke as a gasoline-air mixture enters the cylinder; (*c*) The beginning of the compression stroke. (*d*) At the beginning of the power stroke the spark plug fires, igniting the compressed mixture of gasoline and air. (*e*) At the beginning of the exhaust stroke combustion products are swept out. Note that each cycle involves two complete rotations of the crankshaft.

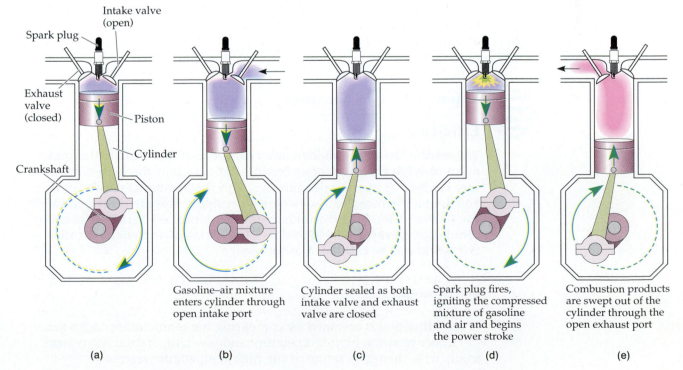

Intake valve (open)
Spark plug
Exhaust valve (closed)
Piston
Cylinder
Crankshaft

Gasoline–air mixture enters cylinder through open intake port

Cylinder sealed as both intake valve and exhaust valve are closed

Spark plug fires, igniting the compressed mixture of gasoline and air and begins the power stroke

Combustion products are swept out of the cylinder through the open exhaust port

(a)　　　(b)　　　(c)　　　(d)　　　(e)

In the language of physics, the exploding hot gas-air mixture is called a *high-temperature reservoir,* and the ambient atmosphere into which the heat of compression is dumped is called a *low-temperature reservoir.* The second law of thermodynamics says that any engine operating between two temperatures must dump some energy in the form of heat into the low-temperature reservoir. You can see how this works in the gasoline engine—heat produced by moving the piston back up has to be dumped. A similar argument can be made for any conceivable engine you could build.

The consequence of this situation is that some of the energy stored in the gasoline can be used to run the car, but some must be dumped into the low-temperature reservoir of the atmosphere. Once that heat energy has gone into the atmosphere, it can no longer be used to run the engine. That energy simply dissipates and is no longer available. Thus this version of the second law tells us that any real engine operating in the world, even an engine in which there is no friction, must waste some of the energy that goes into it.

This version of the second law explains why petroleum reserves and coal deposits play such an important role in the world economy. They are *high-grade* and *nonrenewable* sources of energy that can be used to produce very-high-temperature reservoirs. They are also sources of energy that can be used only once. If these *fossil fuels* are burned to produce a high-temperature reservoir and generate electricity, for example, a large portion of energy must simply be thrown away.

Although the second law sets rather stringent limits on engines that work in cycles, it does not apply to many other uses of energy. No engine is involved if you burn natural gas to heat your home or use solar energy to heat water, for example. Consequently these limits needn't apply. In other words, burning fossil fuels or employing solar energy to supply heat directly can be considerably more efficient than using it to generate electricity.

Science by the Numbers

Efficiency

The second law of thermodynamics can be used to calculate the maximum possible efficiency of an engine. Let's say that the high-temperature reservoir is at a temperature T_{hot}, and the low-temperature reservoir is at a temperature T_{cold} (where all temperatures are measured in the Kelvin scale). Then the maximum theoretical efficiency—the percentage of energy available to do useful work—of any engine in the real world can be calculated as follows:

▶ **In words:**

Efficiency is obtained by comparing the temperature *difference* between the high-temperature and low-temperature reservoirs, with the temperature of the high-temperature reservoir.

▶ **In equation form:**

$$\text{efficiency (in percent)} = \frac{(\text{temperature}_{hot} - \text{temperature}_{cold})}{\text{Temperature}_{hot}} \times 100.$$

▶ **In symbols:**

$$\text{Eff.} = \frac{T_{hot} - T_{cold}}{T_{hot}} \times 100$$

Any loss of energy due to friction in pulleys or gears or wheels in a real machine will make the actual efficiency less than this theoretical maximum. This maximum is actually a very stringent constraint on real engines.

Consider the efficiency of a normal coal-fired generating plant. The temperature of the high-energy steam (the hot reservoir) is about 500 K, while the temperature of the air into which waste heat must be dumped—(the "cold" reservoir) is around room temperature, or 300 K. The maximum possible efficiency of such a plant is given by the second law to be:

$$\begin{aligned}\text{Efficiency (in percent)} &= [(T_{hot} - T_{cold})/T_{hot}] \times 100 \\ &= [(500 - 300)/500] \times 100 \\ &= 40.0\%.\end{aligned}$$

In other words, more than half of the energy produced in a typical coal-burning power plant must be dumped into the atmosphere as waste heat. This fundamental limit is independent of the engineers' ability to make the plant operate efficiently. In fact, engineers do very well in this regard—most generating plants operate within a few percent of the efficiency allowed by the second law of thermodynamics. ■

3. Every Isolated System Becomes More Disordered with Time

The third statement of the second law of thermodynamics is in many ways the most profound. It tells us something about the order of the universe itself. The idea of increasing disorder is perhaps the most familiar way of looking at the second law. It can be stated in colloquial form such as "the world is going to hell in a handbasket" or "things always seem to get worse." For physicists, however, this statement of the second law has a very precise and special meaning.

To understand what this statement means, you have to understand exactly what a physicist means by the terms "order" and "disorder" (Figure 4–8). An ordered system is one in which a number of objects, be they atoms or automobiles, are positioned in a completely regular and predictable pattern. For example, atoms in a perfect crystal or automobiles in a perfect line form highly ordered systems. A disordered system, on the other hand, contains objects that are randomly situated, without any obvious pattern. Atoms in a gas or automobiles after a multi-car pileup on the freeway are good examples of disordered systems.

A more formal definition of order and disorder requires considering the number of different ways a system can be arranged. Consider a parking

Figure 4–8
Highly ordered, regular patterns of objects are less likely to occur than disordered, irregular patterns.

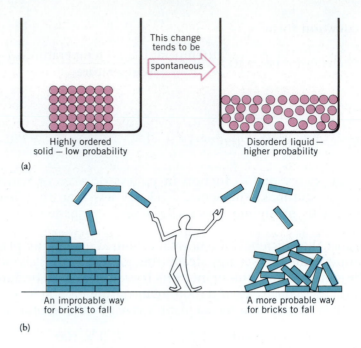

This change tends to be *spontaneous*

Highly ordered solid — low probability

Disorderd liquid — higher probability

(a)

An improbable way for bricks to fall

A more probable way for bricks to fall

(b)

lot with exactly three spaces in it. Suppose you were told that there were exactly three identical cars in that parking lot. In this situation, you would know exactly what state the parking lot was in—each of the spaces would be filled by one of those three cars, and the system is fully ordered.

Now, however, suppose you were told that there was only one car in the parking lot with three spaces. In this case, you couldn't be certain about the state of the parking lot. The car could be in the first parking space, the second, or the third. Thus the car could be in any of three different states. In such a situation, we would say that the parking lot was in a state of higher disorder, or, to use the technical term, higher entropy.

The disorder of a system, then, has a one-to-one correspondence with the number of states that the system could adopt, or, equivalently, with our level of ignorance of the exact state of the system. The term entropy is used to describe this concept. **Entropy** is a measure of the disorder in a physical system. In terms of entropy, the statement of the second law reads:

> **The entropy of an isolated system remains constant or increases.**

In other words, any system left to itself will evolve in the direction of the most disordered state, the state with the maximum uncertainty in the exact conditions of its constituent objects. Without careful chemical controls, atoms and molecules will tend to become more intermixed; without careful driving, automobiles will also tend to become more disordered.

The example we gave at the beginning of the chapter, with the one fast atom in a collection of slow atoms, shows very clearly how such a process works. In the most likely situation, when all of the atoms are in the same low-energy states, the entropy is maximized. A much less probable situation occurs when one of the atoms is in a high-velocity state; the entropy is lower because one part of the system is special. Another way of saying this is that systems tend to avoid states of high improbability.

This concept of probability also explains a number of paradoxes that puzzled scientists around the turn of the century. For example, it is possible that all the air molecules in the room in which you are sitting could suddenly rush over to one side of the room, leaving you in a perfect vacuum. You don't worry about this happening because you feel (rightly) that this event is highly unlikely. In fact, the probability is so low as to make it extremely unlikely you would see it even if you waited the entire lifetime of the universe.

While systems tend to become more disordered, the second law does not require that *every* system must approach a state of lower order. Think about water, a substance of high disorder because water molecules are arranged at random. If you put water into a freezer you get an ice cube, a much more ordered state in which water molecules have formed a regular crystal structure. You have caused a system to evolve to a state of higher order. How can this be reconciled with our statement of the second law?

The answer to this seeming paradox lies in the simple word "isolated." The refrigerator in which you make the ice cubes is not an isolated system because it has a power cord plugged into the wall, and ultimately connected to the generating plant. The isolated system in this case is the refrigerator plus the generating plant. The second law of thermodynamics says that in this particular isolated system, the total entropy must increase. It does not say that the entropy has to increase in all the subparts of the system.

In this example, one part of the system (the ice cube) becomes more ordered, while another part of the system (the generating plant, its burning fuel, and the surrounding air) becomes more disordered. All that the second law requires is that the amount of disorder at the generating plant be greater than the amount of order at the ice cube. As long as this requirement is met, the second law is not violated. In fact, in this particular example the disorder at the generating plant greatly exceeds any possible order that could take place inside your refrigerator.

Science in the Making

The Heat Death of the Universe

The nineteenth-century discovery of the second law of thermodynamics was a gloomy event. The prevailing philosophy of the time was that life, society, and the universe in general were on a never-ending upward spiral of progress. Darwin's 1859 publication of *On the Origin of Species*, which proposed that more complex forms of life could evolve from less complex forms (see Chapter 23), served to reinforce this particular notion of an ever-improving world. In this optimistic climate, the discovery that the energy in the universe was being steadily and irrevocably degraded was very hard for nineteenth-century scientists and philosophers to accept.

In fact, they felt that the second law inevitably meant that all the energy in the universe would eventually be degraded into waste heat, and that everything in the universe would eventually come to the same temperature. They called this the "heat death" of the universe, and they saw it as the ultimate end of the laws of thermodynamics.

This notion even affected the literature and music of the time. For example, here is an excerpt from Algernon Swinburne's "The Garden of Proserpine":

> From too much love of living
> From hope and death set free
> We thank with weak thanksgiving
> Whatever gods there be
> That no man lives forever
> That dead men rise up never
> That even the weariest river
> Flows somewhere safe to sea.

Today, we have a rather different view of how the universe will end, a view that will be discussed in detail in Chapter 18. Suffice to say, however, that Swinburne and his colleagues may have been premature in their gloom (although one could argue that poets of that period seemed to grab at any excuse for being gloomy). ■

Consequences of the Second Law

The Arrow of Time

We live in a world of four dimensions. Three of these—the dimensions that define space—have no obvious directionality. You can go east or west, north or south, and up or down in our universe. But the fourth dimension—time—behaves differently. Time has direction; we can never revisit the past, nor can we foresee the future.

Take one of your favorite home movies or just about any videotape and play it in reverse. Chances are that before too long you'll see something silly—something that couldn't possibly happen, that will make you laugh. Springboard divers fly out of the water and land completely dry on the diving board. From a complete stop golf balls fly off toward the tee. Ocean spray coalesces into smooth waves that recede from shore. Most physical laws, such as Newton's laws of motion or the first law of thermodynamics, say nothing about time. The motions predicted by Newton and the conservation of energy are independent of time. They work just as well if you play a video forward or backward.

But the second law of thermodynamics is different. By defining a *sequence* of events—heat flows from hot to cold; concentrated fuels burn

to produce waste heat; the disorder of isolated systems never spontaneously decreases; you get older—we have established a direction to time.
We experience the passage of events as dictated by the second law. Scientists cannot answer the deeper philosophical question of why we perceive
the arrow of time in only one direction, but through the second law they
can describe the effects of that directionality.

Built-in Limitations of the Universe

The second law of thermodynamics has both practical and philosophical
consequences. It poses severe limits on the way that human beings can
manipulate nature, and on the way that nature itself operates. It tells us
that some things cannot happen in our universe.

At the practical level, it tells us that if we continue to generate electricity by burning fossil fuels or by nuclear fission, we are using up a good
deal of the energy that is locked in those concentrated nonrenewable
resources. These limitations are not a question of sloppy engineering or
poor design, they're simply built into the laws of nature. If you could
design an engine or other device that would extract energy from coal
and oil with higher efficiency than the second law limits, then you could
also design a refrigerator that worked when it wasn't plugged in.

At the philosophical level, the second law tells us that nature has a
built-in hierarchy of more useful and less useful forms of energy. The
lowest or least useful state of energy is the reservoir into which all energy
eventually gets dumped. Once the energy is in that lowest energy reservoir, it can no longer be used to do work. For the Earth, energy passes
through the region that supports life, the *biosphere*, but is eventually lost
as it is radiated into the black void of space.

Does Evolution Violate the Second Law?

Creationists, who argue that life on Earth was created in a single miraculous event, often use the second law of thermodynamics to argue against
theories of life's gradual evolution. These arguments reflect a common
misunderstanding of the second law.

Creationists point out that since life is a highly ordered system—a
system in which trillions of atoms and molecules must occur in exactly
the right sequence—it could not possibly have arisen spontaneously without violating the second law. What they neglect to consider, however, is
that the energy that drives living systems is sunlight, so that the "isolated
system" that the second law speaks of is the Earth's biosphere *plus the
Sun*. All that you need to make the evolution of life consistent with the
second law is that the order observed in living things must be offset by
a greater amount of disorder in the Sun. Once again, as with the earlier
example of the ice cube, this requirement is easily met by the Sun and
the biosphere taken together.

Science cannot yet answer the question of why life arose, nor can it
say whether God was involved in the process. But it can show that the
development of life, as a natural process, is in no way inconsistent with
the universal laws of thermodynamics.

THINKING MORE ABOUT ENTROPY

Aging

Nothing better illustrates the directionality of nature than human aging. It's all very well to talk about collisions of atoms and the making of ice cubes, but when we see the inevitable effects of aging in ourselves and those around us, we come to realize that the second law of thermodynamics has a very real meaning for each of us. In fact, there is probably no older dream in human history than that somehow, some day, someone would find out how to reverse the aging process.

Modern biologists approach the problem of aging by noting that the process of evolution (see Chapter 23) acts to preserve those properties that help an organism survive until it has offspring to which those properties can be transmitted. After that, no particular reason exists from the point of view of a species for an individual to survive any longer. Indeed, throughout most of the history of the human race, few individuals lived past the age of 40, so aging was not perceived to be a significant problem until recently.

Two general types of theory attempt to explain why aging occurs. The first we can call "planned obsolescence." It suggests that the human body is actually designed to self-destruct after a certain time, perhaps to insure that more food and other resources are available for children. The second, which we can call the "accumulated accident" school, holds that the general wear and tear of existence eventually overcomes the body's ability to make repairs, and the system just runs down. In modern language, scientists talk of accumulated damage to DNA (see Chapter 23), the molecule that contains the cell's operating instructions.

At the moment, the evidence seems to support the second of these alternatives. Scientists are starting to identify specific processes that damage DNA, and to document the precise sort of damage that occurs. As always, the research must be done on animals before it can be applied to humans. One group at the University of California at Irvine, for example, produced a strain of fruit flies that live twice as long as normal, and has identified a substance produced in their cells that seems to explain their enhanced lifetime. It is, of course, a long way from fruit flies to human beings, but it's not too much of a stretch to say that sometime we will understand why our bodies age; from that point, it is probably only a small step to being able to do something about it.

What sort of problems do you think society would have to face if life expectancy increased to 120 years? What kinds of changes in education, work habits, and government might you expect to see? Do you think this sort of research should be pursued?

Summary

All objects in the universe are at a *temperature* above *absolute zero* temperature, and thus they hold some heat energy—the kinetic energy of moving atoms. *Heat capacity* defines how much heat energy a substance can hold.

Heat transfer between two objects that are at different temperatures may occur in three ways. *Conduction* involves the transfer of heat energy through the collisions of individual atoms and molecules; *thermal conductivity* is a measure of how easily this energy transfer occurs. *Convection* involves the motion of a mass of fluid in a *convection cell,* in which warmer atoms are physically transported from one place to another. Heat can also be transferred by *radiation*—infrared energy and other forms of light that travel across a room or across the vastness of space until they are absorbed.

The first law of thermodynamics promises that the total amount of energy never changes, no matter how you shift it from one form to another, but the *second law of thermodynamics* places severe restrictions on how you can shift energy. Three different but equivalent statements of the second law underscore these restrictions:

1. Heat will not flow spontaneously from a colder to a hotter body. This first statement places a restriction on the transfer of heat—for example, you have to supply an external source of energy to a refrigerator before it will work.

2. It is impossible to construct a machine that does nothing but convert heat into useful work. This different but equivalent statement of the second law precludes the construction of an engine that operates with 100% *efficiency*. All engines operate on a cycle, and every engine must expend some of its energy returning to its initial state.

3. The *entropy* of an isolated system always increases. The third statement of the second law of thermodynamics introduces the concept of entropy—the tendency of isolated systems to become more disordered with time. This directionality of energy flow in the universe defines our sense of the direction of time.

Key Terms About Heat and the Second Law of Thermodynamics

temperature	heat transfer	convection	second law of thermodynamics
absolute zero	conduction	convection cell	efficiency
heat capacity	thermal conductivity	radiation	entropy

Key Equation

$$\text{efficiency (in percent)} = \frac{(T_{hot} - T_{cold})}{T_{hot}} \times 100.$$

Review Questions

1. Identify the three ways that heat can be transferred and give examples of each.

2. What kind of heat transfer depends only on the collisions between individual atoms and molecules?

3. What kind of heat transfer depends on the bulk motion of large numbers of molecules?

4. By what process can heat be transferred across the vacuum of empty space?

5. What is the difference between temperature and heat?

6. What is heat capacity? Is it the same for all materials?

7. Describe three common temperature scales. What fixed points are used to calibrate them?

8. State the second law of thermodynamics in three different ways.

9. What is the high-temperature reservoir in your car's engine? What is the low-temperature reservoir?

10. What is entropy?

11. To what does the phrase "heat death of the universe" refer?

12. How does hair help keep humans and other mammals warm?

Discussion Questions

1. Why don't all the atoms in the room you're sitting in move to one side, leaving you in a vacuum?

2. Why are there big cooling stacks around nuclear reactors and coal-fired generating plants?

3. Identify three examples of the second law of thermodynamics in action that have occurred since you woke up this morning.

4. Imagine lying on a hot beach on a sunny summer day. In what different ways is heat transferred to your body? In each case what was the original source of the heat energy?

5. "Cogeneration" is a term used to describe systems in which waste heat from electrical generating plants is used to heat nearby homes and businesses with efficiencies much greater than 50%. Does cogeneration violate the second law? Why or why not?

6. Why is a perpetual-motion machine impossible?

7. Sea water is full of moving molecules that possess kinetic energy. Could we extract this energy from sea water? Why or why not?

8. Describe the kinds of heat transfer that occur while you are cooking a meal. Where does the heat energy come from? Where does it end up?

9. Why do some animals roll up into a ball when they are cold?

10. Why are feather beds warm, and why is goose down considered the best filling for a parka?

11. Why do human beings wear clothes? Compare our behavior in this regard to other warm-blooded animals.

Problems

1. Calculate the maximum possible efficiency of a power plant that burns natural gas at a temperature of 600 Kelvin, with low-temperature surroundings at 300 Kelvin. How much more efficient would the plant be if it were built in the Arctic where the low-temperature reservoir is at 250 Kelvin? Why don't we build all power plants in the Arctic?

2. The Ocean Thermal Electric Conversion system (OTEC) is an example of a high-tech electrical generator. It takes advantage of the fact that in the tropics, deep ocean water is at a temperature of 4°C, while the surface is at a temperature around 25°C. The idea is to find a material that boils between these temperatures. The material in fluid form is brought up through a large pipe from the depths, and the expansion associated with its boiling is used to drive an electrical turbine. The gas is then pumped back to the depths, where it condenses back into a liquid and the whole process repeats.
   a. What is the maximum efficiency with which OTEC can produce electricity? (*Hint:* Remember to convert all temperatures to the Kelvin scale.)
   b. Why do you suppose engineers are willing to pursue the scheme, given your answer in (*a*)?
   c. What is the ultimate source of the energy generated by OTEC?

Investigations

1. Research the daily high and low temperatures for the past week in a nearby city and one of its sur-

rounding towns. On average, what is the difference in temperature? What causes this difference?

2. What kind of insulation is installed in your home? What could you do to improve your home's insulation?

3. Get an aluminum cup, a coffee mug, and a plastic drinking cup. Simply by feeling these objects, can you guess their relative thermal conductivities? What experiment might you perform to test your guess?

4. Visit a building supply store and look at the doors and windows they sell. What steps have manufacturers taken to reduce heat flow from homes?

5. Play a home movie or videotape backward. How many violations of the second law of thermodynamics can you spot?

Additional Reading

Atkins, P.W. *The Second Law*. New York: Scientific American Library, 1984.

Morris, R. *Time's Arrow*. New York: Touchstone, 1985.

Electricity and Magnetism

Electricity and magnetism are two different aspects of one force—the electromagnetic force

A Random Walk
Late for Work at the Copy Center

Imagine waking up on a cold, dry winter day and realizing you're late for work. You rush to throw on your clothes and comb your hair with rapid, vigorous strokes. When you look in the mirror you realize that the folds of your shirt are sticking together and your hair is standing on end. Then, at work, you rush to photocopy an important report, only to find the copies sticking to each other, slowing your efforts. Static electricity has struck.

Imagine walking home from work on a hot, humid summer day. The sky is dark and you hear the distant rumble of thunder. Suddenly, a jagged lightning bolt slices the horizon. You hurry to get home before the thunderstorm. Static electricity has, quite literally, struck again.

Believe it or not, the phenemenon that causes your clothes to stick together and your hair to stand on end in the cold, dry winter is exactly the same phenomenon that causes lightning in the hot, humid summer—the force called electromagnetism.

Nature's Other Forces

When the girl touched the electrically charged sphere, her hair became electrically charged as well. Individual hairs repel each other and thus "stand on end."

According to Newton's laws of motion, nothing happens without a force, but the law of gravity that Newton discussed cannot explain many everyday events. How does a refrigerator magnet cling to metal, defying gravity? How does a compass needle swing around to the north? How can static cling wrinkle your shirt, or lightning shatter an old tree? All of these phenomena involve motions that point to the existence of some underlying force that is different from gravity.

Newton may not have known about these forces, but he did give us a method for studying them. First observe natural phenomena and learn how they behave, then organize those observations into a series of natural laws, and finally use those laws to predict future behavior of the physical world. This is the process we have called the scientific method.

In particular, we will find Newton's first law of motion (see Chapter 2) to be very useful in our investigation of nature's other forces. According to this law, whenever we see a change in the state of motion of any material object, we know that a force has acted to produce that change. Thus, whenever we see such a change and can rule out the action of known forces such as gravity, we can conclude that the change must have been caused by a hitherto unknown force. We shall use this line of reasoning to show that electric and magnetic forces exist in the natural world.

Our understanding of the phenomena associated with static electricity and magnetism began in the eighteenth century with a group of scientists in Europe and North America who called themselves "electricians." These researchers were fascinated by the many curious phenomena associated with nature's unseen forces. Their thoughts were not focused on practical applications, nor could they have imagined how their work would help transform the world.

Static Electricity

Various phenomena related to electricity have been known since ancient times. The Greeks knew that if you rub a piece of amber with cat's fur and then touch other objects with the amber, those other objects will be repelled from each other. The same thing happens, they found, if you rub a piece of glass with silk: objects touched with the glass will be repelled from each other. On the other hand, if you bring objects that have been touched with the amber near objects touched with the glass, they are attracted toward each other. Objects that behave in this way are said to possess **electrical charge,** or to be charged.

The force that moves objects toward and away from each other in these simple demonstrations was named **electricity** (from the Greek word for amber). In these simple experiments, the electrical charge doesn't move once it has been placed on an object, so the force is also called **static electricity**.

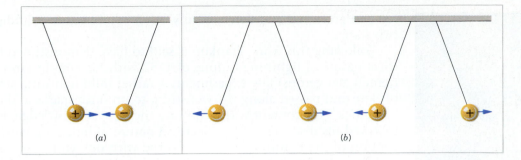

Figure 5–1

(*a*). If the charge on a suspended ball is the same as that on another suspended ball, the two objects will repel each other. (*b*). If charges on the two objects are opposite, they will attract each other.

The electric force is obviously different from gravity. Unlike the electric force, gravity can never be repulsive—when a gravitational force acts between two objects, it always pulls them together. The electrical force, on the other hand, can attract some objects toward each other and push other objects away from each other. Furthermore, the electric force is vastly more powerful than gravity. A simple comb charged with static electricity easily lifts a piece of paper against the gravitational pull of the entire Earth.

Today, we understand that the properties of the electrical force arise from the existence of two kinds of electrical charge (Figure 5–1). We say that objects touched with the same source—be it amber or glass—have the same electrical charge and are repelled from each other. On the other hand, one object touched with amber will have a different electrical charge than a second object touched with glass. This difference is reflected in their behavior—they attract each other.

Science in the Making

Benjamin Franklin and Electrical Charge

The most famous North American "electrician" was Benjamin Franklin (1706–1790), one of the pioneers of electrical science as well as a central figure in the founding of the United States of America. Franklin began his electrical experiments in 1746 with a study of electricity generated by friction. Most scientists of the time thought that electrical effects resulted from the interaction of two different "electrical fluids." Franklin, however, became convinced that all electrical phenomena could be explained as the result of transfer of a *single* electric fluid that could be shifted from one object to another. He realized that objects could have an excess or a deficiency of this fluid, and he applied the names "negative" and "positive" to these two situations. There's no particular significance to this choice of words—he could have chosen

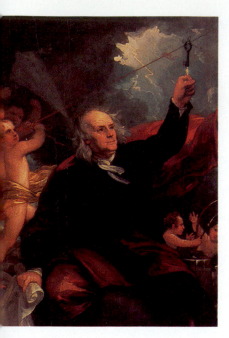

Benjamin Franklin engaged in his famous (and potentially lethal) kite experiment.

red and green, up and down, or any other combination to indicate the two states.

Following this work, Franklin is said to have demonstrated the electrical nature of lightning in June of 1752 with his famous (and extremely dangerous) kite experiment. A rather mild lightning stroke hit his kite and passed along the wet string to produce sparks and an electric shock. Not content with acquiring theoretical knowledge, Franklin followed his discovery of the electrical nature of lightning with the invention of the lightning rod, a metal rod with one end in the ground and the other sticking up above the roof of a building. It carries the electrical charge in lightning bolts into the earth, diverting it away from the building. Lightning rods caught on quickly in the wooden cities of North America and Europe, preventing countless deadly fires. The descendants of Franklin's lightning rods are in widespread use today. ■

The Movement of Electrons

We now understand that there are two kinds of electrical charge, so observations of both attractive and repulsive behavior of charged objects are easy to understand. In modern language, we say that all objects are made up of minute building blocks called atoms, and all atoms are made up of still smaller particles that have electrical charge. As we shall see in Chapter 7, in every atom, negatively charged *electrons* move in orbit around a heavy, positively charged *nucleus* near the center, just as planets orbit the Sun. The electrons and the nucleus have opposite electrical charges, so an attractive force exists between them. This force plays roughly the same role that gravity plays in keeping the solar system together. Most atoms are electrically neutral, because the positive charge of the nucleus cancels the negative charge of the electrons.

Electrons—particularly electrons in outer orbits—tend to be rather loosely bound to their parent nucleus. These electrons may be knocked loose in collisions, and once they're knocked loose they can move around freely through the material.

When electrons are pulled out of a material, they no longer cancel the positive charges in the nucleus. The result is a net excess of positive charge in the object, and we say that the object as a whole has acquired a **positive electric charge.** Similarly, an object acquires a **negative electric charge** when extra electrons are pushed into it—this is what happens to a comb when you run it through your hair on a dry day. Electrons are pushed into the comb, so it acquires a negative charge. Your hair loses electrons, so individual strands become positively charged.

During a thunderstorm, the same phenomenon occurs on a much larger scale, as wind and rain disrupt the normal distribution of electrons in clouds. When a charged cloud passes over a tall tree or tower, the violent electric discharge called lightning may result from the attraction of the positive charges on the ground and negative charges in the cloud (although in the case of lightning, both the positive and negative charges move).

Although historical investigations of electrical charge tended to concentrate on somewhat artificial experiments, we have come to know that electrically charged particles play important roles in many natural systems. Virtually all of the atoms in the Sun, for example, have lost electrons and thus are positively charged. In all advanced life forms (including human beings), charged atoms are routinely moved into and out of cells to maintain the processes of life. As you read these words, for example, positively charged potassium and sodium atoms moving across the membranes of cells in your optic nerve to carry signals to your brain.

Coulomb's Law

The phenomenon of electricity remained something of a mild curiosity until the mid-eighteenth century, when scientists began applying the scientific method to investigate it. One of the first things that had to be done, of course, was to develop a precise statement about the nature of the electrical force. The French scientist Charles Augustin de Coulomb (1736–1806) was most responsible for this work. During the 1780s, at the same time the United States Constitution was being written by Benjamin Franklin and others, Coulomb devised a series of experiments in which he passed different amounts of electrical charge onto objects and then measured the force between them. After repeated careful measurements he discovered that the electrical force was in some ways very similar to the gravitational force that Isaac Newton had discovered a century earlier.

Coulomb summarized his discoveries in a simple relationship known as **Coulomb's law,** which states:

▶ **In words:**

> The force between any two electrically charged objects is proportional to the product of their charges divided by the square of the distance between them.

▶ **In equation form:**

$$\text{force (in newtons)} = k \times \frac{\text{1st charge} \times \text{2nd charge}}{\text{distance}^2}.$$

▶ **In symbols:**

$$F = k \times \frac{q_1 \times q_2}{d^2},$$

where distance d is measured in meters, charge q is measured in a unit called the coulomb (see below), and k is the Coulomb constant, a number that plays the same role in electricity that the gravitational constant G in plays in gravity. Like G, k is a number (9.00×10^9 newton-meter^2/coulomb^2 in one common system of units) that can be determined experimentally, and which turns out to be the same for all charges and all separations of those charges anywhere in the universe.

Coulomb observed that if two electrically charged objects are moved farther and farther away from each other, the force between them gets smaller and smaller, just like gravity. In fact, if the distance between two objects is doubled, the force decreases by a factor of four—the familiar 1/distance^2, or inverse square relationship, that we saw in the law of universal gravitation in Chapter 2. The farther you are from a charge, the less you are affected by it. Coulomb also discovered that the size of the force depends on the product of the charges of the two objects—

double the charge on one object and the force doubles, double the charge on both objects and the force increases by a factor of four. Coulomb's law is a summary of a large number of experiments done on stationary electrical charges. It can also be thought of as a summary of the behavior of static electricity.

In order to come to a result like this, scientists had to define a unit of electrical charge, called the *coulomb* (abbreviated C) after the scientist who did so much of this important work. A coulomb equals the charge on 6.3×10^{18} electrons, a very large number, indeed. When this many electrons have been moved onto an object or removed from it, that object will have one coulomb of electrical charge.

Though you might not realize it, you see the results of Coulomb's law every day in the behavior of ordinary water. Water molecules, which are composed of two hydrogen atoms linked to one oxygen atom, have a negatively charged end and a positively charged end, a phenomenon known as *polarity*. When you sprinkle ordinary salt crystals—the compound sodium chloride—into water, positively charged sodium atoms and negatively charged chlorine atoms are pulled apart by electrostatic forces. The salt dissolves.

The behavior of salt contrasts with that of common plastics, which do not dissolve. Plastics are not made of charged particles, so water cannot exert an electrostatic force and pull them apart.

Science by the Numbers

Two Forces Compared

In Chapter 7 we will examine compelling evidence that the atoms that compose all the materials in our physical surroundings have a definite internal structure. Tiny negatively charged particles called electrons circle in orbits around a positively charged nucleus. Thus, inside an atom, two forces act between the nucleus and an electron: the force of gravity and the electrical force. We can use this fact to get a sense of the relative strength of the two forces.

The simplest atom is hydrogen, in which a single electron circles a single positively charged particle known as a proton (see Chapter 7). The masses of the electron and proton are 9×10^{-31} kg and 1.7×10^{-27} kg, respectively. The charge on the proton is 1.6×10^{-19} C, and the charge on the electron has the same magnitude but is negative. A typical separation of these two particles in an atom is 10^{-10} m.

Given these numbers, what are the values of the electrical and gravitational attractions between these two particles?

Gravity The force of gravity between the two particles will be:

$$\text{force of gravity (in newtons)} = G \times \frac{\text{mass}_1 \text{ (in kg)} \times \text{mass}_2 \text{ (in kg)}}{\text{distance (in m)}^2}$$

$$= (6.7 \times 10^{-11} \text{ m}^3/\text{kg-s}^2)$$

$$\times \frac{(1.7 \times 10^{-27} \text{ kg}) \times (9 \times 10^{-31} \text{ kg})}{(10^{-10} \text{ m})^2}$$

$$= 1.0 \times 10^{-47} \text{ N}.$$

Electricity   The electrical force, on the other hand, is given by Coulomb's law to be

$$
\begin{aligned}
\text{force of static electricity (in newtons)} \\
= k \times \frac{\text{charge}_1 \text{ (in C)} \times \text{charge}_2 \text{ (in C)}}{\text{distance (in m)}^2} \\
= (9 \times 10^9 \text{ N-m}^2/\text{C}^2) \times (1.6 \times 10^{-19} \text{ C}) \times \frac{1.6 \times 10^{-19} \text{ C}}{(10^{-10} \text{ m})^2} \\
= 2.6 \times 10^{-8} \text{ N}.
\end{aligned}
$$

From this simple calculation we can see that, in the atom, the electrical force is many orders of magnitude larger than the gravitational force. This is why, in our discussion of the atom in subsequent chapters, we ignore the effects of gravity completely. ■

The Electric Field

Imagine that you have an electrical charge sitting at a point (Figure 5–2). The charged object could be a piece of lint, an electron, or one of your hairs. If you brought a second charged object to one spot near that piece of lint, the second object would feel a force. If you then moved that second object to another spot near the lint, it would still feel a force, but the force would, in general, be a different magnitude and point in a different direction than at the first spot. In fact, the second charged object would feel a force at every point in space around the piece of lint.

You can make a picture that represents this fact as in Figure 5–2. The arrow at each point around the original charged object represents the force that would be felt by another charged object if that object were brought to the point in space where the arrow originates. The collection of all the arrows that represent these forces is called the **electric field** of the original charged object. We can think of every charged object as being surrounded by such a field, as shown in the figure. Notice that the electrical field is defined as the force that *would* be felt by another charge if that charge were located at a particular point, so that the field is present even if no other charge is in the region.

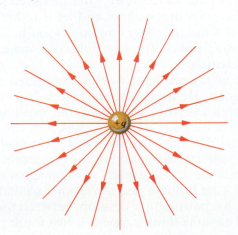

Figure 5–2

An electrical field surrounding a positive charge, $+q$, may be represented by lines of force radiating outward. Any charged object that approaches $+q$ experiences a greater and greater electrical force the closer it gets. Positively charged objects will be repelled, while negatively charged objects will be attracted.

Technically, the electric field at a point is defined to be the force that would be felt by a positive 1-coulomb charge if it were brought to that point. The field is usually drawn so that the direction of the arrow corresponds to the direction of that force, and the length of the arrow corresponds to its magnitude.

Magnetism

Just as electrical phenomena were known to the ancient philosophers, so too were the phenomena we place under the title of magnetism. The first known magnets were naturally occurring iron minerals. If you bring one of these minerals (a common one is magnetite or "lodestone") near a piece of iron, the iron will be attracted to it. You have undoubtedly seen experiments in which magnets were placed near nails, which jumped up and hung from them.

The fact that the nails behave in this way tells you that there must be yet another force in nature, a force different from both electricity and gravity. Electrical attraction doesn't make the nails move, nor is it gravity that causes the nails to jump up. The simple experiment of picking up a nail with a magnet illustrates beyond a shadow of a doubt that there is a **magnetic force** in the universe—a force that can be identified and described by the same methods we used to investigate gravity and electricity.

While electricity remained a curiosity until well into modern times, magnetism was put to practical use very early. The *compass*, invented in China and used by Europeans to navigate the oceans during the age of exploration, is the first magnetic device on record. A sliver of lodestone, left free to rotate, will align itself in a north-south direction. We use compasses so often these days that it's easy to forget how important knowing direction was to early travelers, particularly those who ventured out of sight of land in sailing ships.

In the late sixteenth century the English scientist William Gilbert (1544–1603) conducted the first serious study of magnets. Though revered in his day as a doctor (he was physician to both Queen Elizabeth I and King James I) his most lasting fame came from his discovery that every magnet can be characterized by what he called poles. If you take a piece of naturally occurring magnet and let it rotate, one end of the magnet points north and the other end points south. These two ends of the magnet are called **poles.** The two poles of a magnet are given the labels **north** and **south.**

In the course of his research, Gilbert discovered many important properties of magnets. He learned to magnetize iron and steel rods by stroking them with a lodestone. He discovered that hammered iron becomes magnetic and found that iron's magnetism can be destroyed by heating. He realized that the Earth itself is a giant magnet—a fact that, as we shall see, explains the operation of the compass. Gilbert also documented many of the most basic aspects of the magnetic force. He found that if two magnets are brought near each other so that the north poles are close together, a repulsive force develops between the magnets—they are forced apart. The same thing happens if two south poles are brought

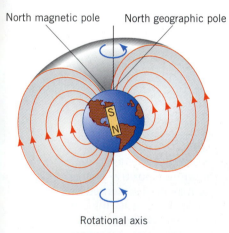

North magnetic pole   North geographic pole

Rotational axis

Figure 5–3

A compass needle and the Earth. Any magnet will twist because of the forces between its poles and those of the Earth.

together. If, however, the north pole of one magnet is brought near the south pole of another magnet, the resulting force is attractive. In this respect, magnetism seems to mimic the eighteenth-century studies of static electricity. William Gilbert's results can be summarized in two simple statements:

> **Every magnet has at least two poles.**
> **and**
> **Like magnetic poles repel each other, while unlike poles attract.**

Once you know that a magnet has two poles, then you can understand how a compass behaves. The Earth itself is a giant magnet, with one pole in Canada and the other pole in Antarctica. If a piece of magnetized iron (for example, a compass needle) is allowed to rotate freely, one of its poles will be attracted to and twist around toward Canada in the north, and the other end will point to Antarctica in the south (see Figure 5–3).

Just as the electric force can be represented in terms of an electric field, so too can the magnetic force be represented in terms of a magnetic field. If a small compass needle is brought near a magnet as shown in Figure 5–4, the forces exerted by the magnet will twist the needle around. The direction of the needle will, in general, be different at different locations around the magnet.

Just as we can imagine any collection of electric charges as being surrounded by an electric field—imaginary lines of force—we can imagine every magnet as being surrounded by an imaginary set of lines that show the orientation of compasses. These lines are drawn so that if a compass were brought to a point in space, the needle would turn and point along the line. The number of lines in a given area is a measure of the strength of the forces exerted on the compass. A collection of lines that map out the directions in which compass needles would point is called a **magnetic field** (Figure 5–5).

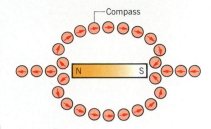

Figure 5–4
A magnetic field. Small magnets placed near a large one orient themselves along the lines of the magnetic field, as shown.

Figure 5–5
(*a*) A bar magnet and its magnetic dipole field. (*b*) Iron filings placed near a bar magnet align themselves along the field.

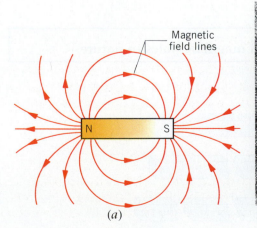

(*a*)

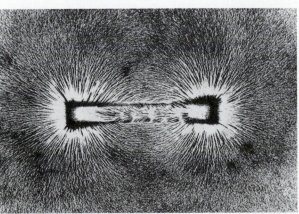

(*b*)

Magnetic Navigation

Humans aren't the only navigators who make use of the Earth's magnetic field. The fact that the Earth is surrounded by a magnetic field is used by many living things to guide their movements. This was first established by scientists at the Massachusetts Institute of Technology in 1975, when they were studying a single-celled bacterium that lived in the ooze at the bottom of nearby swamps. They found that the bacteria incorporated about 20 little chunks of an iron ore called magnetite (the material often found in natural magnets) into their bodies. These chunks were strung out in a line, in effect forming a microscopic compass needle.

Because the Earth's magnetic field dips into the surface in the Northern Hemisphere and rises up out of it in the Southern, that Massachusetts bacterium has a built-in "up" and "down" indicator. In fact, the bacterium uses its internal magnet to allow it to navigate down into the nutrient-rich ooze at the bottom of the pond. Interestingly enough, related bacteria in the Southern Hemisphere follow the field lines in the opposite direction to get to the bottom of their ponds.

Since 1975, similar internal magnets have been discovered in many animals. Some migratory birds, for example, use internal magnets as one of several cues to guide them on flights thousands of miles in length. In one case, the Australian silvereye, there is even evidence that the bird can "see" the magnetic field of the Earth, through a process involving modification of molecules normally involved in color vision.

The Dipole Field

The magnetic field shown in Figure 5–5 plays a very important role in nature. We call this field, which arises whenever a magnet has both a north and a south magnetic pole, a *dipole field* because the bar magnet has two poles. All magnets found in nature have both north and south poles—you never find one without the other. Even if you take an ordinary bar magnet and cut it in two, you don't get a north and a south pole in isolation, but rather two small magnets each with a north and a south pole (Figure 5–6). If you took each of those halves and cut them in half, you would continue to get smaller and smaller dipole magnets. In fact, it seems to be a general rule of nature that:

> **There are no isolated magnetic poles in nature.**

Figure 5–6

Cut magnets. If you break a dipole magnet in two, you get two smaller dipole magnets, not an isolated north or south pole.

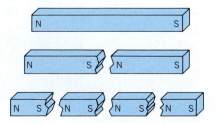

In the language of physicists, a single isolated north or south magnetic pole is called a *magnetic monopole*. Although physicists have conducted extensive searches for monopoles, no experiment has yet found unequivocal evidence for their existence. In the next section we'll see why.

Connections Between Electricity and Magnetism

In our everyday experience, static electricity and magnetism seem to be two unrelated phenomena. Yet scientists in the nineteenth century, probing deeper into the electric and magnetic forces, discovered remarkable connections between the two, a discovery that transformed every aspect of technology.

Science in the Making

Luigi Galvani and Life's Electric Force

You might think that studies of magnets and electric charges have little to do with biology, but scientists of the late eighteenth century discovered remarkable links between life and electricity. Of all the phenomena in nature, none fascinated these scientists more than life, the mysterious force that allowed animals to move and grow. An old doctrine called "vitalism" held that there was a special force found only in living things, and not found in the rest of nature. Luigi Galvani (1737–1798), an Italian physician and anatomist, added fuel to the debate about the nature of life with a series of classic experiments demonstrating the effects of electricity on living things.

His most famous investigations employed an electric spark to induce convulsive twitching in amputated frogs' legs—a phenomenon not unlike a person's involuntary reaction to a jolt of electricity. Later he was able to produce a similar effect simply by poking a frog's leg simultaneously with one fork of copper and one of iron. In modern language, we would say that the electric charge and the presence of the two metals in the salty fluid in the frog's leg led to a flow of electrical charge in the frog's nerves, a process that caused contractions of the muscles. Galvani, however, argued that his experiments showed that there was a vital force in living systems—something he called "animal electricity"—that made them different from inanimate matter. This idea gained some acceptance among the scientific community, but provoked a long debate between Galvani and the Italian physicist Allesandro Volta (1745–1827). Volta argued that Galvani's effects had to do with chemical reactions between the metals and the salty fluids of the frogs' legs. In retrospect, we can see that both of these scientists had part of the truth. Muscle contractions are indeed initiated by electrical signals, even if there is no such thing as animal electricity, and electrical charges can be induced to flow by chemical reactions.

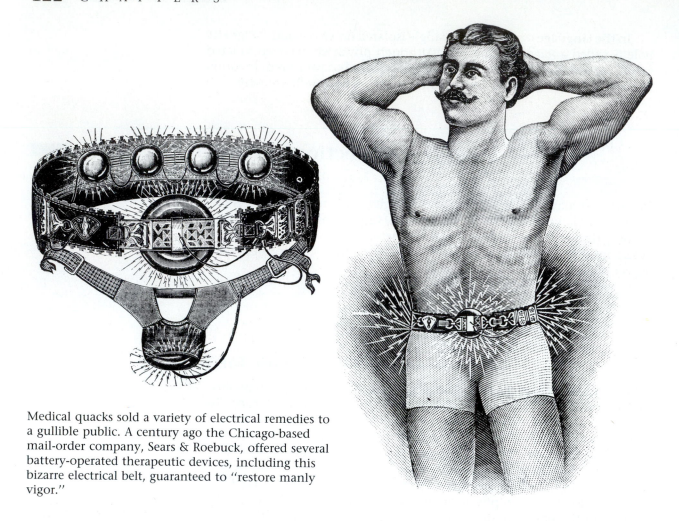

Medical quacks sold a variety of electrical remedies to a gullible public. A century ago the Chicago-based mail-order company, Sears & Roebuck, offered several battery-operated therapeutic devices, including this bizarre electrical belt, guaranteed to "restore manly vigor."

The controversy that surrounded Galvani's experiments had many surprising effects. On the practical side, as we discuss in the text, Volta's work on chemical reactions led to the development of the battery and, indirectly, to our modern understanding of electricity. The notion of animal electricity proved a great boon to medical quacks and con men, and for centuries various kinds of electrical devices were palmed off on the public as cures for almost every known disease.

Finally, in a bizarre epilogue to Galvani's research, other researchers used batteries to study the effects of electric currents on human cadavers. In one famous public demonstration, a corpse was made to sit up and kick its legs by electrical stimulation. Such experiments helped inspire Mary Shelley's famous novel, *Frankenstein*. ■

Batteries and Electric Current

Although we encounter static electricity in our everyday lives, most of our contact with electricity comes from moving charges. In your home, for example, negatively charged electrons move through wires to run all of your electrical appliances. A flow of charged particles is called an **electric current.**

Until the work of Allessandro Volta, scientists could not produce persistent electrical currents in their laboratories, and therefore knew little about them. As a result of his investigations into Galvani's work, Volta developed the first **battery,** a device that converts stored chemical energy in the battery materials into kinetic energy of electrons running through an outside wire.

The first battery was a crude affair, but we now use its descendants to start our cars and run all sorts of portable electronic equipment. Your car battery, a reliable and beautifully engineered device, routinely performs for years before it needs replacing. It is made of alternating plates of two kinds of material, lead and lead oxide, immersed in a bath of dilute sulfuric acid. When the battery is being discharged, the lead plate interacts with the acid, producing lead sulphate (the white crud that collects around the posts of old batteries) and some free electrons. These electrons run through an external wire to the other plates, where they interact with the lead oxide and sulfuric acid to form lead sulphate. The electrons running through the outside wire are what start your car.

When the battery is completely discharged, it consists of plates of lead sulphate immersed in water, a configuration from which no energy can be obtained. Running a current backward through the battery, however, runs all the chemical reactions in reverse and restores the original configuration. We say that the battery has been recharged. Once this is done, the whole cycle can proceed again. In your car, the generator constantly recharges the battery whenever the engine is running, so that it is always ready to use.

Magnetic Effects from Electricity

In the spring of 1820 a strange thing happened during a physics lecture in Denmark. The lecturer, a professor by the name of Hans Christian Oersted (1777–1851), was using a battery to demonstrate some properties of electricity. By chance he noticed that whenever he connected the battery (so that an electric current began to flow), a compass needle on a nearby table began to twitch and move. When he disconnected the battery, the needle went back to pointing north. This accidental discovery led the way to one of the most profound insights in the history of science. Oersted had discovered nothing less than the fact that electricity and magnetism—two forces that seemed as different from each other as night and day—are in fact intimately related to each other. They are two sides of the same coin.

In subsequent studies, Oersted and his colleagues established very clearly that whenever electrical charge flows through a wire, a magnetic field will appear around that wire. A compass brought near the wire will twist around until it points along the direction of that field. This leads to an important experimental finding in electricity and magnetism:

Hans Christian Oersted (1770–1851)

> **Magnetic fields can be created by the motion of electrical charges.**

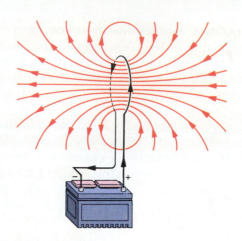

Figure 5–7

A schematic drawing of an electromagnet reveals the principal components, a loop of wire and a source of electric current. When a current flows through the wire loop, a magnetic field is created around it.

Like all fundamental discoveries, the discovery of this law of nature has important practical consequences. Perhaps most importantly, it led to the development of the **electromagnet,** a device composed of a coil of wire that produces a magnetic field whenever electrical charge runs through the wire. Almost every electrical appliance in modern technology uses this device.

The Electromagnet

Electromagnets work on a very simple principle, as illustrated in Figure 5–7. If an electric current flows in a loop of wire, then a magnetic field will be created around the wire, just as Oersted discovered in 1820. That magnetic field will have the shape sketched in the figure—a shape familiar to you as the dipole magnetic field shown in Figure 5–5.

In other words, we can create the equivalent of a piece of lodestone or magnetized piece of iron simply by running electrical current around a loop of wire. The stronger the current—i.e., the more electrical charge we push through the wire—the stronger the magnetic field will be. But unlike a piece of lodestone or a bar magnet, an electromagnet can be turned on and off. To differentiate between these two sorts of magnets, we often refer to magnets made from materials such as iron as permanent magnets.

The electromagnet can be used in all sorts of practical ways, including buzzers, switches, and electric motors. In each of these devices a piece of iron is placed near the magnet. When the current flows in the loops of wire, the iron is pulled toward the magnet. In some cases, the electromagnet can be used to complete an electrical circuit by pulling a switch closed. As soon as the current is turned off in the electromagnet, a spring pushes the iron back, and the current in the larger circuit also shuts off.

You use electromagnetic switches extensively in your household appliances. For example, your home is probably heated by a furnace that is linked to a thermostat on the wall in your living room or hallway. You set the thermostat to a specific temperature. If the temperature in the room falls below the desired temperature, the thermostat responds by using an electromagnet to close a switch, allowing a small current to flow. While the current is flowing in the electromagnet, the switch is closed and the furnace operates, heating the rooms in your house. As

Electromagnets, which can be turned on and off, are ideal for moving scrap iron at a junkyard.

soon as the temperature reaches the level you have set, however, the thermostat stops the current that flows to the electromagnet, the switch opens, and the furnace shuts off. In this way, you can adjust the temperature of your house without having to watch it constantly and run to the basement every time you need to turn on the furnace.

Technology

The Electric Motor

All **electric motors,** which play a key role in modern society, incorporate electromagnets. They all operate by supplying a current to an electromagnet to make it move and generate mechanical power. The motor employs a permanent magnet, as shown in Figure 5–8, and a rotating loop of wire inside the poles of those magnets. The current in the rotating loop adjusts so that when it is oriented as shown in Figure 5–8a, the south pole of the electromagnet lies near the north pole of the permanent magnet, and the north pole of the electromagnet lies near the south pole of the permanent magnet. The attractive force between north and south poles causes the loop of wire to spin. As soon as the loop of wire gets to the position shown in Figure 5–8b, the current reverses so that the south pole of the electromagnet lies just past the south pole of the permanent magnet, and the north pole of the electromagnet lies just past the north pole of the permanent magnet. The repulsive force between these like magnetic poles acts to continue the rotation. By alternating the current in the loop, you can keep a continuous rotational force on the wire, and keep the wire turning. This sort of arrangement operates in an electric hand drill, an electric fan, and the motors that raise and lower the windows in your car.

This simple diagram contains all the essential features of an electric motor, but real electric motors are much more complex than the one

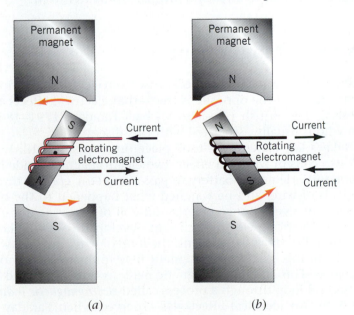

(a)   (b)

Figure 5–8
An electric motor. The simplest motors work by placing an electromagnet that can rotate between two permanent magnets. (a) When the current is turned on, the north and south poles of the electromagnet are attracted to the south and north poles of the permanent magnets. (b) As the electromagnet rotates, the current direction is switched, causing the electromagnet to continue rotating.

we have shown. Typically, they have three or more different electromagnets and at least three permanent magnets, and the alternation of the current direction is somewhat more complicated than we have indicated. By artfully juxtaposing electromagnets and permanent magnets, inventors have produced an astonishing variety of electric motors: fixed speed for your CD player, variable speed for your food processor, reversible motors for power drills and screwdrivers, and specialized motors for many precision industrial uses. ■

Why Magnetic Monopoles Don't Exist

Earlier we stated that there are no isolated magnetic poles in nature—you can't have an isolated north pole or south pole. Electromagnets provide a basis for understanding where this law comes from. Every magnetic field that we know in the universe results from the motion of electrical charges. If you think of an electron in orbit around an atom as described earlier, you'll realize that the orbiting single electron constitutes a "current," in every respect identical to the circle of wire in an electromagnet. The only difference is that the current in the atomic loop consists of a single electron going around and around in an orbit, while the current in the wire consists of many electrons moving around and around in a much larger loop.

The magnetism in permanent magnets can be traced ultimately to the summation of countless current loops made by electrons going around orbits in atoms (see Chapter 7). This fact explains why ordinary magnets can never be broken down into magnetic monopoles. If you break a magnet down to one last individual atom, you still have a dipole field because of the atomic-scale current loop. If you try to break the atom down further, the dipole field will disappear and there will be no magnetism except that associated with the particles themselves. Thus magnetism in nature is ultimately related to the arrangement of electrical charges, rather than to anything intrinsic to matter itself.

■ Electrical Effects from Magnetism

Now that you know magnetic effects arise from electricity, it should come as no surprise that the opposite is true—that electrical effects arise from magnetism. The British physicist Michael Faraday (1791–1867) is the person who is usually associated with this discovery.

Faraday's key experiment took place on August 29, 1831, when he placed two coils of wire—in effect, two electromagnets—side by side in his laboratory. He used a battery to pass an electric current through one of the coils of wire, and he watched what happened to the other coil. Astonishingly, even though the second coil of wire was not connected to a battery, a strong electrical current developed. We now know what happened in Faraday's experiment: the loop of wire, through which current was running, produced a magnetic field in the neighborhood of the second loop. This changing magnetic field, in turn, produced a current in the second loop through a process called *electromagnetic induction* (see Figure 5–9). An identical effect was observed when Faraday waved a

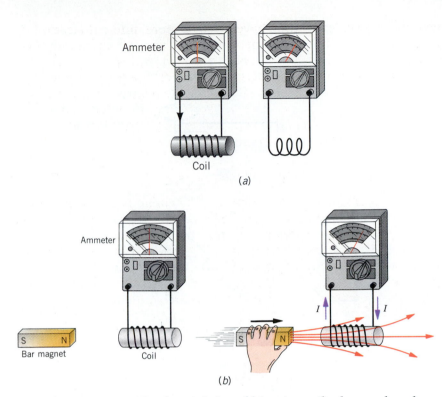

Coil

(a)

Ammeter

Bar magnet

Coil

(b)

Figure 5–9
Electromagnetic induction. (a) When a current flows in the circuit on the left, a current is observed to flow in the circuit on the right, *even though there is no battery or power source in that circuit.* (b) Moving a magnet into the region of a coil of wire causes a current to flow in the circuit, even in the absence of a battery or other source of power.

permanent magnet in the vicinity of his wire coil—he produced an electric current without a battery.

Michael Faraday's research can be summarized by a simple law:

> **Electric fields and electric currents can be produced by changing magnetic fields.**

Figure 5–10 illustrates the **electric generator,** or dynamo, a vital tool of modern technology that demonstrates this effect. Place a loop of electrical wire with no batteries or other power source between the north and south poles of a strong horseshoe magnet. As long as the loop of wire stands still, no current flows in the wire, but as soon as we begin to rotate the loop, a current flows in the wire. This current flows in spite the fact that there is no battery or other power source in the wire.

From the point of view of the electrons in the wire, any rotation changes the orientation of the magnetic field. The electrons sense a changing magnetic field and hence, by Faraday's findings, a current flows in the loop. If we spin the loop continuously, a continuous current flows in it. The current flows in one direction for half of the rotation, then flows in the opposite direction for the other half of the rotation. This extremely important practical device, the electrical generator, followed immediately from Faraday's discovery of electromagnetic induction.

In an electrical generator, some source of energy—water passing over a dam, steam produced by a nuclear reactor or coal-burning furnace, or wind-driven propeller blades, for example—turns a shaft. In your car, the energy to turn the coils in a magnetic field comes from the gasoline

Figure 5–10

An electric generator. As long as the loop of wire rotates, there is a changing magnetic field near the loop and a current flows in the wire.

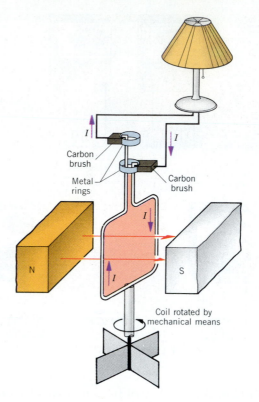

that is burned in the motor. In every generator the rotating shaft links to coils of wire that spin in a magnetic field. Because of the rotation, electrical current flows in the wire and can be tapped off onto external lines. Almost all the electricity used in the United States is generated in this way.

You may have noticed a curious fact about electric motors and generators. In an electric motor, electrical energy is converted into the kinetic energy of the spinning shaft, while in a generator, the kinetic energy of the spinning shaft is converted into electrical energy. Thus motors and generators are, in a sense, exact opposites in the world of electromagnetism.

Because the current in the generating coils flows first one way and then the other, it will do the same thing in the wires in your home. This kind of current, the kind used in household appliances and cars, is called **alternating current,** or **AC,** because the direction keeps alternating. In contrast, chemical reactions in a battery cause electrons flow in one direction only and produce what is called **direct current,** or **DC.**

Science in the Making

Michael Faraday

Michael Faraday, one of the most honored scientists of the nineteenth century, did not come easily to his profession. The son of a blacksmith, he received only a rudimentary education as a member of a small Christian sect. Faraday was apprenticed at the age of 14 to a Lon-

Electricity is generated when steam or water power turns a turbine, which contains coils of wire and magnets.

don book merchant, and he became a voracious reader as well as a skilled bookbinder. Chancing upon the *Encyclopedia Britannica,* he was especially drawn to scientific articles, and he determined then and there to make science his life.

The young Faraday pursued his scientific career in style. He attended a series of public lectures at the Royal Institution by London's most famous scientist, Sir Humphry Davy, a world leader in physical and chemical research. Then, in a bold and flamboyant move, Faraday transcribed his lecture notes into beautiful script, bound the manuscript in the finest tooled leather, and presented the volume to Davy as his calling card. Michael Faraday soon found himself working as Davy's laboratory assistant.

After a decade of work with Davy, Faraday had developed into a creative scientist in his own right. He discovered many new chemical compounds, including liquid benzene, and enjoyed great success with his own popular lectures for the general London public at the Royal Institution. But his most lasting claim to fame was a series of classic experiments through which he discovered a central idea that helped link electricity and magnetism. ■

Michael Faraday
(1791–1867)

Maxwell's Equations

We now understand the basic facts that govern the behavior of electrical and magnetic phenomena in our world. Electricity and magnetism are not distinct phenomena at all, but are simply different manifestations of one underlying fundamental entity—**electromagnetism.** In the 1860s Scottish physicist James Clerk Maxwell (1831-1879) realized that the four very different statements about electricity and magnetism that we have talked about constitute a single coherent description of electricity and magnetism. He also recognized that something was missing from the law that described the production of magnetic fields by electrical phenomena, and he was able to insert the missing piece.

In the end, the four statements that he wrote down have come to be known as *Maxwell's equations,* because he was the first to realize their true import, and because he was the first to manipulate the mathematics to make important predictions—predictions that we will discuss in detail in the next chapter. For reference, the four fundamental laws of electricity and magnetism known as Maxwell's equations are:

1. Coulomb's Law—like charges repel, unlike attract.
2. There are no magnetic monopoles in nature.
3. Magnetic phenomena can be produced by electrical effects.
4. Electrical phenomena can be produced by magnetic effects.

Electric Circuits

Most people come into contact with electrical phenomena through electric circuits in their homes and cars. An **electric circuit** is an unbroken path of material that carries electricity. For example, the electric light that you are using to read this book is part of an electrical circuit that begins at a power plant with an electric generator, many miles away (Figure 5–11). That electricity continues through power lines into your town and is distributed on overhead or underground wires until it finally gets to where you live. There, the circuit of which the light is a part is made up of copper wires that run through the walls of your home. One set of these wires goes first to a circuit breaker (to break the circuit in case of a dangerous overload), then to a switch, and finally to the light

Figure 5–11

Your toaster completes part of an electric circuit that extends all the way back to your local power generating plant.

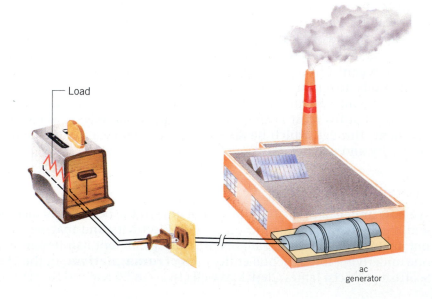

Load

ac
generator

bulb. When you turn the switch to the "on" position, you complete an unbroken path that runs all the way from the generating plant to the light bulb. When electricity flows through the light, the current heats up the filament until it glows. When you put the switch in the "off" position, it's like raising a drawbridge—the current is blocked from flowing into this part of the circuit and none reaches the light. The room becomes dark.

Every circuit consists of three parts: a source of energy, a closed path usually made of metal wire through which the current can flow, and a device that uses the electrical energy—perhaps a motor or light bulb.

The energy source is usually a battery or an electrical generator. Batteries, which are ideal for powering small, portable electric circuits, rely on chemical energy to push a direct current of electrons through wires. Electrical generators, which may be located hundreds of miles from where the electricity is finally used, produce alternating current as we have seen. Your car carries both a battery and a generator under the hood.

One way to think about electrical circuits is to draw an analogy between electrons flowing through a wire and water flowing through a pipe. In the case of the water, we use two quantities to characterize the flow: the amount of water that passes a point each second, and the pressure behind that water. The numbers we use to describe the flow of electrons in a circuit are exactly the same.

The amount of current—the number of electrons—that actually flows in a wire is measured in a unit called the **ampere** or **amp,** named after French physicist Andre-Marie Ampere (1775–1836). One amp corresponds to a flow of one coulomb (the unit of electric charge) per second past a point in the wire:

1 amp of current = 1 coulomb of charge per second.

Electrical current, therefore, is analogous to the current of a river or stream. Typical household appliances use anywhere from about 1 amp (100-watt bulb) to 40 amps (an electric range with all burners and the oven blazing away).

We call the pressure produced by the energy source in a circuit the **voltage,** measured in **volts** (abbreviated V) and named after Alessandro Volta, the Italian scientist who invented the chemical battery. You can think of voltage in circuits much the same way you think of water pressure in your plumbing system. More volts in a circuit mean more "oomph" to the current, just as more water pressure makes the water flow faster. Typically, a new flashlight battery produces 1.5 volts, a fully charged car battery about 15 volts (even though they are called "12-volt" batteries), and ordinary household circuits operate on either 115 or 230 volts.

Wires through which the current flows are analogous to pipes carrying water: the smaller the pipe, the harder it is to push water through it. Similarly, it is harder to push electrons through some wires than others. The quantity that measures how hard it is to push electrons through wires is called **electrical resistance,** and it is measured in a unit called the **ohm.** The higher the resistance, the more of the electron energy that is converted into heat. Ordinary copper wire, for example, has a low resistance, which explains why we use it to carry electricity around our homes. Toasters and space heaters, on the other hand, employ high-resistance wires so that they will glow bright red and give off large amounts of heat when current flows through them. In transmission lines, it is important that as much energy as possible gets from one end of the line to the other; thus we use very thick low-resistance wires.

The mechanism behind electrical resistance is easy to understand. As electrons move through a wire, they collide with atoms. As a result of these collisions, some of the kinetic energy of the electrons is transferred to kinetic energy of the atoms, a process that heats up the wire. In fact, electrons collide so often with atoms that their progress through the wire is surprisingly slow. You can walk faster than an electron can move through the copper wires in your home, for example.

The *load* in any electrical circuit is the "business end"—the place where useful work gets done. The filament of a light bulb, the heating element in your hair dryer, or an electromagnetic coil of wire in an electric motor are typical loads in household circuits. The power used by the load depends both upon how much current flows through it and the size of the voltage. The greater the current and the higher the voltage, the more power is used. A simple equation allows us to calculate the amount of electrical power used.

▶ **In words:**

The power consumed by an electrical appliance is equal to the product of the current and the voltage.

▶ **In equation form:**

power (in watts) = current (in amps) × voltage (in volts).

▶ **In symbols:**

$P = I \times V$.

This equation tells us that both the current and the voltage have to be high for a device to consume high levels of electrical power.

Table 5–1 summarizes some key terms about electrical circuits.

Table 5–1 • Terms Related to Electric Circuits

Term	Definition	Unit	Plumbing Analog
Voltage	Electrical pressure	volt	Water pressure
Resistance	Resistance to electron flow	ohm	Pipe diameter
Current	Flow rate of electrons	amp	Flow rate
Power	Current × voltage	watt	Rate of work done by moving water

EXAMPLE 5–1: Starting Your Car

When you turn on the ignition of your automobile, your 15-volt car battery must turn a 400-amp starter motor. How much power is required to start your car?

▶ **Solution:** We apply the equation for electrical power.

power (in watts) = current (in amps) × voltage (in volts)
= 400 amps × 15 volts
= 6000 watts
= 6 kilowatts

In the early days of automobiles most vehicles were started by a hand crank, which might have required 100 watts of power, a reasonable amount for an adult. Modern, high-compression automobile engines require much more starting power than could be generated by one person. ▲

EXAMPLE 5–2: The Power of Sound

In Example 3–2, we talked about a 250-watt compact disc system. Assuming that this system is plugged into a normal household outlet rated at 115 volts, how much current will flow through the stereo at full power?

▶ **Reasoning:** The power consumed by the CD system is given by

power (in watts) = current (in amps) × voltage (in volts)

▶ **Solution:** We can manipulate this equation to find the current:

$$\text{current} = \frac{\text{power}}{\text{voltage}}$$
$$= \frac{250 \text{ watts}}{115 \text{ volts}}$$
$$= 2.17 \text{ amps}$$

This amount is slightly more current than flows through two 100-watt light bulbs. ▲

Human Body

The Propagation of Nerve Signals

All of your body's movements, from the beating of your heart to the blinking of your eyes, are controlled by nerve impulses. Although nerve signals in the human body are electrical in nature, they bear little resemblance to the movement of electrons through a wire. Nerve cells of the type shown in Figure 5–12 form the fundamental element of the nervous system. A nerve cell consists of a central body with a large number of filaments going out from it. These filaments connect one nerve cell to others. The longest filament is known as the *axon*, and it is along the axon that a nerve signal moves.

The membrane that surrounds the axon is a complex structure, full of channels through which atoms and molecules can move. When the nerve cell is resting, positively charged objects tend to be outside the membrane, negatively charged ones inside. When an electrical signal triggers the axon, the membrane is distorted and, for a short time, positive charges (mainly in the form of sodium) pour into the cell. As soon

Figure 5–12

The nerve cell consists of a central body and a number of filaments. The dendrites receive incoming signals, and the axon conducts outgoing signals away from the cell body. The myelin sheath helps insulate the axon from neighboring electrical interference.

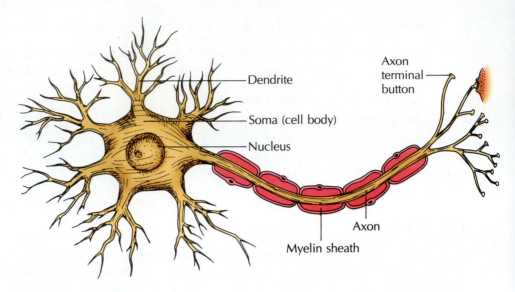

as the inside becomes more positively charged, the membrane changes again and positive charges (this time mainly potassium) move back outside to restore the original charge. This charge disturbance moves down the filament, creating what we call a nerve signal. It's much more complex than a simple migration of electrons.

When the signal reaches the end of one of the filaments, it is transferred to the next cell by a group of molecules called *neurotransmitters* that are sprayed out from the end of the "upstream" cell, and received by special structures on the "downstream" cell. The reception of neurotransmitters causes the cell membrane to deform and a nerve signal starts down *that* cell.

Thus, although the human nervous system is not as simple as an ordinary electrical circuit, it does operate by electrical signals. ∎

THINKING MORE ABOUT ELECTROMAGNETISM

Basic Research

It's hard to imagine modern American society without electricity. We use it to move goods, to communicate with each other, to light and warm our homes, and to manufacture the necessities and amenities of life. Chances are that nothing within sight of you now would be as it is if not for the use of electricity.

Yet the men who gave us this marvelous gift were not, by and large, primarily concerned with developing better lamps or clocks or modes of transportation. In terms of the categories we introduced in Chapter 1, they were doing basic research. Galvani and Volta, for example, were drawn to the study of electricity by their studies of frog muscles that contracted by jolts of electrical charge. Volta's first battery was built to duplicate the organs found in electric fish. Scientific discoveries—even scientific discoveries that bring enormous practical benefit to humanity—can come from unusual and unexpected sources.

What does this tell you about the problem of allocating government research funding? Can you imagine trying to justify funding Galvani's experiments on frogs' legs to a government panel on the grounds that it would lead to something useful? Would a federal research grant designed to produce better lighting systems have produced the battery (and, eventually, the electric light), or would it more likely have led to an improvement in the oil lamp? How much funding do you think should go to offbeat areas (on the chance that they may produce a large payoff) as compared to projects that have a good chance of producing small but immediate improvements in the quality of life?

While you're thinking about these issues, you might want to keep in mind Michael Faraday's response to a question. When asked by a political leader what good his electric motor was, he is supposed to have answered, "What good is it? Why, Mr. Prime Minister, someday you will be able to tax it!"

Summary

Electricity and *magnetism* are phenomena that involve forces quite different from the universal gravitational force that Newton described in the seventeenth century. Nevertheless, Newton's laws of motion provided scientists of the eighteenth and nineteenth centuries a way to describe and quantify a range of intriguing electromagnetic behavior.

The phenomena of *static electricity*, including lightning, static cling, and the small sparks produced when walking across a wool rug on a cold winter day, are caused by the transfer of electrons between objects. An excess of electrons imparts a *negative charge*, while a deficiency causes an object to have a *positive charge*. Objects with like charge experience a repulsive force, while oppositely charged objects attract each other. These observations were quantified in *Coulomb's law*, which states that the magnitude of electrostatic force between any two objects is proportional to the charges of the two objects, and inversely proportional to the square of the distance between them.

Other scientists, investigating the very different phenomenon of magnetism, observed that every *magnet* has a *north and south pole*, and that magnets exert forces on each other. No matter how many times a magnet is divided, each of its pieces will have two poles—there are no isolated magnetic poles. Like magnetic poles repel each other, while opposite poles attract. A compass is a needle-shaped magnet that is designed to point at the poles of the Earth's magnetic field.

Both electric and magnetic forces can be described in terms of force *fields*—imaginary lines that reveal the directions of forces that would be experienced in the vicinity of electrically charged or magnetic objects. Scientists in the nineteenth century discovered that the seemingly unrelated phenomena of electricity and magnetism were actually two aspects of a single *electromagnetic force.*

Hans Oersted found that an electric current passing through a coil of wire produces a magnetic field. The *electromagnet* and *electric motor* were direct results of his work. Michael Faraday discovered the opposite effect when he induced an electric current by placing a wire coil near a magnetic field, thus designing the first *electric generator*, which produced an *alternating current (AC)*. *Batteries*, on the other hand, develop a *direct current (DC)*. All *electric currents* (measured in *amperes*) are characterized by an electric "push" or *voltage* (measured in *volts*) and *electrical resistance* (measured in *ohms*).

James Clerk Maxwell realized that the many independent observations about electricity and magnetism, taken together, constitute a complete description of electromagnetism.

Key Terms About Electricity and Magnetism

electrical charge	poles (north and south)	alternating current (AC)
electricity	magnetic field	direct current (DC)
static electricity	electric current (measured in amperes)	electromagnetism
positive charge		electric circuit
negative charge	battery	ampere, or amps
Coulomb's law	electromagnet	voltage (measured in volts)
electric field	electric motor	electrical resistance (measured in ohms)
magnetic force	electric generator	

Key Equations About Electricity and Magnetism

electrostatic force (in newtons) $= k \times \dfrac{\text{1st charge} \times \text{2nd charge (in coulombs)}}{\text{distance}^2}$

1 coulomb = the charge on 6.3×10^{18} electrons
electric power (in watts) = current (in amps) $\times$ voltage (in volts)
1 ampere of current = 1 coulomb of charge per second

Universal Electrostatic Constant

$k = 9.00 \times 10^9$ newton-meter^2/coulomb^2

Review Questions

1. How do you know there is such a thing as an electrical force?
2. How does the electrical force differ from gravity in its manifestations?
3. What would a modern scientist say that Franklin's "single fluid" was?
4. How can the movement of negative charges like the electron produce a material that has a positive charge?
5. What is an electric field?
6. How do you know there is such a thing as a magnetic force?
7. What was the first practical application of magnetism?
8. What is a dipole field?
9. Under what circumstances can electric charges produce a magnetic field?
10. Describe the working and function of an electromagnet.
11. How does an electromagnet differ from a permanent magnet?
12. How are electromagnets used in an electric motor?
13. How can a single atom produce a magnetic field?
14. Under what circumstances can a magnetic field produce an electric current?
15. How does an electric generator work?
16. What are Maxwell's equations?
17. What does the volt measure? Where do you run across this term in your everyday experience?
18. What does the ampere measure? Where do you run across this term in your everyday experience?
19. What does the ohm measure? Where do you run across this term in your everyday experience?
20. What is the relation between the power an appliance consumes, the voltage across it, and the current through it?

Discussion Questions

1. Your car uses a 15-volt battery, but do you need all of its energy to start the car? (*Hint:* What happens if you leave the lights on for half an hour—can you still start it?)
2. If you took an electric motor and turned it by hand, what do you think would happen in the coils of wire?
3. What would you say was the motivation in Faraday's classic experiment on electromagnetic induction?
4. How did Benjamin Franklin know that lightning had an electrical nature?
5. Why can't you find a magnet monopole by taking an atom apart?
6. Identify five things in your room that would not be possible without discoveries in electromagnetism.
7. How does the first law of thermodynamics apply to electrical circuits?
8. How does the second law of thermodynamics apply to electrical circuits?

Problems

1. Based on electric charges and separations, which of the following atomic bonds is strongest?
   a. A +1 sodium atom separated by 2.0 distance units from a −1 chlorine atom in table salt.
   b. A +1 hydrogen atom separated by 1.0 distance units from a −2 oxygen in water.
   c. A +4 silicon atom separated by 1.5 distance units from a −2 oxygen in glass.
   (*Hint:* You are only interested in the *relative* strengths, which depend only on the relative charges and distances.)

2. When a car battery runs down, it is recharged by running current through it backward. Typically, you might run 5 amps at 12 volts for an hour. How much energy does it take to recharge a battery?

3. Most household circuits have fuses or circuit breakers that open a switch when the current in the circuit exceeds 15 amps. Would the lights go off when you plug in an air conditioner (1 kilowatt), a TV (250 watts), and four 100 watt light bulbs? Why?

4. Energy efficient appliances are important in today's economy. Suppose that a light bulb gave as much light as a 100-watt bulb, but consumed only 20 watts, while costing $2.00 more. If electricity costs 8 cents per kilowatt hour, how long will the bulb have to operate to make up the difference in price?

5. An energy efficient air conditioner draws 7 amps in a standard 115-volt circuit. It costs $40 more than a standard air conditioner that draws 12 amps. If electricity costs 8 cents per kilowatt hour, how long would you have to run the efficient air conditioner to recoup the difference in price?

Investigations

1. Make an inventory of all your electrical appliances. How many watts does each use?

2. Most household circuits have fuses or circuit breakers that open a switch when the current in the circuit exceeds 15 amps. How many of the appliances in (1) could you run on the same circuit without overloading it?

3. Read Jules Verne's *The Mysterious Island*. Do you think it is possible for castaways to build an electrical generator as described in the book?

4. Examine carefully your most recent electric bill. How much power did you use? How much did it cost? Is there a discount for electricity used at off-peak hours? Examine your living place for all the ways in which you use electricity. Plan a strategy for reducing your electric bill by 10% next month. You can reduce consumption by turning off lights and appliances when not in use, installing lower-wattage bulbs, or using electricity during low-rate times.

5. Find out where your electric power is generated. Does your local utility buy additional power from some other place? What kind of fuel or energy is used to drive the turbines? Are there pollution controls that restrict the use of certain kinds of fuels at your local power plant?

6. How many kilowatts of electrical power does a typical commercial power plant generate? How much electricity does the United States use each year? Is this amount going up or down?

7. Take an old electric razor or other small motor-driven appliance and dissect the motor. How many permanent magnets are inside? How many separate coils of wire?

8. Identify the major electrical-circuit components in your automobile. Which require the greatest power?

9. How do electric eels generate electrical shocks?

10. How does an electroencephalogram (EEG) work? How does it differ from an electrocardiogram (EKG)?

11. Many kinds of living things, from bacteria to vertebrates, incorporate small magnetic particles. Investigate the ways living things use magnetism.

12. Read the novel *Frankenstein* (or see the classic 1931 movie with Boris Karloff and Colin Clive). Discuss the ideas about the nature of life that are implicit in the story. Does it represent a realistic picture of scientific research? Why or why not?

Additional Reading

Everitt, C.W.F. *James Clerk Maxwell: Physicist and Natural Philosopher*. New York: Scribners, 1976.

Franklin, Benjamin. *The Autobiography of Benjamin Franklin*. Boston: St Martins Press, 1993.

Pera, Marcello. *The Ambiguous Frog: The Galvani-Volta Controversy on Animal Electricity*. Princeton: Princeton University Press, 1992.

Electromagnetic Radiation

Whenever an electrically charged object is accelerated, it produces electromagnetic radiation—waves of energy that travel at the speed of light.

A Random Walk
A Day at the Beach

Few experiences are more relaxing than a day at the beach. The sight of waves washing ashore, the sound of good music, and the feel of the Sun's rays beating down help us forget about the pressure of exams and term papers. What might surprise you is that underlying all of those familiar experiences at the beach is the phenomenon of waves.

Waves are all around us. Waves of water travel across the surface of the ocean and come crashing against the land. Waves of sound travel through the air when we listen to music. Some parts of the United States suffer from mighty waves of rock and soil called earthquakes. All of these familiar kinds of waves must move through matter.

But the most remarkable waves of all—electromagnetic waves that fill every corner of our environment every moment of our lives—can travel through an absolute vacuum at the speed of light. The sunlight that warms you at the beach and provides virtually all of the energy necessary for life on Earth is transmitted through space by just such a wave. The radio waves that carry your favorite music, the microwaves that heat your dinner, and the X-rays your dentist uses to check for cavities are also types of electromagnetic waves—invisible forms of energy that travel at light speed.

The Nature of Waves

Waves are fascinating, at once familiar and yet somewhat odd. Scientists study waves by observing their distinctive behavior, particularly their unique ability to transfer energy without transferring mass.

Energy Transfer by Waves

Energy can travel in two forms in our everyday world—the particle and the wave. Suppose you have a domino sitting on a table and you want to knock it over, a process that requires transferring energy from you to the domino. One way to proceed would be to take another domino and throw it. From the standpoint of energy, you would say that the muscles

Figure 6–1
You can use a domino to knock over other dominoes in two different ways: (*a*) you can throw a domino, or (*b*) you can trigger a wave of dominoes.

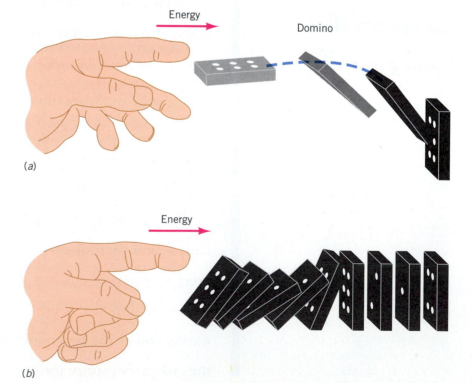

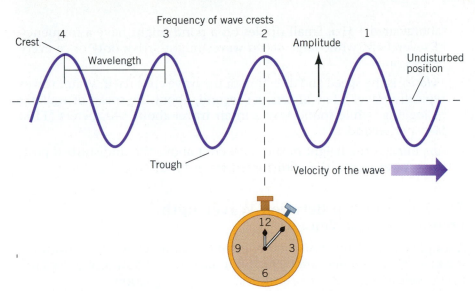

Figure 6–2

A cross-section of a wave reveals the characteristics of wavelength, velocity, and amplitude. Successive wave crests are numbered *1, 2, 3,* and *4*. An observer at the position of the clock records the number of crests that pass by in a second. This is the frequency, which is measured in cycles per second, or hertz.

in your arms imparted kinetic energy to the moving domino, which, in turn, would impart enough of that energy to the standing domino to knock it over (Figure 6–1*a*). We say that the energy transfers by the motion of a solid piece of matter.

Alternatively, you could line up a row of standing dominoes, knock over the first, which would then knock over the second, which in turn would knock over the third, and so on (Figure 6–1*b*). Eventually the falling chain of dominoes would hit the last one and you would have achieved the same goal. In the case of the lined-up dominoes, however, no single object traveled from you to the most distant domino. In the language of physics we say that you started a wave of falling dominoes and the wave is what knocked over the final one. A **wave,** then, is a travelling disturbance; it carries energy from place to place without requiring matter to travel across the intervening distance.

The Properties of Waves

Think about a familiar example of waves. You are standing on the banks of a quiet pond on a crisp autumn afternoon. There's no breeze and the pond in front of you is still and smooth. You pick up a rock and toss it into the middle of the pond. As soon as the rock hits the water a series of ripples moves outward from the point of impact. In cross-section, the ripples have the familiar wave shape shown in Figure 6–2. You can use four measurements to characterize the ripples.

1. **Wavelength** is the distance between crests, or the highest points of adjacent waves. On a pond the wavelength might be only a centimeter or two, while ocean waves may be tens or hundreds of meters between crests.

2. **Frequency** is the number of wave crests that go by a given point every second. A wave that sends one crest by every second (completing one *cycle*) is said to have a frequency of one cycle per second or one **hertz**

(abbreviated 1 Hz). Small ripples on a pond might have a frequency of several Hz, while large ocean waves might arrive only once every few seconds.

3. *Velocity* is the speed and direction of the wave crest itself. Water waves typically travel a few meters per second, about the speed of walking or jogging, while sound waves in air travel about 340 meters (1100 feet) per second.

4. *Amplitude* is the height of the wave crest above the undisturbed position; for example, the undisturbed water level.

The Relationship Between Wavelength, Frequency, and Velocity

A simple relationship exists among three fundamental wave quantities—wavelength, frequency, and velocity. In fact, if we know any two of the three, we can calculate the third from a simple equation.

To understand why this is so, think about waves on water. Suppose you are sitting on a sailboat, as shown in Figure 6–3, watching a series of wave crests passing by. You can count the number of wave crests going by every second (the frequency) and measure the distance between the crests (the wavelength). From these two numbers, the speed of the wave can be calculated.

If, for example, one wave arrives every two seconds and the wave crests are six meters apart, then the waves must be traveling six meters every two seconds—a velocity of three meters per second. You might look out across the water and see a particularly large wave crest that will arrive at the boat after four intervening smaller waves. You would predict that the big wave is 30 meters away (five times the wavelength), and that it will arrive in 10 seconds. That kind of information can be very helpful if you are plotting the best course for an America's Cup yacht race or estimating the path of potentially destructive ocean waves.

This relationship among wavelength, velocity, and frequency can be written in equation form:

Figure 6–3

Waves passing a sailboat reveal how wavelength, velocity, and frequency are related. If you know the distance between wave crests (the wavelength) and the number of crests that pass each second (the frequency), then you can calculate the wave's velocity.

▶ **In words:**

The velocity of a wave is equal to the length of each wave times the number of waves that pass by each second.

▶ **In equation form:**

wave velocity (in m/s) = wavelength (in m) × frequency (in Hz).

▶ **In symbols:**

$v = \lambda \times \omega,$

where λ and ω are the common symbols for wavelength and frequency. This simple equation holds for all kinds of waves.

EXAMPLE 6–1: Seismic Shocks

A pile driver pounds a metal I-beam two times per second, driving it into rock. Each shock sends a wave travelling 6 kilometers per second through the rock. Called *seismic waves*, these kinds of sound waves play a major role in our measurements of the interior of the Earth (see Chapter 15). What is the wavelength of these seismic waves?

▶ **Reasoning:** We know that

wave velocity (in m/s) = wavelength (in m) × frequency (in Hz).

In this case we know the velocity (6 kilometers per second, which equals 6000 meters per second) and the frequency (2 times per second, which equals 2 hertz).

▶ **Solution:** We can rearrange the equation to solve for wavelength.

$$\text{wavelength (in m)} = \frac{\text{velocity (in m/s)}}{\text{frequency (in Hz)}}$$
$$= \frac{6,000 \text{ m/s}}{2 \text{ Hz}}$$
$$= 3,000 \text{ m}$$

These seismic waves, therefore, are quite long—3 kilometers or about 1.8 miles. Nevertheless, they're much shorter than distances in the Earth, and thus travel many times their wavelength before they die out. ▲

The Two Kinds of Waves: Transverse and Longitudinal

Imagine that a chip of bark or a piece of grass is lying on the surface of the pond when you throw a rock. When the ripples go by, that floating object and the water around it move up and down; they do not move to a different spot. At the same time, however, the wave crest moves in a direction parallel to the surface of the water. This means that *the motion of the wave is different from the motion of the medium on which the wave moves.* This kind of wave, where the motion moves perpendicular to the direction of the wave, is called a *transverse wave.*

Figure 6–4

Transverse (*a*) and longitudinal (*b*) waves differ in the motion of the wave relative to the motion of individual particles.

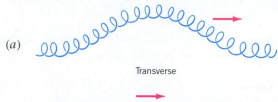

(*a*)

Transverse

(*b*)

Longitudinal

You can observe (and participate in) this phenomenon if you ever go to a sporting event in a crowded stadium where fans "do the wave." Each individual simply stands up and sits down, but the visual effect is of a giant sweeping motion around the entire stadium. In this way waves can move great distances, even though individual pieces of the transmitting medium hardly move at all.

Not all waves are transverse waves like those on the surface of water—we used the example of a pond simply because it is so familiar and can be visualized. Sound is a form of wave that moves through the air. When you talk, for example, your vocal cords move air molecules back and forth. The vibrations of these air molecules set the adjacent molecules in motion, which sets the next set of molecules in motion and so forth. A circular wave moves out from your mouth, a wave that looks very much like ripples on a pond. The only difference is that in the air the wave crest that is moving out is not a raised portion of a water surface, but a denser region of air molecules. In the language of physics, sound is a *pressure wave* or *longitudinal wave*. As a longitudinal wave of sound moves through the air, gas molecules vibrate forward and back *in the same direction as the wave*. This motion is very different from the transverse wave of a ripple in water, where the water molecules move *perpendicular to the direction of the waves* (see Figure 6–4). Note that in either longitudinal or transverse waves, the energy always moves in the direction of the wave.

 ## Science by the Numbers

The Sound of Music

The speed of sound in air is more or less constant for all kinds of sound. The way we perceive a sound wave, therefore, depends on its other properties—wavelength, frequency, and amplitude. For example, we sense the amplitude of a sound wave as loudness—the bigger the amplitude, the louder the sound. Similarly, we hear high-frequency sound waves (sound with short wavelengths) as high pitches, while we perceive low-frequency sound waves (with long wavelength) as low sounds.

You can see one consequence of this contrast when you watch a symphony orchestra. The highest notes are played by small instruments such as the piccolo and violins, while the lowest notes are the domain of the massive tuba and double basses. Each of a pipe organ's thousands of pipes produces a single note. An organ pipe encloses a

Different size instruments in a New Orleans jazz ensemble play in different ranges. The large tuba on the left plays low notes, while the smaller trumpet at center plays in a higher range.

column of air in which a sound wave can travel back and forth, down the length of the pipe over and over again. The number of waves completing this circuit every second—the frequency—defines the pitch that you hear (see Figure 6–5).

The note that we hear as "middle A," the pitch to which most orchestras tune, has a frequency of 440 Hz. Sound travels through air at about 340 meters per second. An organ pipe open at both ends produces a note with a wavelength twice as long as the pipe. The length of an organ pipe air column that plays middle A, therefore, is given by half the wavelength in the equation

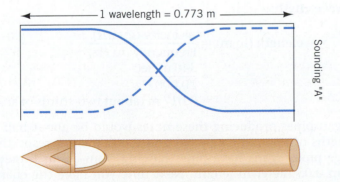

1 wavelength = 0.773 m

Sounding "A"

Figure 6–5

An organ pipe produces a single note. Air in the pipe vibrates and produces a sound wave with a wavelength related to the length of the pipe.

$$\text{wavelength (in m)} = \frac{\text{velocity (in m/s)}}{\text{frequency (in Hz)}}$$
$$= \frac{340 \text{ m/sec}}{440 \text{ Hz}}$$
$$= 0.773 \text{ m (about 2 feet)}$$

The length of the organ pipe is half the wavelength:

$$\text{organ pipe length (in m)} = \frac{\text{wavelength (in m)}}{2}$$
$$= \frac{0.773 \text{ m}}{2}$$
$$= 0.387 \text{ m (about 15 inches)}$$

Notes in the middle range on the pipe organ are thus produced by pipes that are about a half-meter long. ■

EXAMPLE 6–2: The Limits of Human Hearing

The human ear can hear sounds at frequencies from about 20 to 20,000 Hz. What are the longest and shortest wavelengths you can hear? What are the longest or shortest organ pipes you are likely to see?

▶ **Reasoning:** Each organ pipe has a fixed length and produces just one note. We have to calculate the wavelength needed for both the lowest and highest frequency.

▶ **Solution:** The lowest audible note, at 20 Hz, would require a wavelength:

$$\text{wavelength (in m)} = \frac{\text{velocity (in m/s)}}{\text{frequency (in Hz)}}$$
$$= \frac{340 \text{ m/sec}}{20 \text{ Hz}}$$
$$= 17 \text{ m (about 50 feet long)}$$

Similarly, the highest audible note, at 20,000 Hz, is produced by a wavelength:

$$\text{wavelength (in m)} = \frac{\text{velocity (in m/s)}}{\text{Frequency (in Hz)}}$$
$$= \frac{340 \text{ m/sec}}{20,000 \text{ Hz}}$$
$$= 0.017 \text{ m (about two-thirds of an inch)}$$

Organ pipes producing these notes would be about half the wavelengths of approximately 8.5 and 0.009 meters, respectively. Most large pipe organs have pipes ranging from about 8 meters to less than 0.05 meter in length. Next time you have the chance, visit a church or auditorium with a large pipe organ and look at the variety of pipes. Not only are there many different lengths, but there are also many distinctive shapes, each sounding like a different instrument.

Confronted by very low-frequency sounds, we often don't so much hear them as feel them—we sense vibrations in our bodies. You may have experienced this sensation when hearing very low notes on an organ. Some animals (elephants, for example) routinely use sound in the 20–40 Hz range to communicate with each other over long distances. The mating call of the female elephant, for example, is experienced as a vibration by humans, but attracts bull elephants from many miles away. ▲

Use of Sound by Animals

Humans use sound to communicate, of course, as do many other animals. But some animals have refined the use of sound to a much higher level. In 1793, the Italian physiologist Lazzaro Spallanzani did some experiments with bats that established that they use sound to locate their prey. He took bats that lived in the cathedral tower in Pavia, blinded them, and then turned them loose. Weeks later, those bats had fresh insects in their stomachs, proving that they didn't locate food by sight. Similar experiments with bats that were made deaf, however, showed that they could neither fly nor locate insects.

Today, we understand that bats navigate by emitting high-pitched sound waves and then listening for the reflection of those waves off of other objects. By measuring the time it takes for a pulse of sound waves to go out, be reflected, and come back, the bat can determine the distance to surrounding objects, particularly the flying insects that make up its diet. Typically, a bat can detect the presence of an insect up to 10 meters away. In addition, the bat can use the Doppler effect (see below) to tell whether the target is moving.

A bat navigates by emitting high-pitched sounds and listening for their echoes.

In an interesting application of the principle of natural selection (see Chapter 23), some species of moths have developed sophisticated sense organs to hear the sound emitted by bats. Using ears on their thorax or abdomen, these moths can hear the high-pitched sounds emitted by bats, and thus can tell when they are being "seen." When they hear the sound, they take immediate evasive action. In a few cases, the moths have developed an even more sophisticated defense. When a bat approaches, they emit a series of high-pitched clicks that, in effect, "jam" the bat's detection system.

Whales, dolphins, and porpoises use sound echoes as a navigation tool in the ocean, much as bats do in air. Sometimes, however, the sounds that they emit are in the audible range for humans. Perhaps the most famous example of sophisticated use of sound by animals is the songs of humpback whales, which have appeared on innumerable commercial recordings. The functions of these songs are not clear. It appears, however, that all of the whales in a wide area of ocean (the South Atlantic, for example) sing the same song, although some individuals may leave out parts. The songs change every year, but the whales in a given area change their songs together.

Interference

Waves from different sources may overlap and affect each other in the phenomenon called interference. **Interference** describes what happens when waves from two different sources come together at a single point—each wave interferes with the other, and the observed wave amplitude is the sum of the amplitudes of the interfering waves. Consider the common situation shown in Figure 6–6. Suppose you and a friend each throw rocks into a pond at two separate points, labeled *A* and *B* in the figure. The waves from each of these two points travel outward, and eventually will meet at a point like the one labeled *C*. What will happen when the two waves come together?

One easy way to think about what happens is to imagine that each part of each wave carries with it a set of instructions for the water surface—"move down two feet," or "move up one foot." When two waves arrive

Figure 6–6

Two waves originating from different points create an interference pattern. Bright regions correspond to constructive interference, while dark regions correspond to destructive interference.

simultaneously at a point, the surface responds to both sets of instructions. If one wave says to move down three feet and the other to move up one foot, the result will be that the water surface will move down two feet.

Each point on the surface of the water, then, moves a different distance up or down depending on the instructions that are brought to it by the waves from points *A* and *B*. One possible situation is shown in Figure 6–7a. Two waves, each carrying the command "go up one foot" arrive at a point together. The two waves act together to lift the water surface to the highest possible height it can have. This is called *constructive interference* or reinforcement. On the other hand, you could have a situation like the one shown in Figure 6–7b, where the two waves arrive at the point such that one is giving an instruction to go up one foot and the other to go down one foot. In this case, the water surface will not move at all—it is dead still. This is called *destructive interference* or cancellation. And, of course, waves can interfere anywhere between these two extremes.

The most familiar example of destructive interference does not take place with water waves, but with sound waves. Occasionally an auditorium may be designed in such a way that almost no sound can be heard in certain seats. This unfortunate situation results when two waves—for example, one directly from the stage and one bouncing off the ceiling—arrive at those seats in just such a way as to cause partial or total destructive interference. One of the main goals of acoustical design of auditoriums, a field that relies on complex computer modeling of sound interference patterns, is to avoid such problems.

Another place where you see interference, believe it or not, is at service stations on hot afternoons. If you look carefully at the oil slicks on the pavement, you often see a kind of iridescence—a rainbow play of colors on the dark oil surface. In this case two light waves are interfering. One wave is the sunlight that bounces off the top of the oil film, and the other is the wave that goes through the oil film and bounces off the bottom. The two waves come to your eye, and if it happens that the waves corresponding to green light interfere constructively, then that spot in the oil slick will appear green. Other parts of the slick correspond to constructive and destructive interference for other colors, and the result is the iridescent display you see.

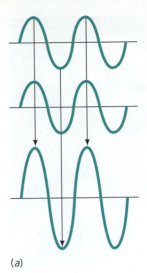

(a)

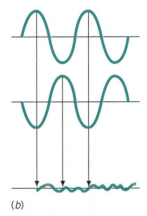

(b)

Figure 6–7
Cross-sections of interfering waves illustrate the phenomena of constructive (*a*) and destructive (*b*) interference.

The Electromagnetic Wave

Have you ever had your teeth X-rayed? Cooked a meal in a microwave oven? Watched television or listened to radio? If so, then you have firsthand experience with the phenomenon of electromagnetic waves.

From the point of view of a physicist, waves are characterized by a mathematical expression called a "wave equation," which describes the movement of the medium for every wave, whether it's a water wave moving through a liquid, a sound wave in air, or a seismic wave causing an earthquake. Physicists have learned over long periods of time that whenever an equation of this or some closely related form appears, a corresponding wave should be seen in nature.

James Clerk Maxwell
(1831–1879)

Soon after Maxwell wrote down the four equations that describe electricity and magnetism (see Chapter 5), he realized that some rather straightforward mathematical manipulation led to yet another equation, one that describes waves. The waves that Maxwell predicted are rather strange sorts of things, and we'll describe their anatomy in more detail later. The important point, however, is that these are waves in which energy is transferred not through matter, but through electric and magnetic fields. It appears from the equations that whenever an electric charge is accelerated, for example, one of these waves is emitted. Maxwell called this phenomenon **electromagnetic waves** or **electromagnetic radiation.** An electromagnetic wave is a self-propagating wave made up of electric and magnetic fields fluctuating together.

The equations also predicted exactly how fast the waves could move—the wave velocity depends only on known constants such as the universal electrostatic constant in Coulomb's equation (see Chapter 5). These numbers are known from experiment, and when Maxwell put the numbers into his expression for the velocity of his new waves, he found a very surprising answer. The predicted velocity of the wave turned out to be 300,000 kilometers per second (186,000 miles per second).

If you just had an "aha!" moment, you can imagine how Maxwell felt. The number that he had gotten, of course, is the well-known speed of light, which means that the waves described by his equation are actually the familiar (but mysterious) waves of light.

This result was astonishing. For centuries scientists had puzzled over the origin and nature of light. Newton and others had discovered natural laws that describe the connections between forces and motion, as well as the behavior of matter and energy. But light remained an enigma. How did radiation from the Sun travel to the Earth? What caused the light produced by a candle? Scholars had been hard pressed to advance beyond the poetry of the third verse of the Old Testament: "And God said, 'Let there be light': and there was light."

There is no obvious reason why static cling, refrigerator magnets, or the workings of an electric generator should be connected in any way to the behavior of visible light. Yet Maxwell discovered that **light** and other kinds of radiation are a type of wave that is generated whenever electrical charges are accelerated.

The Anatomy of the Electromagnetic Wave

A typical electromagnetic wave, shown in Figure 6–8, consists of electric and magnetic fields arranged at right angles to each other and perpendicular to the direction the wave is moving. To understand how the waves work, go back to Maxwell's equations that describe how a changing magnetic field produces an electric field, and vice versa (Chapter 5). At the point labeled *A* in Figure 6–8, the electric and magnetic fields associated with the wave are at maximum strength, but these fields are changing slowly, so both decrease in strength. At point *B*, the fields are at minimum strength, but they are changing rapidly and so begin to increase. Thus, at point *A*, the magnetic and electric fields begin to die out, while at point *B* just the reverse happens. In this way the electromagnetic wave leapfrogs through space, bouncing its energy back and forth between electric and magnetic fields as it goes.

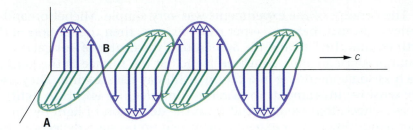

Figure 6–8

A diagram of an electromagnetic wave shows the relationship of the electric field, the magnetic field, and the direction that the wave is moving. *A* and *B* indicate points of maximum and minimum field strength.

Once you understand that the electromagnetic wave has this kind of ping-pong arrangement between electricity and magnetism, you can understand one of the most puzzling points about it—the fact that it can travel through a vacuum. Every other wave we talked about is easy to visualize because the wave moves through a medium. We know that the motion of the wave is not the same as the motion of the medium, but in every other way the medium is there to give the wave solid support. The electromagnetic wave is, in a sense, the logical extension of this idea. It is a wave that has no medium whatsoever, but simply keeps itself going through its own internal mechanisms.

Electromagnetic waves, then, are a form of radiant energy, or radiation, created when electrical charges accelerate. Once they start moving, however, they operate according to their own logic and no longer depend on the source that emitted them.

Science in the Making

The Ether

When Maxwell first proposed his idea of electromagnetic radiation he was not prepared to deal with a wave that required no medium whatsoever. Previous scientists who had studied light, including such luminaries as Isaac Newton, assumed that light must travel through a hypothetical substance called "ether" that permeates all space. Ether, they thought, served as the medium for light, and so Maxwell assumed that ether provided the medium for his electromagnetic waves. In Maxwell's picture, the ether was something like a tenuous Jell-O that filled all of space. An accelerating charge shook the Jell-O at one point, and after that the electromagnetic waves moved outward at the speed of light.

The idea of an ether in one form or another goes back a long time, all the way to the Greeks. It may be true that "nature abhors a vacuum," but the human mind seems to abhor it more, and for most of recorded history people filled the vacuum of space with this imaginary substance. It wasn't until 1887 that two U.S. physicists, Albert A. Michelson (1852–1931) and Edward W. Morley (1838–1923), working at what is now Case Western Reserve University in Cleveland, did a series of experiments that demonstrated that the ether could not be detected. This failure was interpreted by many scientists to mean that the ether did not exist.

The concept of the experiments was very simple. Michelson and Morley reasoned that if an ether really existed, then the motion of the Earth around the Sun and the Sun around the center of our galaxy would produce an apparent ether "wind" at the surface of the Earth, much as someone riding in a car on a still day feels a wind. They used very sensitive instruments to search for interference effects in light waves—effects that would result from the deflection of light by the ether wind. When their experiment turned up no such deflection, they concluded that the ether could not exist.

In 1907 Albert Michelson became the first U.S. scientist to win a Nobel Prize, an honor that recognized his pioneering experimental studies of light. ■

Light

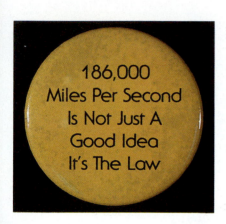

186,000
Miles Per Second
Is Not Just A
Good Idea
It's The Law

Once Maxwell understood the nature of light, his equations allowed him to draw several important conclusions. For one thing, since the velocity of the electromagnetic waves depends entirely on the nature of interactions between electrical charges and magnets, it cannot depend on the properties of the wave itself. This means that all electromagnetic waves, regardless of their wavelength or frequency, have to move at exactly the same velocity. This velocity—**the speed of light**—turns out to be so important in science that we give it a special letter, c. The speed of electromagnetic waves in a vacuum is one of the fundamental constants of nature.

For electromagnetic waves, then, the relation between velocity, wavelength, and frequency takes on a particularly simple form:

$$\text{wavelength} \times \text{frequency} = c$$
$$= 300,000 \text{ km/s } (= 186,000 \text{ miles/s})$$

In other words, if you know the wavelength of an electromagnetic wave, you can calculate its frequency and vice versa.

The Energy of Electromagnetic Waves

Think about how you might produce an electromagnetic wave with a simple comb. Electromagnetic waves are generated any time a charged object is accelerated, so imagine combing your hair on a dry winter day when the comb picks up a static charge. Each time you move the comb back and forth an electromagnetic wave traveling 300,000 kilometers per second is sent out from the comb.

If you wave the electrically charged comb up and down slowly, once every second, you create electromagnetic radiation, but you're not putting much energy into it. You produce a low-frequency, low-energy wave with a wavelength of about 300,000 kilometers. (Remember, each wave moves outward 300,000 kilometers in a second, and that's the separation between wave crests).

If, on the other hand, you could vibrate the comb vigorously—say at 300,000 times per second—you would produce a higher-energy, high-frequency wave with a one-kilometer wavelength. By putting more energy into accelerating the electrical charge, you have more energy in the electromagnetic wave.

Visible light, the first example of an electromagnetic wave known to humans, bears out this particular kind of reasoning. A glowing ember has a dull red color, corresponding to relatively low energy. Hotter, more energetic fires, on the other hand, show a progression of more energetic colors, from the yellow of a candle flame to the blue-white flame of a blowtorch. These colors are merely different frequencies, and therefore different energies, of light; high-frequency light corresponds to a blue color, low-frequency to red.

Red light has a wavelength corresponding to the distance across about 7000 atoms or about 700 nanometers (a nanometer is 10^{-9} meter, about 40 billionths of an inch). Red light is the longest wavelength that the eye can see, and is the least energetic of the visible electromagnetic waves. Violet light, on the other hand, has a shorter wavelength corresponding to the distance across about 4000 atoms, or about 400 nanometers, and is the most energetic of the visible electromagnetic waves. All of the other colors—green, yellow, and so on—have wavelengths and energies between those of red and violet.

EXAMPLE 6–3: Figuring Out Frequency

The wavelength of red light is about 700 nanometers, or 7×10^{-7} m. What is the frequency of a red light wave?

▶ **Reasoning:** We know that for all electromagnetic waves,

$$\text{wavelength} \times \text{frequency} = 300,000 \text{ km/s}$$
$$= 3 \times 10^{8} \text{ m/s}.$$

We want to determine frequency, so we have to rearrange this equation:

$$\text{frequency} = \frac{(3 \times 10^{8} \text{ m/s})}{\text{wavelength}}$$

▶ **Solution:** This means that for red light with a wavelength of 7×10^{-7} m,

$$\text{frequency} = \frac{3 \times 10^{8} \text{ m/sec}}{7 \times 10^{-7} \text{ m}}$$
$$= 0.43 \times 10^{15} \text{ Hz}$$
$$= 4.3 \times 10^{14} \text{ Hz}.$$

(Remember, a hertz equals one cycle per second.) In order to generate red light by vibrating a charged comb you would have to wiggle it more than four hundred trillion (400,000,000,000,000) times per second. ▲

The Doppler Effect

Once waves have been generated, their motion is independent of the source. (It doesn't matter what kind of charged object accelerates to produce an electromagnetic wave: once produced, all such waves behave exactly the same way.) This statement has an important consequence that was discovered in 1842 by the Austrian physicist Christian Johann Doppler (1803–1853). This consequence is called the Doppler effect in his honor. The **Doppler effect** describes the way the frequency of a wave appears to change if there is relative motion between the wave source and the observer.

Let's take sound as an example. Figure 6–9a shows the way a sound wave looks when the source is stationary relative to a listener—when you listen to your radio, for example. In this case everything sounds "normal."

If the source of sound—a racing ambulance, for example—is moving relative to the listener, however, a different situation occurs (Figure 6–9b). Periodically a pulse of high pressure moves away from the source of the sound wave and travels out in a sphere *centered on the spot where the source was located when that particular crest was emitted.* By the time the source is ready to emit other crests, it will have moved. Thus the second sound-wave sphere emitted will be centered at the new location. As the source continues to move, it will emit sound waves centered farther and farther to the right in the figure, producing a characteristic pattern as shown.

To an observer standing in front of the source, the crests appear to be bunched up. Another way of saying this is that to this observer, the frequency of the wave appears to be higher than it would normally be. In the case of a sound wave, this means that the sound will be higher-pitched than it would have been had the source been stationary. On the other hand, if the observer is standing in back of the source, the distance between crests will be stretched out, and the frequency and pitch of the resulting sound will be lower.

You probably have heard the Doppler effect. Think of standing on a highway while cars go by at high speeds. The engine noise appears to be very high-pitched as the car approaches you and then suddenly drops in pitch as the car passes you. This effect is particularly striking at automobile races where cars are moving at very high velocity.

This sort of change of pitch was, in fact, the first example of the Doppler effect to be studied. Scientists hired a band of trumpeters to sit on an open railroad car and blast a single long, loud note as the train whizzed by at a carefully controlled speed. Musicians on the ground determined the pitches they heard as the train approached and as it receded, and they compared those pitches to the actual note the musicians were playing.

The same sort of bunching up and stretching out of crests can happen for any wave, including light. If you are standing in the path of a source of light that is moving toward you, the light you see will be of higher frequency, and hence will look bluer than it would ordinarily. (Remember, blue light has a higher frequency than red light.) We say the light is *blueshifted* (Figure 6–9c).

If, on the other hand, you are standing in back of the moving source, the distance between crests will be stretched out and it will look to you as if the light had a lower frequency. We say that it is *redshifted.* In Chapter 17 we will see that the redshifting of light from distant moving sources is one of the main clues that we have about the structure of the universe.

The Doppler effect also has practical applications much closer to your home. Police radar units send out a pulse of electromagnetic waves that is absorbed by the metal in your car, then reemitted. The waves that come back will be Doppler shifted, and by comparing the frequency of the wave that went out and the wave that comes back, the speed of your car can be deduced. A similar technique is used by meteorologists, who employ Doppler radar to measure wind speed and direction during the approach of potentially damaging storms (see Chapter 16).

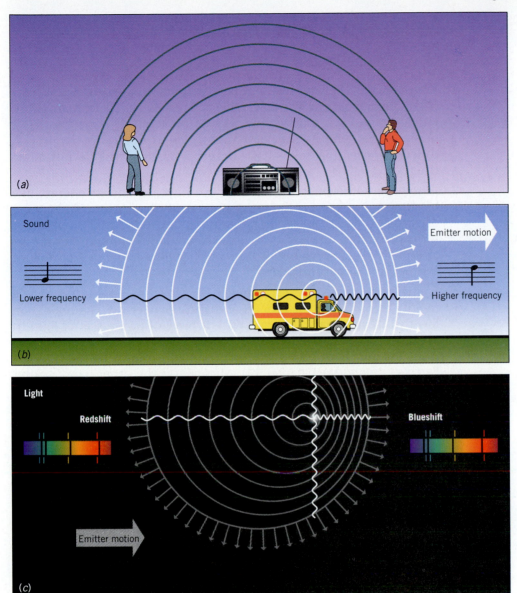

Figure 6–9

The Doppler effect occurs whenever a source of waves is moving relative to the observer of the waves. (*a*) Water waves are compressed in the direction of motion and spread out behind. (*b*) Sound waves seem to increase or decrease in pitch, depending on whether the sound is approaching or receding. (*c*) The Doppler shift for light waves causes a blueshift for approaching sources, and a redshift for receding sources.

Transmission, Absorption, and Scattering

The only way we can know about electromagnetic radiation is to observe its interaction with matter. Our eyes, for example, interact with visible light and send nerve impulses to our brain; impulses that are interpreted as what we "see."

When an electromagnetic wave hits matter it can do one of three things:

1. *Transmission.* The wave will often pass right through matter, as light passes through your window. This process is called **transmission**. Other than slowing down the wave a bit while it is in transit, or perhaps changing its direction slightly as in eyeglasses—a process called *refraction* (Figure 6–10*a*)—transparent materials do not affect the wave.

2. *Absorption.* Other matter may soak up the wave and its energy, like blacktop on a summer day—the process of **absorption**. The energy of absorbed electromagnetic radiation is converted into some other form of energy, usually heat. Black and dark colors, for example, absorb visible light: you've probably noticed how hot black pavement can become on a sunny day.

3. *Scattering.* Alternatively, electromagnetic waves may be absorbed and rapidly reemitted in the process of **diffuse scattering** (Figure 6–10*b*). Most white materials, such as a wall or piece of paper, scatter light in all directions. White objects like clouds and snow, which scatter light from the Sun back into space, play a major role in controlling the Earth's climate (see Chapter 14). Mirrors, on the other hand, scatter light at the same angle as the original wave, a process called *reflection* (Figure 6–10*c*).

All electromagnetic waves are detectable in some way. For each of them to be useful, researchers must find appropriate materials to transmit, absorb, and scatter the waves. For each wavelength there must be instruments that produce the waves and others that detect their presence. While only a very narrow range of electromagnetic waves can be detected by the optical instrument we call the human eye, scientists have devised an extraordinary range of transmitters and detectors to produce and measure electromagnetic radiation that we can't see.

The Electromagnetic Spectrum

A profound puzzle accompanied Maxwell's original discovery that light is an electromagnetic wave. Waves can be of almost any length. Water waves on the ocean for example, range from tiny ripples to globe-spanning tides. Yet visible light spans an extremely narrow range of wavelengths, only about 390 to 710 nanometers. According to the equations that Maxwell derived, electromagnetic waves could exist at any wavelength (and, consequently, any frequency) whatsoever. The only constraint is that the wavelength times the frequency must be equal to the speed of light. Yet when Maxwell looked into the universe, he saw visible light as the only example of electromagnetic waves. It was as if a splendid symphony, ranging from the deep bass of the tuba to the sharp shrill of the piccolo was poised to play, but you could hear only a couple of notes.

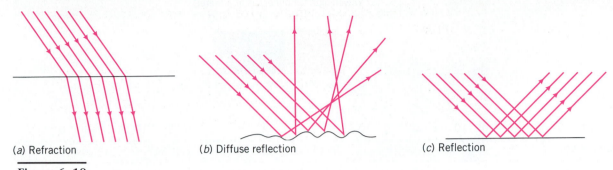

(*a*) Refraction   (*b*) Diffuse reflection   (*c*) Reflection

Figure 6–10

When light interacts with matter it may be (*a*) refracted, (*b*) scattered diffusely, and (*c*) reflected. It may also be absorbed.

In such a situation it would be natural to wonder what had happened to the rest of the waves. Scientists looked at Maxwell's equations, looked at nature, and realized that something was missing. The equations predicted that there ought to be more kinds of electromagnetic waves than light; waves that no one had seen up to that time; waves performing the waltz between electricity and magnetism, but with frequencies and wavelengths different from those of visible light.

These as yet unseen waves would have exactly the same structure as the one shown in Figure 6–8, but they could have either longer or shorter wavelengths than visible light depending on the acceleration of the electric charge that created them. These waves would move at the speed of light, and would be exactly the same as visible light except for the differences in the wavelength and frequency.

Between 1885 and 1889 the German physicist Heinrich Rudolf Hertz (1857-1894), after whom the unit of frequency is named, performed the first experiments that confirmed these predictions. He discovered the waves that we now know as radio. Since that time, all manner of electromagnetic waves have been discovered, from those with wavelengths longer than the radius of the Earth to those with wavelengths shorter than the size of the nucleus of the atom. They include radio waves, microwaves, infrared, visible light, ultraviolet, X-rays, and gamma rays. This entire symphony of waves is called the **electromagnetic spectrum** (see Figure 6–11). Remember that every one of these waves, no matter what its wavelength or frequency, is the result of an accelerating electric charge.

Radio Waves

The **radio wave** part of the electromagnetic spectrum ranges from the longest waves—those whose wavelength is longer than the size of the Earth—to waves a few meters long. The corresponding frequencies—from roughly a kilohertz (a thousand cycles per second) to several hundred megahertz (a million cycles per second)—correspond to the familiar numbers on your radio dial. There are various subdivisions of radio waves, but the most important fact about them is that, like light, they can

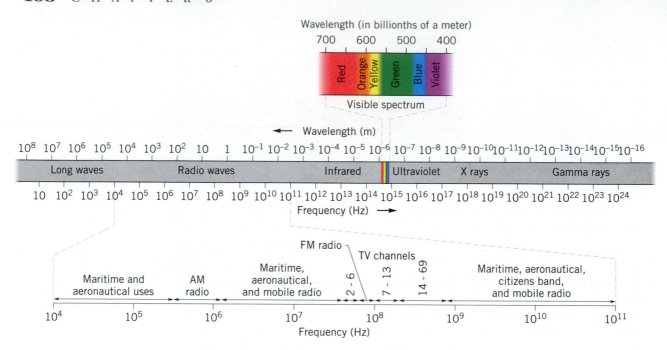

Figure 6–11

The electromagnetic spectrum includes all kinds of waves that travel at the speed of light, including radio, microwave, infrared, visible light, ultraviolet, X-rays, and gamma rays. Note that sound waves, water waves, seismic waves, and other kinds of waves that require matter in order to move travel much slower than light speed.

penetrate long distances through the atmosphere. This makes radio waves very useful in communication systems.

Have you ever been driving at night and picked up a radio signal from a station a thousand miles away? If so, you have had first-hand experience of the ability of radio waves to travel for long distances through the atmosphere. In Chapter 17, we will see how important this fact is for astronomy, where scientists speak of the "radio window" in the atmosphere, a window that allows Earth-based telescopes to monitor radio waves emitted by objects in the sky.

A typical radio wave used for communication can be produced by pushing electrons back and forth rapidly in a tall metal antenna. This acceleration of electrons produces outgoing radio waves, just as throwing a rock in a pond produces outgoing ripples. When these waves encounter another piece of metal (for example, the antenna in your radio or TV set) the electric fields in the waves accelerate electrons in that metal, so that its electrons move back and forth. This constitutes an electric current that electronics in your receiver eventually turn into a sound or a picture.

Most construction materials are at least partially transparent to radio waves. Thus you can listen to the radio even in the basement of most buildings. In long tunnels or deep valleys, however, absorption of radio waves by many feet of rock and soil may limit reception.

In the United States, the Federal Communications Commission (FCC) assigns frequencies in the electromagnetic spectrum for various uses. Each commercial radio station is assigned a frequency (which they use in association with their call letters), as is each television station. All manner of private communication—ship-to-shore radio, civilian band radio, emergency police and fire channels, and so on—need their share of the spectrum as well. In fact, the right to use a part of the electromagnetic spectrum for communications is very highly prized because only so many frequencies exist, and many more people want to use those frequencies than can do so. For example, rights to a single FM radio frequency in Los Angeles recently sold for $110 million.

Technology

AM and FM Radio Transmission

Radio waves carry signals in two ways: AM and FM. Broadcasters can send out their programs at only one narrow range of frequency, a situation very different from music or speech, which use a wide range of frequencies. Thus radio stations cannot simply transform a range of sound-wave frequencies into a similar range of radio-wave frequencies. Instead, the information to be transmitted must be impressed in some way on the narrow frequency range of your station's radio waves.

This problem is similar to one you might experience if you had to send a message across a lake with a flashlight at night. You could adopt two strategies. You could send a coded message by turning the flashlight on and off, thus varying the brightness (the amplitude) of the light. Alternatively, you could change the color (the frequency) of the light by passing blue or red filters in front of the beam.

Radio stations also adopt these two strategies (see Figure 6–12). All stations begin with a carrier wave of fixed frequency. AM radio sta-

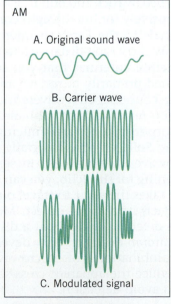

(*a*) Amplitude modulation

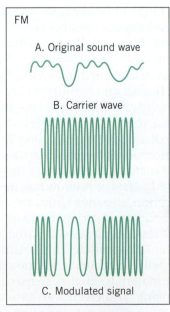

(*b*) Frequency modulation

Figure 6–12

AM (amplitude modulation) and FM (frequency modulation) transmission differ in the way that a sound wave (*a*) is superimposed on a carrier wave of constant amplitude and frequency (*b*). The carrier wave can be varied, or modulated, to carry information (*c*) by altering its amplitude or its frequency.

tions typically broadcast at frequencies between about 530 and 1600 kilohertz (thousands of hertz), while the carrier frequencies of FM radio stations range from about 88 to 110 megahertz (millions of hertz).

The process called *amplitude modulation*, or *AM*, depends on varying the strength (or amplitude) of the radio's carrier wave according to the sound signal to be transmitted (Figure 6–12*a*). Thus, the shape of the sound wave is impressed on the radio's carrier wave signal. When this signal is taken into your radio, the electronics are designed so that the original sound signal is recovered and used to run the speakers. The original sound signal is what you hear when you turn on your radio. The AM frequencies are better able to scatter off the layers of the atmosphere, and so can be heard over greater distances.

Alternatively, you can slightly vary the frequency of the radio's wave according to the signal you want to transmit, a process called *frequency modulation*, or *FM*, as shown in Figure 6–12*b*. A radio that receives this particular signal will unscramble the changes in frequency and convert them into electrical signals that run the speakers so that you can hear the original signal. TV broadcasts, which use carrier frequencies about a thousand times higher than FM radio, typically send the picture on an AM signal, and the sound on an FM signal at a slightly different frequency. ■

Microwaves

Microwaves are electromagnetic waves whose wavelengths range from about 1 meter (a few feet) to 1 millimeter (a thousandth of a meter, or about 0.04 inch). There are no absolute boundaries to the different kinds of electromagnetic waves, and the wavelengths of microwaves and radio waves overlap. The longer wavelengths of microwaves travel easily through the atmosphere, like their cousins in the radio part of the spectrum, though most microwaves are absorbed by rock and building materials. Therefore, microwaves are used extensively for line-of-sight communications. Most satellites broadcast signals to the Earth in microwave channels, and these waves also commonly carry long-distance telephone calls and television broadcasts. The satellite antennas that you see on private homes and businesses are designed primarily to receive and use microwave transmissions, as are the large cone-shaped receivers attached to the microwave relay towers situated on many hills or tall buildings.

The distinctive transmission and absorption properties of microwaves make them ideal for use in aircraft radar. Solid objects, especially those made of metal, reflect most of the microwaves that hit them. By sending out timed pulses of microwaves and listening for the echo, you can judge the direction, distance (from the time it takes the wave to travel out and back), and speed (from the Doppler effect) of a flying object. Modern military radar is so sensitive that it can detect a single fly at a distance of a mile. To counteract this sensitivity, aircraft designers have developed planes with "stealth" technology—combinations of microwave-absorbing materials, angled shapes that reduce the apparent cross-section of the plane, and electronic jamming to avoid detection.

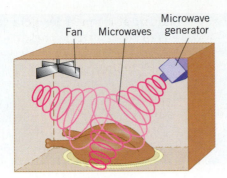

Fan   Microwaves   Microwave generator

Figure 6–13
Every microwave oven contains a device that generates microwaves by accelerating electrons, and walls that scatter the microwaves until they are absorbed, usually by water molecules that get hot from the absorbed energy.

Technology

Microwave Ovens

The same kind of waves that are used for phone calls, television broadcasts, and radar can be used to cook your dinner in an ordinary microwave oven. This type of oven works as shown in Figure 6–13. A special electronic device in the oven accelerates electrons rapidly and produces the microwave radiation, which is a form of energy. These microwaves are guided into the main cavity of the oven, which is composed of material that scatters microwaves. Thus the wave energy remains inside the box until it is absorbed by something.

It turns out that microwaves are absorbed quickly by water molecules. This means that the energy used to create microwaves is carried by those waves to food inside the oven, where it is absorbed by water and converted into heat. This absorption of microwave energy results in a very rapid rise in temperature, and rapid cooking. Paper and glass, which don't contain water molecules, are not heated by microwaves. Despite the different applications, from the point of view of the electromagnetic spectrum there is no fundamental difference between the microwaves used for cooking and those used for communication. ■

Infrared Radiation

Infrared radiation includes wavelengths of electromagnetic radiation that extend from a millimeter down to about a micron (10^{-6} meter, or less than a ten-thousandth of an inch). Our skin, which absorbs infrared radiation, provides a crude kind of detector. You feel infrared radiation when you put your hands out to a warm fire or the cooking element of an electric stove. Infrared waves are what we feel as heat radiation.

Warm objects emit infrared radiation, and this fact has been used extensively in both civilian and military technology. Infrared detectors are used to guide air-to-air missiles to the exhaust of jet engines in enemy aircraft, and infrared detectors are often used to "see" human beings and

Figure 6–14

A photograph using infrared film reveals how fast heat energy escapes from different houses. This "false-color" image is coded so that white is hottest, followed by red, pink, blue, and black.

warm engines at night. Similarly, many insects (such as mosquitoes and moths) and other nocturnal animals (including opossums and some snakes) have developed sensitivity to infrared radiation; thus they can "see" in the dark.

Infrared detection is also used to find heat leaks in homes and buildings (Figure 6–14). If you take a picture of a house at night using film that is sensitive to infrared radiation, places where heat is leaking out will show up as bright spots on the film. This information can be used to correct the loss and thus conserve energy. In a similar way, Earth scientists often monitor volcanoes with infrared detectors. The appearance of a new "hot spot" may signal an impending eruption.

Visible Light

All of the colors of the rainbow are contained in **visible light,** whose wavelengths range from red light at about 700 nanometers down to violet light at about 400 nanometers (Figure 6–15). From the point of view of

Figure 6–15

A glass prism separates white light into the visible spectrum.

the larger universe, the visible world in which we live is a very small part of the total picture (see Figure 6–10).

Our eyes distinguish several different *colors*, but these portions of the electromagnetic spectrum have no special significance except in our perceptions (Figure 6–16). In fact, the distinct colors that we see—red, yellow, green, and blue—represent very different-sized slices of the spectrum. The red and green portions of the spectrum are rather broad, spanning more than 50 nanometers of frequencies; we thus perceive many different wavelengths as red or blue. In contrast, the yellow part of the spectrum is quite narrow, encompassing wavelengths from only about 570 to 590 nanometers.

Why should we be able to distinguish such a restricted range of wavelengths as the yellow part of the spectrum? The Sun's light is especially intense in this part of the spectrum, so some biologists suggest that our eyes evolved to be especially sensitive to these wavelengths; to take maximum advantage of the Sun's light. Our eyes are ideally adapted for the light produced by our Sun during daylight hours. Animals such as owls and cats that hunt at night have eyes that are more sensitive to infrared wavelengths—radiation that makes warm living things stand out against the cooler background.

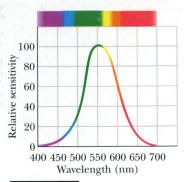

Figure 6–16

The relative sensitivity of the human eye differs for different wavelengths. Our perception peaks at wavelengths that we perceive as yellow, though the colors we see have no special physical significance.

The Human Body

The Eye

The light detector with which we are most familiar is one we carry around with us all the time—the human eye. The eye is a marvelously complex organ, turning incoming electromagnetic radiation into images through the use of a combination of physical and chemical processes (see Figure 6–17).

Light waves enter the eye through a clear lens whose thickness can be changed by a sheath of muscles around it. The direction of the waves is changed by refraction in the lens so that they are focused at receptor cells located in the retina at the back of the eye. There the light is absorbed by two different kinds of cells, called *rods* and *cones* (the names come from their shape, not their function). The rods are sensitive to light and dark, including low levels of light; they give us night vision. Three kinds of cones, sensitive to red, blue, and green light, allow us to see colors.

The energy of incoming light triggers complex changes in molecules in the rods and cones, initiating a series of reactions that eventually leads to a nerve signal that travels along the optic nerve to the brain (see Chapter 5).

The question of how such a complex organ could develop is a very interesting one. The question of why it should respond to the band of radiation given off most copiously by the Sun is another. It turns out that both of these questions can be answered quite elegantly in terms of the principle of natural selection, which we shall discuss in detail in Chapter 24. ■

Figure 6–17

A cross-section of the human eye reveals the path of light, which enters through the protective cornea and travels through the colored iris. The pupil changes the size of the aperture through which light passes, thus controlling the amount of light entering the eye. Muscles move the eye and change the shape of the lens, which focuses light onto the retina, where the light's energy is converted into nerve impulses. These signals are carried to the brain along the optic nerve.

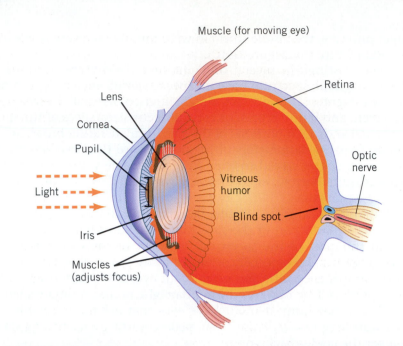

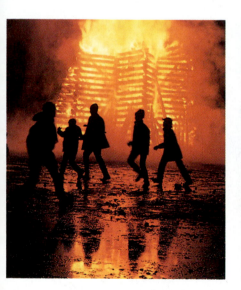

A variety of chemical reactions produce light energy. (*a*) Fire. (*b*) Light sticks. (*c*) A firefly.

Ultraviolet Radiation

At wavelengths shorter than visible light, we begin to find waves of high frequency and therefore high energy and potential danger. The wavelengths of **ultraviolet radiation** range from 400 nanometers down to about 100 nanometers (the size of a hundred atoms placed end-to-end) in length. The energy contained in longer ultraviolet waves can cause a chemical change in skin pigments, a phenomenon known as tanning. This lower-energy portion of the ultraviolet is not particularly harmful by itself.

Shorter-wavelength (higher-energy) ultraviolet radiation, on the other hand, carries more energy—enough that this radiation can damage skin cells, causing sunburn and skin cancer in human beings. The wave's energy is absorbed by cells and can cause extensive damage to DNA (see Chapter 23). The ability of ultraviolet radiation to kill living cells is used by hospitals to sterilize equipment and kill unwanted bacteria.

The Sun produces intense ultraviolet radiation in both the longer and shorter wavelengths. Fortunately, our atmosphere absorbs much of the harmful short wavelengths and thus shields living things. Nevertheless, if you spend much time outdoors under a bright Sun you should protect exposed skin with a sunblocking chemical, which is transparent (colorless) to visible light, but reflects harmful ultraviolet rays.

The energy contained in both long and short ultraviolet wavelengths can be absorbed by atoms, which in special materials may subsequently emit a portion of that absorbed energy as visible light. (Remember, both visible light and ultraviolet light are forms of electromagnetic radiation, but visible light has longer wavelengths, and therefore less energy, than ultraviolet radiation.) This phenomenon, called *fluorescence*, provides the so-called "black light" effects so popular in stage shows and night clubs. The post office commonly "tags" its postage stamps with invisible fluorescent ink, which enables automated machines with ultraviolet lights to locate and cancel stamps with great efficiency.

X-rays

X-rays are electromagnetic waves that range in wavelength from about 100 nanometers down to 0.1 nanometer, smaller than a single atom. These high-frequency (and thus high-energy) waves can penetrate several centimeters into most solid matter, but are absorbed to different degrees by all kinds of materials. This fact allows X-rays to be used extensively in medicine to form visual images of bones and organs inside the body. Bones and teeth absorb X-rays much more efficiently than skin or muscle,

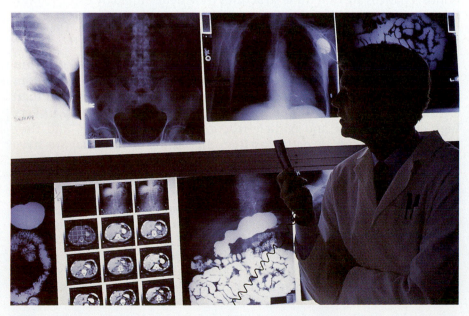

A physician examines medical x-rays. Internal structures are revealed because bones and different tissues absorb x-rays to different degrees.

so a detailed picture of inner structures emerges. X-rays are also used extensively in industry to inspect for defects in welds and manufactured parts.

The X-ray machine in your doctor's or dentist's office is something like a giant light bulb with a glass vacuum tube. At one end of the tube is a tungsten filament that is heated to very high temperature by an electrical current, just like in an incandescent light bulb. At the other end is a polished metal plate. X-rays are produced by applying an extremely high voltage—negative on the filament and positive on the metal plate—so electrons stream off the filament and smash into the metal plate at high velocity. The sudden deceleration of the negatively charged electrons releases a flood of high-energy electromagnetic radiation—what we call X-rays—that travel from the machine to you at the speed of light.

X-ray Diffraction

Have you ever noticed that when light glances off a phonograph record or a CD at an angle, you see a play of colors? Tiny grooves in the record or CD that are spaced at distances comparable to the wavelength of light create a phenomenon called *diffraction*, a special kind of scattering of electromagnetic waves in which the waves are strongly reinforced in certain directions by the process of interference (see above). X-ray diffraction works in exactly this way.

It turns out that planes of atoms in a crystal are spaced at roughly the same distance apart as the wavelength of X-rays—about 10 nanometers, or a hundred-millionth of an inch. Because of this coincidence, crystals diffract X-rays. X-ray crystallographers use the information contained in diffracted X-rays to determine the spacing and positions of atoms in a crystal (see Chapter 10). They often record diffracted X-rays on film, just like a medical doctor. Chemists and mineralogists often use crystal-structure models with balls and sticks to represent atoms and their bonds; chances are that these structures were determined by using X-rays.

Gamma Rays

The highest energies in the electromagnetic spectrum are called **gamma rays.** Their wavelengths range from slightly less than the size of an atom (about 0.1 nanometer, or 10^{-10} meter) to the size of a nucleus (less than a trillionth of a meter). Gamma rays are normally emitted only in very high-energy nuclear and particle reactions (see Chapter 12), and they are not as common in nature as the other kinds of rays that we have talked about.

Gamma rays have many uses in medicine. Some types of medical diagnosis involve giving the patient a radioactive chemical that emits gamma rays. If that chemical concentrates at places where bone is actively healing, for example, then doctors can monitor the healing by locating the places where gamma rays are emitted. The gamma ray detectors used in this specialized form of nuclear medicine are both large (to capture the energetic waves) and expensive. Doctors also use gamma rays for the treatment of cancer in humans. In these treatments, high-energy gamma rays are directed at tumors or malignancies that cannot be removed surgically. If the gamma ray energy is absorbed in those tissues, the tissues will die and the patient has a better chance to live.

Gamma rays are also used extensively in astronomy because many of the interesting processes going on in our universe involve bursts of very high energy and, hence, the emission of gamma rays.

THINKING MORE ABOUT ELECTROMAGNETIC RADIATION

Is ELF Radiation Dangerous?

We usually think of electromagnetic radiation in terms of relatively high-frequency waves, such as radio waves and microwaves at millions or billions of hertz, but Maxwell's equations tell us that *any* accelerated charge will emit waves of electromagnetic radiation, not just those that are shaken rapidly. In particular, the electrons that move back and forth in wires to produce the alternating current in household wiring generate electromagnetic radiation. Every object in which electric power flows, from power lines to toasters, is a source of this weak, *extremely low frequency (ELF) radiation*.

For more than a century human beings in industrialized countries have lived in a sea of weak ELF radiation, but until recently no questions were raised about whether that radiation might have an effect on human health. In the late 1980s, however, a series of books and magazine articles created a minor sensation by claiming that there is evidence that exposure to ELF radiation causes some forms of cancer, most notably childhood leukemia.

Scientists tended to downplay these claims. They pointed out that the electric fields due to power lines at the location of a cell are a thousand times smaller than those due to natural causes (such as electrical activity in nearby cells). They pointed out that age-corrected cancer rates in the United States (with the exception of lung cancer, which is caused primarily by smoking) have remained constant or dropped over the last 50 years, though exposure to ELF radiation has increased enormously.

They also questioned the validity of studies claiming correlations with cancer, arguing that when a more detailed analysis of results was conducted, the correlation disappeared.

Activists who suspected a lurking danger in power lines argued that the effect they were trying to measure was very small, and therefore easy to lose in the mass of carcinogens to which Americans are exposed. Opponents countered that numerous health scares in the past have proved to be utterly without basis, and predicted that ELF will play out the same way. They pointed out that original studies of problems like the link between smoking and lung cancer gave unambiguous answers from the start.

This situation is typical of encounters at the border between science and public health. Preliminary data indicate a possible health risk in the environment but do not prove that the risk is real. Settling the issue by further study takes years, and in the meantime people have to make decisions about what to do. In addition, as in the case of ELF radiation, the cost of removing the risk is often very high.

Suppose you were a scientist who had shaky evidence that some common food—bread, for example, or a familiar kind of fruit—could be harmful. What responsibility would you have to make your results known to the general public? If you stress the uncertainty of your results and no one listens, should you make sensational (perhaps unsupported) claims to get people's attention?

Summary

Waves provide a way to transfer energy from one place to another through a medium without matter actual traveling across the intervening distance. Every wave can be characterized by its *wavelength,* a velocity, an amplitude, and a *frequency* (measured in cycles per second, or *hertz*). Transverse waves, such as swells on the ocean, occur when the medium moves perpendicular to the direction of the waves. Longitudinal waves, such as sound, occur when the medium moves in the same direction as the wave.

Two waves can interact with each other, causing constructive or destructive *interference.* The observed frequency of a wave depends on the relative motion of the wave's source and the observer, a phenomenon known as the *Doppler effect.*

The motion of every wave can be described by a characteristic wave equation. James Clerk Maxwell recognized that simple manipulation of his equations that describe electricity and magnetism pointed to the existence of *electromagnetic waves* or *electromagnetic radiation,* alternating electric and magnetic fields that can travel through a vacuum at *the speed of light.* This discovery solved one of the oldest mysteries of science, the nature of *light.* While visible light was the only kind of electromagnetic radiation known to Maxwell, he predicted the existence of other kinds with longer and shorter wavelengths. Soon thereafter a complete *electromagnetic spectrum* of waves, including *radio waves, microwaves, infrared radiation, visible light, ultraviolet radiation, X-rays,* and *gamma rays,* was recognized.

Electromagnetic radiation can interact with matter in three ways: it can be *transmitted, absorbed,* or *scattered.* We use these properties in countless ways every day—radio and TV, heating and lighting, microwave ovens, tanning salons, medical X-rays, and more. Much of science and technology during the past 100 years has been an effort to find new and better ways to produce, manipulate, and detect electromagnetic radiation.

Key Terms About Electromagnetic Radiation

wave	the speed of light, *c*	microwaves
wavelength	Doppler effect	infrared radiation
frequency (measured in hertz)	electromagnetic spectrum	visible light
interference	transmission	ultraviolet radiation
electromagnetic wave, or electromagnetic radiation	absorption	X-rays
light	diffuse scattering	gamma rays
	radio waves	

Key Equation About Waves

wave velocity (m/s) = wavelength (m) × frequency (Hz)

$$\text{wavelength (in m)} = \frac{\text{velocity (in m/s)}}{\text{frequency (in Hz)}}$$

1 hertz = 1 cycle/second

For light: wavelength (in m) × frequency (in Hz) = *c*

Constant: Speed of Light

$c = 300{,}000 \text{ km/s} = 3 \times 10^8 \text{ m/s}$

Review Questions

1. Identify the two different forms in which energy can be transferred.

2. What three characteristics can be used to describe a wave? How are these three related?

3. What are some everyday examples of waves?

4. Describe the difference between a longitudinal wave and a transverse wave.

5. What happens when two different waves overlap?

6. What causes an electromagnetic wave?

7. What features do all electromagnetic waves have in common?

8. Why did Maxwell think there were kinds of electromagnetic radiation other than visible light?

9. What was the first kind of electromagnetic radiation other than visible light to be discovered? Who discovered it? How was he honored by other scientists?

10. What three things can happen when electromagnetic radiation hits a piece of matter?

11. What are some uses of microwaves?

12. Give one important difference between the effects of long and short wavelengths of ultraviolet light.

13. What are some uses of gamma rays?

14. What kinds of electromagnetic radiation can you detect with your body?

15. What are similarities and differences between water waves and light waves?

16. Describe the Doppler effect.

17. Identify an object that absorbs (*a*) radio waves; (*b*) microwaves; (*c*) visible light.

Discussion Questions

1. Describe the major differences between a sound wave and a radio wave.

2. If a tree falls in a forest, what kinds of waves are created? Where did the energy that produced those waves come from?

3. Propose an experiment to test whether lead is transparent to radio waves.

4. "The wave" in a football stadium is an example of a transverse wave. How would you instruct fans to create a longitudinal "wave"?

5. Imagine that you are looking into a store window on a sunny day. You see yourself reflected in the glass as you look at the items for sale. Describe the different ways that the sunlight is interacting with matter.

6. How did Michelson and Morley demonstrate that ether does not exist? What was their hypothesis?

7. What do you think causes the beautiful iridescent sheen of a butterfly wing?

8. Why do people wear light-colored clothing in summer and dark-colored clothing in winter?

9. Why might a car radio emit static when you drive through an area with tall buildings?

10. What characteristic of X-rays makes them well-suited for medical diagnosis? Are X-rays fundamentally different from other kinds of electromagnetic radiation?

Problems

1. An organ pipe is 3.1 meters long. What is the frequency of the sound it produces? Extra credit: To what pitch does that frequency correspond?

2. An ocean liner experiences broad waves, called swells, with a frequency of one every 10 seconds (0.1 Hz) and a wavelength of 220 feet. Assume the waves are moving due east. If the liner maintains a speed of 15 miles per hour, will it have a smoother trip going east or west? Why?

3. Radio and TV transmissions are being emitted into space, so "Star Trek" episodes are streaming out into the universe. The nearest star is 9.5×10^{17} meters away. If civilized life exists on a planet near this star, how long will they have to wait for the next episode?

Investigations

1. Visit a local hospital and see how many types of electromagnetic radiation are used on a regular basis. From radio waves to gamma waves, how are the distinctive characteristics of absorption and transmission for each segment of the spectrum used at the facility?

2. What frequencies of electromagnetic radiation are used for communications by the emergency response groups (police, fire, ambulance) in your community? What are the corresponding wavelengths of these signals?

3. Examine a microwave oven or, better yet, obtain an old broken oven that you can dissect. Locate the source of microwaves. Which materials in the oven transmit microwaves? Which ones scatter microwaves? Do you think any of the components absorb microwaves? Why?

4. In large metropolitan areas a license to broadcast electromagnetic waves at an AM frequency may change hands for millions of dollars. (*a*) Why is electromagnetic "real estate" so valuable? Investigate how frequencies are divided up and who regulates the process. Should individuals or corporations be allowed to "own" portions of the spectrum, or to buy and sell pieces of it? (*b*) Currently the only portions of the electromagnetic spectrum that are regulated by national and international law are the longer wavelengths, including radio and microwave. Why are the shorter wavelengths, including infrared, visible light, ultraviolet, and X-ray wavelengths, not similarly regulated?

5. Whales communicate over distances of thousands of miles. Investigate how they do this. What kind of wave is used? What is its speed? Its frequency? Its wavelength? Listen to a recording of humpback whale songs.

6. Find out how sonar works. Compare it to the use of sound by bats. Discuss the analogy between the defenses of moths and the use of radar detectors by motorists.

Additional Reading

Lewis, Thomas. *Empire of the Air: The Men Who Made Radio.* New York: Edward Burlingmane Books, 1991.

Sobel, Michael I. *Light.* Chicago: University of Chicago Press, 1989.

The Atom

All of the matter around us is made of atoms—the chemical building blocks of our world.

A Random Walk
Breathing Lessons

Imagine being exposed to a substance so corrosive that it reacts violently with almost anything it touches. Upon contact with it, shiny metals become pitted and stained, while many everyday objects burst into flame or even explode at the slightest spark. To the best of our knowledge, this dangerous substance exists on only one planet, Earth. The substance is free oxygen.

Oxygen is a chemical element, one of about a hundred basic building blocks that make up all the objects around us. Over 20% of the air around you is oxygen. Unlike most other elements, however, oxygen is starved for electrons. Oxygen gas attacks many other elements, greedily taking their electrons and, in the process, releasing a lot of energy. Once an oxygen atom seizes extra negative electrons, however, it becomes electrically charged and may become locked into a chemical compound. Glass, china, and almost all the rocks at the Earth's surface are more than half oxygen atoms, tightly bonded into durable solids. The water you drink is almost 90% oxygen by weight.

With every breath you take, oxygen from the air enters your body and sustains life by participating in chemical reactions that release the energy you use to grow and move and think. Without an extremely active element like oxygen, these reactions couldn't take place. Our lives depend on the fact that oxygen is very nasty stuff.

The Smallest Pieces

Think about the paper in this book. You could take a page from the book and cut it in half. Then you could cut the half in half, and then cut half of the half of the half, and so on. If paper is really smooth and continuous, there would be no end to this process; no smallest piece of paper that couldn't be cut further. But is the paper really made that way?

The Greek Atom

In about 530 B.C., a group of Greek philosophers, the most famous of whom was a man named Democritus, gave this question some serious thought. Democritus argued (purely on philosophical grounds) that if you took the world's sharpest knife and started slicing chunks of matter, you would eventually come to a smallest piece—a piece that could not be divided further. He called this smallest piece the "atom," which translates roughly as "that which cannot be divided." He argued that all material was formed from these atoms, and that the atoms are eternal and unchanging, but that the relationship between the atoms is constantly shifting.

These Greek philosophers did not really engage in science—their reasoning had none of the interplay between observation and hypothesis that characterizes the scientific method. It wasn't until the beginning of the nineteenth century that the modern theory of atoms was born.

The Beginning of an Atomic Theory

Modern atomic theory is generally attributed to an English meteorologist, John Dalton (1766–1844). In 1808, Dalton published a book called *New System of Chemical Philosophy,* in which he argued that the new knowledge being gained by chemists about materials provided evidence, in and of itself, that matter was composed of atoms. The argument went something like this: Chemists knew that most materials can be broken down into simpler chemicals. If you burn wood, for example, you get carbon dioxide, water, and all sorts of materials in the ash. If you use an electric current to break down water, you get two gases, hydrogen and oxygen. In addition, no matter how much of a material you break down, it always breaks down in the same proportion. Water, for example, always yields one part hydrogen to eight parts oxygen by weight.

On the other hand, a few materials could not be broken down into

Figure 7–1
Atoms may be envisioned as solid balls that stack together to form crystals, like fruit at the supermarket. Atomic models are often drawn with spheres, though we now know that atoms are not solid objects.

other things. Chemists could heat wood to get charcoal (essentially pure carbon), for example; but, try as they might, they couldn't break the carbon down any further. Those materials that could not be broken down further were called chemical **elements.**

The hypothesis that we now call *atomism* was very simple. Dalton suggested that for each chemical element there was a corresponding species of indivisible objects called **atoms.** He borrowed the name from the Greeks, but very little else. Two or more atoms stuck together form a **molecule**—the same term applies to any cluster of atoms that can be isolated, whether it contains two atoms or a thousand. Molecules make up most of the different kinds of material we see around us. Water, for example, forms from one oxygen atom and two hydrogen atoms (giving the familiar H_2O composition).

In Dalton's view, atoms were truly indivisible—he thought of them as little bowling balls (see Figure 7–1). In Dalton's world, then, indivisible atoms provide the fundamental building blocks of all matter.

This picture was not only intellectually attractive, but it also conformed to much of what chemists had discovered by Dalton's time. For example, it explained why chemical elements couldn't be broken down—they were made from a single type of atom that was, by definition, indivisible. It also explained the regular proportions of elements in compounds, such as the one-to-eight ratio of hydrogen to oxygen in water. Because of these successes, scientists rapidly accepted the atomic picture, which became a part of the way we think about the world. Even though we know today that atoms are made up of even smaller parts, we still use the old name "atom" for historical reasons.

Discovering Chemical Elements

Describing and isolating chemical elements provided a major challenge for chemists of the nineteenth century. In 1800 fewer than 30 elements had been isolated, not enough to establish any systematic trends in their chemical behavior. In the early 1800s a new process called *electrolysis*, facilitated by Volta's invention of the battery (see Chapter 5), allowed many new elements to be separated by means of electric current. More than two dozen more elements thus were discovered in the first half of the nineteenth century.

Dmitri Mendeleev's invention of the first periodic table of the elements in 1869 (see Chapter 1) capitalized on these discoveries and pointed the way to even more new elements. That original table listed several dozen elements on the basis of their atomic weights (in rows from the upper left) and by groups with distinctive chemical properties (in columns). What Mendeleev could not know was that his table revealed much about the underlying structure of atoms and their electrons.

In the early 1800s, the list of known chemical elements was rapidly expanding, but contained only a few dozen entries. Today, the periodic table lists 109 elements, of which 92 appear in nature and the rest have been produced artificially. Most of the materials we encounter in everyday life are not elements, but compounds of two or more elements bound together. Table salt, plastics, stainless steel, paint, window glass, and soap are all made from a combination of elements.

Nevertheless, we do have experience with a few chemical elements in our everyday lives:

helium A light gas that has many uses besides filling party balloons and blimps. In liquid form, helium is used to maintain superconductors at low temperatures (see Chapter 10).

carbon Pencil lead, charcoal, and diamonds are all examples of pure carbon. The differences between these materials has to do with the way the atoms of carbon are linked together, as we discuss in Chapter 10.

aluminum A lightweight metal used for many purposes. The dull white surface of the metal is actually a combination of aluminum and oxygen, but if you scratch the surface, the shiny material underneath is pure elemental aluminum.

copper The reddish metal of pennies and pots. Copper wire provides a cheap and efficient conductor of electricity.

gold A soft, yellow, dense, and highly valued metal. For thousands of years the element gold has been coveted as a symbol of wealth. Today it coats critical electrical contacts in spacecraft and other sophisticated electronics.

Table 7–1 • Important Terms Related to Atoms

Element:	A chemical substance that cannot be broken down further.
Atom:	The smallest particle that retains its chemical identity.
Molecule:	Any collection of two or more atoms bound together.
Electron:	An atomic particle with negative charge and low mass.
Nucleus:	The small, massive central part of an atom.
Proton:	Positively charged nuclear particle.
Neutron:	Electrically neutral nuclear particle.
Ion:	An electrically charged atom.

Although we know of more than 90 different elements in nature, living systems are constructed almost entirely from just half a dozen. Most of the atoms in your body, for example, are hydrogen, carbon, oxygen, or nitrogen, with smaller but important roles played by phosphorus and sulfur. Why these particular elements dominate the structure of living things is discussed at length in Chapters 20, 21, and 23, but for the moment we simply note that the atoms that make up our bodies are a small subset of the collection of atoms in nature.

The Structure of the Atom

Dalton's idea of the atom as a single indivisible entity was not destined to last. In 1897, English physicist Joseph John Thomson (1856–1940) unambiguously identified a particle called the **electron,** an object that has a negative electrical charge and is much smaller and lighter than even the smallest atom known. Since there was no place from which a particle such as the electron could come other than inside the atom, Thomson's discovery provided incontrovertible evidence for what people had suspected for a long time—that atoms, themselves, are not the fundamental building block of matter, but rather are made up of things that are smaller and more fundamental still. Table 7–1 summarizes some of the important terms related to atoms.

The Atomic Nucleus

The most important discovery about the structure of the atom was made by New Zealand–born physicist Ernest Rutherford (1871–1937) and his coworkers in Manchester, England, in 1911. The basic idea of the experiment is sketched in Figure 7–2. The experiment started with a piece of radioactive material—matter that sends out energetic particles (see Chapter 11). For our purposes, you can think of radioactive materials as sources of tiny subatomic "bullets." The particular material Rutherford used produced bullets that scientists had named *alpha particles,* which are thousands of times heavier than electrons. By arranging the apparatus as shown, Rutherford produced a stream of these subatomic bullets moving toward the right in the figure. In front of this stream, he placed a thin foil of gold.

The experiment was designed to measure something about the way atoms were put together. At the time, people believed that the small, negatively charged electrons were scattered around the entire atom, more or less like raisins in a bun. Rutherford was trying to shoot bullets into the "bun" and see what happened.

What the experiment revealed was little short of astonishing. Almost all the subatomic bullets either passed right through the gold foil unaffected or were scattered through very small angles. This result is easy to interpret: it means that most of the heavy alpha particles passed through spaces in between gold atoms, and that those that hit the gold atoms were only moderately deflected by the relatively low-density material in them.

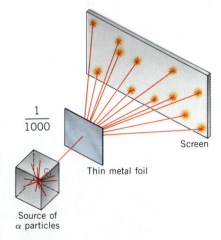

$\frac{1}{1000}$

Screen

Thin metal foil

Source of α particles

Figure 7–2

In Rutherford's experiment a beam of radioactive particles was scattered by atomic nuclei in a piece of gold foil. A lead shield protected researchers from the radiation.

Figure 7–3

The Nuclear Regulatory Commission uses a highly stylized atomic model as its logo.

One alpha particle in a thousand, however, was scattered through a large angle; some even bounced straight back. After almost two years of puzzling over these extraordinary results, Rutherford concluded that a large part of each atom's mass is located in a very small, compact object at the center—what he called the **nucleus.** About 999 times out of 1000 the alpha particles either missed the atom completely or went through the low-density material in the outer reaches of the atom. About 1 time out of 1000, however, the alpha particle hit the nucleus and was bounced through a large angle.

You can think of the Rutherford experiment in this way: If the atom were a large ball of mist or vapor taller than a skyscraper, and the nucleus were a bowling ball at the center of that sphere of mist, then most bullets shot at the atom would go right through. Only those that hit the bowling ball would be bounced through large angles. In this analogy, of course, the bowling ball plays the part of the nucleus, while the mist is the domain of the electrons.

As a result of Rutherford's work, a new picture of the atom emerged—one that is very familiar to us. Rutherford described a small, dense, positively charged nucleus sitting at the atom's center, while light, negatively charged electrons circle it, like planets orbiting the Sun. Indeed, Rutherford's discovery has become an icon of the modern age, adorning everyday objects from postage stamps to bathroom cleaners (see Figure 7–3).

Later on, physicists discovered that the nucleus itself is made up primarily of two different kinds of particles (see Chapter 11). One of these carries a positive charge and is called a *proton*. The other, whose existence was not confirmed until 1932, carries no electrical charge and is called a *neutron*.

For each positively charged proton in the nucleus of the atom, there is normally one negatively charged electron in orbit. Thus the electrical charges of the electrons and the protons cancel out, and every atom is electrically neutral. In some cases, atoms either lose or gain electrons. In this case, they acquire an electrical charge and are called *ions*.

Why the Rutherford Atom Couldn't Work

The picture of the atom that Rutherford developed is intellectually appealing, particularly because it recalls to us the familiar stately orbits of our solar system. We have already learned enough about the behavior of nature, however, to know that the atom that is described in the text could not possibly exist in nature. Why do we say this?

We learned in Chapter 2 that an object traveling in a circular orbit is constantly being accelerated—it is not in uniform motion because it is continually changing direction. Furthermore, we learned in Chapter 5 that any accelerated electric charge must give off electromagnetic radiation, as a consequence of Maxwell's equations. Thus, if an atom was of the Rutherford type, the electrons moving in their orbits would constantly be giving off energy in the form of electromagnetic radiation. This energy, according to the first law of thermodynamics, would have to come from somewhere (remember conservation of energy!), so the electrons would gradually spiral in toward the nucleus as they gave up their energy to electromagnetic radiation. Eventually, the electrons would have to fall into the nucleus and the atom would cease to exist in the form we know.

In fact, if you put in the numbers, the life expectancy of the Rutherford atom turns out to be less than a second. Given the fact that many atoms have survived billions of years, since almost the beginning of the universe, this calculation poses a serious problem for the simple orbital model of the atom.

When Matter Meets Light

Almost from its inception, the Rutherford model of the atom encountered difficulties. Some of the problems involved its violations of fundamental physical laws as described above, while others were more mundane—the Rutherford atom simply did not explain all the behavior of atoms that scientists knew about. The first decades of the twentieth century was a period of tremendous ferment in the sciences as people scrambled to find a new way of describing the nature of atoms.

The Bohr Atom

In 1913 Niels Bohr (1885–1962), a young Danish physicist working in England, produced the first model of the atom that avoided the kinds of objections encountered by Rutherford's model. The **Bohr atom** is very strange: it does not match well with our intuition about the way things

Niels Bohr (1885–1962) with Aage Bohr, one of his five sons. Both won Nobel Prizes in Physics.

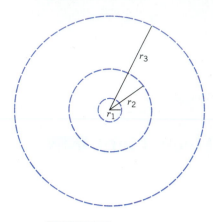

Figure 7–4

A schematic diagram of the Bohr atom reveals three of the possible locations for electrons.

"ought to be" in the real world. The only thing in the Bohr atom's favor was that it worked.

The Bohr atom began with an educated guess; a hunch, if you will. The young scientist was deeply immersed in studying the way that atoms interact with light and other forms of electromagnetic radiation. He realized that one way of explaining what he saw in the laboratory was that electrons circling the nucleus, *unlike planets circling a sun,* could not maintain their orbits at just any distance from the center. He suggested that there were only certain orbits—he called them "allowed orbits"—located at specified distances from the center of the atom in which an electron could exist for long periods of time *without giving off radiation.* Bohr's picture of the atom is shown in Figure 7–4. The idea is that the electron can exist at a distance r_1 from the nucleus, or at a distance r_2 from the nucleus, or at a distance r_3 from the nucleus, and so on. As long as the electron remains at one of those distances, its energy is fixed. The electron cannot ever, at any time, be in orbit any place between these allowed distances.

One way to think about the Bohr atom is to imagine what happens when you climb a flight of steps. You can stand on the first step or you can stand on the second step. It's very hard, however, to imagine what it would be like to stand somewhere between two steps. In just the same way, an electron can be in the first allowed orbit, or in the second allowed orbit, but it can't be in between.

In terms of energy, both the steps and the electrons in an atom have a very simple pictorial description (see Figure 7–5). Each time you change steps in your home, your gravitational potential energy changes. Similarly, each time an electron changes levels, its energy changes.

An electron in an atom can be in any one of a number of allowed orbits, and each allowed orbit is at a different energy. You can see this by noting that you would have to exert a force over a distance to move an electron from one allowed orbit to another, just as your muscles have to exert a force to get you up a flight of stairs. The allowed energy levels of an atom, then, occur as a series of steps as shown in the figure. The

Figure 7–5

Stairs provide an analogy to energy changes associated with electrons in the Bohr atom.

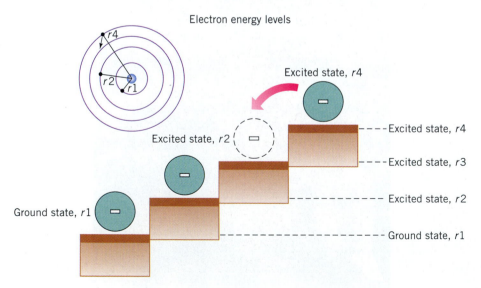

electron in the lowest energy level is called the *ground state,* while all energy levels above the ground state are called *excited states.*

Photons: Particles of Light

The major feature of the Bohr atom is that an electron in a higher energy level can move down into an available lower energy level. This process is analogous to that by which a ball at the top of a flight of stairs can bounce down the stairs under the influence of gravity.

Assume that an electron is in an excited state, as shown in Figure 7–6. The electron can move to the lowest state, as shown, but if it does, something must happen to the extra energy. Energy can't just disappear. This realization was Bohr's great insight. The energy that's left over when the electrically charged electron moves from an upper state to a lower state is emitted by the atom in the form of a single packet of electromagnetic radiation–a particle-like unit of light called a **photon.** Every time an electron jumps from a higher to a lower energy level, a photon moves away at the speed of light.

The concept of a photon raises a perplexing question: Is light—Maxwell's electromagnetic radiation—a wave or a particle? We will explore this puzzle at some length in Chapter 8, once we have learned more about the behavior of atoms.

The interaction of atoms and electromagnetic radiation provides the most compelling evidence for the Bohr atom. If electrons are in excited states, and if they make transitions to lower states, then photons are emitted. If we look at a group of atoms in which these transitions are

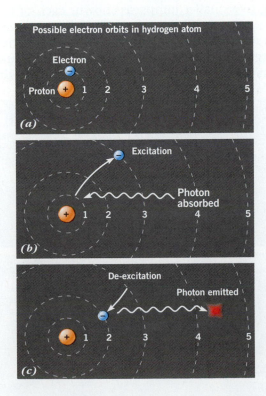

(a) (b) (c)

Figure 7–6
Electrons may jump between the energy levels shown in (a) and in the process (b) absorb or (c) emit energy in the form of a photon.

occurring, we will see light or other electromagnetic radiation. Thus, when you look at the flame of a fire or the glowing heating coil of an electric stove in your kitchen, you are actually seeing photons that were emitted by electrons jumping between allowed states in that material's atoms.

Not only does the Bohr atom give us a picture of how matter emits radiation, it also provides an explanation for how matter absorbs radiation. Start with an electron in a low-energy state, perhaps its ground state. If a photon arrives that has just the right amount of energy so that it can raise the electron to a higher energy, the photon can be absorbed and the electron will be pushed up to an excited state. Absorption of light is thus nothing more than the mirror image of light emission.

Our picture of the interaction of matter and radiation is exceedingly simple, but two key ideas are embedded in it. For one thing, when an electron moves from one allowed orbit to another, it cannot ever, at any time, be at any place in between. This rule is built into the definition of an allowed orbit. This means that the electron must somehow disappear from its original orbit and reappear in its final orbit *without ever having to traverse any of the positions in between*. This process, called a **quantum leap** or **quantum jump,** cannot be visualized, but it is something that seems to be fundamental in nature—an example of the "quantum weirdness" of nature at the atomic scale that we'll discuss in Chapter 8.

The second point to make is that if an electron is in an excited state, it can, in principle, get back down to the ground state in a number of different ways. Look at Figure 7–7. An electron in the upper energy level can move to the ground state by making one large jump and emitting a single photon with large energy. Alternatively, it can move to the ground state by making two smaller jumps, as shown. Each of these jumps will emit photons of somewhat less energy. The energies emitted in the two different jumps will generally be different from each other, but the sum of the two energies will equal that of the single large jump. We would expect, if we had a large collection of atoms of this kind, that some

Figure 7–7

An electron can jump from a higher to lower energy level in a single quantum leap (*a*), or by multiple quantum leaps (*b*).

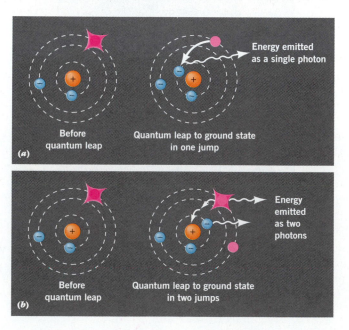

(*a*) Before quantum leap — Quantum leap to ground state in one jump — Energy emitted as a single photon

(*b*) Before quantum leap — Quantum leap to ground state in two jumps — Energy emitted as two photons

electrons would make the large leap while others make the two smaller ones. Thus, when we look at a collection of these atoms, we would see three different energies of photons.

In Chapter 6 we saw that the energy of electromagnetic radiation is related to its frequency, and, in the visible range, to the color of the light we perceive. If we look at a collection of atoms of the type shown in Figure 7–7, we would observe three different frequencies, one corresponding to each of the quantum leaps illustrated. If all of these frequencies were in the visible range, we would see three different colors.

Another subtle point about the Bohr atom arises if you ask the question, "How do electrons get to the excited states in the first place?" Obviously, it requires energy to lift an electron from the ground state to any excited state. This energy has to come from somewhere. We have already mentioned one possibility; that the atom will absorb a photon of just the right frequency to raise the electron to a higher energy level. There are other possibilities, however. If the material is heated, for example, atoms will move fast, gain kinetic energy, and undergo energetic collisions. In these collisions, an atom can absorb energy, and then use that energy to move electrons to a higher state. This explains why materials often glow when they are heated.

Science by the Numbers

All Possible Quantum Leaps

How many different frequencies might there be in an atom that has four allowed energy levels? An electron in the highest energy state could make one transition to the ground state, giving us one photon of high energy. Alternatively, it could jump to the next lowest state (one more photon), from which point it could either jump to the ground state directly (yet another photon) or make two successive jumps to get to the ground state (two more photons). Finally, the electron could jump directly to the first excited state (one additional photon) and from there to the ground state. All intermediate quantum jumps are included in the jumps just described for the electron in the highest energy state. If you add up all of these possible transitions you will find that this atom gives off photons of six different energies. ■

An Intuitive Leap

As we pointed out above, Bohr first proposed his model of the atom based on an intuition guided by experiments and ideas about the behavior of things in the subatomic world. In some ways the Bohr model was completely unlike anything we experience in the macroscopic world—indeed, the model seemed to some a little bit "crazy." It took two decades for scientists to develop a theory called quantum mechanics that showed why electrons can exist only in allowed orbits and not in between. We will discuss this justification for the Bohr atom in Chapter 8, but it should be remembered that the justification occurred long after the initial hypothesis. The Bohr atom was accepted by physicists because it worked—because it explained what they saw in nature and allowed them to make predictions about the behavior of real matter.

How could Bohr have come up with such a strange picture of the atom? He was undoubtedly guided by some of the early work that would lead to the theory of quantum mechanics (see Chapter 8). In the end, however, this is an unsatisfactory explanation. Many people at the time studied the interactions of atoms and light, but only Bohr was able to make the leap of intuition to his description of the atom. This insight, like Newton's realization that gravity might extend to the orbit of the Moon, remains one of the great intuitive achievements of the human mind.

▇ Spectroscopy

Whenever energy is added to a system with many atoms in it, electrons in some atoms are pushed to excited states. As time goes by, these electrons will make quantum leaps down to the ground state, giving off photons as they do. If some of those photons are in the range of visible light, the source will appear to glow.

You may not realize it, but you have looked at such collections of atoms all your life. Common mercury vapor street lamps contain bulbs filled with mercury gas. When the gas is heated, electrons are moved up to excited states. When they jump down they emit photons that give the lamp a bluish-white color. Other types of street lights, often used at freeway interchanges, use bulbs filled with sodium atoms. When sodium is excited, the most frequently emitted photons lie in the yellow range, so the lamps look yellow.

Yet another place where you can see photons emitted directly by single quantum leaps is in Day-Glo colors—the vivid colors often used

The elements (*a*) sodium, (*b*) potassium, and (*c*) lithium impart distinctive colors to a flame.

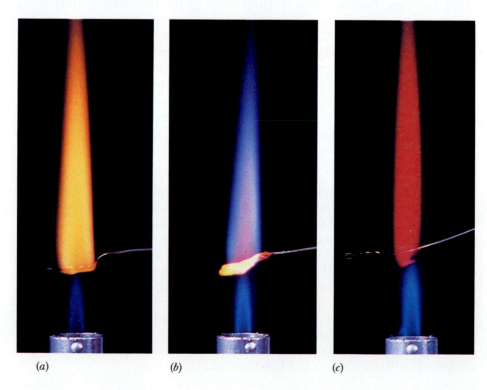

(*a*)   (*b*)   (*c*)

in sports clothing and for other purposes. From these examples, you can draw two conclusions: (1) quantum leaps are very much in evidence in your everyday life, and (2) different atoms give off different characteristic photons.

The second of these two facts is extremely important for scientists. If you think about the structure of an atom, the idea that different atoms have different characteristic photons shouldn't be too surprising. Electron energy levels depend on the electrical attraction between the nucleus and the electrons, just as the orbits of the planets depend on the gravitational attraction between the planets and the Sun. Different nuclei have different numbers of protons, so electrons circling them are in different orbits. In fact, the arrangement of the allowed energy levels is different in each of the hundred or so different chemical elements. Since the energy and frequency of the photons emitted by an atom depend on the differences in energy between the allowed levels, each chemical element gives off a distinct set of characteristic photons.

You can think of the collection of characteristic photons emitted by each chemical element as a kind of "fingerprint"—something that is distinctive for that chemical element and none other. This feature opens up a very interesting possibility. The total collection of photons emitted by a given atom is called its **spectrum,** a characteristic fingerprint that can be used to identify chemical elements even when they are very difficult to identify by any other means.

In practice, the identification process works like this: Light from the atoms is spread out by being passed through a prism (Figure 7–8). Each possible quantum jump corresponds to light at a specific wavelength, so each type of atom produces a set of lines, as shown in Figure 7–9. This spectrum is the atomic fingerprint.

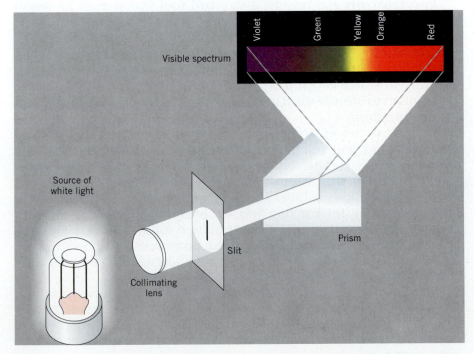

Figure 7–8
A glass prism spreads out the colors of the visible spectrum.

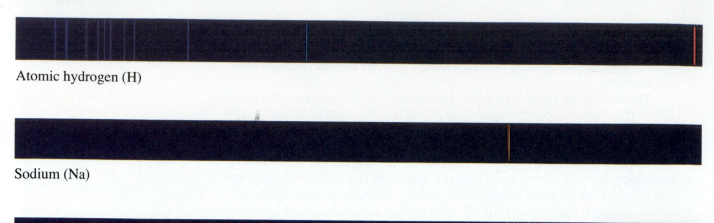

Atomic hydrogen (H)

Sodium (Na)

Neon (Ne)

Figure 7–9

Line spectra, shown here for hydrogen, sodium, and neon, provide distinctive fingerprints for elements and compounds.

The Bohr picture suggests that if an atom gives off light of a specific wavelength and energy, then it will also *absorb* light at that wavelength. The emission and absorption processes, after all, involve quantum jumps between the same two energy levels but in different directions. Thus if white light shines through a material containing a particular kind of atom, certain wavelengths of light will be absorbed. Observing that light on the other side of the material, you will see certain colors missing. The dark areas corresponding to the absorbed wavelengths are called absorption lines. This set of lines will be as much an atomic fingerprint as the set of colors that the atoms emit. And although the use of visible light is very common, it should be obvious that these arguments hold for radiation in any part of the electromagnetic spectrum.

Spectroscopy has become a standard tool that is used in almost every branch of science. Astronomers use emission spectra to find the chemical composition of distant stars, and they study absorption lines to determine the chemical composition of interstellar dust and the atmospheres of the outer planets. Spectroscopic analysis is also used in manufacturing to search for impurities on production lines, and by police departments to identify small traces of unknown materials when conducting investigations.

In a classic set of experiments in the early 1940s, scientists used spectroscopy to work out in detail how chemical reactions governed by protein enzymes (see Chapter 20) proceed in cells. In these experiments, a fluid containing the materials undergoing the chemical reactions was allowed to flow down a tube. The farther down the tube the fluid was, the farther along the reaction was. By measuring spectra at different points along the tube, scientists were able to follow the change in the behavior of electrons as the chemical reactions went along. In this way, part of the enormously complex problem of understanding the chemistry of life was unraveled.

More recently, scientists have begun to develop instruments that can use the principles of spectroscopy to identify pollutants emitted by automobile tailpipes as the cars drive by. If they are successful, we will have a major new tool in our battle against air pollution and acid rain (see Chapter 25).

Science in the Making

The Story of Helium

You have probably experienced helium—you may even have used it to inflate party balloons. Helium gas turns out to be a very interesting material, not only for its properties (it's so light that it floats in air), but because of the history of its discovery.

The word helium refers to *helios,* the Greek word for Sun, because helium was first discovered in spectral lines in the Sun in 1868 by the English scientist Joseph Norman Lockyer (1836–1920). Helium is very rare in the atmosphere of the Earth, and before Lockyer's discovery scientists were not even aware of its existence. Following the discovery, there was a period of about 30 years when astronomers accepted the fact that the element helium existed in the Sun, but were unable to find it on the Earth.

This supposition led to a very interesting problem. Could it be that there were chemical elements on the Sun that simply did not exist on our own planet? If so, it would call into question our ability to understand the rest of the universe, for the simple reason that if we don't know what an element is and can't isolate it in our laboratories, then we can never really be sure that we understand its properties.

The existence of helium on Earth wasn't confirmed until 1895, when Lockyer identified its spectrum in a sample of radioactive material. ■

Technology

The Laser

The Bohr atom provides an excellent way of understanding the workings of one of the most important devices in modern science—the laser. The word *laser* is an acronym for *l*ight *a*mplification by *s*timulated *e*mission of *r*adiation. At the core of every laser is a collection of atoms—a crystal of ruby, perhaps, or a gas enclosed in a glass tube. The term "stimulated emission" refers to a process that goes on when light and these atoms interact. If an electron is in an excited state, as shown in Figure 7–10, and one photon of just the right energy passes nearby, the electron may be stimulated to make the jump to a lower energy state, thus releasing a second photon. By "just the right energy" for the first photon, we mean a photon whose energy corresponds to the energy gap between two electron energy levels in the atom.

Furthermore, the stimulated atom emits photons in a special way. Remember that light is a form of electromagnetic radiation that can be described as a wave. In a laser the crests of all the emitted photon waves line up exactly with the crests of the first photon, and the signal is enhanced by constructive interference. In the language of the physicists, we say that the photons are "coherent." Thus in stimulated emission you have one photon at the beginning of the process and two coherent photons at the end.

Figure 7–10
Lasers produce a beam of
light when one photon
stimulates the emission of
other photons.

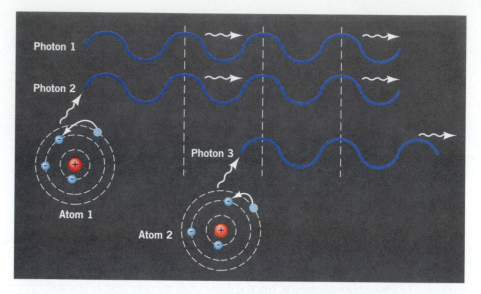

Now suppose that you have a collection of atoms where most of
the electrons are at the excited state, as shown in Figure 7–11. If a single photon of the correct frequency enters this system from the left
and moves to the right, it will pass the first atom and stimulate the
emission of a second photon. You will then have two photons moving
to the right. As these photons encounter other atoms, they, too, stimulate emission so that you have four photons. It's not hard to see that
light amplification in a laser will happen very quickly, cascading so
that soon there is a flood of photons moving to the right through the
collection of atoms. Energy added to the system from outside continuously returns atoms to their excited state—a process called *pumping*—
so that more and more coherent photons can be produced.

In a laser, the collection of excited atoms is bounded on two sides
by mirrors so that photons moving to the right hit the mirror, are reflected, and make another pass through the material, stimulating even
more emission of photons as they go. If a photon happens to be lined
up exactly perpendicular to the mirrors at the end of the laser, it will
continue bouncing back and forth. If its direction is off by even a
small angle, however, it will eventually bounce out through the sides
of the laser and be lost. Thus only those photons that are exactly
aligned will wind up bouncing back and forth between the mirrors,
constantly amplifying the signal. Aligned photons will traverse the laser millions of times, building up an enormous cascade of coherent
photons in the system.

Figure 7–11
The action of a laser. Electrons in the laser's atoms
are continuously "pumped"
into an excited state by an
outside energy source, and
the beam of photons is released when the electrons
return to their ground state.

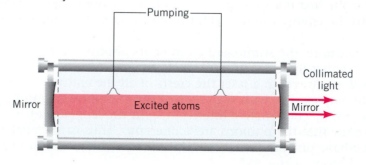

(a)

(b)

The mirrors are designed to be partially reflective—perhaps 95% of the photons that hit the mirror are reflected back into the laser, and 5% are allowed to leak out. Those photons that leak out form the familiar laser beam, which is made up of intense, coherent light.

Laser beams have found thousands of uses in science and industry since their development in the 1960s. Low-power lasers are ideal for optical scanners, such as the ones in your supermarket checkout line, and they make ideal light pointers for lectures and slide shows. The fact that the beam of light travels in a straight line makes the laser invaluable in surveying over long distances—indeed, modern subway tunnels are routinely surveyed by using lasers to provide a straight line underground. Lasers are also used to detect movement of seismic faults in order to predict earthquakes (see Chapter 15). In this case, a laser is directed across the fault, so that small motions of the ground are easily measured. Finely focused laser beams have revolutionized delicate procedures such as eye surgery. Much more powerful lasers can transfer large amounts of energy. They are often used as cutting tools in factories, as well as implements for performing some kinds of surgery. The military has also adopted laser technology, both in targeting and range finders, and in designs for futuristic energy beam weapons.

From the point of view of science, however, lasers are important because they enable us to make extremely precise and detailed measurements of the properties and structures of atoms. Almost all modern studies of the atom depend in one way or another on the laser. ∎

Lasers are now used extensively in industry and medicine. (a) A laser beam facilitates delicate eye surgery. (b) Robot arms use lasers to complete precise welding on an automobile assembly line.

The Periodic Table of the Elements

Dmitri Mendeleev, the Russian scientist who noticed the regularity or periodicity in the known chemical elements (see Chapter 1), related that periodicity to each element's atomic weight. Today, each element is assigned an integer, called the *atomic number,* which defines the sequence of elements in the table. The atomic number corresponds to the number of protons in the atom or, equivalently, the number of electrons in orbit around the nucleus. If you arrange the elements as shown in Figure 7–12, with elements getting progressively heavier as you read from left to right and top to bottom as in a book, then elements in the same vertical column have very similar chemical properties. Mendeleev called this discovery the **periodic table of the elements,** and you may have seen versions of it in your high school science classroom.

Periodic Chemical Properties

The most striking characteristic of the periodic table is the similarity of elements in any given column. In the far left-hand column of the table, for example, are highly reactive elements called alkali metals (lithium, sodium, potassium, etc.). Each of these soft, silvery elements forms compounds (called *salts*) by combining in a one-to-one ratio with any of the elements in the seventh column (fluorine, chlorine, bromine, etc.). Water dissolves these compounds, which include common table salt, or sodium chloride.

The elements in the second column, including beryllium, magnesium, and calcium, are called the *alkaline earth metals,* and they too display similar chemical properties. All of these elements, for example, combine with oxygen in a one-to-one ratio to form colorless, solid compounds with very high melting temperatures.

Elements in the far right-hand column (helium, neon, argon, and so on), by contrast, are all colorless, odorless gases that are almost impossible to coax into any kind of chemical reaction. These *noble gases* find applications when ordinary gases are too reactive. Helium lifts blimps, because the only other lighter-than-air gas is hydrogen, which is explosive. Argon fills incandescent light bulbs, because nitrogen or oxygen would react with the hot filament.

In the late nineteenth century scientists knew that the periodic table "worked"—it organized the 63 elements known at that time and implied the existence of others—but they had no idea *why* it worked. Their faith in the periodic table was buttressed by the fact that when Mendeleev first wrote it down, there were holes in the table—places where he predicted elements should go, but for which no element was known. The ensuing search for the missing kinds of atoms produced the elements we now call scandium (in 1876) and germanium (in 1886).

Why the Periodic Table Works

With the advent of Bohr's atomic model and its modern descendants, we finally have an understanding of why the periodic table works. We now realize that the pattern of elements in the periodic table reflects the arrangement of electrons in their orbits.

The atom is largely empty space. When two atoms come near enough to each other to undergo a chemical reaction—a carbon atom and an oxygen atom in a burning piece of coal, for example—the outermost electrons meet each other first. We will see in Chapter 9 that these outer-

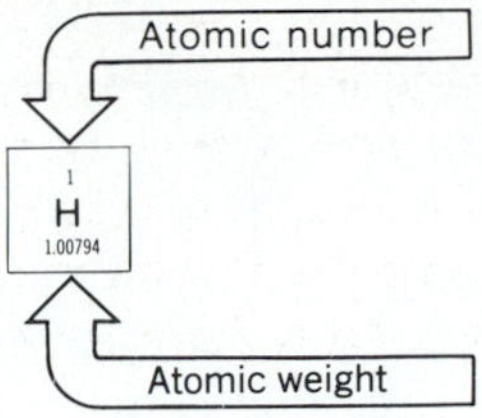

Figure 7–12

Periodic table of the elements:

| | | Atomic number → H 1.00794 ← Atomic weight | | | | | | | | | | | | | | | |

	IA (1)	IIA (2)	IIIB (3)	IVB (4)	VB (5)	VIB (6)	VIIB (7)	VIII (8)	VIII (9)	VIII (10)	IB (11)	IIB (12)	IIIA (13)	IVA (14)	VA (15)	VIA (16)	VIIA (17)	0 (18) Noble gases
1	1 H 1.00794																	2 He 4.00260
2	3 Li 6.941	4 Be 9.01218											5 B 10.81	6 C 12.011	7 N 14.00674	8 O 15.9994	9 F 18.99840	10 Ne 20.1797
3	11 Na 22.98977	12 Mg 24.3050											13 Al 26.98154	14 Si 28.0855	15 P 30.97376	16 S 32.066	17 Cl 35.4527	18 Ar 39.948
4	19 K 39.0983	20 Ca 40.078	21 Sc 44.9559	22 Ti 47.88	23 V 50.9415	24 Cr 51.9961	25 Mn 54.9380	26 Fe 55.847	27 Co 58.93320	28 Ni 58.69	29 Cu 63.546	30 Zn 65.39	31 Ga 69.723	32 Ge 72.61	33 As 74.92159	34 Se 78.96	35 Br 79.904	36 Kr 83.80
5	37 Rb 85.4678	38 Sr 87.62	39 Y 88.90585	40 Zr 91.224	41 Nb 92.90638	42 Mo 95.94	43 Tc 98.9072	44 Ru 101.07	45 Rh 102.90550	46 Pd 106.42	47 Ag 107.8682	48 Cd 112.411	49 In 114.82	50 Sn 118.710	51 Sb 121.75	52 Te 127.60	53 I 126.90447	54 Xe 131.29
6	55 Cs 132.90543	56 Ba 137.327	57 *La 138.9055	72 Hf 178.49	73 Ta 180.9479	74 W 183.85	75 Re 186.207	76 Os 190.2	77 Ir 192.22	78 Pt 195.08	79 Au 196.96654	80 Hg 200.59	81 Tl 204.3833	82 Pb 207.2	83 Bi 208.98037	84 Po 208.9824	85 At 209.9871	86 Rn 222.0176
7	87 Fr 223.0197	88 Ra 226.0254	89 †Ac 227.0278	104 Unq 261.11	105 Unp 262.114	106 Unh 263.118	107 Uns 262.12	108 Uno (265)	109 Une (266)									

Periods

* Lanthanides:

58 Ce 140.115	59 Pr 140.90765	60 Nd 144.24	61 Pm 144.9127	62 Sm 150.36	63 Eu 151.965	64 Gd 157.25	65 Tb 158.92534	66 Dy 162.50	67 Ho 164.93032	68 Er 167.26	69 Tm 168.93421	70 Yb 173.04	71 Lu 174.967

† Actinides:

90 Th 232.0381	91 Pa 231.0359	92 U 238.0289	93 Np 237.0482	94 Pu 244.0642	95 Am 243.0614	96 Cm 247.07003	97 Bk 247.0703	98 Cf 251.0796	99 Es 252.083	100 Fm 257.0951	101 Md 258.10	102 No 259.1009	103 Lr 260.105

The periodic table of the elements. The weights of the elements increase from left to right. Each vertical column groups elements with similar chemical properties.

most electrons, in fact, govern the chemical properties of materials. We have to understand the behavior of these electrons if we want to understand the periodic table.

To do this, we need to know one more curious fact about electrons. Electrons are particles that obey what is called the *Pauli exclusion principle,* which says that no two electrons can occupy the same state at the same time. One analogy is to compare electrons to cars in a parking lot—each car takes up one space, and once a space is filled, no other car can go there. Electrons behave just the same way—once an electron fills a particular niche in the atom, no other electron can occupy the same niche. A parking lot can be full long before all the actual space in the lot is taken up with cars, because the driveways and spaces between cars must remain empty. So, too, a given Bohr orbit can be filled with electrons long before all the available space is filled.

In fact, it turns out that there are only two spaces that an electron can fill in the lowest, innermost Bohr orbit. One of these spaces corresponds to a situation in which the electron "spins" clockwise on its axis, the other to a situation in which it "spins" counterclockwise on its axis. When we start to catalog all possible chemical elements in the periodic table, then, we have element one (hydrogen) with a single electron in the lowest orbit, and element two (helium) with two electrons in that same orbit. After this, if we want to add one more electron it has to go into the second electron orbit because the lowest orbit is completely filled up. This situation explains why only hydrogen and helium appear in the first row in the periodic table.

Adding a third electron yields an atom with two electrons in the innermost orbit, and a single electron in the second electron orbit. This element is lithium, the one just below hydrogen in the first column of the periodic table. Lithium and hydrogen both have a single electron in

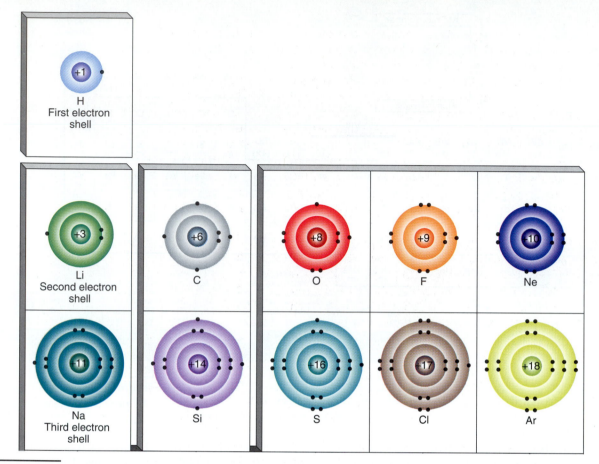

Figure 7–13

A representation of electrons in a number of common atoms.

their outermost electron orbit, so they have very similar chemical properties.

The second electron orbit has room for eight electrons, a fact reflected in the eight elements of the periodic table's second row, from lithium with three electrons to neon with ten. Neon appears directly under helium, and we expect these two gases to have similar chemical properties because both have completely filled outer electron orbits.

Thus a simple counting of the states available to electrons in the lowest two orbits explains why the first row in the periodic table has two elements in it and the second row eight. By similar (but somewhat more complicated) arguments, you can show that the Pauli exclusion principle requires that the next row of the periodic table have 8 elements, the next 18, and so on. (See Figure 7–13.) Thus, with an understanding of the structure of the atom, the seemingly mysterious regularity that Mendeleev found among the chemical elements becomes an example of the basic laws of nature at work.

Counting Atoms

The periodic table arranges elements in order of increasing number of protons, which corresponds closely to the order of increasing mass. The mass of atoms is generally given in atomic mass units, defined to be one-twelfth the mass of a carbon atom with 6 protons and 6 neutrons. Think of this unit, then, as roughly the mass of a proton or neutron. How do chemists go from this definition of mass in terms of atoms to the kind of mass you would read on a laboratory balance?

A carbon atom has 12 atomic mass units, an iron atom has 56. Thus there will be just as many atoms in 12 grams of carbon as there are in 56 grams of iron. In the language of the chemist, we say that each of these samples contains one "gram molecular weight," or one *mole,* of atoms. A mole is thus a (very large) number of atoms. In fact, a mole of any material is defined as the number of atoms or molecules contained in the number of grams that is numerically equal to the material's mass in atomic mass units. Thus 12 grams of carbon atoms, 16 grams of oxygen atoms, 32 grams of oxygen molecules, and 18 grams of water molecules all constitute one mole of those materials.

The number of atoms in a mole is known as Avogadro's number, after Amadeo Avogadro, the eighteenth-century Italian scientist who first determined it. There are about 6.02×10^{23} atoms or molecules in a mole. This number is huge—if you had a mole of pennies, for example, there would be enough money to give every man, woman, and child in the United States a grand total of 24 trillion dollars!

THINKING MORE ABOUT ATOMS

Are Atoms Real?

Throughout the nineteenth century, physicists and chemists came to realize that matter really behaved *as if* it were made up of atoms. The question remained, however, whether the atoms were real or simply useful fictions. After all, when you look at the paper on this page you don't see atoms. In a sense, this debate mirrors the current argument over whether even smaller components of matter called quarks (see Chapter 12) are real or not.

By the end of the nineteenth century, however, an important series of observations were made that argued for the reality of atoms. If you look at very small objects—grains of pollen suspended in a liquid, for example—they seem to jitter around. Over long periods of time each grain tends to wander in a random fashion, changing directions all the time—a phenomenon called *Brownian motion.* It is clear from Newton's laws of motion that something in the water has to be causing the tiny movements by exerting a force, but what could it be?

In 1905 Albert Einstein demonstrated that individual collisions with large numbers of wa-

ter molecules were driving the motion of the pollen grains. Einstein realized that a small object suspended in liquid would be constantly bombarded by moving atoms. At any given moment, there will, purely by chance, be more atoms hitting on one side than the other. The object will be pushed toward the side with fewer collisions. A moment later, however, more atoms will be hitting another surface, and the object will change direction. Over time, Einstein argued, atomic collisions would produce precisely the sort of erratic motion that you see in the microscope.

Einstein used the mathematics of statistics to make a number of predictions about how fast and how far the suspended grains would move, based on the hypothesis that the motion was due to collisions with real atoms—predictions that other scientists could test in the laboratory. The French physicist Jean Baptiste Perrin (1870–1942) published the results of his careful experiments on Brownian motion in 1909. His results agreed with Einstein's calculations. With this work, the question of the reality of atoms was laid to rest for good. Hypothetical mental constructs can't collide with solid objects and make them move.

Until recently, all evidence for atoms was indirect though compelling; no one had ever "seen" an atom. In the late 1980s, however, scientists devised instruments called scanning probe microscopes, with which it is possible to "photograph" atoms and thus provide direct visual evidence of their reality (Figure 7–14).

When in the chain of historical events would you have been willing to believe that atoms are real? When Dalton explained the existence of elements? When Einstein explained Brownian motion? When you were shown a picture like the one in Figure 7–14? Never? What does it take to make something "real"? And, finally, does it make a difference to science whether atoms are real or not?

Figure 7–14
This electronic image representing individual atoms was taken with an instrument called a scanning probe microscope. The "mountains" correspond to individual xenon atoms that were placed on a "plain" of nickel. The image reveals the first time an atomic structure had been "hand built."

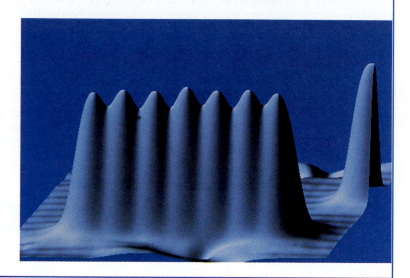

Summary

All the solids, liquids, and gases around us are composed of about 100 different *elements*. *Atoms,* the building blocks of our chemical world, combine in groups of two or more called *molecules.* For thousands of years atoms were discussed purely as hypothetical objects, but studies of Brownian motion early in the twentieth century and recent imaging of individual atoms in new kinds of microscopes have confirmed the existence of these tiny particles.

Each atom contains a massive central *nucleus* made from positively charged protons and electrically neutral neutrons. Around the nucleus orbit *electrons,* swift negatively charged particles that have only a small fraction of the mass of protons and neutrons. Early models of this kind of atom treated electrons like planets orbiting around the Sun. Those models were flawed, however, because each electron, constantly accelerating, would have to emit electromagnetic radiation continuously. Niels Bohr proposed an alternate model in which electrons rested in various energy levels, much as you can stand on different levels of a flight of stairs.

Electrons in the *Bohr atom* can shift to a higher energy level by absorbing the energy of heat or light. They can also drop into a lower energy level, and in the process release a *photon,* an individual electromagnetic wave. These changes in electron energy level are called quantum leaps or quantum jumps. *Spectroscopic* studies of the light emitted or absorbed by atoms—the atom's *spectrum*—reveal the nature of each atom's electron energy levels.

Each of an atom's energy levels, or electron orbits, holds only a limited number of electrons. The innermost electron orbit will accommodate only 2 electrons, the second and third orbits contain up to 8 electrons, while the fourth and fifth orbits hold 18 electrons. This electronic structure is reflected in the organization of the *periodic table of elements,* which lists all the elements in rows corresponding to each orbit, and in columns corresponding to elements with similar numbers of outer electrons and thus similar chemical behavior.

Key Terms About Atoms

element	electron	photon	spectroscopy
atom	nucleus	quantum leap, or quantum jump	periodic table of the elements
molecule	Bohr atom	spectrum	

Review Questions

1. What experiment first demonstrated the existence of atoms?

2. What three particles make up every atom? What are the major differences among these particles?

3. How does the nucleus differ from electrons?

4. What is a molecule?

5. Why did some alpha particles bounce backward off atoms in Rutherford's experiment?

6. Describe the main features of the Bohr atom.

7. What is a quantum leap? What can cause an electron to undergo such a change?

8. How do an emission spectrum and an absorption spectrum differ in appearance?

9. Cite three examples of everyday objects with vivid emission spectra.

10. How can scientists on Earth determine the elements that occur in stars?

11. Describe the workings of a laser.

12. In what ways are all the elements in a given column of the periodic table similar?

Discussion Questions

1. In what ways are atoms like the solar system of planets orbiting around the Sun? In what ways are they different?

2. The leaves of a tree are bright green. What do you think a leaf's absorption spectrum might look like?

3. Rutherford's experiment involved firing nucleus-sized "bullets" at atoms of gold. He found that one atom in a thousand bounced backward. What might Rutherford have seen if atoms were completely uniform in mass? What if electrons were the more massive particle? (*Hint:* What happens when a bowling ball collides with a ping-pong ball?)

4. Based on your knowledge of Newton's laws of motion, the laws of thermodynamics, and the nature of electromagnetic radiation, explain why the Rutherford model of the atom couldn't work.

5. People sometimes use the phrase "quantum leap" to signify a major breakthrough or stunning advance. Knowing what you do about the Bohr atom, is that an appropriate meaning? Why or why not?

6. When you shine invisible ultraviolet light (black light) on certain objects, they glow with brilliant colors. How might this behavior be explained in terms of the Bohr atom?

7. Why do different lasers have different-colored beams?

8. What does it mean to say the periodic table was useful because it "worked"? How does this relate to the scientific method? Describe an imaginary discovery that might have invalidated the periodic table.

9. If you were told that fluorine is an extremely reactive element (that is, it combines readily with other elements) what other elements could you guess were also extremely reactive? Why?

10. Space probes often carry compact spectrometers among their scientific hardware. What kind of spectroscopy might scientists use to determine the surface composition of cold, outer planets that orbit the Sun? How might they use spectroscopy to determine the atmospheric composition of these planets?

Problem

1. If the electrons in an atom can occupy any one of five different energy levels, how many lines might appear in that atom's spectrum?

Investigations

1. Simple hand-held spectroscopes are available in many science labs. Look at the spectra of different kinds of light bulbs: an incandescent bulb, a fluorescent bulb, a halogen bulb, and any other kinds available to you. What differences do you observe in their spectra? Why?

2. Place pieces of transparent materials between a strong light source and the spectrometer described in Investigation 1. Does the spectrum change? Why?

3. Why do colors look different when viewed indoors, under fluorescent light, and outdoors in sunlight? How might you devise an experiment to quantify these differences?

Additional Reading

Perrin, Jean. *Atoms.* Woodbridge, Connecticut: Oxbow Press (1990) (translation of the 1913 French edition).

von Baeyer, Hans Christian. *Taming the Atom.* New York: Random House (1992).

Quantum Mechanics

At the subatomic scale everything is quantized. Any measurement at that scale significantly alters the object being measured.

A Random Walk
Take Me Out to the Ball Game

Imagine yourself at a big-league ball game under the lights of a great stadium. Cheering fans fill the stands, roving vendors sell their food and drink, and the pitcher and batter play out their classic duel. The pitcher stares the batter down, winds up, and hurls a fast ball. But the batter is ready and pounces on the pitch. The ball leaps off the bat with a sharp crack. And then all the lights go out.

Where is the ball? You can't see it, but given a videotape replay of the bat hitting the ball you could determine the exact position and velocity of the ball at the time the lights went out. You could, at least in principle, calculate the subsequent path of the ball and predict where it would eventually land. We know how objects behave in our everyday world, even if we can't see them.

But the behavior of particles at the scale of electrons is different. Early in this century scientists realized that you can never know the exact position and the exact velocity of an electron simultaneously. For the first time the scientific community was faced with the fact that there are some things we cannot know about our physical universe. Despite these limits, we can and do use electrons in thousands of products, from VCRs to microwave ovens, thanks to the advances of quantum mechanics, as we will discuss. Given the importance of things such as fax machines, video games, electronic fuel-injection systems, and the myriad other devices that depend on the microelectronics industry, it's very difficult to imagine modern civilization operating without this branch of science.

The World of the Very Small

In Chapter 7 we saw that when an electron moves between energy levels and emits a photon, it is said to make a *quantum leap*. The term quantum mechanics refers to the theory that describes this event and other events at the scale of the atom. The word quantum comes from the Latin word for "bundle," while mechanics, as we saw in Chapter 2, is the study of the motion of material objects. **Quantum mechanics,** then, is the branch of science that is devoted to the study of the motion of objects that come in small bundles, or quanta. We have already seen that material inside the atom comes in little bundles—tiny pieces of matter we call electrons circle in orbits around another little bundle of matter we call the nucleus. In the language of physicists, the atom's matter is said to be *quantized*.

Electric charge is also quantized—every electron has a charge of exactly -1, while every proton has a $+1$ charge. We have seen that photons emitted by an atom can have only certain values of energy, so that energy levels in the atom and emitted energy are quantized. In fact, inside the atom, in the world of the submicroscopically small, *everything* comes in quantized bundles.

Our everyday world isn't like this at all. Although we have been told since childhood that the objects around us are made up of atoms, for all intents and purposes we experience matter as if it were smooth, continuous, and infinitely divisible. Indeed, for almost anything we want to do in the physical world, the idea of matter existing in continuous form works as well as anyone would want.

The quantum world is foreign to our senses. All of the intuition that we have built up about the way the world operates—all of the "gut feelings" we have about the universe—comes from our experiences with large-scale objects made up of apparently continuous material. If it should turn out (as it does) that the world of the quantum does not match our intuition, we should not be surprised. We have never dealt with this kind of world, so we have no particular reason, based on observations or experience, to believe that it should behave one way or the other.

This warning may not make you feel much better as you learn just how strange and different the quantum world really is, but it might help you come to intellectual grips with a most fascinating part of our physical universe.

Measurement and Observation in the Quantum World

Every measurement in the physical world incorporates three essential components:

1. A sample—a piece of matter to study.
2. A source of energy—light or heat or kinetic energy that interacts with the sample.
3. A detector to observe and measure that interaction.

When you look at a piece of matter such as this book, you can see it because light bounces off the book and comes to your eye, a very sophisticated detector (see Chapter 6). When you examine a piece of fruit at the grocery store you apply energy by squeezing it to detect if it feels too ripe.

Many professions employ sophisticated devices to make their measurements. Air traffic controllers reflect radio waves off airplanes to determine their positions, oceanographers bounce sound waves off deep-ocean sediments to map the seafloor, and dentists pass X-rays through your teeth and gums to look for cavities. In our everyday world we assume that such interactions of matter and energy do not change the objects being measured in any appreciable way. Microwaves don't alter an airplane's flight path, nor do sound waves disturb the topography of the ocean's bottom. And while prolonged exposure to X-rays can be harmful, the dentist's brief exploratory X-ray photograph has no obvious immediate effects on the tooth. Our experience tells us that a measurement can usually be made on a macroscopic object—something large enough to be seen without a microscope—without altering that object, because the energy of the probe is much less than the energy of the object.

The situation is rather different in the quantum world. If you want to "see" an electron, you have to bounce energy off it so that the information can be carried to your detectors. But nothing at your disposal can interact with the electron without simultaneously affecting it. You can bounce a photon off it, but in the process the electron's energy will change. You can bounce another particle off it, but the electron will recoil like a pool ball. No matter what you try, the energy of the probe is too close to the energy of the thing being measured. The electron cannot fail to be altered by the interaction.

Many everyday analogies illustrate the process of measurement in the quantum world. It's like trying to detect bowling balls by bouncing other bowling balls off them. The act of measurement in the quantum world poses a dilemma analogous to trying to discover if there is a car in a tunnel when the only means of finding out is to send another car into the tunnel and listen for a crash. With this technique you can certainly discover whether the first car is there. You can probably even find out where it is by measuring the time it takes the probe car to crash. What you *cannot* do, however, is assume that the first car is the same after the interaction as it was before. In the same way, nothing in the quantum world can be the same after the interaction associated with a measurement as it was before.

In principle, this argument would apply to any interaction, whether it involves photons and electrons or photons and bowling balls. As we demonstrate in the "Science by the Numbers" in this chapter, however, the effects of the interaction for large-scale objects are so tiny that they can simply be ignored, while in the case of interactions at the atomic level, they cannot. This fundamental difference between the quantum and macroscopic worlds is what makes quantum mechanics quite different from the classical mechanics of Isaac Newton. Remember that every experiment, be it on planets or fruit or quantum objects, involves interactions of one sort or another. The consequences of small-scale interactions make the quantum world different, not the fact that a measurement is being made. In other words, contrary to some popular accounts of the field, the presence of an experimenter is not what makes the quantum world different, but rather the disruptive nature of quantum interactions themselves.

The Heisenberg Uncertainty Principle

In 1927 a young German physicist, Werner Heisenberg (1901–1976), put the idea of limitations on quantum-scale measurements into precise mathematical form. His work, which was one of the first results to come from the new science of quantum mechanics, is called the *Heisenberg uncertainty principle* in his honor. The central concept of the uncertainty principle is simple:

> **At the quantum scale, any measurement significantly alters the object being measured.**

Suppose, for example, you have a particle such as an electron in an atom and want to know where it is *and* how fast it's moving. The uncertainty principle tells us that it is impossible to measure both the position and the velocity with infinite accuracy at the same time.

The reason for this state of affairs, of course, is that every measurement changes the object being measured. Just as the car in the tunnel could not be the same after the first measurement was made on it, so too will the quantum object change. The result is that as you measure one property such as position more and more exactly, your knowledge of a property such as velocity gets fuzzier and fuzzier.

The uncertainty principle doesn't say that we cannot know a particle's location with great precision. It is possible, at least in principle, for the uncertainty in position to be zero, which would mean that we know the exact location of a quantum particle. In this case, however, the uncertainty in the velocity has to be infinite. Thus, at the point in time when we know exactly where the particle is, we have no idea whatsoever how fast it is moving. By the same token, if we know exactly how fast the quantum particle is moving, we cannot know where it is. It could, quite literally, be in the room with us or it could be in China.

In practice, every quantum measurement involves trade-offs. We accept some fuzziness in the location of the particle and some fuzziness in the knowledge of the velocity, playing the two off against each other to get the best solution to whatever problem it is we're working on. We cannot have precise knowledge of both at the same time, but we can know either one as accurately as we like at any time.

Let's look a little more closely at the differences between the world of our intuition and the quantum world. In the former, we assume that measurement doesn't affect the thing being measured, so that we can have exact simultaneous knowledge of both the position and the velocity of an object such as a car or a baseball. In the quantum world, as Heisenberg taught us, we cannot.

Heisenberg put his notion into a simple mathematical relationship, which is a complete and exact statement of the **uncertainty principle.**

▶ **In words:**

> The error or uncertainty in the measurement of an object's position, times the error or uncertainty in that object's velocity, must be greater than a constant (Planck's constant) divided by the object's mass.

▶ **In equation form:**

> (uncertainty in position) $\times$ (uncertainty in velocity) $> \dfrac{h}{\text{mass}}$,
>
> where h is a number known as *Planck's constant* (see below).

▶ **In symbols:**

> $$\Delta x \times \Delta v > \frac{h}{m}.$$

This equation is a precise, shorthand way of saying that you can never know both the position and velocity of an object with perfect accuracy. The difference between our everyday world and the world inside the atom hangs on the question of the numerical value of h/m, the numbers on the right side of Heisenberg's equation. In SI units (see Appendix A), Planck's constant, h, has a value of 6.63×10^{-34} joule-seconds.

The important point about the Heisenberg relationship is not the exact value of the number, h/m, but the fact that the number is greater than zero. Look at it this way. If you make more and more precise measurements about the location of a particle, you determine its position more and more exactly, and the uncertainty in position, Δx, must get smaller and smaller. In this situation, it follows that the uncertainty in velocity, Δv, has to get bigger and bigger. In fact, we can use the uncertainty principle to calculate exactly how much uncertainty there must be in our knowledge of velocity for a given uncertainty in position, and vice versa.

Science by the Numbers

Uncertainty in the Newtonian World

The best way to understand why we do not have to worry about the uncertainty principle in our everyday life is to calculate the uncertainty in measurements in two separate situations: large objects and very small objects.

1. Small Uncertainties with Large Objects A moving automobile with a mass of 1000 kilograms is located in an intersection that is 5 meters across. How precisely can you know how fast the car is traveling?

▶ **Reasoning:** We can solve this problem by noting that if the car is somewhere in an intersection 5 meters across, then the uncertainty in position of the car is about equal to 5 meters. Thus we know the car's mass and uncertainty in position, and we can calculate the uncertainty in velocity from the uncertainty equation:

$$\text{(uncertainty in position)} \times \text{(uncertainty in velocity)} > \frac{h}{\text{mass}}.$$

First, we must rearrange this equation to solve for uncertainty of velocity:

$$\begin{aligned}\text{(uncertainty in velocity)} &> \frac{h/\text{mass}}{\text{uncertainty in position}}\\ &> (6.63 \times 10^{-34}\,\text{J-s}/1000\,\text{kg})/(5\,\text{m})\\ &> (6.63 \times 10^{-37})/5\,\text{J-s/kg-m}\\ &> 1.33 \times 10^{-37}\,\text{m/s}.\end{aligned}$$

Thus, the uncertainty in the velocity of the automobile is greater than 1.33×10^{-37} m/s (note that the unit J-s/kg-m is equivalent to m/s; see Problem 2 at the end of the chapter). This uncertainty is extremely small. Theoretically, we could know the velocity of the car to an accuracy of 37 decimal places! In practice, however, we have no method of measuring velocities with present or foreseeable human technology to an accuracy remotely approaching this. The uncertainty is, for all practical purposes, indistinguishable from zero. Therefore, for objects with significant mass such as automobiles, the effects of the uncertainty principle are totally negligible. The equation confirms our experience that Newtonian mechanics works perfectly well in dealing with everyday objects.

2. Large Uncertainty with a Small Object Contrast the above example with the uncertainty in velocity of an electron in an atom, located within an area about 10^{-10} meters on a side. To what accuracy can we measure the velocity of that electron?

The mass of an electron is 9.11×10^{-31} kg. If we take the uncertainty in position to be 10^{-10} meters, then according to the uncertainty principle:

$$\text{(uncertainty in velocity)} > \frac{h/\text{mass}}{\text{(uncertainty in position)}}$$

$$> \frac{(6.63 \times 10^{-34}\,\text{J-s})/(9.11 \times 10^{-31}\,\text{kg})}{10^{-10}\,\text{m}}$$

$$> 7.3 \times 10^6\,\text{m/s}.$$

This uncertainty is very large indeed. The mere fact that we know that an electron is somewhere in an atom means that we cannot know its velocity to within a million meters per second.

For ordinary-sized objects such as cars and bowling balls, whose mass is measured in kilograms, the number on the right side of the uncertainty relation is so small that we can treat it as being zero. Only when the masses get very small, as they do for particles such as the electron, does the number on the right get big enough to make a practical difference. ■

Probabilities

The uncertainty principle has consequences that go far beyond simple statements about measurement. In the quantum world we must radically change the way that we describe events. Consider an everyday example in which the uncertainties are much larger (but easier to picture) than those associated with Heisenberg's equation. Think of the example we used at the beginning of this chapter: a batter hitting a ball during a nighttime baseball game, with the stadium lights going out at the moment of impact. Where will the ball be in five seconds?

If you were an outfielder, this would be more than a philosophical question. You would need to know where to go to make your catch, even in the dark. In a Newtonian world, you would have no problem in doing this. If you knew the position and velocity of the ball at the instant the lights went out, some simple calculations would tell you exactly where the ball would be at any time in the future.

If you were a quantum outfielder in an atom-sized ball field, on the other hand, you would have a much harder time of it. You couldn't know both the position and velocity of the quantum ball when the lights go out—at best you could put some bounds on them. You might, for example, be able to say something like "It's somewhere inside this three-foot circle and traveling between 30 and 70 feet per second." This means that when you have to guess where it would be in five seconds, you wouldn't be able to do so with any precision. If you were thinking in Newtonian terms, you would have to say that the ball could be 147 feet from the plate (if it were traveling 30 feet per second and located at the back of the three-foot circle), 353 feet from the plate (if it were traveling 70 feet per second and located at the front of the circle), or anyplace in between. The best you could do would be to predict the likelihood, or **probability,** that the ball would be anywhere in the outfield, and you could present these probabilities on a graph like the one shown in Figure 8–1.

This example shows that the uncertainty principle requires a description of quantum-scale events in terms of probabilities. Just like the baseball in our example of the darkened stadium, there must be uncertainties

Figure 8–1

The position of a "quantum baseball" cannot be determined precisely. Instead, you can predict only the probabilities of the ball being various distances from home plate, as discussed in the text. The most likely location is at the peak of the curve, but the ball could be anywhere else.

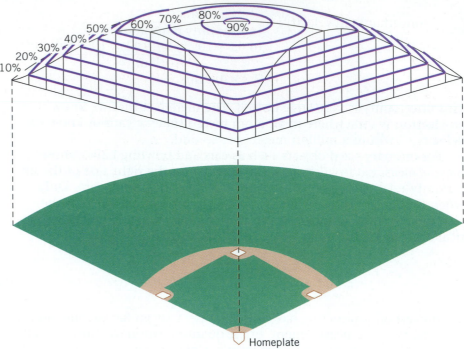

in the position and velocity for every quantum object when we first start observing it, and hence there will be uncertainties at the end— uncertainties that can be dealt with by reporting probabilities.

This result is extremely important. It tells us that we cannot think of quantum events in the same way that we think of normal events in our everyday world. In particular, we have to rethink what it means to talk about concepts such as regularity, predictability, and causality at the quantum level.

Wave-Particle Duality

Quantum mechanics is sometimes called *wave mechanics* because it turns out that quantum objects sometimes act like particles, and sometimes like waves. This dichotomy is known as the problem of "wave-particle duality," and it has puzzled some of the best minds in science. To understand it, think about how particles and waves behave in our macroscopic world.

The Double-Slit Test

Energy travels either as a wave or as a particle in our everyday world (see Chapter 6). Particles transfer energy through collisions, while waves transfer energy through collective motion of the media or electromagnetic fields. Every aspect of the everyday world can be neatly divided into particles or waves, and many experiments can be used to determine whether something is a particle or wave. The most famous of these experiments uses a double-slit apparatus, which consists of a barrier that has two slits in it (Figure 8–2). If particles such as baseballs are thrown from the left side, a few will make it through the slits but most will bounce off. If you were standing on the other side of the barrier you would expect to see the baseballs coming through more or less in the two places shown, accumulating in two piles behind the barrier. You wouldn't expect to see many particles (baseballs) at the spots between the slits.

If, however, waves of water were coming from the left side, you would expect to see the results of constructive and destructive interference (see Chapter 6). Rather than the two piles of baseballs, we would see perhaps half a dozen regions of high waves beyond the barrier, interspersed with regions of still water.

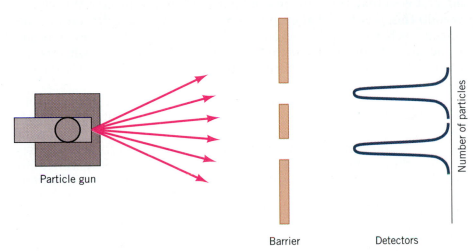

Particle gun

Barrier   Detectors

Number of particles

(a)

Figure 8–2

(*a*) The two-slit experiment may be used to determine whether something is a wave or a particle. A stream of particles striking the barrier will accumulate in the two regions directly behind the slits. (*b*) When waves converge on two narrow slits, however, constructive and destructive interference results in a series of peaks.

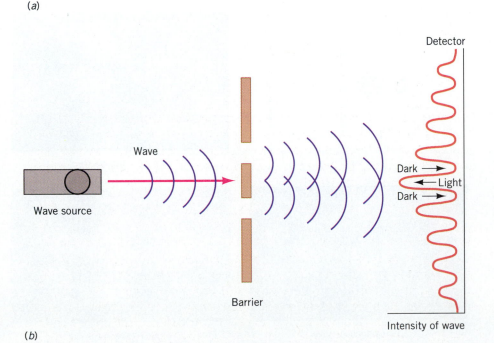

Detector

Wave

Wave source

Dark →
← Light
Dark →

Barrier

Intensity of wave

(b)

Now, let's use the same arrangement to see whether light behaves as a particle or a wave. In Chapter 7 we learned that light is emitted in discrete bundles of energy called *photons*. On the one hand, photons behave like particles in the sense that they can be localized in space. You can set up experiments in which a photon is emitted at one point, then received somewhere else after an appropriate lapse of time, just as a baseball is "emitted" by a pitcher and "received" later by a catcher (Figure 8–2a). If, on the other hand, you shine light—a flood of photons—on the two-slit apparatus, you will definitely get an interference pattern on the right (Figure 8–2b). In that experiment, photons act like waves. The big question: How can photons sometimes act like waves and sometimes act like particles?

You can make the problem more puzzling still by setting up the apparatus so that only one photon at a time comes through the slits. If you do this, you find that each photon arrives at a specific point at the screen—behavior you would expect of a particle. If you allow photons to accumulate over long periods of time, however, they will arrange themselves into an interference pattern characteristic of a wave (Figure 8–3).

You could do a similar series of experiments with any quantum object—electrons, for example, or even atoms. They all exhibit the properties of both particles and waves, depending on what sort of experiment is done. If you perform an experiment that tests the particle properties of these things, they look like particles. If you perform an experiment to test their wave properties, they look like waves. Whether you see quantum objects as particles or waves seems to depend on the experiment that you do.

Figure 8–3

When electrons or photons (light particles) pass through a two-slit apparatus one at a time (a), they cause 100 single spots on a photographic film (b). As the number of electrons increases to 3000 (c), and then to 70,000 (d), a wave-like interference pattern emerges. The bright areas are places where constructive interference occurs, and the dark areas correspond to destructive interference.

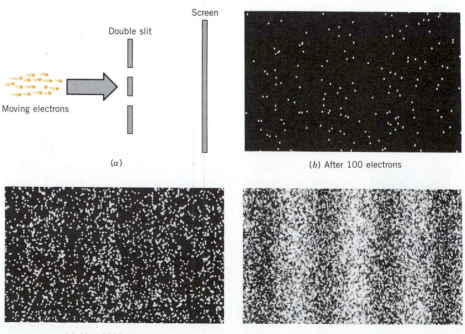

(a)

(b) After 100 electrons

(c) After 3000 electrons

(d) After 70 000 electrons

Some experimenters have gone so far as to try to "trick" quantum particles such as electrons into revealing their true identity by using modern fast electronics to decide whether a particle- or wave-type experiment is being done *after* the quantum object is already on its way into the apparatus. Scientists who do these experiments find that the quantum object seems to "know" what experiment is being done, because the particle experiments always turn up particle properties, while the wave experiments always turn up wave properties.

At the quantum level the objects that we talk about are neither particles nor waves in the classical sense. In fact, we can't really visualize them at all, because we have never encountered anything like them in our everyday experience. They are a third kind of object, neither particle nor wave, but exhibiting the properties of both. If you persist in thinking about them as if they were baseballs or surf coming into a beach, you will quickly lose yourself in confusion.

It's a little bit like finding someone who has seen only the colors red and green in her entire life. If she has decided that everything in the world has to be either red or green, she will be totally confused by seeing the color blue. What she has to realize is that the problem is not in nature, but in her assumption that everything has to be either red or green.

In the same way, the problem of wave-particle duality arises from our assumption that everything has to be either a wave or a particle. If we allow ourselves the possibility that quantum objects are things that we have never encountered before, and that they therefore might have unencountered properties, the puzzle vanishes. It only vanishes, however, if we agree that we won't try to draw a picture of them or pretend that we can actually visualize what they are.

Technology

The Photoelectric Effect

When photons strike some materials, their energy can be absorbed by electrons, which are shaken loose from their home atoms. If the material in question is in the form of a thin sheet, then when light strikes one side, electrons are observed coming out of the other. This phenomenon is called the *photoelectric effect*.

One aspect of the photoelectric effect played a major role in the history of quantum mechanics. The time between the arrival of the light and the appearance of the electrons is extremely short—far too short to be explained by the relatively gentle action of a wave nudging the electrons loose. In fact, it was Albert Einstein who pointed out that the explanation of this rapid response depended on the particle-like nature of the photon. He argued that the interaction between the light and the electron is something like the collision between two billiard balls, with one ball shooting out instantly after the collision. It was this work, which led to our modern concept of the photon, that was the basis of Einstein's Nobel Prize in 1921.

The conversion of light energy into electrical current is used in many familiar devices. In your camera, for example, a photoelectric device measures the amount of light available. The amount of electric current produced by the light is used to determine how wide to open the lens and what the shutter speed should be. In telephone systems that use fiber optics—glass fibers that act like pipes for visible light—light signals strike sophisticated semiconductor devices (see Chapter 10) and shake loose electrons. These electrons form a current that ultimately drives the diaphragm in your telephone and produces the sound that you hear. In CAT scans (see "The Human Body" section in this chapter), photoelectric devices convert X-ray photons into electrical currents whose strength can be used to produce a picture of a patient's internal organs. As all of these examples show, an understanding of the way that objects interact in the quantum world can have enormous practical consequences. ■

The Human Body

The CAT Scan

Ordinary X-ray photographs depend on the differences in density (and therefore in the ability to absorb X-rays) of the various materials in the body. In these photographs, the X-rays make one pass through, in one direction only, to produce the familiar pictures. They cannot produce a three-dimensional image of the interior of the body, nor can they produce sharp images of organs whose densities are not significantly different from the densities of their surroundings. These shortcomings are overcome by a different X-ray technique known as computerized axial tomography, or CAT.

The easiest way to visualize a CAT scan is to imagine dividing the body into slices perpendicular to the backbone, with each slice being a millimeter or so in width. The material in each slice is probed by successive short bursts of X-rays, lasting only a few milliseconds each, that cross the slice in different directions. Each part of the slide is thus traversed by many different X-ray bursts. Each burst of X-rays contains the same number of photons when it starts, and the number that go all the way through the body (i.e., that are not absorbed by material along their path) is measured by a photoelectric device (see the "Technology" section above).

Once all the data on a given slice have been obtained, a computer works out the density of each point of the body and produces a detailed cross-section along that particular slice. A complete picture of the body (or a specific part of it) can then be built up by looking at successive slices. ■

Wave-Particle Duality and the Bohr Atom

Treating electrons as waves helps explain why only certain orbits are allowed in atoms (see Chapter 7). Every quantum object displays a simple relationship between its speed (when we think of it as a particle) and its wavelength (when we think of it as a wave). It turns out that for electrons,

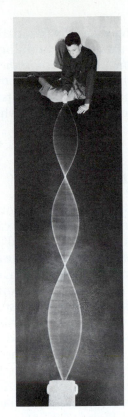

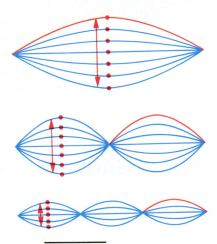

Figure 8–4

A vibrating string adopts a regular pattern, known as a standing wave. These photos and diagrams illustrate fixed patterns with 1/2, 1, and 3/2 wavelengths.

protons, and other quantum objects, a faster speed always corresponds to a shorter, more energetic wavelength (or a higher frequency).

If you think of an electron as a particle, then you can treat its motion around an atom's nucleus in the same Newtonian way you treat the motion of the Earth in orbit around the Sun. That is, for any given distance from the nucleus, the electron must have a precise velocity to stay in a stable orbit. Provided it is moving at this velocity, it will stay in that orbit just as the Earth stays in a stable orbit around the Sun. Any faster and it must adopt a higher orbit; any slower and it will move closer to the nucleus.

If we now choose to think about the electron as a wave, however, a different set of criteria can be used to decide how to put the electron into its orbit. A wave on a straight string (on a guitar, for example) vibrates only at certain frequencies that depend on the length of the string (Figure 8–4). These frequencies correspond to fitting 1/2, 1, and 3/2 wavelengths on the string in the figure. Now imagine bending the guitar string around into a circular orbit. In this case, you will only be able to fit certain standing waves in the orbit, as shown in Figure 8–5.

You can now ask a simple question: Are there any orbits for which the wave and particle descriptions are consistent? Are there, in other words, orbits for which the velocity of the electron (when we think of it as a particle) is appropriate to the orbit, while at the same time the electron wave (when we think of it as a wave) fits onto the orbit, given the relation between wavelength and velocity?

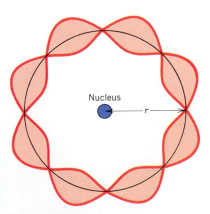

Figure 8–5

An electron in orbit about an atom adopts a standing wave like a vibrating string. This illustration shows a standing wave with four wavelengths fit into the orbit's circumference.

When you do the mathematics, you find that the only orbits that satisfy these twin conditions are the Bohr orbits. In other words, *the only orbits that are allowed in the atom are those for which it makes no difference whether we think of the electron as a particle or a wave.* In a sense, then, the wave-particle duality exists in our minds, and not in nature—nature has arranged things so that what we think doesn't matter.

Quantum Weirdness

The fact that quantum objects behave so differently from objects in our everyday experience causes many people to worry that nature has somehow become "weird" at the subatomic level. The description of particles in terms of a wave defies our common sense. Situations in which a photon or electron seems to "know" how an apparatus will be arranged before the arranging is done seem wrong and unnatural. Many people, scientists and nonscientists alike, find the conclusions of quantum mechanics to be quite unsettling.

Albert Einstein argued that the reason quantum mechanics was so strange was that it was not complete. He felt that there must be some way of getting at the missing information about quantum objects—in effect, some way of measuring them without disturbing them. This way of looking at quantum mechanics eventually came to be called the "hidden variable theory." It holds that the variables we see, such as position and velocity, are not adequate to describe nature at the atomic scale. If only we could get all the information about quantum objects, Einstein believed, then we would be able to describe the quantum-scale universe exactly, just as we do classical objects. The weirdness would simply disappear. In effect, Einstein argued that we end up having to describe nature in terms of probabilities simply because we don't know enough about the quantum world.

Decades of research have not borne out Einstein's ideas. In fact, in the early 1960s physicists discovered that there are experiments that will give different results depending on whether something like an electron or photon is described by a wave function or by hidden variables. When these experiments are done, the results are unambiguous: the quantum mechanical predictions are correct. The success of quantum mechanical predictions means that we have to live with the description of the subatomic world in terms of probabilities. We can't get out of it simply by dreaming that some time in the future things will be different. The quantum world is what it is, and there's no way around it. If you want to play the quantum game, it seems, you have to play by the quantum rules.

This conclusion has led philosophers to think a great deal about what quantum mechanics tells us about the world. Some have argued, for example, that the advent of quantum mechanics means that ordinary causality no longer holds—that we can no longer predict how a physical system will evolve in time. For these philosophers, the comforting clockwork Newtonian world has vanished forever.

Most scientists (the authors included) do not hold with this view. The success of quantum mechanics is proof enough that there is a correct way to describe an atomic-scale system, and if you ignore this fact you can get into a lot of trouble. Newtonian notions like position and velocity just aren't appropriate for the quantum world, which must be described from the beginning in terms of waves and probabilities. Quantum mechanics, then, becomes a way of predicting how subatomic objects change in time. If you know the state of an electron now, you can use quantum mechanics to predict the state of that electron in the future. This process is identical to the application of Newton's laws of motion in the macroscopic world. The only difference is that in the quantum world the "state" of the system is a probability.

In the view of most working scientists, quantum mechanics is a marvelous tool that allows us to do all sorts of experiments and build all manner of new and important pieces of equipment. The fact that we can't visualize the quantum world in terms that are familiar to us seems a small price to pay for all the benefits we receive.

Science in the Making

A Famous Interchange

Many people are disturbed by the fact that nature must be described in terms of probabilities at the subatomic level. Many scientists were also disturbed when quantum mechanics was first developed in the early twentieth century. Even Albert Einstein, one of the founders of quantum mechanics, could not accept what it was telling us about the world. He spent a good part of the last half of his life trying to refute it. His most famous statement from this period was, "I cannot believe that God plays dice with the universe."

Confronted one too many times with this aphorism, Einstein's life-long friend and colleague Neils Bohr is supposed to have replied, "Albert, stop telling God what to do." ■

THINKING MORE ABOUT QUANTUM MECHANICS

Uncertainty and Human Beings

The ultimate Newtonian view of the universe was the concept of the Divine Calculator (see Chapter 2). This mythical being, given the position and velocity of every particle in the universe, could produce an exact prediction of every future state of those particles. The difficulty with this concept, of course, is that if the future of the universe is laid out with clockwork precision, it allows no room for human action. No one can make a choice about what he or she will do, because that choice is already determined and exists (in the mind of God or the Divine Calculator) before it is made.

Quantum mechanics gives us one way to get out of this particular bind. Heisenberg tells us that, although we might be able to predict the future if we knew the position and velocity of every particle exactly, we can never actually get those two numbers. The Divine Calculator in a quantum world is doomed to wait forever for the input data with which to start the calculation.

One area where the uncertainty principle is starting to play a somewhat unexpected role is in the old philosophical argument about the connection between the mind and the brain. The brain is a physical object, an incredibly complex organ that processes information in the form of nerve impulses. (A more detailed description of the workings of the brain is given in Chapter 10.) The problem: What is the connection between the physical reality of the brain—the atoms and structures that compose it—and the consciousness that we all experience?

Many scientists and philosophers have argued that there really is no more to either brain or mind than physical structures. They have traditionally run into a problem, however, because if the brain is purely a physical object, its future states should be predictable. What to do with human creativity, with human diversity, if this is the case?

Recently, scientists (most notably Roger Penrose of Cambridge University) have argued that quantum mechanics can introduce a kind of unpredictability, an unprogrammability, that squares better with our perceptions of our own minds. How could quantum mechanics do this? Is there anything about the probabilities of quantum objects that might make it difficult (or even impossible) to make precise predictions of the future state of the brain?

Summary

Matter and energy at the atomic scale come in discrete packets called quanta. The rules of *quantum mechanics,* the laws that allow us to describe and predict events in the quantum world, are disturbingly different from Newton's laws of motion.

At the quantum scale, unlike our everyday experience, any measurement of the position or velocity of a particle causes the particle to change in unpredictable ways. The mere act of measurement alters the thing being measured. Werner Heisenberg quantified this situation in the Heisenberg *uncertainty principle*, which states that the uncertainty in position of a particle multiplied times the uncertainty in its velocity must be greater than a small positive number. Unlike the Newtonian world, you can never know the exact position *and* velocity of a quantum particle.

These uncertainties preclude us from describing atomic-scale particles in the classical way, with precise position and velocity throughout time. Instead, quantum descriptions are given in terms of *probabilities* that an object will be in one state or another. Furthermore, quantum objects are not simply particles or waves, a dichotomy familiar to us in the macroscopic world. They represent something completely different from our experience, incorporating properties of both particles and waves.

Key Terms About Quantum Mechanics

quantum mechanics   uncertainty principle   probability

Key Equation

$$(\text{uncertainty in position}) \times (\text{uncertainty in velocity}) > \frac{h}{\text{mass}}$$

Review Questions

1. What does "quantum mechanics" mean?
2. Give three examples of properties that are quantized at the scale of an electron.
3. What are the three essential parts of every physical measurement?
4. In what way is a measurement at the quantum scale of an electron different from a measurement at large scales of everyday objects?
5. State Heisenberg's uncertainty principle.
6. Under what circumstances can you know the position of an electron with great accuracy?

7. Why is quantum mechanics sometimes called "wave mechanics"?

8. What is "wave-particle duality"?

Discussion Questions

1. Identify the sample, source of energy, and detector in the following experiments:
   a. Measuring the speed of a tennis ball.
   b. Observing bacteria in a microscope.
   c. Testing the electrical resistance of a piece of wire.
   d. Determining the force required to bend an iron bar.

2. Which properties of electrons are particle-like? Which are wavelike?

3. What role does probability play in description of subatomic events?

4. Sketch a possible probability diagram for the final resting position of a golf ball on a driving range. Assume that the golf tee is the starting point and that an average drive is 250 feet.

5. In Chapter 2 we discussed the fact that chaotic systems are, for all practical purposes, unpredictable. How does this sort of unpredictability differ from that associated with quantum mechanics?

6. Present an argument in terms of the wave nature of the electron that shows that electrons in Bohr orbits cannot emit radiation and spiral in toward the nucleus, as they might be expected to do on the basis of Maxwell's equations. (*Hint:* See "Why the Rutherford Atom Couldn't Work," in Chapter 7.)

7. How does light behave like a particle? How does it behave like a wave?

8. If you threw baseballs through a large two-slit apparatus, would you produce a diffraction pattern? Why or why not?

Problems

1. A ball (mass 0.1 kg) is thrown with a speed between 20.0 and 20.1 meters/sec. How accurately can we determine its position?

2. In "Science By the Numbers" we converted the unit, $\frac{\text{J-s}}{\text{kg-m}}$, to the unit of velocity (m/s) without comment. Demonstrate the equivalence of these two units.

3. An atom of iron (mass 10^{-25} kg) travels at a speed between 20.0 and 20.1 meters/sec. How accurately can we determine its position? Is the accuracy in either of the above situations attainable? How does it compare to the size of an atom? Of a nucleus?

Investigations

1. Look up the doctrine of predestination in an encyclopedia. Does it have a logical connection to the notion of the Divine Calculator? Which came first historically?

2. Werner Heisenberg was a central, and ultimately controversial, figure in German science of the 1930s and 1940s. Read a biography of Heisenberg. Discuss how his early work in quantum mechanics influenced his prominent scientific role in Nazi Germany.

3. What changes in artistic movements were taking place during the period 1900 (just before the discoveries of quantum mechanics) and the mid-twentieth century? Are there any connections between the artistic and scientific movements of those times?

4. Some people interpret the Heisenberg uncertainty principle to mean that you can never really know anything for certain. Would you agree or disagree?

Additional Reading

Cassidy, David. *The Life and Times of Werner Heisenberg.* New York: W. H. Freeman, 1992.

Casti, John L. *Paradigms Lost.* New York: William Morrow (1989).

Mermin, David. "Is the Moon Really There When Nobody Looks?", *Physics Today.* April, 1985, pp 38–47.

Penrose, Roger. *The Emperor's New Mind.* New York: Oxford University Press, 1992.

Trefil, James. *Quantum Physics Works.* Smithsonian Magazine, August, 1987.

Atoms in Combination: The Chemical Bond

Atoms bind together in chemical reactions by the rearrangement of electrons.

A Random Walk
Throwing Things Away

Think about all the things you've thrown away during the past month. Every day you toss out aluminum cans, plastic wrappers, glass jars, food scraps, and lots of paper. From time to time you also discard used batteries, disposable razors, dirty motor oil, worn-out shoes and clothes, even old tires or broken furniture. What happens to all that stuff after it becomes trash?

More and more, communities try to recycle much of their waste. Plastic, glass, aluminum, and newspaper, for example, can be reprocessed and turned into new products and packaging. Old oil can be reprocessed and tires retreaded. But most of the things you throw out must be put into landfills, where, we hope, they will eventually break down into soil.

The situation that faces our society is more than a little ironic. Everything with which you have contact in your daily life—everything you throw away—is made from collections of atoms bonded together. While they are in the store and for as long as we use them, we want these products and their packaging to

9

last and keep looking new. But as soon as we throw them out, we'd like our disposable materials to fall apart and disappear. One way to achieve this end is to engineer biodegradable materials—paper, plastics, fabrics, and other goods designed to break apart when thrown away.

But what holds materials together in the first place? Why do certain atoms, when brought close together, develop an affinity and stick to each other? How do the molecules that play such an important role in our life retain their identity? And how can we design new materials that will fall apart when their useful lives are over? The answers lie in the nature of the chemical bond.

Electron Shells and Chemical Bonding

Think about how two atoms might interact. You know that the atom is mostly empty space, with a tiny, dense nucleus surrounded by swift electrons. If two atoms approach each other, their outer electrons—the "border guards," if you will—encounter each other first. Whatever holds two atoms together thus involves primarily those outer electrons. In fact, the outer electrons play such an important role in determining how atoms combine that they are given the special name of *valence electrons* (see Chapter 7). Chemical bonding often involves an exchange or sharing of valence electrons, and the number of electrons in an atom's outermost orbit is called its *valence*. Chemists often express the importance of the number of outer electrons by saying that valence represents the combining power of a given atom.

It turns out that by far the most stable arrangement of electrons—the electron configuration of lowest energy—is a completely filled outer shell. It is a fact of chemical life that different electron shells hold different numbers of electrons, which gives rise to the structure of the periodic table of the elements (see Chapter 7). A glance at the periodic table tells us that atoms with a total of 2, 10, 18, or 36 electrons (the atoms that appear in the extreme right-hand column) have completely filled shells. Atoms with this many electrons in their outermost orbits are inert gases, which do not combine readily with other materials. Indeed, helium, neon, and argon, with atomic numbers 2, 10, and 18, and thus completely filled electron shells, are the only common elements ordinarily found as isolated atoms.

Every object in nature tries to reach a state of lowest energy, and atoms are no exception. Atoms that do not have filled electron shells are more likely to react with other atoms to produce a state of lower energy. You are familiar with this kind of process in many other natural systems. If you put a ball on top of a hill, for example, it will tend to roll down to the bottom, creating a system of lower gravitational potential energy. Similarly, a compass needle tends to align itself spontaneously with the Earth's magnetic field, thereby lowering its magnetic potential energy. In exactly the same way, when two or more atoms come together the electrons tend to rearrange themselves to minimize the chemical potential energy of the entire system. This situation may require that they exchange or share electrons, and, as often as not, that process involves creating completely filled or "closed" electron shells.

The **chemical bond** is the name we give to the attraction that results from the redistribution of electrons that leads to a more stable configuration between two or more atoms.

Most atoms adopt one of three simple strategies to achieve a filled shell: they give away electrons, accept electrons, or share electrons.

If the bond formation takes place spontaneously, without outside intervention, energy will be released in the reaction. The burning of wood or paper (once their temperature has been raised high enough) is a good example of this sort of process, and the heat you feel when you put your hands toward a fire derives ultimately from the chemical potential energy that is given off as electrons are reshuffled. Alternatively, atoms may be pushed into new configurations by adding energy to systems. Much of industrial chemistry, from the smelting of iron to the synthesis of plastics, operates on this principle.

Types of Chemical Bonds

Atoms link together by three principal kinds of chemical bonds—ionic, metallic, and covalent—all of which involve redistributing electrons between atoms. In addition, three types of attractions—polarization, van der Waals interactions, and hydrogen bonding—result from shifts of electrons within their atoms or groups of atoms. Each type of bonding or attraction corresponds to a different way of rearranging electrons and each produces distinctive properties in the materials it forms.

Ionic Bonds

We have seen that atoms with filled outer electron shells are particularly stable. By the same token, atoms with only one electron or one missing electron in their outer shells are particularly reactive—in effect, they are "anxious" to fill their outer shells. Such atoms tend to form **ionic bonds**—chemical bonds in which the electrical force between two oppositely charged ions holds the atoms together.

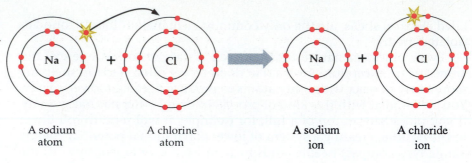

A sodium   A chlorine   **A sodium**   **A chloride**
atom   atom   **ion**   **ion**

Figure 9–1

Sodium, a highly reactive element, readily transfers its single valence electron to chlorine, which is one electron shy of a filled shell. The result is the ionic compound sodium chloride—ordinary table salt. In these diagrams, electrons are represented as dots in shells. The inner shell can contain no more than two electrons, while the second and third shells are limited to eight electrons.

Ionic bonds often form as one atom gives up an electron while another receives it. Sodium (a soft, silvery white metal), for example, has 11 electrons in an electrically neutral atom, 2 in the lowest orbit, 8 in the next, and a single electron in its outer shell. Sodium's best bonding strategy, therefore, is to lose one electron. The seventeenth element, chlorine (a yellow-green toxic gas), on the other hand, is one electron shy of a filled shell. Highly corrosive chlorine gas will react with almost anything that can give it an extra electron. When you place sodium in contact with chlorine gas, the result is predictable: in a fiery reaction, each sodium atom donates its extra electron to a chlorine atom (see Figure 9–1).

In the process of this vigorous electron exchange, atoms of sodium and chlorine become electrically charged—they become ions. Neutral sodium has 11 positive protons in its nucleus, balanced by 11 negative electrons in orbit. By losing an electron, sodium becomes an ion with one unit of positive charge, shown as Na^+ in Figure 9–1. Similarly, neutral chlorine has 17 protons and 17 electrons. The addition of an extra negative electron creates a chloride ion with one unit of negative charge, shown as Cl^- in the figure. The mutual electrical attraction of positive sodium and negative chloride ions is what forms the ionic bonds between sodium and chlorine. The resulting compound, sodium chloride, or common table salt, has properties totally different from either sodium or chlorine.

Under normal circumstances sodium and chlorine ions will lock together into a crystal, a regular arrangement of atoms such as the one shown in Figure 9–2. Alternating sodium and chloride ions form an elegant repeating structure in which each Na^+ is surrounded by six Cl^-, and vice versa.

Ionic bonds may involve more than a single electron transfer. The twelfth element, magnesium, for example, donates two electrons to oxygen, which has eight electrons. In the resulting compound, MgO (magnesium oxide) both atoms have stable filled shells of 10 electrons, and the ions, Mg^{2+} and O^{2-}, form a strong ionic bond. Ionic bonds involving the negative oxygen ion O^{2-} and positive ions such as calcium (Ca^{2+}), magnesium (Mg^{2+}), silicon (Si^{4+}), and iron (Fe^{2+}), are found in many everyday objects: in most rocks and minerals, in ceramics such as china and glass, and in bones and egg shells.

Ionic bonds in these compounds can be very strong, but only in certain ways. You can picture how this works by thinking about Tinkertoys. Tinkertoy structures can be quite strong; when assembled, it is difficult to break one apart by just pushing in the directions of the sticks.

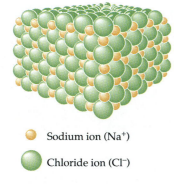

○   Sodium ion (Na^+)

●   Chloride ion (Cl^-)

Figure 9–2

The atomic structure of a sodium chloride crystal consists of a regular pattern of alternating sodium and chloride ions.

But Tinkertoy bonds break easily by twisting or snapping the sticks. In the same way, ionic bonds hold atoms together, but if the atoms should for some reason become displaced, the bond can't hold them very well. As a consequence, ionic-bonded materials such as rock, glass, or egg shells are usually quite brittle. These materials are strong in the sense that you can pile lots of weight on them. But once they shatter—once the ionic bond has been broken—they can't be put back together again.

EXAMPLE 9–1: Ionic Bonding of Three Atoms

Calcium chloride, which plays an important role as an industrial desiccant (a substance that speeds drying), is an ionic-bonded compound with one part calcium to two parts chlorine ($CaCl_2$). How are the electrons arranged in this compound?

▶   **Reasoning:**   From the periodic table (see Chapter 7), calcium and chlorine are elements 20 and 17, respectively. Calcium, therefore, has 18 electrons (2 + 8 + 8) in inner shells, and 2 valence electrons. Chlorine has 10 electrons (2 + 8) in its inner shells and 7 electrons in the outer one, meaning that it is 1 electron short of a filled outer shell (see Figure 9–3).

▶   **Solution:**   Calcium has two electrons to give, and chlorine seeks one electron, to achieve stable filled outer shells. Therefore, calcium gives one electron to each of two chlorine atoms, and the resulting Ca^{2+} ion attracts two Cl^- ions to form $CaCl_2$.   ▲

Metallic Bonds

Atoms in an ionic bond transfer electrons directly—electrons are on "permanent loan" from one atom to another. Atoms in a metal also give up electrons, but they use a very different bonding strategy. In the

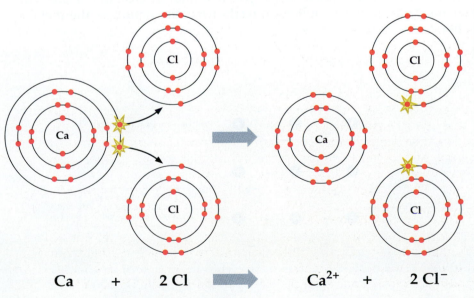

$$Ca \quad + \quad 2\,Cl \quad \Longrightarrow \quad Ca^{2+} \quad + \quad 2\,Cl^-$$

Figure 9–3
Calcium and chlorine neutral-atom electron configurations.

metallic bond, electrons are redistributed so that they are shared by all the atoms as a whole.

Sodium metal, for example, is made up entirely of individual sodium atoms. All of these atoms begin with 11 electrons, but they release 1 to achieve the more stable 10-electron configuration. The extra electrons move away from their parent atoms to float around the metal, forming a kind of sea of negative charge. In the negative electron sea, the positive sodium ions adopt a regular crystal structure, as shown in Figure 9–4.

You can think of the metallic bond as one in which each atom shares its outer electron with all the other atoms in the system. Picture the free electrons as a kind of loose glue in which the metal atoms are placed. In fact, the idea of a metal as being a collection of marbles (the ions) in a sea of stiff gluelike liquid provides a useful analogy.

Metals, characterized by their shiny luster and ability to conduct electricity, are formed by almost any element or combination of elements in which large numbers of atoms share electrons to achieve a more stable electron arrangement. Some metals, such as aluminum, iron, and copper, are familiar from everyday experience. But many elements can form into a metallic state when the conditions are right, including some that we normally think of as gases, such as hydrogen or oxygen at very high pressure. In fact, the great majority of chemical elements are known to occur in the metallic state. In addition, two or more elements can combine to form a metal *alloy,* such as brass (a mixture of copper and zinc) or bronze (an alloy of copper and tin). Modern specialty steel alloys often contain more than half a dozen different elements in carefully controlled proportions.

The special nature of the metallic bond explains many of the distinctive properties we observe in metals. If you attempt to deform a metal by pushing on the marble-and-glue bonding system, things will gradually rearrange themselves and come to some new configuration. It is very hard to shatter a metallic bond just by pushing or twisting, because the atoms are able to rearrange themselves. Thus when you hammer on a piece of metal, you leave indentations but do not break it, in sharp contrast to what happens when you hammer on a ceramic plate. This malleability of metals follows directly from the nature of the metallic bond.

Figure 9–4

Metallic bonding, in which a bond is created by the sharing of electrons between individual metal atoms.

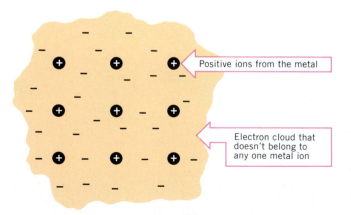

Positive ions from the metal

Electron cloud that doesn't belong to any one metal ion

In Chapter 10, we will examine more closely the electrical properties of materials held together by the metallic bond. We will see that this particular kind of bond produces materials through which electrons—electrical current—can flow.

Covalent Bonds

In the ionic bond, one atom donates electrons to another on more-or-less permanent loan. In the metallic bond, on the other hand, atoms share some electrons throughout the material. In between these two types of bond is the extremely important **covalent bond,** in which well-defined clusters of neighboring atoms, called **molecules,** share electrons. These strongly bonded groups may consist of anywhere from two to many millions of atoms.

The simplest covalently bonded molecules contain two atoms of the same element, such as the diatomic gases hydrogen (H_2), nitrogen (N_2), and oxygen (O_2). In the case of hydrogen, for example, each atom has a single electron, a relatively unstable situation. Two hydrogen atoms can pool their electrons, however, to create a much more stable two-electron arrangement. The two hydrogen atoms must remain close to each other for this sharing to continue, so a chemical bond is formed, as shown in Figure 9–5. Similarly, two oxygen atoms, each with eight electrons, share pairs of electrons.

Hydrogen, oxygen, nitrogen, and other covalently bonded molecules have lower chemical potential energy than isolated atoms because electrons are shared. This lower chemical potential energy means these molecules are less likely to react chemically than the isolated atoms.

The most fascinating of all covalently bonded elements is carbon, the basis of all life on Earth. Carbon, with two electrons in its inner shell and four in its outer shell, presents a classic case of a half-filled shell. When carbon atoms approach each other, therefore, a real question arises as to whether they ought to accept or donate four electrons to achieve a more stable arrangement. You could imagine, for example, a situation where some carbon atoms give four electrons to their neighbors, while

Figure 9–5

Two hydrogen atoms become an H_2 molecule by sharing each of their electrons in a covalent bond. This bonding may be represented schematically in a dot diagram (*a*), or by the merging of two atoms with their electron clouds (*b*).

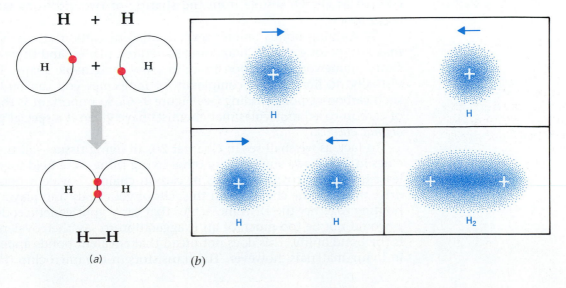

(a)   (b)

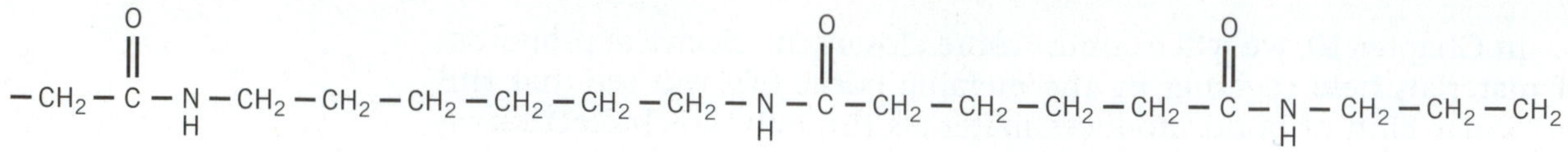

(a)

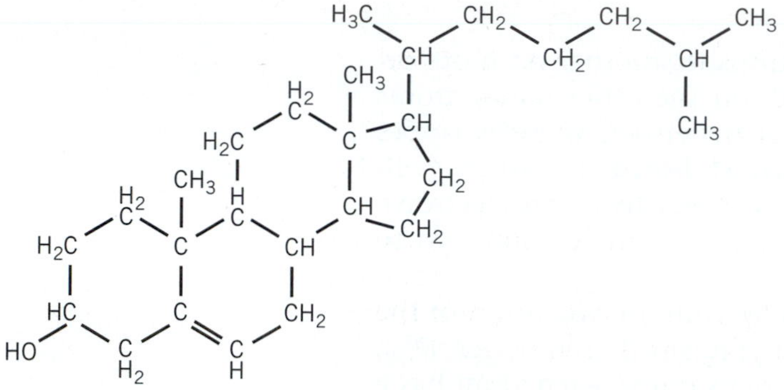

(b)

Figure 9–6

Carbon-based molecules may adopt almost any shape. The molecules may consist of long straight chains of carbon atoms that lead to fibrous materials such as nylon (a), or they may incorporate complex rings and branching arrangements that form lumpy molecules such as cholesterol (b).

other carbon atoms accept four electrons, to create a compound with extremely strong ionic bonds between C^{4+} and C^{4-}. Alternatively, carbon might become a metal in which every atom releases four electrons into an extremely dense electron sea. But neither of these things happens.

In fact, the strategy that lowers the energy of the carbon-carbon system the most is for the carbon atoms to share their outer electrons. Once bonds between carbon atoms have formed, the atoms have to stay close to each other for the sharing to continue. Thus the bonds generated are just like the bond in the case of hydrogen. The case of carbon is unusual, however, because the shape of its electron orbits allows a single carbon atom to form covalent bonds with up to four other atoms by sharing one of its four valence electrons with each. A *single bond* (shown as H—H) is formed when only one electron is shared, while a *double bond* (shown as O=O) results from the sharing of two electrons with one other atom.

By forming bonds among several adjacent carbon atoms, you can make rings, long chains, branching structures, planes, and three-dimensional frameworks of carbon in almost any imaginable shape. There is virtually no limit to the complexity of molecules you can build from such carbon-carbon bonding (see Figure 9–6). So important is the study of carbon-based molecules that chemists have given it a special name—*organic chemistry.*

In fact, as we shall see in Chapter 20, all living tissue—all the molecules in your body and in every other living thing—are held together at least in part by covalent bonds in carbon chains. Covalent bonds also drive much of the chemistry in the cells of your body and play a role in holding together the DNA molecules that carry your genetic code. Thus it would not be too much of an exaggeration to say the covalent bond is the bond of life. This does not mean that covalent bonds appear only in living materials, however. The transistors in the microchip that runs your computer are made from covalently bonded silicon, an element, like carbon, that has four electrons in its outer orbit. Like ionic and metallic bonds, covalent bonds can vary in strength. They hold together human tissue, but they also hold together the diamond, the hardest substance known.

Polarization and Hydrogen Bonds

Ionic, metallic, and covalent bonds form strong links between individual atoms, but molecules also experience forces that hold one to another. In many molecules the electrical forces are such that, although the molecule by itself is electrically neutral, one part of the molecule has more positive or negative charge than another. In water, for example, the electrons tend to spend more time around the oxygen atom than around the hydrogens. This uneven electron distribution has the effect of making the oxygen side of the water molecule more negatively charged, and the two "Mickey Mouse ears" of the hydrogen atom more positively charged (see Figure 9–7). Atom clusters of this type, with a positive and negative end, are called *polar molecules*.

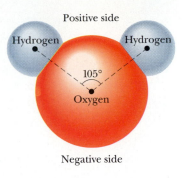

Figure 9–7

The water molecule and its polarity.

The electrons of an atom or molecule brought near a polar molecule such as water will tend to be pushed away from the negative side and shifted toward the positive side. Thus the side of an atom facing the negative end of a polar molecule will become slightly positive itself. This subtle electron shift, called *polarization*, in turn will give rise to an electrical attraction between the negative end of the polar molecule and the positive side of the other molecule. The electron movement thus creates an attraction between the atom and the molecule, even though the atoms and molecules in this scheme all may be electrically neutral.

One of the most important consequences of forces due to polarization is the ability of water to dissolve many materials. Water, made up of strongly polar molecules, exerts forces that tend to pull ions like Na^+ and Cl^- out of crystals.

A process related to the forces of polarization leads to the **hydrogen bond,** a weak bond that may form after a hydrogen atom links to an atom of certain other elements (oxygen or nitrogen, for example) by a covalent bond. Because of the kind of rearrangement of electrical charge described above, the hydrogen may become polarized and develop a slight positive charge, which attracts another atom to it. You can think of the hydrogen atom as a kind of bridge in this situation, causing a redistribution of electrons that, in turn, holds the larger atoms or molecules together. Individual hydrogen bonds are weak, but in many molecules they occur repeatedly and therefore play a major role in determining the molecule's shape and function. Note that while all hydrogen bonds require hydrogen atoms, not all hydrogen atoms are involved in hydrogen bonds.

Hydrogen bonds are common in virtually all biological substances, from everyday materials such as wood, plastics, silk, and candle wax, to the complex structures of every cell in your body. As we shall see in Chapter 21, hydrogen bonds in every living thing link the two sides of the DNA double helix together, although the sides themselves are held together by covalent bonds. Ordinary egg white is made from molecules whose shape is determined by hydrogen bonds, and when you heat the material—when you fry an egg, for example—hydrogen bonds are broken and the molecules rearrange themselves so that instead of a clear liquid you have a white gelatinous solid.

van der Waals Forces

The hydrogen bond exists because atoms or molecules can become polarized—their electrons can shift to one side or another and thus create local electrical charges. In the molecules we've discussed so far, that electrical charge is more or less permanently locked into polar molecules in a fixed or static arrangement. Another attractive bond, called the van der Waals force, results from the polarization of electrically neutral atoms or molecules; in this case the atoms and molecules are not themselves polar.

When two atoms or molecules are brought near each other, every part of one atom or molecule feels an electrical force exerted on it by all parts of the other. For example, an electron in one atom will be repelled by the electrons, but attracted to the nucleus, of an adjacent atom. The net result of these forces exerted on the electron may be a temporary shift of the electron. The same thing happens to every electron in any nearby atom or molecule, and the net result is that every electron is constantly shifting because of the presence of others.

While it may not seem obvious that this mutual, dynamic deformation might give rise to a net attractive force, that is exactly what happens in some molecular materials. In these compounds, even if all the molecules are neutral and nonpolar, the sum of attractive forces wins out over repulsive forces and weak bonds are formed. This weak force that binds two atoms or molecules together is called **van der Waals force.**

If you take a piece of clay and rub it between your fingers, your fingers pick up a slick coating of the material, even though the clay crumbles easily. The reason is that the clay is made up of sheets of atoms. Within each sheet, atoms are held together by strong ionic and covalent bonds. One sheet is held to another, however, by comparatively weak van der Waals forces. This situation is not unlike the way a stack of photocopying paper will stick together on a dry day. Each sheet of paper is strong, but the stack of paper is stuck together by much weaker electrostatic forces. It is easy to pull the stack apart, but very difficult to rip the stack in two. When you crumble clay in your fingers, therefore, you are breaking weak van der Waals bonding between layers, but preserving the stronger bonds that hold each layer together. The clay stains on your hands are thin sheets of atoms, held together by ionic and covalent bonds, and the crumbling you feel is the breaking of the van der Waals bonds.

Many other examples of van der Waals forces can be seen in your everyday life. If you rub talcum powder on your body, for example, you use a pure white material not unlike like the clay discussed above—that is, material in which strongly bonded sheets are held together by van der Waals forces. Similarly, when you write with a "lead" pencil, van der Waals–bonded layers of graphite—a form of carbon—are transferred from pencil to paper (see Figure 9–8). As you draw the pencil across the paper, you break the van der Waals forces and leave behind sheets of graphite in a dark streak across the paper. Van der Waals forces also hold together molecules in many everyday liquids and soft solids, from candle wax and soap to Vaseline and other petroleum products.

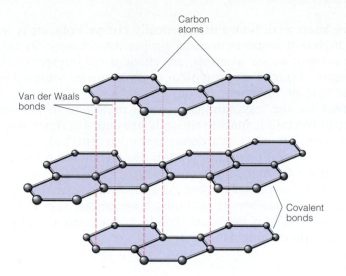

Carbon
atoms

Van der Waals
bonds

Covalent
bonds

Figure 9–8
Graphite, a form of carbon that serves as the "lead" in your pencil, contains layers of carbon atoms strongly linked to each other by covalent bonds (represented by solid lines). These layers are held to each other by much weaker van der Waals bonds (represented by dashed lines).

States of Matter

So far, we have been talking about the way that the limited number of chemical elements in the periodic table can be locked together to give materials with many different properties. But chemical bonds alone cannot explain everything about the variety of materials in our world. All of our experience comes from interactions with large groups of atoms. Depending on how these groups of atoms are organized, they may take on many different forms. These different modes of organization, called the **states of matter,** include gases, plasmas, liquids, and solids.

Gases

A **gas** is any collection of atoms or molecules that expands to take the shape of and fill the volume available in its container. Most common gases, including those that form our atmosphere, are invisible, but the force of a gust of wind is proof that matter is involved. The individual particles that comprise a gas may be isolated atoms such as helium or neon, or small molecules such as nitrogen (N_2) or carbon dioxide (CO_2).

If we could magnify an ordinary gas a billion times, we would see these particles randomly flying about, bouncing off each other and anything else they contact. The gas pressure that inflates a basketball or tire is a consequence of these countless collisions. If we pump in more air, or heat the tire, the number of collisions, and thus the pressure, will go up.

Plasma

At the extreme temperatures characteristic of the Sun, high-energy collisions between atoms may strip off electrons, creating a new state of matter—a **plasma**—in which positive nuclei move about in a sea of electrons. Such a collection of electrically charged objects is something like a gas, but it displays unusual properties not seen in other states of matter. Plasmas, for example, are efficient conductors of electricity. Furthermore, though they are gaslike, plasmas can be confined in a strong magnetic field or "magnetic bottle."

Plasmas are the least familiar state of matter to us, yet more than 99.9% of all the visible mass in the universe exists in this form. Not only are most stars composed of a dense hydrogen- and helium-rich plasma mixture, but several planets, including the Earth, have regions of thin plasma in their outer atmospheres.

Some gradations exist between gas and plasma. Partially ionized gases in neon lights or fluorescent light bulbs, for example, have a small fraction of their electrons in a free state. While not a complete plasma, these ionized gases do conduct electricity.

Liquids

Any collection of atoms or molecules that has no fixed shape but maintains a fixed volume is called a **liquid.** Other than water and biological fluids, there are few naturally occurring liquids on Earth. Water, by far the most abundant liquid on the Earth's surface, is a dynamic force for geological change (see Chapter 16), and water-based solutions are essential to all life.

At the molecular level, liquids behave something like a container full of sand grains. The grains fill whatever volume they are poured into, freely flowing over each other without ever taking on a fixed shape. Attractive forces between individual atoms or molecules hold the liquid together. At the surface of the liquid, these attractive forces act to prevent atoms or molecules from escaping. In effect, they pull the surface in, giving rise to *surface tension,* the property that causes small quantities of the liquid to form beads or droplets.

Solids

Solids include all materials that possess a more or less fixed shape and volume. In all solid materials the chemical bonds are both sufficiently strong and directional to preserve a large-scale external form. In detail, however, solids adopt several quite different kinds of atomic structures.

In **crystals,** groups of atoms occur in a regularly repeating sequence, the same atom or atoms appearing over and over again in a predictable way. A crystal structure can be described by first determining the size and shape of the tiny boxlike unit that repeats, and then recording the exact type and position of every atom that appears in the box. In common salt (see Figure 9–9), for example, the box is a tiny cube less than a billionth of a meter on an edge, and each box contains sodium atoms at the cube center and corners, and chlorine atoms at the center of every face. The regular atomic structure of crystals often leads to large single crystals with beautiful flat faces.

Common single crystals include grains of sand and salt, computer chips, and the gemstones in jewelry. Most crystalline solids, however, are composed of numerous interlocking crystal grains. The two most important groups of these types of materials in our everyday life are

Figure 9–9

Individual salt crystals reveal a characteristic cube shape.

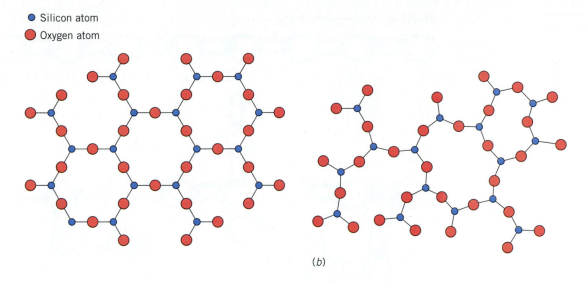

● Silicon atom
● Oxygen atom

(b)

metals and alloys that are characterized by metallic bonds, and most *ceramics,* a broad class of hard, durable solids, including bricks, concrete, pottery, porcelain, and numerous synthetic abrasives, as well as teeth and bones and most rocks and minerals.

Glasses, in contrast to crystals, are solids with predictable local environments for most atoms, but no long-range order to the atomic structure (see Figure 9–10). In most common window and bottle glasses, for example, the most abundant elements, silicon and oxygen, form a strong three-dimensional framework. Every silicon atom is surrounded by a fixed number of oxygen atoms, and most oxygens are linked to two silicons. If you were placed on any atom in a piece of glass, chances are you could predict the next-door atoms. Nevertheless, glasses have no regularly stacked boxes of structure. Travel more than two or three atom diameters from any starting point, and there is no way that you could predict whether you'd find a silicon or an oxygen atom.

Plastics form a third important class of solid materials, defined primarily by their ability to be molded or formed into virtually any desired shape. Though almost unknown a few decades ago, plastics have become our most versatile commercial materials, providing an extraordinary range of uses: clear films for lightweight packaging, dense castings for durable machine parts, thin strong fibers for clothing, colorful moldings for toys, and many others. While most plastics are employed in packaging and construction, they also serve as paints, inks, glues, sealants, foam products, and insulation. New kinds of tough, resilient plastics have revolutionized many sports with products such as high-quality bowling and golf balls, and durable football and ice hockey helmets, not to mention a host of completely new products from Frisbees to roller blades.

Plastics, along with numerous biological molecules such as animal hair, plant cellulose, cotton, and spider's webs, are examples of **polymers**—solids composed of extremely long and large molecules that are formed from numerous smaller molecules, like links forming a chain. The atomic structure of these materials, therefore, is essentially one-dimensional, with predictable repeating sequences of atoms along the

Figure 9–10
The arrangement of atoms in a crystal (*a*) is regular and predictable over distances of thousands of atoms. The arrangement of atoms in a glass (*b*), is regular only on a local scale, but irregular over a distance of three or four atoms.

Most plastics can be recycled, thus saving petroleum (the raw material for all plastics) and reducing the volume of solid waste.

Figure 9–11

Branching and unbranching polymers form from small molecules, just like long chains can form from individual links.

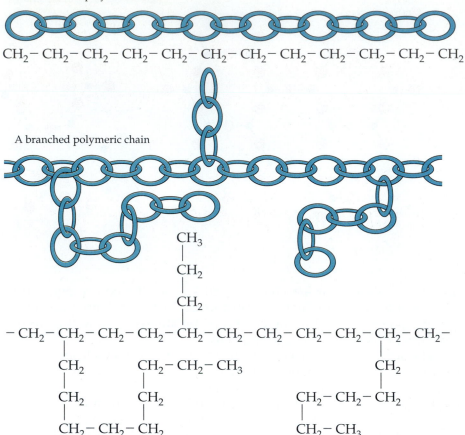

An unbranched polymeric chain

$$CH_2 - CH_2 - CH_2 - CH_2 - CH_2 - CH_2 - CH_2 - CH_2 - CH_2 - CH_2 - CH_2 - CH_2$$

A branched polymeric chain

$$CH_3$$
$$|$$
$$CH_2$$
$$|$$
$$CH_2$$
$$|$$
$$- CH_2 - CH_2 - CH_2 - CH_2 - CH_2 - CH_2 - CH_2 - CH_2 - CH_2 - CH_2 - CH_2 -$$
$$|||$$
$$CH_2CH_2 - CH_2 - CH_3CH_2$$
$$|||$$
$$CH_2CH_2CH_2 - CH_2 - CH_2$$
$$|||$$
$$CH_2 - CH_2 - CH_2CH_2 - CH_2 - CH_2CH_2 - CH_3$$

polymer chain (see Figure 9–11). Plastics consist of complex intertwined polymer strands, much like the strands of fiberglass insulation. When heated, these strands slide across each other to adopt new shapes. When cooled, the plastic fiber mass solidifies into whatever shape is available.

 ## Technology

Liquid Crystals and Your Hand Calculator

Almost every material on Earth is easily classified as a solid, a liquid, or a gas, but scientists have recently synthesized an odd intermediate state of matter, called *liquid crystals*. These materials have quickly found their way into many kinds of electronic devices, including the digital display of your pocket calculator. The distinction between a liquid and a crystal is one of atomic-scale order: atoms are disordered in a liquid, and ordered in a crystal. But what happens in the case of a liquid formed from very long molecules? Like a box of uncooked spaghetti, in which the individual pieces are mobile but well-oriented, these molecules may adopt a very ordered arrangement even in the liquid form.

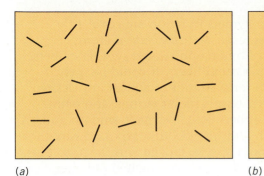

(a)   (b)

Figure 9–12

Schematic diagram of a liquid crystal. Under normal circumstances the elongated molecules are randomly oriented, but in an electric field the molecules align in an orderly pattern.

Liquid crystal displays are used in most pocket calculators.

If the molecules are polar, they may behave like tiny compass needles. Under ordinary circumstances these molecules will occur in random orientations, as in a normal liquid (Figure 9–12*a*). Under the influence of an electric field, however, the molecules may adopt a partially ordered structure, in which the long molecules line up side-by-side (Figure 9–12*b*). This change in structure may even change some of the liquid's physical properties—its color or light-reflecting ability, for example. This phenomenon is now widely used in liquid crystal displays in watches and computers, in which electric impulses align molecules in select regions of the screen to provide a rapidly changing visual display.

Are liquid crystals found in nature? Every cell membrane is composed of a double layer of elongated molecules, called *lipids* (see Chapter 21). Many scientists now believe that these "lipid bilayers" originated in the primitive ocean as molecules similar to today's liquid crystals. ■

Science in the Making

The Discovery of Nylon

Nature's success in making strong, flexible fibers inspired scientists to try the same thing. American chemist Wallace Carothers (1896–1937) began thinking about polymer formation while a graduate student in the 1920s. At the time no one was sure how natural fibers formed, or what kinds of chemical bonds were involved. Carothers wanted to find out.

The chemical company Du Pont took a gamble by naming Carothers head of its new "fundamental research" group in 1928. No pressure was placed on him to produce commercial results, but within a few years his team had developed the synthetic rubber neoprene, and by the mid-1930s they had learned to make a variety of extraordinary polymers, including nylon, the first human-made fiber. Carothers also demonstrated conclusively that polymers in nylon are covalently bonded chains of small molecules, each with six carbon atoms.

Du Pont made a fortune out of nylon and related synthetic fibers. Nylon had many advantages over natural fibers. It could be melted and squeezed out of spinnerets to form strands of almost any desired size: threads, rope, surgical sutures, tennis racket strings, and paint-brush bristles, for example. These fibers could be made smooth and straight like fishing line, or rough and wrinkled like wool, to vary the texture of fabrics. Nylon fibers could also be kinked with heat, to provide permanent folds and pleats in clothing. The melted polymer could even be injected into molds to form durable parts such as tubing or zipper teeth. And nylon was cheap—formed, as Du Pont advertised, from "air, water, and coal."

Sadly, Wallace Carothers did not live to see the impact of his extraordinary discoveries. Suffering from increasingly severe bouts of depression and convinced that he was a failure as a scientist, Carothers took his own life in 1937, just a year before the commercial introduction of nylon. ■

Changes of State

Place a tray of liquid water in the freezer and it will turn to solid ice. Heat a pot of water on the stove and it will boil away to a gas. These everyday phenomena are examples of **changes of state**—transitions among the solid, liquid, and gas states. *Freezing* and *melting* involve changes between liquids and solids, while *boiling* and *condensation* are changes between liquids and gases. In addition, some solids may transform directly to the gaseous state by *sublimation*.

Temperature induces these transitions by changing the speed at which molecules vibrate. An increase in the temperature of ice to above 0°C (32°F), for example, causes molecular vibrations to increase to the point that individual molecules jiggle loose and the crystal structure starts to break apart. A liquid forms. Then, above 100°C (212°F), individual water molecules move fast enough to break free of the liquid surface and form a gas. These transformations require a great deal of energy, because a great many chemical attractions must be broken to change from a solid to a liquid, or from a liquid to a gas. Thus a pot of water may reach boiling temperature fairly quickly, but it takes a long time to break all the attractions between water molecules and boil the water away. By the same token, a glass of ice water will remain at zero degrees for a long time, even on a warm day, until enough energy has been absorbed to break all the ice attractions. Only after the last bit of ice is gone can the water temperature begin to rise.

In the past few decades scientists have discovered that changes in pressure also induce changes of state. Many common solids and liquids become gaseous at low or vacuum pressures, while all known gases and liquids become solid at modest pressures equivalent to a few tens of thousands of atmospheres. Researchers have found that at even higher pressures, equivalent to several million times the pressure of the Earth's atmosphere, many nonmetallic elements and salts become metallic.

Chemical Reactions and the Formation of Chemical Bonds

Atoms or smaller molecules come together to form larger molecules, and larger molecules break up, in processes that we call **chemical reactions.** Chemical reactions are an integral part of our daily lives. When we take a bite of food, light a match, wash our hands, or drive a car we initiate chemical reactions. And every moment of every day, countless chemical reactions in every cell of our bodies sustain life.

All chemical reactions involve rearrangement of the atoms in elements and compounds, as well as rearrangement of electrons to form chemical bonds. Thus, while there are countless millions of known chemical reactions, many of the most familiar ones can be grouped into a few broad categories based on how atoms and electrons are distributed.

Chemical Reactions and Energy: Rolling Down the Chemical Hill

Before we start to discuss the way that atoms can combine to form molecules, it would be a good idea to pause for a moment to think about why these reactions take place at all. The fundamental reason, as so often happens with natural phenomena, has to do with energy.

Consider, for example, one of the electrons in the neutral sodium atom shown in Figure 9–1. This electron is moving in orbit around the nucleus, so it has kinetic energy. In addition, the electron possesses potential energy because it is a certain distance from the positively charged nucleus. You can see this in several ways. You can notice that if the electron were at rest, rather than moving in its orbit, it would fall in toward the nucleus, just as the Earth would fall in toward the Sun in the analogous situation. Thus, just by virtue of its position, the electron is capable of doing work. This, of course, is the way potential energy was defined in Chapter 3. Alternatively, you could imagine starting with the electron close to the nucleus and ask what you would have to do to bring it to the position of the orbit. You would obviously have to exert a force (to overcome the attraction of the electron to the nucleus), and you would have to exert that force over a distance. Thus, just as lifting this book up against the force of gravity gives it gravitational potential energy, lifting the electron up against the pull of the nucleus gives it electrical potential energy.

Finally, the electron has an additional component of potential energy because of the electrical repulsion between it and all of the other electrons in the atom. This is analogous to the small contribution to the Earth's potential energy from the other planets. The sum of these three energies— the kinetic energy associated with orbital motion, the potential energy associated with the nucleus, and the potential energy associated with the other electrons—is the total energy of the single electron in its orbit.

The sum of the energies of all of the electrons is the total energy of the atom. For the isolated sodium atom in Figure 9–1, the total energy is the sum of the energy of the 11 electrons; for the chlorine atom, it is the sum of the energies of the 19 electrons. The total energy of the sodium-chlorine system is, then, the sum of the individual energies of the two atoms.

Now ask what happens to the energy of the sodium-chlorine system after the ionic bond has formed. The force on each individual electron is now different than it was before. For one thing, the number of electrons in each atom has changed; for another, the atoms are no longer isolated, so electrons in the sodium can exert forces on electrons in the chlorine and vice versa. Consequently, the orbits of all the electrons will shift a little due to the formation of the bond. This means that each electron will find itself in a slightly different position with regard to the nucleus, will be moving at a slightly different speed, and will experience a slightly different set of forces than it did before. The total energy of each electron will be different after the bond forms, the total energy of each atom will be different, and the total energy of the system will be different.

Whenever two or more atoms come together to form chemical bonds, the total energy of the system will be different after the bonds form than it was before. Two possibilities exist: either the final energy is less than the initial energy, or the final energy is greater than the initial energy.

The reaction that produces sodium chloride from sodium and chlorine is an example of the first kind of reaction. The total energy of all the electrons in the system is less after the atoms have come together than it was before. The difference in the energies is given off during the reaction in the form of heat, light, and sound (there is an explosion). A chemical reaction in which the final energy of the electrons is less than the initial energy, and which therefore gives off energy in some form, is said to be *exothermic*.

Many examples of exothermic reactions occur in everyday life. The energy that moves your car is given off by the chemical combination of gasoline and oxygen in the car's engine. If your kitchen stove runs on natural gas, the energy to cook your food comes from a similar reaction with oxygen. The chemical reactions in the battery that runs your Walkman also produce energy, although in this case some of the energy is in the form of kinetic energy of electrons in a wire.

If the final energy of the electrons in a reaction is greater than the initial energy, then you have to supply energy to make the chemical reaction proceed. Such reactions are said to be *endothermic*. The chemical reactions that go on when you are cooking (frying an egg, for example, or baking a cake) are of this type. You can put the ingredients of a cake together and let them sit for as long as you like, but nothing will happen until you turn on the oven and supply energy in the form of heat. When the energy is available, electrons can move around and rearrange their chemical bonds. The result: a cake where before there was only a mixture of flour, sugar, and other materials.

As we saw earlier, you can think of chemical reactions as being analogous to a ball lying on the ground. If the ball happens to be at the top of a hill, it will lower its potential energy by rolling down the hill, giving up the excess energy in the form of frictional heat. If the ball is at the bottom of the hill, you have to do work on it to get it to the top.

In the same way, exothermic reactions correspond to systems that "roll down the hill," going to a state of lower energy and giving off excess energy in some form. Endothermic reactions, on the other hand, have to be "pushed up the hill," and hence absorb energy from their surroundings.

Although we have discussed the energies involved in chemical reactions in terms of two atoms coming together to form a single molecule, the arguments we've used are equally applicable to reactions in which two or more molecules or atoms are produced in the final state, or in which more than two atoms or molecules come together to start the reaction. The important question is whether the sum of the energies of all the electrons in the beginning is greater than or less than the sum of the energies of all the electrons at the end.

Common Chemical Reactions

Many millions of chemical reactions take place in the world around us. Some occur naturally, and some occur as the result of human design or intervention. In your everyday world, however, you are likely to see a few types of chemical reactions over and over again. Let's examine a few of these common reactions in more detail.

1. Oxidation and Reduction   Perhaps the most distinctive chemical feature of our planet is the abundance of a highly reactive gas, oxygen, in our atmosphere. This trait leads to many of the most familiar chemical reactions in our lives. **Oxidation** includes any chemical reaction in which an atom (like oxygen, or any other atom that will accept electrons) accepts electrons while combining with other elements. The atom that transfers the electrons is said to be *oxidized*. Rusting, in which iron metal combines with oxygen to form a reddish iron oxide, is a common consequence of gradual oxidation. Burning or *combustion* is a much more rapid oxidation, in which oxygen combines with carbon-rich materials to produce carbon dioxide and other by-products that often pollute.

Hydrocarbons—chemical compounds of carbon and hydrogen—provide the most efficient fuels for combustion, with only carbon dioxide and water (hydrogen oxide) as products. This reaction is

▶ **In words:**

   hydrocarbon + oxygen → carbon dioxide + water + energy,

   where the symbol → should be read "reacts to form." This reaction may also be written as

▶ **In chemical notation:**

   $C_nH_{2n+2} + O_2 \rightarrow CO_2 + H_2O + \text{energy}.$

If you heat or cook with natural gas, then you are using an oxidation reaction to generate energy in your home. The term "natural gas" refers to a compound that chemists call methane. A methane molecule consists of a single carbon atom covalently bonded to four hydrogen atoms, as shown in the drawing in Figure 9–13. The oxidation reaction involved in the burning of methane is written

$$CH_4 + 2O_2 \rightarrow CO_2 + 2H_2O.$$

Figure 9–13

In pictorial form, an oxidation reaction involves the transfer of electrons to oxygen atoms. When natural gas (CH_4) burns, it combines with two oxygen molecules (O_2) to form a molecule of carbon dioxide (CO_2) and two molecules of water (H_2O).

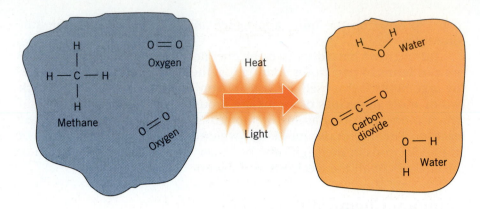

The opposite of oxidation is **reduction,** a chemical reaction in which electrons are transferred to an atom from other elements. The atom that receives the electrons is said to be *reduced*. Thousands of years before scientists discovered oxygen, primitive metalworkers had learned how to reduce metal ores by smelting. In the iron smelting process (typical of the industry carried out at hundreds of North American furnaces 250 years ago), iron makers heated a mixture of ore (iron oxide) and lime (calcium oxide) in an extremely hot charcoal fire. The lime lowered the melting temperature of the entire mixture, which then reacted to produce iron metal and carbon dioxide.

A material that has been oxidized has lost electrons to oxygen or some other atom. A material that has been reduced, on the other hand, has gained electrons. You may find the mnemonic OIL RIG ("oxidation is loss, reduction is gain") useful in keeping the two straight.

Oxidation and reduction are essential to life, and define the principal difference between plants and animals. As we shall see in Chapter 22, animals take in carbon-based molecules in their food and allow it to be oxidized in their cells. Carbon dioxide is released as a by-product. Plants take in carbon dioxide and use the energy in sunlight to reduce it, releasing oxygen as a by-product.

2. Precipitation-Solution Reactions Water and many other liquids have the ability to dissolve solids. You can observe such *solution reactions* when you put salt or sugar into water. Allow a salt solution to become too concentrated, however, and the opposite reaction—*precipitation*—will occur. The next time you are at the ocean, scoop up a handful of salt water and, as the water evaporates, watch as tiny salt crystals precipitate on your palm.

3. Acid-Base Reactions Acids are common substances, used by people for thousands of years. The word has even entered our everyday vocabulary—we may refer to someone with an "acid" wit when we mean something sharp and corrosive. Acids corrode metals and have a sour taste. For our purposes, we can make a technical definition of an acid as follows: An acid is any material that, when put into water, produces positively charged hydrogen ions (i.e., protons) in the solution. Lemon juice, orange juice, and vinegar are examples of common weak acids, while sulfuric acid (used in car batteries) and hydrochloric acid (used in industrial cleaning) are strong acids.

Bases are another class of corrosive materials. They taste bitter and generally feel slippery between your fingers. For our purposes, we can define a base as any material that, when put into water, produces negatively charged OH^- ions. This ion, consisting of an oxygen-hydrogen system that has an extra electron, is called the hydroxide ion. Most antacids (e.g., milk of magnesia) are weak bases; cleaning fluids containing ammonia are the strongest of bases we most often encounter, although still relatively weak. Most common drain cleaners are examples of strong bases.

Although the common definitions of acids and bases involve taste and feel, you shouldn't try these tests yourself. Many acids and bases are *extremely* dangerous and corrosive—for example, battery acid (H_2SO_4) and lye (NaOH).

When acids and bases are brought together in the same solution, the H^+ and the OH^- ions come together to form water, and we say that the substances neutralize each other. For instance, if we mix hydrochloric acid and lye, we will see the following reaction:

$$HCl + NaOH \rightarrow H_2O + NaCl.$$

The formation of water removes both the positively charged hydrogen ion and the negatively charged hydroxide ion from solution, and the other parts of the original molecules come together to form a new material. Molecules formed by the neutralization of an acid and a base are called salts.

If you take an antacid when you have have an upset stomach, you are running a neutralization reaction in your body. Ordinary over-the-counter antacids contain bases such as aluminum hydroxide [$Al(OH)_3$] or sodium bicarbonate ($NaHCO_3$), which react with some of the acid in the stomach. These products do not neutralize all the stomach's hydrochloric acid—only enough of it to alleviate the symptoms.

The definition of acids and bases in terms of whether they add positively charged protons or negatively charged hydroxide groups to water leads to a simple way of measuring the strength of a solution. Although you might not think so at first glance, pure distilled water always contains some protons and hydroxide groups. A small number of water molecules are always being broken up, and at the same time elsewhere in the liquid protons and hydroxide groups come together to form new molecules of water. In fact, in pure water there are almost exactly 10^{-7} moles of positively charged particles per liter. Acids contain more positive charges than this, while bases contain fewer.

This fact is used to set the scale for measuring acids and bases. Pure water is defined to have a pH ("*power of H*ydrogen") of 7. An acid solution that has more positive charges—a concentration of 10^{-6} moles per liter, for example—will have a lower pH (a pH of 6 in this example). A base that has a lower concentration of positive charges—10^{-10} moles per liter, for example—will have a higher pH (a pH of 10 in this example). Some common pH values:

stomach acid	1.0–3.0
normal rainwater	5.6
pure water	7.0
human blood	7.3–7.5
household ammonia	11.0

4. Polymerization and Depolymerization The molecular building blocks of most common biological structures are small (see Chapter 21), consisting of at most a few dozen atoms. Yet most biological molecules are huge, with up to millions of atoms in a single unit. How can these small building blocks yield the large structures characteristic of living things? The answer lies in a process known as polymerization.

A *polymer* is a large molecule that is made by linking smaller, simpler molecules together repeatedly to build up a complex structure. The word comes from the Greek *poly-* (many) and *meros* (parts). In spider's webs, growing hair, clotting blood, and a thousand other processes, living systems have mastered the art of combining small molecules into long chains.

Polymerization reactions include all chemical reactions that form long strands from small molecules. Synthetic polymers usually begin in liquid form, with relatively small molecules that freely move past their neighbors. The common plastic polyethylene, for example, begins as a gas with molecules containing just six atoms (two carbons and four hydrogens), while the liquid that makes a common nylon contains molecules with six atoms of carbon, eleven of hydrogen, one of nitrogen, and

Figure 9–14

Polymers may form by condensation (*a*) or addition polymerization reactions (*b*).

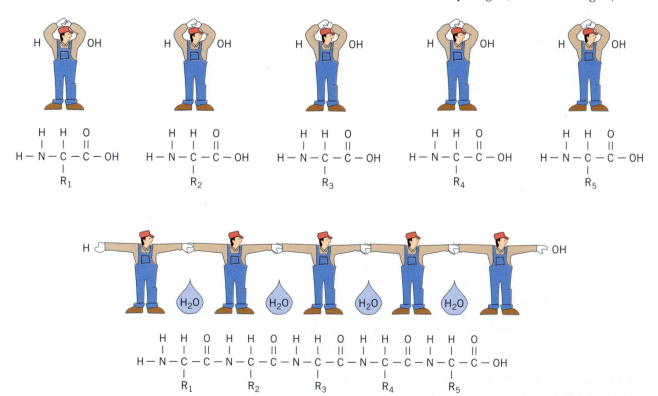

(*a*)

one of oxygen—a combination that chemists write as $C_6H_{11}NO$. Polymers form from the liquid when the ends of these molecules begin to link up. In the case of nylon, the polymer forms by a *condensation reaction* in which each new polymer bond forms by the release of a water molecule (Figure 9–14*a*). Polyethylene, on the other hand, forms by *addition polymerization,* in which the basic building blocks are simply joined end to end (Figure 9–14*b*).

Polymers play an enormous role in our lives. The useful properties of these varied materials are related to the shapes of the molecules and the way they come together to form materials. In polyethylene, for example, long-chain molecules wrap themselves together into something like a plate of hairy spaghetti. It's hard for water molecules to penetrate into this material, so it is widely used in packaging. The "plastic" on prepackaged fruit or meat at your supermarket may be made from polyethylene. A closely related polymer is polyvinyl chloride, whose basic building block, vinyl chloride, is just an ethylene molecule in which one of the four hydrogens has been replaced by a chlorine atom. Because the chlorine atom is bigger than hydrogen, the molecules of the polymer are lumpy and cannot pack too closely together. Commercial PVC, widely used in water and sewer pipes, contains other kinds of molecules that move between the polymers, lubricating the system and making the resulting material highly flexible. Credit cards are also made from this material. Other common polymers include polypropylene (artificial turf), polystyrene ("foam" cups and packaging), and Teflon (nonstick cookware).

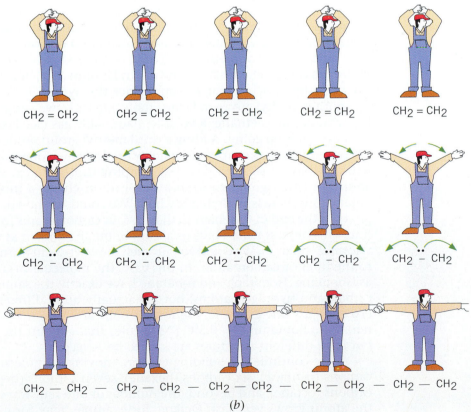

$CH_2 = CH_2$ $CH_2 = CH_2$ $CH_2 = CH_2$ $CH_2 = CH_2$ $CH_2 = CH_2$

$CH_2 - CH_2$ $CH_2 - CH_2$ $CH_2 - CH_2$ $CH_2 - CH_2$ $CH_2 - CH_2$

$CH_2 - CH_2 - CH_2 - CH_2 - CH_2 - CH_2 - CH_2 - CH_2 - CH_2 - CH_2$

(*b*)

Many polymers are extremely long-lasting, a situation that presents a growing problem in an age of diminishing landfills. Nevertheless, most polymers are not permanently stable. Given time, they will decompose into smaller molecules. This breakdown of a polymer into short segments is called *depolymerization*.

Perhaps the most familiar depolymerization reactions occur in your kitchen. Polymers cause the toughness of uncooked meat and the stringiness of many raw vegetables. We cook our food, in part, to break down these polymers. Chemicals such as meat tenderizers and marinades can also contribute to depolymerization and can improve the texture of some foods.

Not all depolymerization is desirable. Museum curators are painfully aware of the breakdown process, which affects leather, paper, textiles, and other historic artifacts made of organic materials. Storage in an environment of low temperature, low humidity, and an inert atmosphere (preferably without oxygen) may slow the depolymerization process, but there is no known way to repolymerize old brittle objects.

Building Molecules: The Hydrocarbons

As an example of how a wide variety of materials can be made by putting the same molecular building blocks together in different ways, let's start with the methane molecule shown in Figure 9–13 and build a family of molecules known to chemists as alkanes. Alkanes are flammable materials (either gases or liquids) that burn readily and are often used as fuels. Most of the components of the gasoline in your car, for example, are members of this family. Alkanes are one example of **hydrocarbons,** molecules made completely from hydrogen and carbon atoms.

You can think of methane as being composed of a carbon and three hydrogens (what chemists call the methyl group) and a fourth hydrogen. We begin by noticing that we can replace the hydrogen in the methane by another methyl group to form a molecule with two carbons in it. This larger molecule is ethane, a volatile, flammable gas (see Figure 9–15).

You can keep going. Adding a third methyl group produces propane, a three-carbon chain. Propane is widely used as a fuel for portable stoves—you may have used it on your last camping trip. The next step is to add another methyl group to form a four-carbon chain, a molecule called butane. But there is an ambiguity here. We could, as shown, add the new group at the end of the chain so that all four carbon atoms form a straight line. This process would give us a form of butane known as n-butane (or "normal" butane). We could equally well, however, add the methyl group to the interior carbon atom. In this case, the molecule would be known as isobutane. Isobutane and n-butane have exactly the same numbers of carbon and hydrogen atoms, but are actually quite different materials. (To give just one example, the former boils at $-11.6°C$ while the latter remains a liquid up to $-0.5°C$.) Molecules that contain the same atoms but have different structures are said to be *isomers*.

As we continue the building process, moving to molecules with five carbons (pentane), six carbons (hexane), seven carbons (heptane), eight carbons (octane), and beyond, the number of different ways to assemble the atoms grows very fast. Octane, for example, may form 18 different isomers. Some of these variants have long chains, others are branched.

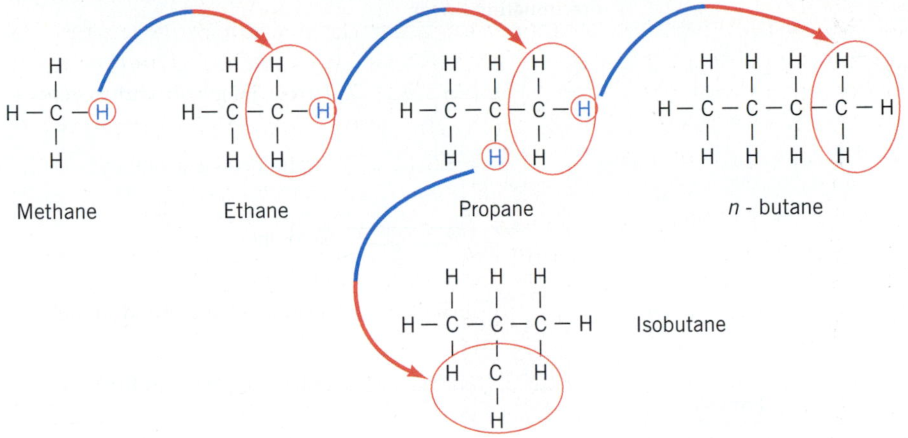

Methane   Ethane   Propane   *n* - butane

Isobutane

Figure 9–15
Building up the alkane series by adding methyl groups to methane. The first three members of the group are methane, ethane, and propane.

As we shall see, these structural differences play an important role in a molecule's usefulness as an automotive fuel.

All other things being equal, the carbon chain length affects whether the alkane is a solid or liquid: the longer the chain, the higher the temperature at which that material can remain a solid. If carbon chains are straight, then alkanes with a half dozen or so carbons are liquid, but those with more than 10 are soft solids. Good quality paraffin candle wax, which melts only near a hot flame of a wick, for example, is composed primarily of chains with 20 to 30 carbon atoms. The presence of branches in the chain, however, makes it more difficult for the molecules to pack together efficiently. One consequence of branching is that the melting points are generally lowered compared to those of straight alkanes.

Technology

Refining Petroleum

Deep underground are vast lakes of a thick, black liquid called *petroleum,* derived from many kinds of transformed molecules of former life forms. Petroleum is an extremely complex mixture of organic chemicals, as much as 98% molecules of hydrogen and carbon (mostly in the form of hydrocarbons), with about 2% of other elements. Engineers must separate this mixture into much purer fractions through the process of *distillation.*

Hydrocarbons with different numbers and arrangements of carbon atoms have very different boiling temperatures. The key to distillation, then, is to boil off and collect different kinds of molecules successively, according to their boiling points. The most *volatile* hydrocarbon—the one with the lowest boiling point—is simple methane (CH_4), or natural gas. At the opposite extreme are very-long-chain hydrocarbons with dozens of carbons, as in the molecules that comprise hard waxes, asphalt, and tar.

Modern chemical plants bristle with tall cylindrical towers that distill petroleum. Engineers pump crude oil into a tower, which is heated

Figure 9–16

A distillation column in a chemical plant separates hydrocarbons into fractions useful as gases, gasoline, kerosene, heating oil, lubricating oils and paraffin, asphalt, and tar.

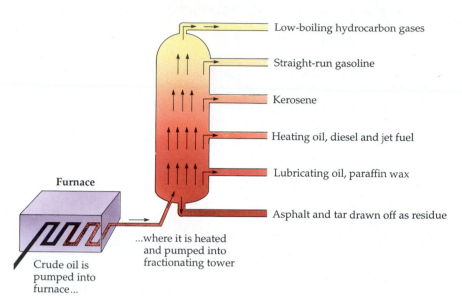

Fractionating tower

Low-boiling hydrocarbon gases

Straight-run gasoline

Kerosene

Heating oil, diesel and jet fuel

Lubricating oil, paraffin wax

Asphalt and tar drawn off as residue

Furnace

Crude oil is pumped into furnace...

...where it is heated and pumped into fractionating tower

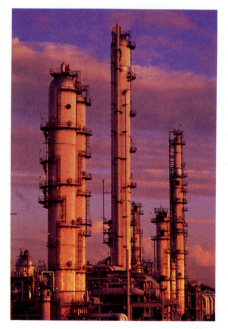

Modern chemical plants bristle with tall distillation columns, in which petroleum products are purified.

from below to create a temperature gradient up the tower (see Figure 9–16). At various levels of the tower, useful petroleum products such as gasoline or heating oil are recovered and sent to other parts of the chemical plant for further refining and processing.

The gasoline you buy at a service station is usually rated in "octanes." The octane rating of a gasoline is based on its ability to stand high compression in a cylinder without igniting. A fuel mixture that ignites while the piston is still moving up and compressing the gas in the cylinder will cause the engine to knock, a highly undesirable quality. In general, the more highly branched a molecule is, the better it will perform as far as knocking is concerned.

A particular isomer of octane that has five carbons in a row and the other on the sides turns out to have very good antiknock properties. This isomer is called isooctane, and an octane rating of 100 for any fuel mixture means that it is as good as pure isooctane. At the opposite extreme, n-heptane is an isomer with seven carbons in a straight chain that produces knocking all the time. An octane rating of zero means the fuel mixture is as bad as pure n-heptane. The octane rating of a fuel, then, is simply a statement that it performs as well as a particular mixture of isooctane and n-heptane. A fuel rated at 95 octane, for example, performs as well as a mixture of 95% isooctane and 5% n-heptane. ■

The Human Body

The Clotting of Blood

Whenever you get a cut that bleeds, your blood begins a remarkable and complex sequence of chemical reactions called clotting. Normal blood, a liquid crowded with cells and chemicals that distribute nutrients and energy throughout your body, flows freely through the circulatory system. When that system is breached and blood escapes, however, the damaged cells release a chemical called prothrombin.

Prothrombin itself is inactive, but other blood chemicals convert it into the active chemical thrombin. The thrombin breaks apart other normally stable chemicals that are always present in blood, and thus produces small molecules that immediately begin to polymerize. The new polymer, called fibrin, congeals quickly and forms a tough fiber net that traps blood cells and seals the break in minutes.

Clotting chemistry differs depending on the nature of the injury and the presence of foreign matter in the wound. Biologists have discovered more than a dozen separate chemical reactions that may occur during the process. A number of diseases and afflictions may occur if some part of this complex chemical system is not functioning properly. Hemophiliacs lack one of the key clotting chemicals and so may bleed continuously from small cuts. Some lethal snake venoms, on the other hand, work by inducing clotting in a closed circulatory system. ■

Science by the Numbers

Balancing Chemical Equations

When chemical reactions occur, atoms are rearranged and new molecules are formed. In this process, however, the total number of atoms of each species must be the same after the reaction as it was before. The requirement that no atoms be created or destroyed in the chemical reaction is usually cast as the requirement that the chemical equation that describes the reaction be "balanced."

Your car battery, for example, contains plates of lead (Pb) and lead dioxide (PbO_2) immersed in a solution of sulfuric acid (H_2SO_4). After the battery has been discharged, both the lead and the lead dioxide have been converted into lead sulfate ($PbSO_4$), and the sulfuric acid has been converted into water (H_2O). If you wrote a chemical reaction that included all these substances, you would get

$$Pb + PbO_2 + H_2SO_4 \rightarrow PbSO_4 + H_2O.$$

This equation, however, is not balanced. There are two lead atoms on the left side, for example, and only a single lead atom on the right. The only way to balance the equation—to make sure that there are as many atoms at the end as at the beginning—is to write

$$Pb + PbO_2 + 2H_2SO_4 \rightarrow 2PbSO_4 + 2H_2O.$$

As you can see, there are now two atoms of lead on each side of the equation, two SO_4 groups, four hydrogens, and two oxygens. This balanced equation represents the reshuffling of atoms that goes on when your battery discharges. It tells us that two molecules of sulfuric acid will be used for each molecule of lead, and that two molecules of water and two of lead sulfate will be produced at the end.

Every chemical reaction, no matter how complex, must balance: you must end with the same number of atoms with which you began. ■

THINKING MORE ABOUT ATOMS IN COMBINATION

Life Cycle Costs

Every month, chemists around the world develop thousands of new materials and bring them to market. Some of these materials do a particular job better than those they replace, some do jobs that have never been done before, and some do jobs more cheaply. All of them, however, share one property—when the useful life of the product of which they are part is over, they will have to be disposed of in a way that is not harmful to the environment. Until very recently, engineers and planners had given little thought to this problem.

Think about the battery in your car, for example. The purchase price covered the cost of mining the lead in its plates, pumping the oil that was made into its plastic case, assembling the final product, and so on. When that battery runs down, all of these materials have to be dealt with responsibly. If you throw the battery into a ditch somewhere, the lead may wind up in nearby streams and wells.

One way of dealing with this sort of problem, of course, is to recycle materials—pull the lead plates out of the battery, process them, and then use them again. But even in the best system, some materials can't be recycled, either because they have become contaminated with other materials through use, or because we don't have technologies capable of doing the recycling. These materials have to be disposed of in a way that isolates them from the environment. The question becomes, "Who pays?"

Traditionally, in the United States, the person who does the dumping—in effect, the last user—must see to the disposal. In some European countries, however, a new concept is being introduced. Called "life cycle costing," this notion is built around the proposition that once a manufacturer uses a material, he or she owns it forever and is responsible for its disposal. The cost of a product such as a new car, then, has to reflect the fact that someday that car may be abandoned and the manufacturer will have to pay for its disposal.

Life cycle costing obviously increases the immediate price of commodities, contributing to inflation in the process. What do you think the proper trade-off is in this situation? How much extra cost should be imposed up front compared to eventual costs of disposal?

Summary

Atoms link together by *chemical bonds,* which form when a rearrangement of electrons lowers the potential energy of the electron system—particularly by the filling of outer electron shells. *Ionic bonds* lower chemical potential energy by the transfer of one or more electrons to create atoms with filled shells. The positive and negative ions created in the process bond together through electrostatic forces. In *metals,* on the other hand, isolated electrons in the outermost shell wander freely throughout the material and create *metallic bonds.* *Covalent bonds* occur when adjacent atoms, or groups of atoms called *molecules,* share bonding electrons. *Hydrogen bonding* and *van der Waals forces* are special cases involving distortion of electron distributions to create electrical polarity—regions of slightly positive and negative charge that can bind together.

Atoms combine to form several different *states of matter. Gases* are composed of atoms or molecules that can expand to fill any available volume. *Plasmas* are ionized gases in which electrons have been stripped from the atoms. *Liquids* have a fixed volume but no fixed shape. *Solids* have fixed volume and shape. Solids include *crystals,* with a regular and repeating atomic structure; *glasses,* with a nonrepeating structure; and *plastics,* which are composed of intertwined chains of molecules called *polymers.* The various states of matter can undergo *changes of state,* such as freezing, melting, or boiling with changes in temperature or pressure.

Chemical bonds form during *chemical reactions,* which may involve the synthesis or decomposition of chemical elements or compounds. Reactions in which materials lose electrons to atoms such as oxygen are called *oxidation* reactions. In the opposite reaction, called *reduction,* electrons are moved onto atoms.

All life depends on *polymerization* reactions, in which small molecules link together to form long polymer fibers such as synthetic polyesters, vinyl, cellophane, and other plastics, and natural hair, silk, plant fiber, and skin. *Hydrocarbons,* widely used as fuels, are chainlike molecules of carbon and hydrogen atoms. High temperatures and certain chemicals can cause the breakdown of polymers—depolymerization—which is often a key objective in cooking.

Key Terms About Chemical Bonding

chemical bond	ionic bond	metallic bond	metal
covalent bond	molecule	hydrogen bond	van der Waals forces
states of matter	gas	plasma	liquid
solid	crystal	glass	plastic
polymer	change of state	chemical reaction	oxidation
reduction	polymerization	hydrocarbon	

Review Questions

1. What is a valence electron?

2. What is a filled shell? How many electrons do the first few shells hold?

3. Describe the ionic bond. Give an example of a material that uses it.

4. Describe the metallic bond. What properties of metal follow from the properties of the bond?

5. What is an alloy? Give an example.

6. Describe the covalent bond. Give an example of a material that uses it.

7. Describe the hydrogen bond. Explain how it affects the properties of water.

8. Describe the van der Waals force. Give an example of a material that uses it.

9. What is a gas? A liquid? A solid?

10. What is a plasma? Where are plasmas found in the solar system?

11. How is a glass constructed from its atoms? A plastic?

12. Define freezing and melting, boiling and condensation, and sublimation.

13. What is oxidation? Reduction? Polymerization? Give an example of all three.

14. What are the names and properties of the first few alkanes?

Discussion Questions

1. Identify objects around you that use the three kinds of chemical bonding or the three kinds of attraction. Which combine two or more kinds?

2. Classify the solid objects around you as ceramics, metals, glass, and plastic.

3. Based on your experience today, list as many everyday chemical reactions as you can. Try to classify these reactions according to their type (i.e., polymerization, oxidation, etc.)

4. The chemistry of the planet Mars is quite similar to Earth except that there is almost no water. Which common chemical reactions that occur on Earth would you not expect to see on Mars?

5. Cooks often tenderize meat by soaking it in a liquid such as lemon juice or vinegar for several hours. What chemical reaction do you think is taking place in the meat? How is this reaction analogous to heating in an oven?

6. During icy winter conditions we often throw salt on sidewalks and streets. What occurs when salt comes in contact with ice? What reaction is involved? How do you think the melting point of salt water compares to that of pure water?

Investigations

1. Look around your home and school and list the variety of plastic objects. What strategy might you develop to recycle plastics? Note the numbers surrounded by a triangle on many disposable plastics. What do the numbers mean?

2. What materials were used for the construction of buildings, furniture, and transportation devices in the United States 200 years ago? What modern technologies would be difficult or impossible with just those materials?

3. Why is it that materials in landfills don't break down into their constituent chemical parts?

4. Dissect a disposable diaper. How many kinds of materials can you identify? What are the key properties of each? What kind of chemical bonding might contribute to the distinctive properties of these materials? Investigate the arguments for and against using disposable diapers.

5. We often refer to drinking water as being "soft" or "hard," based on the kind of impurities present in the water. Which kind do you have in your community? What kind of chemical reaction can take place if water is too hard? How can you prevent that reaction from occurring?

Additional Reading

Snyder, Carl. *The Extraordinary Chemistry of Ordinary Things*. New York: John Wiley and Sons, 1992.

Yergin, Daniel. *The Prize: The Epic Quest for Money, Oil, and Power*. New York: Simon and Schuster, 1990.

Properties of Materials

A material's properties result from its constituent atoms and the arrangements of chemical bonds that hold those atoms together.

A Random Walk with James Trefil

When I was a kid, I used to have to sit around and listen while the grown-ups reminisced about what a great car the Model T had been. It was dependable, I was told, and, most importantly, no matter what went wrong it could be fixed by anyone with a wrench and a pair of pliers. Needless to say, I didn't give much credibility to all this.

Later, as a student, I acquired the first of a long string of Volkswagen Beetles. Now let me tell you, my friends, *that* was a sweet car! There were never any problems with the cooling system, for the simple reason that there wasn't any—the engine was cooled by the air flowing by. And almost any repair could be made by someone with reasonable mechanical ability and a set of tools. While in graduate school, I spent many happy hours under my car, adjusting this or that.

But I never work on my cars any more. When I look under the hood, all I see is a complex array of computers and microchips—nothing a person can get a wrench around. Yet the car I drive today, provided everything is working, is much more user friendly than my old Volkswagen, and infinitely easier to operate than my parent's Model T. I can start it just by turning a key, for example, rather than getting in front of the car and turning a crank. The flow of gasoline to the cylinders is regulated by a small onboard computer, rather than by a clumsy mechanical carburetor.

This personal story about cars turns out to be a pretty good allegory for the way in which the science of materials has developed in the twentieth century. In the beginning, industry turned out big, relatively simple things that were easy to understand and work with—iron wheels for railroads, steel springs for car suspensions, wooden chairs and tables for the home. Today, industry turns out items that perform the same jobs better, but that are made from new kinds of materials such as plastics, composites, and semiconductors. Instead of manipulating large chunks of material, we now control the way atoms fit together. Like modern cars, modern materials do their job well, but they cannot be made (or, usually, repaired) by a simple craftsman working with simple tools.

So while the materials we use are becoming better at what they do and easier for us to use, it becomes harder and harder for us to understand what those materials are. I might have been able to fix my Volkswagen myself, but there is no way I can look under the hood and shift atoms around in my modern car's microchip. In a sense, the improved performance of modern materials has been bought at the price of our ability to understand them. To a large extent the emphasis of modern materials science has shifted away from manipulating large blocks of stuff, which are readily available to our senses, to manipulating atoms in ever more complex ways. And, of course, we can't see or taste or feel atoms.

Materials and the Modern World

The materials people use, perhaps more than any other single feature of a culture, define the technical sophistication of a society. We speak of primitive groups as Stone Age societies, and recognize Iron Age and Bronze Age cultures as progressively more advanced. Given that perspective, in what age are we living now?

Take a moment and look around your room. How many different kinds of materials do you see? The lights and windows employ glass, a brittle, transparent material. The walls may be made out of gypsum, a chalklike mineral that has been compressed in a machine and placed between sheets of heavy paper. Your chair probably incorporates metal, possibly along with wood and woven fabric.

Many of these materials would have been familiar to Americans 200 years ago, when almost everything was made from less than a dozen common substances: wood, stone, pottery, glass, animal skin, natural fibers, and a few basic metals such as iron and copper. But thanks to the

A typical room is filled with high-tech materials: synthetic fibers, specialized glass, colorful plastics, metal alloys, and cosmetics.

discoveries of chemists, the number of everyday materials has increased a thousandfold in the past two centuries. Steel transformed the nineteenth-century world with railroads and skyscrapers, while aluminum provided a lightweight metal for hundreds of applications. The development of rubber, synthetic fibers, and a vast array of other plastics affected every kind of human activity from industry to sports. Brilliant new pigments enlivened art and fashion, while new medications cured many ailments and prolonged lives. And in our electronic age, the discovery of semiconductor and superconductor materials has changed life in the United States in ways that our eighteenth-century ancestors could not have imagined.

How can chemists devise so many different materials? They succeed in part because materials display so many different properties: color, smell, hardness, luster, flexibility, heat capacity, solubility in water, texture, melting point, strength—the list goes on and on. Each new material holds the promise of doing some job better or cheaper than any other. We rely on superhard abrasives and fine lubricants, tenacious glues and efficient solvents, flexible fibers and rigid plastics—a million products for a million uses.

Based on our understanding of atoms and their chemical bonding (see Chapter 9) we now realize that the properties of every material depend on three essential features:

1. The kind of atoms that make it up
2. The way those atoms are arranged
3. The way the atoms are bonded together

In this chapter, we will look at different properties of materials and see how they relate to their atomic architecture. We will examine the strength of materials—how well they resist outside forces. We will look at the ability of materials to conduct electricity, and whether or not they are magnetic. And, finally, we will describe what are perhaps the most important new materials in modern society, the semiconductor and the microchip, to see how the ability to arrange atoms into new materials can lead to dramatic changes in human society.

The Strengths of Materials

Have you ever carried a heavy load of groceries in a plastic bag? You can cram a bag full of bottles and cans and lift it by its thin handles without fear of breakage. How can something as light, flexible, and inexpensive as a piece of plastic be so strong?

Strength is the ability of a solid to resist changes in shape. Strength is one of the most immediately obvious material properties, and it bears a direct relationship to chemical bonding. A strong material must be made from strong chemical bonds. By the same token, a weak material, like a defective chain, must have weak links between atoms. While no type of bond or attraction is universally stronger than the other kinds, van der Waals forces are generally the weakest. Any material with van der Waals forces will have at least one particularly soft direction of bonding, and you will probably be able to pull the material apart with your hands. You experience this softness whenever you use baby powder, graphite lubricant, or soap.

By contrast, the materials we normally think of as strong and durable, such as rocks, glass, and ceramics, are usually held together primarily by ionic bonds. Next time you see a building under construction, look at the way beams and girders link diagonally to form a rigid framework. Chemical bonds in strong materials do the same thing. A three-dimensional network of ionic bonds in these materials holds them together like a framework of steel girders.

The strongest materials we know, however, incorporate long chains and clusters of carbon atoms held together by covalent bonds. The extraordinary strength of natural spider webs, synthetic Kevlar (used to make bulletproof vests), diamonds, and your plastic shopping bag all stem from the strength of covalent carbon bonds.

The Nature of Strength

Every material is held together by the bonds between its atoms. When an outside force is applied to a material, the atoms shift around, the bonds stretch and compress, and a force is generated inside the material to oppose the force imposed from the outside. The strength of a material is thus related to the size of the force it can generate when it is pushed.

Material strength is not a single property because there are different ways of placing an object under stress. Scientists and engineers recognize three very different kinds of strength when characterizing materials:

1. A material's ability to withstand crushing (*compressive strength*)
2. Its ability to withstand pulling apart (*tensile strength*)
3. Its ability to withstand twisting (*shear strength*).

Your everyday experience will convince you that these three properties are often quite independent. A loose stack of bricks, for example, can withstand crushing—you can pile tons of weight on it without having the stack collapse. But the stack of bricks has little resistance against twisting. Indeed, it can be toppled by a child. A rope, on the other hand, is extremely strong when pulled, but has little strength under twisting or crushing.

The point at which a material stops resisting external forces and begins to deform permanently is called its *elastic limit*. We see examples of this phenomenon every day. When you break an egg, crush an aluminum can, snap a rubber band, or fold a piece of paper, you exceed an elastic limit and permanently change the object. When the materials in your body exceed their elastic limit, the consequences can be catastrophic. Our bones may break if put under too much stress, while arteries under too high pressure may rupture in an aneurysm.

The strengths of a material reflect its arrangements of chemical bonds. Think about how you might design a structure out of Tinkertoys that would be strong under crushing, pulling apart, or twisting. The strongest arrangement would have lots of short sticks with triangular patterns. Diamond is exceptionally strong under all three kinds of stress because

The strength of materials is vital to many activities. Bones and muscles provide a weightlifter's strength.

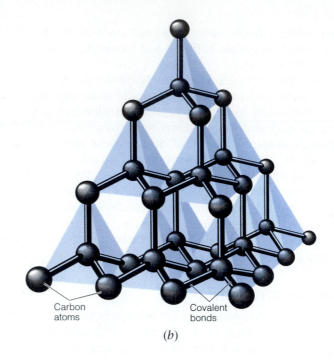

Carbon atoms

Covalent bonds

(a)

(b)

Figure 10–1

The girder framework of a skyscraper (*a*) and the diamond crystal structure (*b*) are both strong because of numerous cross-connections. In a building the connections are steel girders, while in diamond they are carbon-carbon covalent bonds.

of its three-dimensional framework of strong carbon-carbon bonds (see Figure 10–1). Glass, ceramics, and most rocks, which also feature rigid frameworks of chemical bonds, are relatively strong. Many plastics like the one in your shopping bag, however, have strong bonds in only one direction (a common feature of materials called *polymers*) and thus are strong when stretched, but have little strength when twisted or crushed. Materials with layered atomic structures, like a stack of paper, on the other hand, are generally strong when squeezed, but quite weak under other stresses.

Thus the strength of a material—the way it resists or fails to resist outside forces—depends on the kind of atoms in it, the way they are arranged, and the kind of chemical bonds that hold the atoms together.

Composite Materials

Have you driven by a building site recently, or past a place where highway repairs were going on? If so, you almost certainly saw people using a class of materials made by combining two or more substances into a **composite material.** In these kinds of materials, the strength of one of the constituents is used to offset the weakness of another, resulting in a material whose strengths are greater than any of its components.

One of the most common examples of a composite material is plywood, which is a glued composite of thin wood layers with alternating grain direction. Thin sheets of wood tend to break easily along the grain, making them inadequate for many structural uses. In plywood, however, the weakness of one sheet along the grain is compensated for by the strength of the neighboring sheets. Not only is plywood much stronger than a solid board of the same dimension, but it can also be produced from much smaller trees by slicing layers off a rotating log, like removing paper from a roll.

Reinforced concrete provides another common composite material in which steel rods (with great tensile strength) are embedded a concrete mass (with great compressive strength). A similar strategy is used in fiberglass, formed from a cemented mat of glass fibers, and new carbon fiber composites that are providing extraordinarily strong and lightweight structural materials for industry and sports applications.

The modern automobile features a wide variety of composite materials. Windshields are layered to resist shattering and reduce sharp edges in a collision. Tires are intricately formed from rubber and steel belting for strength and durability. Car upholstery commonly mingles natural and artificial fibers, while dashboards often employ complexly laminated surfaces. The bodies of many cars are formed from a fiberglass or other molded lightweight composite. And, as we shall see, all of a modern automobile's electronics, from radio to ignition, depend on semiconductor composites of extraordinary complexity.

Electrical Properties of Materials

Of all the properties of materials, none are more critical to our world than those that control the flow of electricity. Glance around you and tally up the number of electrical devices nearby. Chances are your list will quickly grow to several dozen. Almost every aspect of our technological civilization depends on electricity, so scientists have devoted a good deal of attention to materials that are useful in electrical systems. (See Chapter 5 for a review of electricity and magnetism.) If the job at hand is to send electrical energy from a power plant to a distant city, for example, we need a material that will carry that electrical energy without much loss. If, on the other hand, the job is to put a covering over a wall switch so that we will not be endangered by electricity when we turn on the light, we want a material that will not conduct electricity at all. In other words, a large number of different kinds of materials contribute to any electrical circuit.

Conductors

Any material capable of carrying electrical current—that is, any material through which electrons can flow freely—is called a **conductor.** Metals are the most obvious example of conductors, but there are many others. Water, for example, contains ions of hydrogen (H^+) and hydroxyl (OH^-), as well as dissolved ions of various kinds of minerals. These ions normally carry a charge, and they, too, are free to move if they become part of an electrical circuit. As a result, handling electrical appliances when you are standing on a wet surface or sitting in a bathtub is extremely dangerous. The electricity can be conducted through the water to your body.

We can find out if a material will conduct electricity by making it part of an electrical circuit and seeing if current will flow through it. You already know that metals such as copper conduct electricity—in fact, they almost certainly are carrying electricity through the walls of the building in which you are now sitting.

The arrangement of a material's electrons determines its ability to conduct electricity. In the case of metals, you will recall, some electrons are bonded fairly loosely and shared by all the atoms. If a material like that is made part of an electrical circuit—if, for example, you connect it across the poles of a battery—those electrons are free to move in response. They will flow toward the positive pole of the battery and away from the negative pole.

As we saw in Chapter 5, the motion of electrons in electrical currents is seldom smooth. Under normal circumstances, electrons moving through a metal will collide continuously with the much heavier ions in that metal. In each of those collisions electrons lose some of the energy they have gotten from the battery, and that energy is converted to the faster vibration of ions. We perceive these vibrations as heat, and the property by which materials drain the energy away from a current is called **electrical resistance.** Even very good conductors have some electrical resistance. (The inverse of electrical resistance is *electrical conductivity,* or the ease with which a material allows electrons to flow. Resistance and conductivity are thus two different ways of describing the same property.)

Insulators

Many materials incorporate chemical bonds in which few electrons are free to move in response to the "push" of an electric field. In rocks, ceramics, and many organic materials such as wood and hair, for example, the electrons are bound tightly to one or more atoms by ionic or covalent bonds (see Chapter 9). It takes considerable energy to pry electrons loose from those atoms—energy that is normally much greater than the energy supplied by a battery or an electrical outlet. These materials will not conduct electricity (unless subjected to an extremely high voltage, which can pull the electrons loose). If they are made part of an electrical circuit, no electricity will flow through them. We call these materials **insulators.**

The primary use of insulators in electrical circuits is to channel the flow of electrons and to keep people from touching wires that are carrying current. The shields on your light switches and household power outlets, as well as the shields and casings for most car batteries, for example, are made from plastic—a reasonably good insulating material that has the added advantages of low cost and flexibility. Similarly, electrical workers use protective rubber boots and gloves when working on dangerous power lines. In the case of high-power lines, glass or ceramic components are used to isolate the current because of their strength and superior insulating properties.

Semiconductors

Many materials in nature are neither good conductors nor perfect insulators. We call such materials **semiconductors.** As the name implies, a semiconductor will carry electricity but will not carry it very well. Typically, the resistance of silicon is a million times higher than the resistance of a conductor such as copper. Nevertheless, silicon is not an insulator, because some of its electrons do flow in an electric circuit. In Figure 10–2, we sketch a silicon crystal as it appears under normal circumstances to see why it behaves this way.

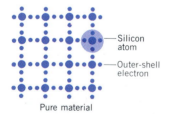

Silicon atom

Outer-shell electron

Pure material

Figure 10–2

A normal silicon crystal displays a regular pattern of silicon atoms. Some of its electrons are shaken loose by atomic vibrations, and these electrons are free to move around and conduct electricity.

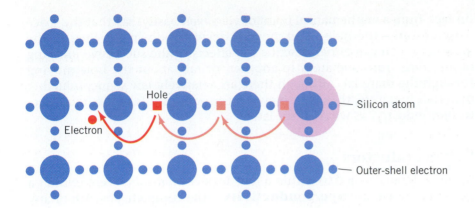

Figure 10–3
A hole in a semiconductor
is produced when an elec-
tron is missing. Holes can
move, just like electrons. As
an electron moves to fill a
hole, it creates another hole
where it used to be.

In this crystal, all the electrons are taken up in the covalent bonds that hold each silicon atom to its neighbor. At absolute zero, in fact, no electrons would be free to move around the material at all. (Remember, absolute zero is the lowest possible temperature, at which no heat energy can be removed from an object.) At room temperature, however, the silicon atoms vibrate, and a few of the electrons that go into making the covalent bonds are shaken loose—think of them as picking up a little of the vibrational energy of the atoms. These *conduction electrons* are free to move around the crystal. If the silicon is made part of an electrical circuit, a modest number of conduction electrons are free to move through the solid.

When a conduction electron is shaken loose, it leaves behind a defect in the silicon crystal—the absence of an electron. This missing electron is called a *hole*. Just as electrons can move around in response to electrical charges, so too can holes (see Figure 10–3).

The motion of holes in semiconductors is something like what you see in a traffic jam on a crowded expressway. A space will open up between two cars, after which one car moves up to fill the space, then another car moves up to fill that space, and another car moves up to fill that space, and so on. You could, of course, describe this sequence of events as the successive motion of a number of cars. But you could just as easily

As cars in a traffic jam
move slowly forward,
"holes" in the traffic can be
described as moving
backward.

(in fact, from a mathematical point of view *more* easily) say that the space between cars—the hole—had moved backwards down the line. In the same way, you can either describe the effects of the successive jumping of electrons from one atom to another, or talk about the hole moving through the material. Although there are relatively few semiconducting materials in nature, they have played an enormous role in the microelectronics industry, as we shall see later.

Superconductors

Some materials cooled to within a few degrees of absolute zero exhibit a property known as **superconductivity**—the complete absence of any electrical resistance. Below some very cold critical temperature, electrons in these materials are able to move without surrendering any of their energy to the atoms. This phenomenon, discovered in Holland in 1911, was not understood until the 1950s. Today, superconducting technologies provide the basis for a billion dollar a year industry worldwide. The principal reason for this success is that once a material becomes superconducting and is kept cool, current will flow in it forever. This behavior means that if you take a loop of superconducting wire and hook it up to a battery to get the current flowing, the current will continue to flow even if you take the battery away.

In Chapter 5, we learned that current flowing in a loop creates a magnetic field. If we make an electromagnet out of superconducting material and keep it cold, the magnetic field will be maintained at no energy cost except for the refrigeration. Indeed, superconductors provide strong magnetic fields much more cheaply than any conventional copper wire electromagnet because they don't heat up from electrical resistance. Superconducting magnets are used extensively in many applications where very high magnetic fields are essential—in particle accelerators (see Chapter 11), in magnetic resonance imaging systems for medical diagnosis, and perhaps eventually in everyday transportation.

How is it that a superconducting material can allow electrons to pass through without losing energy? The answer in at least some cases has to do with the kind of electron-ion interactions that occur. At very low temperatures, heavy ions in a material don't vibrate very much, and can be thought of as being fixed in more or less one place. As a fast-moving electron passes between two positive ions, the ions are attracted to the electron and start to move toward it. By the time the ions respond, however, the electron is long gone. Nevertheless, when the ions move close together, they create a region in the material with a more positive electrical charge than normal. This region attracts a second electron and pulls it in. Thus the two electrons can move through the superconducting material something like the way two bike racers move down a track, with one running interference for the other.

At the very low temperatures at which a material becomes superconductive, electrons hook up in pairs, and the pairs start to interlock like links of a complex, tangled matrix. While individual electrons are very light, the whole collection of interlocked electrons in a superconductor is quite massive. If one electron encounters an ion, the electron can't easily be deflected. In fact, to change the velocity of any electron—something you would have to do to get energy from it—you would have to change the velocity of *all* the electrons. Since this can't be done, no energy is given up in such collisions, and electrons simply move through the material together. If the temperature is raised, though, the ions will vibrate more vigorously and will no longer be able to form the delicate minuet required to produce the electron pairs. Thus, above the critical temperature, superconductivity breaks down.

Science in the Making

High-Temperature Superconductors

Until the mid-1980s, all superconducting materials had to be cooled in liquid helium, a refrigerant that boils at a few degrees above absolute zero, because none of these materials was capable of sustaining superconductivity above about 20 Kelvins. Acting on a hunch, scientists Karl Alex Müller and George Bednorz of IBM's Zurich, Switzerland, research laboratory began a search for new superconductors. Traditional superconductors are metals and alloys, but Bednorz and Müller decided instead to focus on oxides—chemical compounds that contain oxygen, such as most rocks and ceramics. It was an odd choice, for oxides usually make the best electrical insulators, although occasionally an oxide will conduct electricity.

Working with little encouragement from their peers and no formal authorization from their employers, the scientists spent many months mixing chemicals, baking them in an oven, and testing for superconductivity, all with no success. All they seemed to be doing was wasting their time and IBM's resources, but they continued doggedly.

The breakthrough came on January 27, 1986, when a small black wafer of baked chemicals was found to become superconductive at greater than 30 degrees above absolute zero—a temperature that shattered the old record and ushered in the era of "high-temperature" (though still extremely cold) superconductors. Their compound of copper, oxygen, and other elements seemed to defy all conventional wisdom, and it began a frantic race to study and improve the novel material.

Today, oxide materials closely related to those first described by Bednorz and Müller have been found to superconduct at temperatures as high as 160 degrees above absolute zero (see Figure 10–4). It may soon be possible to make commercially useful electrical devices out of these materials, a prospect that would lower the costs of using superconducting magnets.

Perhaps equally important, high-temperature superconductors have taken superconductivity from the domain of a few low-temperature laboratories and brought it into classrooms around the world. As a new generation of scientists grows up with this new kind of superconductor, exciting new ideas and inventions are sure to be found. Within the next generation we may have electric motors that rotate a million revolutions per minute on superconducting bearings, superconducting electrical storage facilities that reduce our energy bills, and magnetically levitated trains that travel at jet speeds between cities. ■

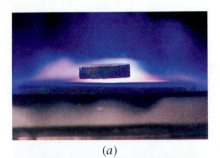

(a)

(b)

Figure 10–4

(*a*) A magnet floats "magically" above a black disk made from a new "high-temperature" superconductor. The clouds in the background form above the cold liquid-nitrogen refrigerant. (*b*) This phenomenon has been used in Japan to float high-speed trains above their tracks.

Magnetic Properties of Materials

The magnets that lie at the heart of most electrical motors and generators, though critical to almost everything we do, are not much evident in our everyday lives. Similarly, we are usually unaware of the magnets that drive our stereo speakers, telephones, and other audio systems. Even refrigerator magnets and compass needles are so common that we take them for granted. But why do some common materials such as iron display strong magnetism, while other substances seem to be unaffected by magnetic fields?

In Chapter 5, we learned that one of the fundamental laws of nature is that every magnetic field is due, ultimately, to the presence of electric currents. In particular, the movement of electrons in the orbits around atoms can be thought of as a small electrical current, so each electron in an atom acts like a little electromagnet. An atom with many electrons, then, can be thought of as being composed of many small electromagnets, each with different strength and pointing in a different direction. The total magnetic field of the atom arises by adding together the magnetic fields of all the tiny electron electromagnets.

It turns out that many atoms have magnetic fields that closely approximate the dipole type (originally shown in Figure 5–4) so that each atom in the material can be thought of as a tiny dipole magnet (Figure 10–5). If this is the case, it is not hard to see how a solid material like a piece of lodestone could have a magnetic field. It is somewhat harder to understand why most materials do *not* have magnetic fields.

In Figure 10–6a, we show the orientation of atomic magnets in a typical material. They point in random directions, so at a place outside the material, their effects tend to cancel. An observer looking at the material will measure no magnetic field, and a compass placed outside the material will not be deflected. This ordinary situation explains how materials made up of tiny magnets can, as a whole, be nonmagnetic.

A few materials in nature, including iron, cobalt, and nickel metals, do not show a random scrambling of atomic magnets. This effect, called **ferromagnetism,** has to do with the details of the sizes of atoms and the separations between them. In a material such as iron, for example, we often find a situation like the one shown in Figure 10–6b. Forces between atoms in these materials tend to favor situations in which the neighboring atoms line up with each other into small magnetic *domains*—regions a few thousand atoms across. Thus, in a normal piece of iron, atoms within a specific domain will all be lined up pointing in the same direction, but the orientation of domains is random. Again, someone standing outside this material will not measure a magnetic field, because the magnetic fields due to different domains cancel each other.

In special cases, as when a ferromagnetic material such as iron cools from very high temperature in the presence of a strong magnetic field, all of the neighboring domains may line up and thus reinforce each other. Only when most of the magnetic domains line up (as shown in Figure 10–6c) do you get a material that exhibits an external magnet field—the arrangement that occurs in permanent magnets.

The ultimate cause of the magnetic field is the movement of individual electrons around individual atoms. These motions must be aligned in the

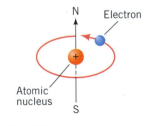

Figure 10–5

The dipole magnetic field of atoms takes the same form as that of larger magnets.

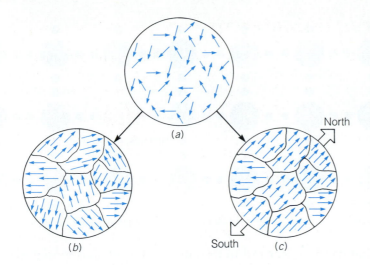

Figure 10–6
Different magnetic behavior in materials. (*a*) Nonmagnetic materials have random orientations of atomic spins. (*b*) Ferromagnetic materials with randomly oriented domains are not magnetic. (*c*) A permanent magnet has uniformly oriented atomic spins.

material, however, in order to form a permanent magnet. This alignment requires the creation of magnetic domains within the material, as well as the lining up of these domains parallel to each other.

Microchips and the Information Revolution

Every material has hundreds of different physical properties. We have already seen how strength, electrical conductivity, and magnetism all result from the properties of individual atoms and how those atoms bond together. We could continue in this vein for many more chapters, examining optical properties, elastic properties, thermal properties, and so on. But such a treatment would miss another key idea about materials: New materials often lead to new technologies that change society.

Of all the countless new materials discovered in the last century, none has transformed our lives more than silicon-based semiconductors. From personal computers to auto ignitions, portable radios to sophisticated military weaponry, microelectronics are a hallmark of our age. Indeed, semiconductors have fundamentally changed the way we manipulate a society's most precious resource—information.

The key to this revolution is our ability to fashion complex crystals atom-by-atom from silicon, a material that is produced from ordinary beach sand.

Doped Semiconductors

The element silicon by itself is not a very useful substance in electrical circuits. What makes silicon useful, and what has driven our modern microelectronic technology, is a process known as doping. **Doping** is the addition of a minor impurity to an element or compound. The idea behind silicon doping is simple. When silicon is melted before being made into circuit elements, a small amount of some other material is added to it. One common additive, for example, is phosphorus, an ele-

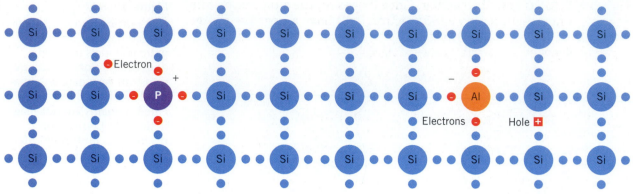

(*a*) Phosphorus-doped silicon *n*-type semiconductor

(*b*) Aluminum-doped silicon *p*-type semiconductor

Figure 10–7

(*a*) Phosphorus-doped silicon *n*-type semiconductors and (*b*) aluminum-doped silicon *p*-type semiconductors are formed from silicon crystals with a few impurity atoms.

ment that has five electrons in its outermost orbit, as opposed to the four of silicon.

When the silicon crystallizes to form the structure shown in Figure 10–7, the phosphorus is taken into the crystalline structure. Of the five outer electrons in each phosphorus atom, however, only four are needed to make bonds in the crystal. The fifth electron is not locked in at all. In this situation, it does not take long for the extra electron to be shaken loose and wander off into the body of the crystal. This action has two important consequences: (1) there are conduction electrons in the material that do not come from the creation of holes in the lattice, and (2) the phosphorus ion that has been left behind has a positive charge. A semiconductor prepared in this way is said to be an *n-type semiconductor*, because the moving charge is a negative electron.

Alternatively, silicon can be doped with an element such as aluminum, which has only three electrons in its outer orbit. In this case, when the aluminum is taken into the crystal structure, there will be a missing electron—a hole—capable of carrying electricity. The hole need not stay with the aluminum, of course, but is free to move around within the

A technician wears protective clothing to maintain the ultraclean environment required in a semiconductor factory.

semiconductor as described earlier. Once it does so, the aluminum atom, which has now acquired an extra electron, will have a negative charge. This type of semi-conductor is called *p-type,* because a hole—a missing negative electron—acts as a positive charge.

Diodes

You can understand the basic workings of a microchip by conducting an experiment in your mind. Imagine taking a piece of n-type semiconductor and placing it against a piece of *p*-type semiconductor. As soon as the two types of material are in contact, you would expect electrons to diffuse from the *n*-type semiconductor over into the *p*-type, while holes would diffuse back the other way. Over a period of time, however, you might expect that most of the electrons in the semiconductor boundary region would actually fall into the holes (provided that there were equal numbers of each), so that eventually both free electrons and free holes would disappear. On one side of the boundary between the two semiconductors there would be a region where negative aluminum ions—ions locked into the crystal structure by the doping process—had acquired an extra electron. Conversely, on the other side of the boundary, you would see an array of positive phosphorus ions, each of which had lost an electron, but which were nonetheless locked into the crystal structure.

A semiconducting device like this—formed from one *p* and one *n* region—is called a **diode** (see Figure 10–8). Once a diode is constructed, a permanent electrical field tends to push electrons across the boundary in only one direction, from the *n*-type side to the *p*-type side. As electrons are pushed "with the grain" in the diode—from negative to positive— the current flows through normally. When the current is reversed, however, the electrons are blocked from going through by the presence of the built-in electric field. Thus the diode acts as a one-way gate, allowing the electrical current through in only one direction.

The semiconductor diode has many uses in technology. One use, for example, can be found in almost any electronic device that is plugged into a wall outlet. As we saw in Chapter 5, electricity is sent to homes in the form of alternating current—AC. It turns out, however, that most home electronics such as televisions and stereos require direct current— DC. A semiconductor diode can be used to convert that alternating current into direct current by blocking off half of it. In fact, if you examine the insides of almost any electronic gear, the power cord leads directly to a diode and other components that convert pulsing AC into steady DC, as shown in Figure 10–9.

Figure 10–8
The semiconductor diode consists of a negatively charged *n* region attached to a positively charged *p* region. Electrons in this diode can flow easily from the negative to the positive region, but will not flow the opposite way. The result is a one-way valve for electrons.

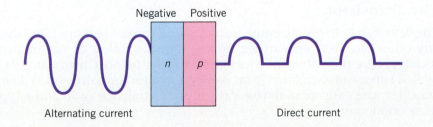

Negative   Positive

n   *p*

Alternating current   Direct current

Figure 10–9
A diode converts alternating current to direct current in most electronic devices. Half of the alternating current passes through the diode, but the other half is blocked.

Barstow Solar One converts energy to electricity through an array of photovoltaic cells at this plant in California's Mojave Desert.

Technology

Photovoltaic Cells and Solar Energy

Semiconducting diodes may play an important role in the energy future of the United States, through the use of a device called the photovoltaic cell. A photovoltaic cell is nothing more than a large semiconductor diode. A thin layer of *n*-type material overlays a thicker layer of *p*-type. Sunlight striking the top *n*-type layer shakes electrons loose from the crystal structure. These electrons are then accelerated through the *n-p* boundary and pushed out into an external circuit. Thus, while the sun is shining, the photovoltaic cell acts in the same way as a battery. It provides a constant push for electrons and moves them through an external circuit. If large numbers of photovoltaic cells are put together, they can generate enormous amounts of current.

Photovoltaic cells enjoy many uses today. Your hand calculator, for example, may very well contain a photovoltaic cell that recharges the batteries (it's the small dark band just above the buttons). Photovoltaic cells are also used in regions where it is hard to bring in traditional electricity—to pump water in remote sites, for example, or to provide electricity in wilderness areas of national parks. ■

The Transistor

The device that drives the entire information age, and perhaps more than any other has been responsible for the transformation of our modern society, is the transistor. Invented just two days before Christmas, 1947, by Bell Laboratory scientists John Bardeen, Walter Brattain, and William Shockley, the early **transistor** was simply a sandwich of *n*- and *p*-type semiconductors.

In one kind of transistor, two *p*-type semiconductors form the "bread" of a sandwich, while the *n*-type semiconductor is the "meat." Another kind of transistor uses the *npn* configuration. Both kinds are shown in Figure 10–10. Electrical leads connect to each of the three semiconductor regions of the transistor. An electrical current goes into the region called the *emitter,* the thin slice of semiconductor in the middle is called the *base,* and the third semiconductor section is the *collector.*

The transistor, then, has two built-in electrical fields—one at each *n-p* junction. The idea of the transistor is that a small amount of electrical charge run into or out of the base can change these electrical fields—in effect, opening and closing the gates of the transistor. The best way to think of the transistor is to make an analogy to a pipe that carries water. The electrical current that flows from emitter to collector is like water that flows through the pipe, and the base is like a valve in the pipe. A small amount of energy applied to turning the valve can have an enormous effect on the flow of water. In just the same way, a small amount of charge run onto the base can have an enormous effect on the current that runs through the transistor.

In your tape deck, for example, small electrical currents are created when the magnetized tape is run past the tape heads. This small current can be fed into the base of a transistor, and can thus be impressed upon the much larger current that is flowing from the emitter to the collector. A device that takes a small current and converts it into a large one is called an *amplifier* (see Figure 10–11). The amplifier in your tape deck takes the small signal created by the tape and converts it into the much larger current that runs the speakers.

As important as the transistor's amplifying properties are, probably its most important use has been as a switch. If you run enough negative charge onto the base, it can repel any electrons that are trying to get

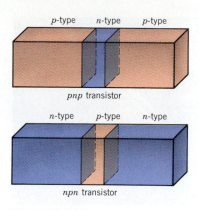

Figure 10–10

A *pnp* and an *npn* transistor.

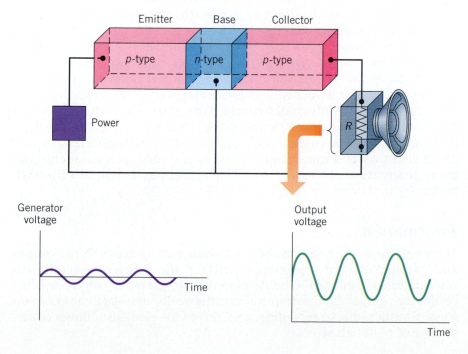

Figure 10–11

A transistor acting as an amplifier. A small amount of energy, supplied by a power source such as a wall outlet to one region of a transistor, can be amplified because electrons flow easily between charged regions.

Figure 10–12

A microchip incorporates many complex circuits built into a single piece of silicon. For comparison, the eye of an ordinary sewing needle is shown in the background.

through. Thus moving electrical charge onto the base will shut off the flow of current through the transistor, while running the electrical charge off the base will turn the current back on. The transistor, then, operates as an electron switch in much the same way as the faucet on your sink acts as a water switch. The fact that a transistor can be operated as a switch means that it can be used as the basic working element of the computer—arguably the most important device developed in the twentieth century.

Microchips

Individual diodes and transistors still play a vital role in modern electronics, but these simple devices have been largely replaced by much more complex arrays of *p*- and *n*-type semiconductors, called **microchips** (see Figure 10–12). Microchips may incorporate hundreds or thousands of transistors in one *integrated circuit,* specially designed to perform a specific function. An integrated circuit microchip lies at the heart of your pocket calculator or microwave oven control, for example. Similarly, arrays of integrated circuits store and manipulate data in your personal computer, and they regulate the ignition in all modern automobiles.

The first transistors were bulky things, about the size of a golf ball, but today a single microchip can integrate hundreds of thousands of these devices. California's Silicon Valley has become a well-known center for the design and manufacture of these tiny integrated circuits.

The production of hundreds of thousands of transistors on a single silicon chip requires exquisite control of atoms. One way of going about it is to put a thin wafer of silicon into a large heated vacuum chamber. Around the edges of the chamber is an array of small ovens, each of which holds a different element, such as aluminum or phosphorus. The side ovens are heated in carefully controlled sequence and opened to allow small amounts of other elements—the dopants—to be vaporized and enter the chamber along with silicon.

If you wanted to make a *p*-type semiconductor, for example, you could mix a small amount of phosphorous with the silicon in the chamber and let it deposit out on the silicon plate at the bottom. Typically, a device called a mask is put over the silicon chip so that the *p*-type semiconductor is deposited only in designated parts of the chip. Then the vapor is cleared from the chamber, a new mask is put on, and another layer of material is laid down. In this way, a complex three-dimensional structure can be built up at a microscopic scale. In the end, each microchip has many different transistors in it, connected exactly as designed by the engineers.

Information

The single most important use of semiconducting devices is in the storage and manipulation of information. In fact, the modern revolution in information technology—the development of arrays of interconnected computers, global telecommunications networks, vast data banks of personal statistics, digital recording, and the credit card—is a direct consequence of materials science.

While it may appear strange to say so, almost all the things we normally consider as conveying information—the printed or spoken word, pictures, or music, for example—can be analyzed in terms of their information content and manipulated by the microchips we've just discussed. The term "information," like many words, has a precise meaning when it is used in the sciences, a meaning that is somewhat different from colloquial usage. In its scientific context, information is measured in a unit that is called the *binary* dig*it*, or **bit.**

You can think of the bit as the two possible answers to any simple question: yes or no, on or off, up or down. A single transistor being used as a switch, for example, can convey one bit of information—it is either on or off. Any form of communication contains a certain number of bits of information. As we shall see shortly, the computer is simply a device that stores and manipulates this kind of information.

Let's begin by asking a very simple question. What is the information content of a single letter of the alphabet? From the point of view of an information theorist, the answer is simply the minimum number of questions with "yes or no" answers that are required to identify a letter of the alphabet unambiguously. Let's see how you might go about asking such questions. Below are five questions you might ask to specify the letter E.

1. Is it in the first half of the alphabet? (yes)
2. Is it in the first six letters of the alphabet? (yes)
3. Is it in the first three letters of the alphabet? (no)
4. Is it D? (no)
5. Is it E? (yes)

From this simple example, you can see that you need the answers to five questions—five bits of information—to specify unambiguously a single letter of the alphabet. (You might be lucky and guess the answer in fewer questions, but five questions are always enough to pinpoint any one of the 26 letters.) We would say, then, that the information content of a single letter of the alphabet is five bits.

As a matter of fact, there are $2 \times 2 \times 2 \times 2 \times 2 = 32$ different things that can be specified using five bits of information. Thirty-two items is not enough, however, to handle all the numbers, capitalization, punctuation marks, and other symbols. Six bits of information are capable of specifying $2 \times 2 \times 2 \times 2 \times 2 \times 2 = 64$ different things, and you could argue that all the things you'd need to specify on a printed page could be included in those 64. Thus the information content of a single ordinary printed symbol is six bits. If you were using switches to store the information on a normal printed page, you would need six of them lined up in a row to specify each symbol.

The average word is six letters long, so the information content of a typical word is 36 bits. The average printed page of a novel contains about 500 words, corresponding to an information content of almost 20,000 bits. A 200-page book in this scheme thus contains about 4 million bits, or 4 megabits, of information.

Historically, switches in computers were lumped together in groups of eight. Such a group is capable of storing eight bits of information, or one **byte.** In terms of this unit, a 200-page book would contain 500,000 bytes, or half a megabyte.

Science by the Numbers

Is a Picture Really Worth a Thousand Words?

Pictures and sounds can be analyzed in terms of information content, just like words. Your television screen, for example, works by splitting the picture into small units called pixels. In North America, the picture is split up into 525 segments on the horizontal and vertical axes, giving a total of about 275,000 pixels (in rounded numbers) for one picture on the TV screen. Your eye integrates these dots into a smooth picture. Every color can be thought of as a combination of the three colors red, green, and blue, and it is usual to specify the intensity of each of these three colors by a number that requires 10 bits of information to be recorded. (In practice, this means that the intensity of each color is specified on a scale of about 1 to 1000.) Thus each pixel requires 30 bits to define its color. The total information content of a picture on a TV screen, then, is

$$275,000 \text{ pixels} \times 30 \text{ bits} = \text{about 8 million bits.}$$

Thus it requires about 8 megabits, or 1 megabyte, to specify a single frame on a TV picture. In passing, we should note that a TV picture typically changes 30 times a second, so the total flow of information on the TV screen may exceed 200 million bits per second.

It would appear, then, that a picture is not only worth a thousand words, but much more. In fact, if a word contains 36 bits of information, then the picture will be worth 8 million bits per picture divided by 36 bits per word, which equals about 220,000 words per picture.

The old saying, if anything, underestimates the truth! ■

Computers

A **computer** is a machine that stores and manipulates information. The information is stored in the computer in transistors, with groups of transistors acting as switches that carry information. In principle, if you had a machine with a few million transistors in it, you could store the text for this entire book. In practice, however, computers do not normally work in this way. They have a *central processing unit (cpu)* in which transistors store and manipulate relatively small amounts of information at any one time. When the information is ready to be stored—for example, when you have finished working on a text in a word processor, or writing a program to perform a calculation—it is removed from the cpu and stored elsewhere. It might, for example, be stored in the form of magnetically oriented particles on a floppy disc or a hard drive. In these cases, a bit

In less than a quarter century the computer has evolved from a specialized research aid to an essential tool for business and education.

of information is no longer a switch that is on or off, but a bit of magnetic material that has been oriented either "north pole up" or "north pole down."

A typical computer can store hundreds of millions of bits (hundreds of megabytes) in magnetic form, and can hold an additional tens of millions of bits of information in its cpu. The ability to store information in this way is extremely important in modern society. As just one example, think about the last time you made an airline reservation. You called in, and your travel agent got into communication with the airline's computer. Stored in strings of bits within that computer are the flights, the seating assignments, the ticket arrangements, and often the address and phone number of every passenger who will be flying on the particular day when you want to fly. When you change your reservation, make a new one, or perform some other manipulation, the information is taken out of storage, brought to the central processing unit, manipulated by changing the exact sequence of bits, and then put back into storage. This process—the storage and manipulation of vast amounts of data—forms the very fabric of our modern society.

The Human Body

The Computer and the Brain

When computers first came into public awareness, there was a general sense that we were building a machine that would in some way duplicate the human brain. Fields like *artificial intelligence* were sold (some would say oversold) on the basis of the idea that they would soon be able to perform all those functions that we normally think of as being distinctly human. In fact, this has not happened. The reason has to do with the difference between the basic unit of the computer, which is the transistor, and the basic unit of the brain, which is the nerve cell.

The transmission of electrical signals between the brain's axons is fundamentally different than in normal electrical circuits (see Chapter 5). This difference in signal transmission alone, however, does not make a brain so different from a computer. A computer normally performs a sequential series of operations—that is, a group of transistors takes two numbers, adds them together, feeds that answer to another group of transistors that performs another manipulation, and so on. Some computers are now being designed and built that have some parallel capacity—machines in which, for example, the addition and other manipulations are done in parallel rather than one after the other. Nevertheless, the natural configuration of computers is to have each transistor hooked to, at most, a couple of others.

A nerve cell in the brain, however, operates in quite a different manner. A typical nerve cell has on the order of a thousand projections from it. These projections are called *dendrites* and, in general, each dendrite on a nerve cell in the brain is connected to a different neighboring nerve cell in the brain. Thus each of the trillions of cells in the brain is connected to a thousand other cells. Whether a nerve cells decides to fire—whether or not the signal moves out along the axon—depends in an unknown way on a complex integration of all the signals that come into that cell from a thousand other cells.

This complex arrangement means that the brain is a system that is highly interconnected—more interconnected than any other system known in nature. In fact, if the brain has trillions of cells and each cell has a thousand connections, there will be on the order of 1,000,000,000,000,000 connections between brain cells. Building a computer of this size and level of connectedness is at present totally beyond the capability of technology. ■

THINKING MORE ABOUT THINKING

Thinking Machines

One of the questions that often intrigues people when they think about complex computers is whether or not a computer can be built that would in some way mimic or replace the human brain. Computers have been built that can add faster or remember more than any single human being, but these specific abilities by themselves do not seem to be crucial in developing a machine that thinks. The real question is whether or not a machine will be designed that is, by general consensus, regarded as "alive" or "conscious."

British mathematician Alan Turing proposed a test that was designed to address this question. The so-called "Turing test" operates this way: A group of human beings sit in a room and interact with something through some kind of computer terminal. They might, for example, type questions into a keyboard and read answers on a screen. Alternatively, they could talk into a microphone and hear answers played back to them by some kind of voice synthesizer. These people are allowed to ask the hidden "thing" any questions they like. At the end of the experiment, they have to decide whether or not they are talking to a machine or a human being. If they can't tell the difference, the machine is said to have passed the Turing test.

As of this date, no machine has passed the test, although there have been occasional contests in Silicon Valley in which machines were put through their paces. But what if a machine did actually pass? Would that mean we have invented a truly intelligent machine? John Searle, a philosopher at the University of California at Berkeley, has recently challenged the whole idea of the Turing test as a way of telling if a machine can think by proposing a paradox he calls the "Chinese room."

The Chinese room works like this: An English-speaking person sits in a room and receives typed questions from a Chinese-speaking person in the adjacent room. The English-speaking person does not understand Chinese, but has a large manual of instructions. The manual might say, for example, that if a certain group of Chinese characters are received, then a second group of Chinese characters should be sent out. The English-speaking person could, at least in principle, pass the Turing test if the instructions were sufficiently detailed and complex. Obviously, however, the English speaker has no idea of what he or she is doing with the information that comes in or goes out. Thus, argues Searle, the mere fact that a machine passes the Turing test tells you nothing about whether it is aware of what it is doing.

Do you think a machine that can pass the Turing test must be aware of itself? Do you see any way around Searle's argument for the Chinese room? What moral and ethical problems might arise if human beings could indeed make a machine that everyone agreed was aware and conscious?

Summary

All materials, from building supplies and fabrics to electronic components and food, have properties that arise from the kinds of constituent atoms and the ways those atoms are bonded together. The high *strength* of materials such as stone and synthetic fibers relies on interconnected networks of ionic or covalent bonds, while many soft and pliable materials such as soap and graphite incorporate weak van der Waals forces. *Composite materials,* such as plywood, fiberglass, and reinforced concrete, merge the special strengths of two or more materials.

The electrical properties of materials also depend on the kinds of constituent atoms and the bonds they form. *Electrical resistance,* a material's resistance to the flow of an electric current, for example, depends on the mobility of bonding electrons. Metals, which are characterized by loosely bonded outer electrons, make excellent *conductors,* while most materials with tightly held electrons in ionic and covalent bonds are good electrical *insulators.* Materials such as silicon that conduct electricity, but not very well, are called *semiconductors.* At very low temperatures some compounds lose all resistance to electricity and become *superconductors.*

Magnetic properties also arise from the collective behavior of atoms. While most materials are nonmagnetic, *ferromagnets* have domains in which electron spins are aligned parallel to each other.

New materials play important roles in modern technology. Semiconductors, in particular, are vital to the modern electronics industry. Semiconductor material, usually silicon, is modified by *doping* with small amounts of another element. Phosphorous doping, for example, adds a few extra negative electrons to produce an *n*-type semiconductor, while aluminum doping provides positive holes in *p*-type semiconductors. Devices formed by juxtaposing *n*- and *p*-type semiconductors act as switches and valves for electricity. A *diode* joins single pieces of *n*- and *p*-type material, for example, to act as a one-way valve for current flow. *Transistors,* which incorporate a *pnp* or *npn* semiconductor sandwich, act as amplifiers or switches for current. *Microchips* can combine up to thousands of *n* and *p* regions in a single integrated circuit that performs complex operations.

Semiconductor technology has revolutionized the storage and use of information. All kinds of information can be reduced to a series of simple "yes-no" questions, or *bits.* Eight-bit words, called *bytes,* are the basic information unit of most modern *computers.*

Key Terms About Material Properties

strength	insulator	doping	bit
composite materials	semiconductor	diode	byte
conductor	superconductivity	transistor	computer
electrical resistance	ferromagnet	microchip	

Review Questions

1. Define three different kinds of strengths. Give examples of materials that are strong in each of these modes.

2. What kinds of bonds are in the strongest materials we know? Why are these bonds so strong?

3. What three atomic factors determine the strength of a material?

4. Diamonds and graphite are both made from carbon atoms. Why are diamonds so much stronger?

5. What is a composite material? Give an example.

6. What is an electrical insulator? Give an example.

7. What is an electrical conductor? Give an example.

8. What is a superconductor? What advantages are gained when they are used to make a magnet?

9. What is a semiconductor? Give an example.

10. Explain how holes can move in a semiconductor.

11. How do electrons behave in the atoms of a ferromagnet? How might a natural ferromagnet form?

12. If the magnetic fields of individual atoms in a material are arranged randomly, is the material magnetic? Why or why not?

13. How can you create a permanent magnet?

14. What is a semiconductor diode? How do diodes convert AC into DC?

15. What is a transistor? What are the base, emitter, and collector, respectively?

16. What is an integrated circuit? How might one be made?

17. What is a bit of information? A byte of information?

18. What is a computer?

19. How many connections are there between cells in the brain? How does this compare to the connections in a large computer?

20. What is the Turing test?

Discussion Questions

1. How do the principles of physics and chemistry both come into play when developing new materials?

2. From the point of view of atomic architecture, how would you design a material that has great strength when pulled apart? Great strength when twisted?

3. How can you explain electrical resistance in terms of the second law of thermodynamics?

4. Why do workers often wear rubber gloves when dealing with electricity?

5. Was the research conducted by Bednorz and Müller on superconductivity an example of "big science" or "little science"?

6. Now that the cold war between the United States and the former Soviet Union is over, the United States government recently announced plans to share its hitherto secret research on strong yet lightweight materials with U.S. automakers. Do you agree with this plan? What benefits and risks do you see?

7. Identify five different materials that were invented in the twentieth century and that you use in your daily life.

8. Identify 10 objects in your home that use semiconductors. What other kinds of materials with special electrical properties are found in all of these 10 objects?

9. How would your life change if you had none of your electrical appliances?

Problems

1. There is an effort in the world today to convert television into so-called high definition TV (HDTV). In HDTV, the picture is split up into as many as 1100 by 1100 (as opposed to 525 by 525) pixels. What is the information content of an HDTV picture? What is the information content that must be transmitted each second in an HDTV broadcast?

2. Construct a set of "yes or no" questions to specify any letter of the alphabet or digit from zero to nine.

3. Construct a set of "yes or no" questions to specify any of the 50 United States.

4. Estimate the total amount of information contained in the printed words in this chapter.

Investigations

1. Why does a magnet become demagnetized when you repeatedly hit it with a hammer? What other ways can you destroy a permanent magnet? Why?

2. Shortly after the discovery of high-temperature superconductivity, many newspapers and TV shows ran features on how these new materials would change society. In what ways might superconductivity change society? Historically, what other new materials have caused significant changes in human societies?

3. Imagine that you are a science fiction writer. Concoct a description of a new material with unique (but plausible) properties and describe how that material might change a society.

4. Research the status of magnetically levitated trains, like the one now operating in Japan. How does it operate? How fast might it go? How soon might such a train operate in North America?

5. Read the book or watch the movie *The Man in the White Suit*. What unique material properties are described, and how is the new technology received by society?

6. Visit a sports equipment store. Learn about the new materials that are used in tennis rackets, football helmets, and sports clothing.

7. Write a short story in which a new material with unique properties plays a central role.

8. What kinds of materials do surgeons use to replace broken hip bones? What are the advantages of this material?

9. Until recently, plastic surgeons used silicone-filled implants for breast enlargement and other procedures. Intensive research is now underway to understand the effects of silicone on the human body, due to numerous adverse reactions to these implants. Investigate the nature of silicone and this recent research. What are the results of these investigations? Originally, why did scientists think silicone would be a safe substance?

Additional Reading

Dennett, Daniel. *Consciousness Explained.* New York: Little, Brown, 1991.

Gordon, J.E. *The New Science of Strong Materials: Or why you don't fall through the floor.* Princeton: Princeton University Press, 1976.

Gordon, J.E. *Structures, or Why Things Don't Fall Down.* New York: Plenum, 1978.

Hazen, Robert M. *The Breakthrough: The race for the superconductor.* New York: Summit Books, 1988.

Rossotti, Helen *Why Things Aren't Grey.* Princeton: Princeton University Press, 1987.

Vrainich, Joseph. *Supertrains.* New York: St. Martin's Press, 1991.

The Nucleus of the Atom

Nuclear energy depends on the conversion of mass.

A Random Walk
A Book Full of Energy

Right now you are sitting down reading a heavy book. How much energy are you holding in your hands? You could drop the book, releasing a little bit of gravitational potential energy (enough to hurt your toe, anyway). You could burn the book (we don't encourage this experiment) and generate heat energy for half an hour or so. But your book holds vastly more energy than that.

If you could tap all the energy stored in this book, you could generate all the electricity used by a very large city in a year. That's a lot of energy—it corresponds to the output of three large power plants operating continuously, burning tons of coal day and night. How can all that energy be stored in such a small container?

The answer: mass is itself a form of energy, as we saw in Chapter 3. We usually don't observe this principle in operation in our daily lives, but inside the nucleus of the atom, where matter is packed densely, the relationship between mass and energy is of the utmost importance.

Empty Space, Explosive Energy

Each grain of sand at the beach contains enough energy, in the form of its mass, to provide all the energy you use in a year.

Imagine that you are holding a basketball, while 25 kilometers (about 15 miles) away a few grains of sand whiz around. All of the vast intervening space—enough to house a fair-sized city—is absolutely empty. No rocks or soil, no grass or trees, not even any air or water fills the void. In some respects, that's what an atom is like, though on a much smaller scale, of course. The nucleus is the basketball, and the grains of sand represent the electrons (remember, though, that electrons in an atom display characteristics of both particles and waves). The atom, with a diameter 100,000 times that of its nucleus, is almost all empty space. (Refer to Chapter 7 for a review of the structure of an atom.)

The previous chapters explored the properties of atoms in terms of their electrons. Chemical reactions, the way a material handles electricity, and even the very shape and strength of objects all depend on the way that electrons in different atoms interact with each other. In terms of our analogy, all of the properties of the atoms that we have studied so far result from things that are taking place 25 kilometers from where the basketball-sized nucleus is sitting. The incredible emptiness of the atom is a key to understanding two important facts about the relation of the atom to its nucleus.

1. *What goes on in the nucleus of an atom has almost nothing to do with the atom's chemistry, and vice versa.* In other words, the chemical bonding of an atom's electrons has virtually no effect on what happens to the nucleus. In most situations you can regard the orbiting electrons and the central nucleus as two separate and independent systems. This is why physicists instead of chemists usually study the nucleus.

2. *The energies available in the nucleus are much greater than those available among electrons.* The particles inside the nucleus are tightly locked in. It takes a great deal more energy to pull them out than it does to remove an electron from an atom.

The enormous energy we can get from the nucleus follows from the equivalence of mass and energy (which we will discuss in more detail in Chapter 13). This relationship is defined in Einstein's most famous equation.

▶ **In words:**

Mass is a form of energy. When mass is converted into energy, the amount of energy produced is enormous—equal to the mass of the object multiplied by the square of the speed of light.

▶ **In equation form:**

Energy = mass × (speed of light)^2,

▶ **In symbols:**

$E = mc^2$.

Remember that the constant c, the speed of light, is a very large number (3×10^8 meters per second), and that large number is squared in Einstein's equation to give an even larger number. Thus even very small amounts of mass are equivalent to very large amounts of energy, as shown in the calculation in the "Science by the Numbers" in this chapter.

Einstein's equation tells us that a given amount of mass can be converted into a specific amount of energy in any form, and vice versa. Technically speaking, this statement is true for any process involving energy. When hydrogen and oxygen combine to form water, for example, the mass of the water molecule is infinitesimally less than the sum of the masses of the original atoms. This missing mass has been converted to binding energy in the molecule. Similarly, when an archer draws a bow, the mass of the bow increases by a tiny amount because of the increased elastic potential energy in the bent material.

The change in mass of objects in everyday events like these is so small that it is customarily ignored, and we speak of the various forms of energy without thinking about their mass equivalents. In nuclear reactions, however, we cannot ignore the mass effects. A nuclear reactor, for example, can transform fully 20% of the mass of a proton into energy in each reaction by a process we will soon discuss. Thus nuclear reactions can convert significant amounts of mass into energy, while chemical reactions, which involve only relatively small changes in electrical potential energy, involve only infinitesimal changes in mass. This difference explains why an atomic bomb, which derives its destructive force from nuclear reactions, is so much more powerful than conventional explosives, such as dynamite, and conventional weapons, which depend on chemical reactions in materials such as TNT.

Science by the Numbers

Mass and Energy

On the average, each person in the United States uses energy at the rate of about one kilowatt-hour each hour. In effect, each individual uses the energy equivalent of a toaster going full blast all the time. How much mass would have to be converted completely to energy to produce your year's supply of energy?

First, we have to calculate the number of hours in a year:

$$24 \text{ hours/day} \times 365 \text{ days/year} = 8760 \text{ hours/year}.$$

The total energy used by one person is thus about 8760 kilowatt-hours every year.

In Appendix A, we find that one kilowatt-hour of energy is the same as 3.6 million joules, so every year each of us uses

$$8760 \times 3.6 \times 10^6 \text{ joule}$$
$$= 3.15 \times 10^4 \times 10^6 \text{ joule}$$
$$= 3.15 \times 10^{10} \text{ joule}.$$

In order to calculate the mass that is equivalent to this large amount of energy, we need to put this energy into the Einstein equation, which we can rewrite as

$$\text{mass} = \frac{\text{energy}}{(\text{speed of light})^2}$$

Written in this form, the number we seek (the mass) is expressed in terms of two numbers we already know. The speed of light, c is 3×10^8 meter/second, so we find:

$$
\begin{aligned}
\text{mass} &= \frac{3.15 \times 10^{10}\,\text{joule}}{(3 \times 10^8\,\text{m/sec})^2} \\
&= \frac{3.15 \times 10^{10}}{9 \times 10^{16}\,\text{m}^2/\text{sec}^2} \\
&= 3.5 \times 10^{-7}\,\text{joule-second}^2/\text{meter}^2 \\
&= 3.5 \times 10^{-7}\,\text{kilograms}
\end{aligned}
$$

In the last step we have to remember that a joule is defined as a kilogram-meter^2/second^2, so the units "joule-second^2/meter^2" in the answer above are exactly the same as kilograms (see Appendix A). Our year's energy budget could be satisfied by a mass that weighs less than a millionth of a kilogram, or about the mass of a very fine sand grain —if you could unlock that energy! ∎

The Organization of the Nucleus

Like the atom, the nucleus is not indivisible but is made up of smaller pieces, most importantly the proton and the neutron. Approximately equal in mass, the proton and neutron can be thought of as the primary building blocks of the nucleus.

The **proton** (from Latin for "the first one") has a positive electrical charge of $+1$ and was the first of the nuclear constituents to be discovered and identified. The number of protons determines the electrical charge of the nucleus. An atom in its electrically neutral state will have as many negative electrons in orbit as protons in the nucleus. Thus the number of protons in the nucleus determines the chemical identity of an atom.

When people began studying nuclei, however, they quickly found that the mass of a nucleus is significantly greater than the sum of the mass of its protons. In fact, for most atoms the nucleus is more than twice as heavy as its protons. What did this observation imply? Scientists realized that the additional mass had to be in the nucleus, and we now realize that this extra mass is supplied by a particle with no electric charge called the **neutron** (for "the neutral one"). The neutron has approximately the same mass as the proton. Thus a nucleus with equal numbers of protons and neutrons will have twice the mass of the protons alone.

The mass of a proton or a neutron is about 2000 times the mass of the electron. Therefore, almost all of the mass of the atom is contained within the protons and neutrons in its nucleus. You can think of things this way: Electrons give an atom its size, but the nucleus gives an atom its mass.

Element Names and Atomic Numbers

The most important fact in describing any atom is the number of protons in the nucleus—the **atomic number**—which defines which *element* you are dealing with. All atoms of gold (atomic number 79) have exactly 79 protons, for example. In fact, the name gold is simply a convenient shorthand for "atoms with 79 protons." Every element has its own atomic number: all hydrogen atoms have just 1 proton, carbon atoms must have 6 protons, and so on. The periodic table of the elements that we discussed in Chapter 7 can be thought of as a chart in which the number of protons in the atomic nucleus increases as we read from left to right.

The central importance of the proton in defining an atom is not surprising. The fixed number of positively charged protons dictates the number of electrons (and thus the chemical properties) of an electrically neutral atom.

Isotopes and the Mass Number

Each element has a fixed number of protons, but the number of neutrons is not so constrained. Think about what that means. Two atoms with the same number of protons may have different numbers of neutrons. Such atoms are said to be **isotopes** of each other, and they have different masses. The total number of protons and neutrons is the *mass number*.

Every element exists in several different isotopes, each with a different number of neutrons. The most common isotope of carbon, for example, has 6 neutrons, so it has a mass number of 12 (6 protons + 6 neutrons); it is usually written ^{12}C or carbon-12, and is called "carbon twelve." Other isotopes of the carbon nucleus, such as carbon-13 with 7 neutrons, and carbon-14 with 8 neutrons, are heavier than normal carbon-12, but they have the same number of electrons and, therefore, the same chemical identity. A neutral carbon atom that has 8 neutrons in its nucleus, for example, still must have 6 electrons in orbit to balance the required 6 protons, but it has a mass number of 14.

The complete set of all the isotopes—every known combination of protons and neutrons—is often illustrated on a graph that plots number of protons versus number of neutrons (see Figure 11–1). Several features

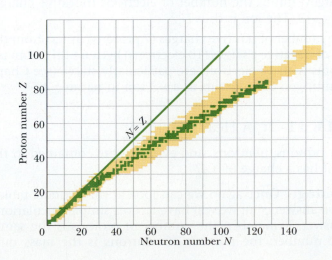

Figure 11–1

A chart of the isotopes. Stable isotopes appear in green, while radioactive isotopes are in beige. Each of the approximately 2000 known isotopes has a different combination of protons (Z on the vertical scale) and neutrons (N on the horizontal scale). Isotopes of the light elements (toward the bottom left of the chart) have similar numbers of protons and neutrons and thus lie close to the diagonal $N = Z$ line at 45 degrees. Heavier isotopes (on the upper right part of the chart) tend to have more neutrons than protons and thus lie well below this line.

are evident from this graph. First, every chemical element has many known isotopes—in some cases, dozens of them. Close to 2000 isotopes have been documented, compared to the hundred or so different elements. This plot also reveals that the number of protons is not generally the same as the number of neutrons. While many light elements, up to about calcium (with 20 protons), often have nearly equal numbers of protons and neutrons, heavier elements tend to have more neutrons than protons. This fact plays a key role in the phenomenon of radioactivity, as we shall see.

EXAMPLE 11–1: Inside the Atom

We find an atom with 7 protons and 8 neutrons in its nucleus and 10 electrons in orbit.

1. What element is it?
2. What is its mass number?
3. What is its electrical charge?
4. How is it possible that the numbers of protons and electrons are different?

▶ **Reasoning:** We can find the first three answers by looking at the periodic table, but we will refer back to Chapter 9 and the discussion of stable electron orbits for the last answer.

▶ **Solution:**

1. The element name depends on the number of protons, which is 7. A glance at the periodic table reveals that element number 7 is nitrogen.
2. Next, we calculate the mass number, which is the sum of protons and neutrons: $7 + 8 = 15$. This isotope is nitrogen-15.
3. The electrical charge equals the number of protons (positive charges) minus the number of electrons (negative charges): $7 - 10 = -3$. The ion is thus N^{-3}.
4. The number of positive charges (7 protons) differs from the number of negative charges (10 electrons) because this atom is an ion. Atoms with 10 electrons are particularly stable (see Chapter 10), so nitrogen often occurs as a -3 ion in nature. ▲

EXAMPLE 11–2: A Heavy Element

How many protons, neutrons, and electrons are contained in the atom ^{238}U when it has a charge of $+4$?

▶ **Reasoning:** Once again we can look at the periodic table for the first two answers, but we will have to do a simple calculation for the last answer. Remember, the number of protons is the same as the atomic number; the number of neutrons is the mass minus the number of protons; and we compare the number of protons and the +4 charge to determine the number of electrons.

▶ **Solution:** From the periodic table the element U, uranium, is element number 92, so it has 92 protons.

The number of neutrons is simply the mass number, 238, minus the number of protons: $238 - 92 = 146$ neutrons.

The number of electrons is equal to the number of protons minus the charge on the ion, which in this case is +4. Thus there are $92 - 4 = 88$ electrons in orbit in this case. ▲

The Strong Force

In Chapter 5 we learned that one of the fundamental laws of electricity is that like charges repel each other. If you think about the structure of the nucleus for a moment, you will realize that the nucleus is made up of a large number of positively charged objects (the protons) in close proximity to each other. Why doesn't the electrical repulsion between the protons push them apart and disrupt the nucleus completely?

The only way the nucleus can be stable is for there to be an attractive force capable of balancing or overcoming the electrical repulsion at the incredibly small scale of the nucleus. You can get some idea of the magnitude of this unknown force by thinking about the strength of the electrical repulsion between protons that it must overcome. If two protons in a nucleus were scaled up to the size of basketballs, and if the charges on them were raised proportionately, then even if you encased those two basketball-sized protons in a room-sized block of solid steel they would be blasted apart almost instantaneously. The electrostatic repulsion would be so strong that it would quite literally tear the steel apart as the protons moved away from each other.

Much of the effort of physicists in the twentieth century has gone into understanding the nature of this force that holds the nucleus together. Whatever the force is, it must be vastly stronger than gravity or electromagnetism, the only two forces we've encountered up to this point. For obvious reasons it is called the **strong force.** The strong force must operate only over very short distances, distances characteristic of the size of the nucleus, because our everyday experience tells us that the strong force doesn't act on large objects. Both with respect to its magnitude and its range, then, the strong force is somehow confined to the nucleus. In this respect, the strong force is unlike both electricity and magnetism.

The strong force has another distinctive feature. If you weigh a dozen apples and a dozen oranges, their total weight is simply the sum of the individual pieces of fruit. But this is not true of protons and neutrons in the nucleus. The mass of the nucleus is always slightly less than the sum of the masses of the protons and neutrons. One way of thinking about the generation of the strong force that holds the nucleus together is to imagine that, when protons and neutrons come together, some of their mass is converted into the energy that binds them together. We know this must be true, because it requires energy to pull most nuclei apart. This so-called "binding energy" varies from one nucleus to another. The iron nucleus is the most tightly bound of all the nuclei. This fact will become important in Chapter 17 when we discuss the death of stars.

■ Radioactivity

A safety officer in protective clothing uses a Geiger counter to examine waste for radioactivity.

The vast majority of atomic nuclei in objects around you—more than 99.999% of the atoms in our everyday surroundings—are stable. In all probability, the nuclei in those atoms will never change to the end of time. But some kinds of atomic nuclei are not stable. Uranium-238, the most common isotope of the rather common element uranium, for example, has 92 protons and 146 neutrons in its nucleus. If you put a block of uranium-238 on a table in front of you and watched for a while, you would find that a few of the uranium nuclei in that block would disintegrate spontaneously. One moment there would be a normal uranium atom in the block, and the next moment there would be fragments of smaller atoms and no uranium. At the same time, fast-moving particles would speed away from the uranium block into the surrounding environment. This process of spontaneous change, and the associated emission of energetic particles, is called **radioactivity** or **radioactive decay.** The emitted particles themselves are referred to as *radiation*. The term radiation used in this sense is somewhat different from the electromagnetic radiation that we introduced in Chapter 6. In this case, radiation refers to whatever comes out from the spontaneous decay of nuclei, be it electromagnetic waves or actual particles with mass.

What's Radioactive?

All but a tiny fraction of the atoms around you are stable, but most everyday elements have at least a few isotopes that are radioactive. Carbon, for example, is stable in its commonest isotopes carbon-12 and carbon-13; but carbon-14, which constitutes about a millionth of the carbon atoms in living things, is radioactive. A few elements such as uranium, radium, and thorium have no stable isotopes at all. Even though most of our surroundings are composed of stable isotopes, a quick glance at the chart of isotopes (Figure 11–1) reveals that most of the 2000 or so known natural and laboratory-produced isotopes are unstable and undergo radioactive decay of one kind or another.

Science in the Making

Becquerel and Curie

The nature of radioactivity was discovered, more or less by accident, by Antoine Henri Becquerel (1852–1908) in 1896. Becquerel had obtained samples of radioactive minerals—natural compounds that incorporate uranium and other radioactive elements. Apparently by chance, he put some of these samples in a drawer of his desk along with an unexposed photographic plate and a metal key. When he developed the photographic plate sometime later, the silhouette image of the key was very clearly visible on it. From this photograph he concluded that some as-yet unknown form of radiation had traveled from the mineral sample to the plate. The key seemed to have absorbed the radiation and blocked it off, but the radiation that got through delivered enough energy to the plate to cause the chemical reactions that normally go into photographic development. Becquerel knew that whatever had exposed the plate must have originated in the minerals and must have traveled at least as far as the plate.

Becquerel's discovery was followed by an extraordinarily exciting time for chemists, who began an intensive effort to isolate and study the elements from which the radiation originated. The leader in the field we now call radiochemistry was also one of the best known scientists of the modern era, Marie Sklodowska Curie (1867–1934). Born in Poland and married to Pierre Curie, a distinguished French scientist, she conducted her pioneering research in France, often under extremely difficult conditions because of the unwillingness of her colleagues to accept her. She worked with tons of exotic uranium-bearing minerals from mines in Bohemia, and she isolated minute quantities of previously unknown elements such as radium and polonium. One of her crowning achievements was the isolation of 22 milligrams of pure radium chloride, which became an international standard for measuring radiation levels. She also pioneered the use of X-rays for medical diagnosis during World War I. For her work she became the first scientist to be awarded two Nobel prizes, one in physics and one in chemistry. She also was one of the first scientists to die from prolonged exposure to radiation, whose harmful effects were not known at that time. Her fate, unfortunately, was shared by many of the pioneers in nuclear physics. ■

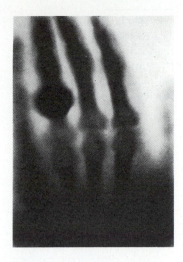

The first x-ray photograph of human bones was taken by Henri Becquerel, who exposed his wife's hand with her wedding ring.

The Kinds of Radioactive Decay

Physicists who studied radioactive rocks and minerals soon discovered three different kinds of radioactive decay, each of which changes the nucleus in its own characteristic way, and each of which plays an important role in modern science and technology. These three kinds of radioactivity were dubbed alpha, beta, and gamma radiation to emphasize that they were unknown and mysterious when first discovered. We retain the names today even though we know perfectly well what they are.

1. Alpha Decay   Some radioactive decays involve the emission of a relatively large and massive particle composed of two protons and two neutrons. Such a particle is exactly the same as the nucleus of a helium-4 atom. It is called an alpha particle, and the process by which it is emitted is called **alpha decay.** (An alpha particle is often represented in equations and diagrams by the Greek letter α.)

The nature of alpha decay was discovered by Ernest Rutherford, the discoverer of the nucleus, in the first decade of the twentieth century. His simple and clever experiment, sketched in Figure 11–2, began with a small amount of radioactive material known to emit alpha particles in a sealed tube. After a number of months, careful chemical analysis revealed the presence of a small amount of helium in the tube—helium that hadn't been present when the tube was sealed. From this observation Rutherford concluded that alpha particles must be associated with the helium atom. Today we would say that Rutherford observed the emission of the helium nucleus in radioactive decay, followed by the acquisition of two electrons to form an atom of helium gas.

The Curie family, with Marie Sklodowska, Pierre, and their child, Irene. Both parents received the Nobel Prize in Chemistry in 1911 for isolating radium and polonium. Their child received the 1935 Prize with her husband, Frederic Joliot-Curie.

Figure 11–2
The Rutherford experiment led to the identification of the alpha particle, which is the same as a helium nucleus.

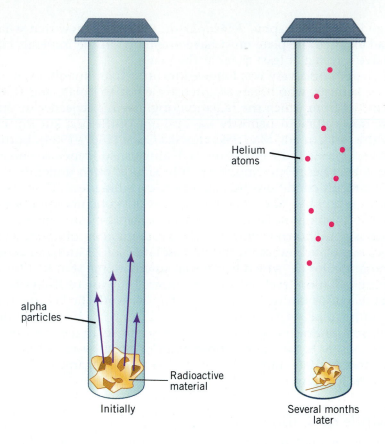

Helium atoms

alpha particles

Radioactive material

Initially

Several months later

Rutherford received the Nobel prize in chemistry for his chemical studies and his work in sorting out radioactivity. He is one of the few people in the world who made his most important contributions to science—in this case the discovery of the nucleus—*after* he received the Nobel Prize.

When the nucleus emits an alpha particle, it loses two protons and two neutrons. This means that the daughter nucleus will have two fewer protons than the original. If the original nucleus is uranium-238 with 92 protons, for example, the daughter nucleus will have only 90 protons, which means that it is a completely different chemical element called thorium. The total mass of the new atom will be 234, so alpha decay causes uranium-238 to transform to thorium-234. The thorium nucleus with 90 protons can accommodate only 90 electrons in its neutral state. This means that, soon after the decay, two of the original complement of electrons will wander away, leaving the daughter nucleus with its allotment of 90. The process of alpha decay reduces the mass and changes the chemical identity of the decaying nucleus.

Radioactivity is nature's "philosopher's stone." The philosopher's stone, according to medieval alchemists, was supposed to turn lead into gold. The alchemists never found their stone because almost all of their work involved what we today would call chemical reactions; that is, they were trying to change one element into another by manipulating electrons. Given what we now know about the structure of atoms, we realize that they were approaching the problem from the wrong end. If you really want to change one chemical element into another, you have to manipulate the nucleus, precisely what happens in the process of radioactivity.

When the alpha particle leaves the parent nucleus, it typically travels at very high speed (often at an appreciable fraction of the speed of light). This means it has a lot of kinetic energy. This energy, like all nuclear energy, comes from the conversion of mass: the mass of the daughter nucleus and the alpha particle, added together, is somewhat less than the mass of the parent uranium nucleus. If the alpha particle is emitted by an atom that is part of a solid body, then it will undergo a series of collisions as it moves from the parent nucleus into the wider world. In each collision it will share some of its kinetic energy with other atoms. The net effect of the decay as far as the surrounding materials is concerned is that the kinetic energy of the alpha particle is eventually converted into heat, and the material warms up. Much of the Earth's interior heat comes from exactly this kind of energy transfer. As we shall see in Chapter 15, this heat is responsible for most of the major surface features of the Earth.

2. Beta Decay   The second kind of radioactive decay, called **beta decay,** involves the emission of an electron. (Beta decay and the electron it produces are often denoted by the Greek letter β). The simplest kind of beta decay that can be observed is for a single neutron. If you put a collection of neutrons on the table in front of you, they would start to disintegrate, with about half of them disappearing in the first ten minutes or so. The most obvious products of this decay are a proton and an electron—both particles that carry an electrical charge and are therefore very easy to detect. This production of a positive and negative particle from a neutral one does not change the total electrical charge of the entire system.

In the 1930s, when beta decay of the neutron was first seen in the laboratory, the experimental equipment available at the time easily detected and measured the energies of the electron and proton. Scientists looking carefully at beta decay were troubled to find that the process appeared to violate the law of conservation of energy, as well as some other important conservation laws in physics. When they added up the mass and kinetic energies of the electron and proton after the decay, they amounted to less than the mass tied up in the energy of the original neutron. If only the electron and proton were given off, the conservation law of energy would be violated.

Rather than face this possibility, physicists at the time followed the lead of Wolfgang Pauli (see Chapter 7) and postulated that another, undetected particle had to be emitted in the decay—a particle that they could not detect at the time, but that carried away the missing energy and other properties. It wasn't until 1956 that physicists were able to detect this missing particle—the *neutrino,* or "little neutral one"—in the laboratory. This particle has no electric charge, travels at the speed of light, and, if stopped, would have no mass. Today, at giant particle accelerators (see Chapter 13) neutrinos are routinely produced and used as probes in other experiments.

When beta decay takes place inside a nucleus, one of the neutrons in the nucleus is converted into a proton, an electron, and a neutrino.

The lightweight electron and the neutrino speed out of the nucleus, while the proton remains. The electron that comes off in beta decay is *not* one of the electrons that was originally circling the nucleus in a Bohr orbit. The electrons emitted from the nucleus come out so fast that they are long gone from the atom before any of the electrons in orbit have time to react. The new atom has a net positive charge, however, and eventually may acquire a stray electron from the environment.

The net effect of a beta decay is that the daughter nucleus has approximately the same mass as the parent (it has the same total number of protons and neutrons), but has one more proton and one less neutron. It is therefore a different element than it was before. Carbon-14, for example, undergoes beta decay to become an atom of nitrogen-14. If you place a small pile of carbon-14 powder—it would look like black soot—in a sealed jar and come back in 20,000 years, most of the powder will have disappeared and the jar will be filled with colorless, odorless nitrogen gas. Beta decay, therefore, is a transformation in which the chemical identity of the atom is changed, but its mass is virtually the same before and after. (Remember: the electron and neutrino that are emitted are extremely lightweight and make almost no difference in the atom's total mass.)

What force in nature could cause an uncharged particle like the neutron to fly apart? The force is certainly not gravitational attraction between masses. It is not the electromagnetic force that causes oppositely charged particles to fly away from each other. And beta decay seems to be quite different from the strong force that holds protons together in the nucleus. In fact, beta decay is an example of the operation of the fourth fundamental force in nature, the *weak force*.

3. Gamma Radiation   The third kind of radioactivity, called **gamma radiation,** is different in character from alpha and beta decay. (Gamma decay and gamma radiation are often denoted by the Greek letter γ.) A "gamma ray" is simply a generic term for a very energetic photon—electromagnetic radiation. In Chapter 6 we saw that all electromagnetic radiation comes from the acceleration of charged particles, and that is what happens in gamma radioactivity.

When an electron in an atom shifts from a higher energy level to a lower one, we know that a photon will be emitted, typically in the range of visible light. In just the same way, the particles in a nucleus can shift between different energy levels. These shifts, or nuclear quantum leaps, involve energy differences thousands or millions of times greater than

Table 11–1 • Types of Radioactive Decay

Type of decay	Particle emitted	Net change
alpha	alpha particle	new element with two less protons, two less neutrons
beta	electron	new element with one more proton, one less neutron
gamma	photon	same element, less energy

those of orbiting electrons. When particles in a nucleus undergo shifts from higher to lower energy levels, some of the emitted gamma radiation is in the range of the X-rays, while others are even more energetic.

Gamma rays are emitted from a nucleus any time the protons and neutrons inside reshuffle. Neither the protons nor the neutrons change their identity, so the daughter atom has the same mass, the same isotope number, and the same chemical identity as the parent. Nevertheless, highly energetic radiation is produced by this process.

Moving Down the Chart of the Isotopes

The three kinds of radioactivity, summarized in Table 11–1, affect the nuclei of isotopes. Gamma radiation changes the energy of the nucleus without altering the number of protons and neutrons—the isotope doesn't change. Alpha and beta radiation result in a new element by changing the numbers of protons and neutrons. Think about how alpha and beta radiation might be represented on the chart of the isotopes (Figure 11–1). Alpha decay removes two protons and two neutrons, so we shift diagonally down and to the left two squares on the chart. Alpha decay thus provides a way to move from heavier to lighter elements.

Beta decay, on the other hand, results in one less neutron and one more proton. On the chart of the isotopes we move diagonally up to the left by one square. Beta decay provides a way to reduce the ratio of neutrons to protons, which is generally greater for heavier elements. Combinations of beta and alpha decays, therefore, provide a means to move from heavier radioactive isotopes to lighter stable isotopes.

Radiation and Health

Why is nuclear radiation so dangerous? Alpha, beta, and gamma radiation all carry a great deal of energy—enough energy to damage the molecules that are essential to the workings of your cells. The most common type of damage is *ionization,* the stripping away of one or more of an atom's electrons. An ionized atom cannot bond in its normal way, and any structures that depend on that atom—a cell wall or a piece of genetic material, for example—will be damaged. Prolonged exposure to ionizing radiation can so disrupt an organism's cells that it will die.

It takes a great deal of radiation to cause sickness or death. Only in unusual circumstances, such as the aftermath of nuclear weapons used on the Japanese cities of Hiroshima and Nagasaki at the end of World War II, or the nuclear reactor accident at Chernobyl in Ukraine in 1986, do people die shortly after exposure. There are, however, possible long-term effects of exposure to lower levels of radiation. While such exposure will not cause death in the short term, it may affect a body's ability to replace old cells and can increase the risk of cancer. Radiation may also damage the genetic material that carries the coded information required to reproduce. Birth defects are tragic consequences of nuclear radiation that may not appear until decades after exposure to radiation.

No one can escape high-energy radiation entirely. Radioactive decay of uranium and other elements in rocks and soils, as well as cosmic radiation from space, bombard us all the time. Radioactive elements even occur in our bones and tissues. Certain kinds of radiation—medical X-rays, for example, and radioactive isotopic tracers for diagnosis—have saved countless lives. Human beings have always lived in a radioactive environment. Nevertheless, it is wise to minimize exposure to unnecessary sources.

The Human Body

Nuclear Tracers

Radioactive materials are useful in medicine and industry because all radioactive isotopes are also chemical elements. The chemistry of atoms is governed by their electrons, which operate independently of the nucleus. The radioactive properties of a material are totally unrelated to the chemical properties. This means that an isotope of a particular chemical that has a radioactive nucleus will undergo the same chemical reactions as the stable isotopes of the same element. If a radioactive isotope of iodine or phosphorus is injected into your bloodstream, for example, it will collect at the same places in your body as stable iodine or phosphorus.

Medical scientists can use this fact to study the functions of the human body and to make diagnoses of diseases and abnormalities (see Figure 11–3). Iodine, for example, concentrates in the thyroid gland, so instruments outside the body can view the workings of your thyroid by examining the decay products of the radioactive isotope of iodine that are injected into the bloodstream and concentrate there. Similarly, phosphorus compounds are attracted to bone. Doctors can examine a healing bone fracture with radioactively enhanced phosphorus compounds injected into the bloodstream. In fact, such radioactive or nuclear tracers are used extensively in medicine, in the earth sciences, in industry, and in all kinds of scientific research to follow the exact chemical progressions of different elements. Small amounts of radioactive material will produce measurable signals as they move through a system, allowing scientists and engineers to trace their pathways. ■

Figure 11–3

Radioactive tracers at work. The patient has been given a radioactive tracer that concentrates in the bone and emits radiation that can be measured on a film. The dark spot in the front part of the skull indicates the presence of a bone cancer.

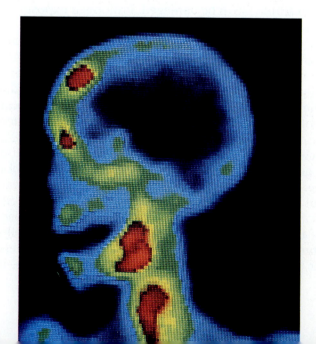

Half Life

A single nucleus of an unstable isotope, left to itself, will decay spontaneously at a specific point in time. That is, the original nucleus will persist up until a specific time, then radioactive decay will occur, and from that point on you will see only the fragments of the decay.

Watching a single nucleus undergo decay is like watching one kernel in a batch of popcorn. Each kernel will pop at a specific time, but all the kernels don't pop at the same time. But even though you can't predict when any one kernel will pop, you can predict the time during which the popping will go on. A collection of radioactive nuclei behaves in an analogous way. Some nuclei decay almost as soon as you start watching; others persist for much longer times. The percentage of nuclei that decay in each second after you start watching remains more or less the same.

Physicists use the term **half-life** to describe the average time it takes for half of a batch of radioactive isotopes to undergo decay. If you have 100 nuclei at the beginning of your observation and it takes 20 minutes for 50 of them to undergo radioactive decay, for example, then the half-life of that nucleus is 20 minutes. If you were to watch that sample for another 20 minutes, however, the nuclei would not all have decayed. You would find that you had about 25 nuclei at the end, then at the end of another 20 minutes you would most likely have 12 or 13, and so on.

Think about this process. Saying that a nucleus has a half-life of an hour does *not* mean that all the nuclei will sit there for an hour, at which point they will all decay. The nuclei, like the popcorn in our example, decay at different times. The half-life is simply an indication of how long, on the average, it will be before an individual nucleus decays.

Radioactive nuclei display a wide range of half-lives. Some nuclei, such as uranium-227, are so unstable that they persist only a tiny fraction of a second. Others, such as uranium-238, have half-lives that range into the billions of years, comparable to the age of the Earth. Between these two extremes you can find a radioactive isotope that has almost any half-life you wish.

We do not yet understand enough about the nucleus to be able to predict half-lives. On the other hand, the half-life is a fairly easy number to measure and therefore can be determined for any nucleus. The fine print on most charts of the isotopes (expanded versions of Figure 11–1) usually includes the half-life for each radioactive isotope.

Radioactive decay of atoms is something like popping kernels of popcorn. You can predict, on average, how many decays or pops will occur, but you can't say for sure when any single atom or kernel will change.

Radiometric Dating

The phenomenon of radioactive decay has provided scientists who study the Earth and human history with one of their most important methods of determining the age of materials. This remarkable technique, which depends on our knowledge of the half-life of radioactive materials, is called **radiometric dating.**

The best-known radiometric dating scheme involves the isotope carbon-14. Every living organism takes in carbon during its lifetime. At this moment, your body is taking the carbon in your food and converting it to tissue, and the same is true of all other animals. Plants are taking in carbon dioxide from the air and doing the same thing. Most of this carbon—about 99%—is in the form of carbon-12, while perhaps 1% is

The Shroud of Turin, with its ghostly image of a man, was dated by carbon-14 techniques to centuries after the death of Christ.

carbon-13. But a certain small percentage, no more than one carbon atom in every million, is in the form of carbon-14, a radioactive isotope of carbon with a half life of about 5700 years.

As long as something is alive, the carbon-14 in its tissues is constantly renewed in the same small proportion that is found in the general environment. When a living thing dies, however, it stops taking in carbon of any form. From the time of death on, therefore, the carbon-14 in the tissues is no longer replenished. Like a ticking clock, carbon-14 disappears atom by atom to form an ever-smaller percentage of the total carbon. We determine the approximate age of a bone, piece of wood, cloth, or other object by carefully measuring the fraction of carbon-14 that remains, and comparing it to the amount of carbon-14 that we know must have been in that material when it was alive. If the material happens to be a piece of wood taken out of an Egyptian tomb, for example, we have a pretty good estimate of how old the artifact is and, probably, when the tomb was built.

Carbon-14 dating often appears in the news when a reputedly ancient artifact is shown to be from more recent times. In a highly publicized experiment, the Shroud of Turin, a fascinating cloth artifact reputed to be involved in the burial of Jesus, was shown by carbon-14 techniques to date from the twelfth century A.D.

Carbon-14 dating has been instrumental in mapping human history over the last several thousand years. When an object is more than about 70,000 years old, however, the amount of carbon-14 left in it is so small that this dating scheme cannot be used. To date rocks and minerals that are millions of years old, scientists must rely on similar techniques that use radioactive isotopes of much greater half-life. Among the most widely used radiometric clocks in geology are those based on the decay of potassium-40 (half-life of 1.25 billion years), uranium-238 (half-life of 4.5 billion years), and rubidium-87 (half-life of 49 billion years). In these cases we measure the total number of atoms of a given element, together with the relative percentage of a given isotope, to determine how many radioactive nuclei were present at the beginning. Most of the ages that we will discuss in the chapters on the earth sciences and evolution are derived ultimately from these radiometric dating techniques.

Geologists collect rock samples for radiometric dating.

Decay Chains

When a parent nucleus decays, the daughter nucleus will not necessarily be stable. In fact, in the great majority of cases, the daughter nucleus is as unstable as the parent. The original parent will decay into the daughter, the daughter will decay into a second daughter, on and on, perhaps for dozens of different radioactive events. Even if you start with a pure collection of atoms of the same isotope of the same chemical element, nuclear decay will guarantee that eventually you'll have many different chemical species in the sample. A series of decays of this sort is called a *decay chain*. The sequence of decays continues until a stable isotope appears. Given enough time, all of the atoms of the original element will eventually decay into that stable isotope.

To get a sense of a decay chain, consider the example we used at the beginning of this chapter—uranium-238, with a half-life of approximately 4.5 billion years. Uranium-238 decays by alpha emission into thorium-234, another radioactive isotope. Thorium-234 undergoes beta decay (half-life of 24.1 days) into protactinium-234 (half-life of about seven hours), which in turn undergoes beta decay to uranium-234. After three radioactive decays we are back to uranium, albeit a lighter isotope with a 247,000-year half-life.

The rest of the uranium decay chain is shown in Figure 11–4. It follows a long path through eight different elements before it winds up as stable lead-208. Given enough time, all of the uranium-238 now in

Figure 11–4

The uranium-238 decay chain. The nuclei in the chain decay by both alpha and beta emission until they reach lead-208, a stable isotope. Some isotopes may undergo either alpha or beta decay, as indicated by splits in the chain. Nevertheless, all paths arrive eventually at lead-208 after 14 decay events.

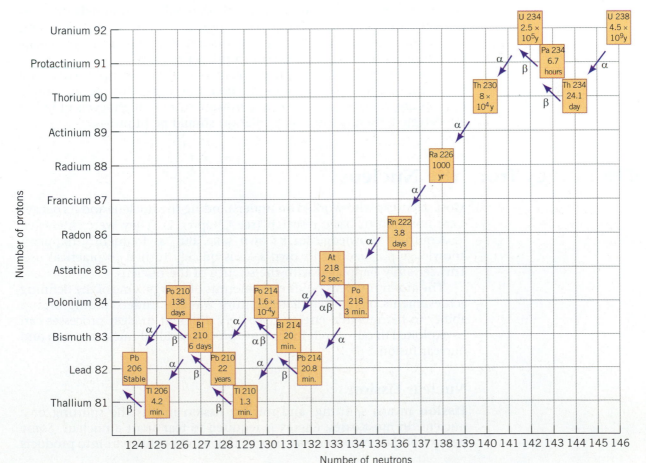

the Earth will eventually be converted into lead. Since the Earth is only about 4.5 billion years old, however, there's only been time for about half of the original uranium to decay, so at the moment (and for the foreseeable future) we can expect to have all the members of the uranium decay chain in existence on the Earth.

Indoor Radon

The uranium-238 decay chain is not an abstract concept, of interest only to theoretical physicists. In fact, the health concern over indoor radon pollution is a direct consequence of the uranium decay chain. Uranium is a fairly common element—about two grams out of every ton of rocks at the Earth's surface are uranium. The first steps in the uranium-238 decay chain produce thorium, radium, and other elements that remain sealed in ordinary rocks and soils. The principle health concern arises from the production of radon-222, about halfway along the path to stable lead. Radon is a colorless, odorless, inert gas that does not chemically bond to its host rock.

As radon is formed, it seeps out of its mineral host and moves into the atmosphere, where it undergoes alpha decay (half-life of about four days) into polonium-218 and a dangerous sequence of short-lived, highly radioactive isotopes. Historically, radon atoms were quickly dispersed by winds and weather, and they posed no serious threat to human health. In our modern age of well-insulated, tightly sealed buildings, however, radon gas can seep in and build up, occasionally to hundreds of times normal levels, in poorly ventilated basements. Exposure to such high radon levels is dangerous because each radon atom will undergo at least five more radioactive decay events in just a few days.

The solution to the radon problem is relatively simple. First, any basement or other sealed-off room should be tested for radon. Simple test kits are available at your local hardware store. If high levels of radon are detected, then the area's ventilation should be improved.

Energy from the Nucleus

Most scientists who worked on understanding the nucleus and its decays were involved in basic research (see Chapter 1). They were interested in acquiring knowledge for its own sake. But, as frequently happens, knowledge pursued for its own sake is quickly turned to practical use. This certainly happened with the science of the nucleus.

The atomic nucleus holds vast amounts of energy. One of the defining achievements of the twentieth century was the understanding of and ability to harness that energy. Two very different nuclear processes can be exploited in our search for energy: processes called nuclear fission and nuclear fusion.

Nuclear Fission

Fission means splitting, and nuclear fission means the splitting of a nucleus. In most cases, energy is required to tear apart a nucleus. Some heavy isotopes, however, have nuclei that can be split apart into products that have less mass than the original. From such nuclei, energy can be obtained from the mass difference.

The most common nucleus from which energy is obtained by fission is uranium-235, an isotope of uranium that constitutes about 0.7% of the uranium in the world. If a neutron hits uranium-235, the nucleus splits into two roughly equal-sized large pieces and a number of smaller fragments. Among these fragments will be two or three more neutrons. If these neutrons go on to hit other uranium-235 nuclei, the process will be repeated and a *chain reaction* will begin, with each split nucleus producing the neutrons that will cause more splittings. By this basic process large amounts of energy can be obtained from uranium.

The device that allows us to extract energy from nuclear fission is called the **nuclear reactor** (see Figure 11–5). The uranium in a reactor contains mostly uranium-238, but it has been processed so it contains much more uranium-235 than it would if it were found in nature. This uranium is stacked in long fuel rods, about the thickness of a lead pencil, surrounded by a metallic protector. Typical reactors will incorporate many thousands of fuel rods. Between the fuel rods is the *moderator,* a fluid, usually water, whose function is to slow down neutrons that leave the rods.

The nuclear reactor works like this: A neutron strikes a uranium-235 nucleus in one fuel rod. The decay products, which include several fast-moving neutrons, leave that particular fuel rod, are slowed down by the moderator, and enter another fuel rod. Because they are moving slowly, they are more likely to initiate fissions in that rod. A chain reaction in a reactor proceeds by steps in which neutrons move from one fuel rod to another. In the process, the energy released by the conversion of matter goes into heating the fuel rods and the water. The water is pumped to another location in the nuclear plant, where it is used to produce steam.

The steam is used to run a generator to produce electricity as described in Chapter 5 (see Figure 5–9). In fact, the only significant difference

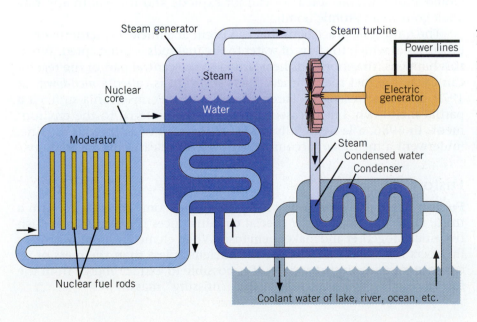

Figure 11–5
A nuclear reactor, shown here schematically, produces heat that converts water to steam.
The steam powers a turbine, just as in a conventional coal-burning plant.

The nuclear power plant at Three Mile Island, near Harrisburg, Pennsylvania, had to shut down after suffering a partial meltdown. Safety measures insured that no radioactive material was released into the environment.

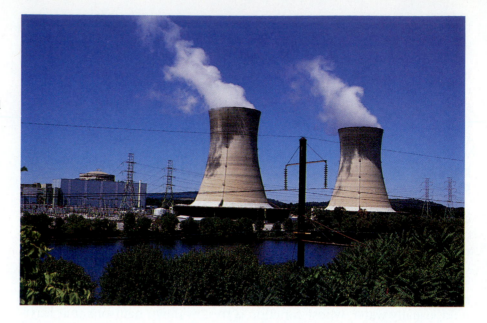

between a nuclear reactor and a coal-fired generating plant is the way in which steam is made. In the former, the energy to produce steam comes from the conversion of mass in uranium nuclei; in the latter, it comes from the burning of coal.

Nuclear reactors must keep a tremendous amount of nuclear potential energy under control while confining dangerously radioactive material. Modern reactors are thus designed with numerous safety features. The water that is in contact with the uranium, for example, is sealed in a self-contained system and does not touch the rest of the reactor. Another built-in safety feature is that nuclear reactors cannot function without the presence of the moderator. If there should be an accident in which the water was evaporated from the reactor vessel, the chain reaction would shut off. Thus a reactor cannot explode and it is *not* in any way analogous to an atomic bomb.

The most serious accident that can occur at a nuclear reactor involves processes in which the flow of water to the fuel rods is interrupted. When this happens, the enormous heat stored in the central part of the reactor can cause the fuel rods to melt. Such an event is called a *meltdown*. In 1979, a nuclear reactor at Three Mile Island in Pennsylvania suffered a partial meltdown, but released no radioactive material to the environment. In 1986, a less carefully designed reactor at Chernobyl, Ukraine, underwent a meltdown accompanied by large releases of radioactivity.

Fusion

Fusion refers to a process in which two nuclei come together to form a third, larger nucleus. Under special circumstances it is possible to push two nuclei together and make them fuse. When elements with low atomic numbers fuse, the mass of the final nucleus is less than the mass of its constituent parts. In these cases, it is possible to extract energy from the fusion reaction by conversion of that "missing" mass.

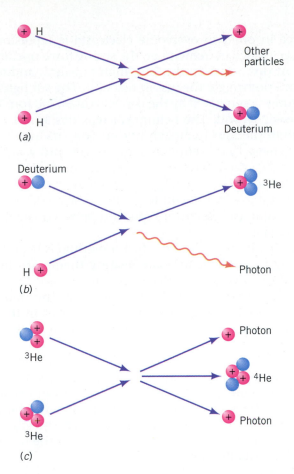

Figure 11–6
A fusion reaction releases energy as nuclei combine. Hydrogen nuclei enter into a multi-step process whose end product is a helium nucleus.

The most common (and arguably the most important) fusion reaction involves the combination of hydrogen nuclei into a helium nucleus (see Figure 11–6). (Remember that the nucleus of the ordinary hydrogen nucleus is a single proton, with no neutron. Thus we use the term "hydrogen nucleus" and "proton" interchangeably.) This reaction powers the Sun and other stars and thus is ultimately responsible for all life on Earth.

You cannot just put hydrogen in a container and expect it to form helium, however. Two positively charged protons must collide with tre-

The Princeton Tokomak is a ring-shaped magnetic chamber. Hydrogen plasma is confined in this ring during attempt to achieve nuclear fusion reactions.

mendous force in order to overcome electrostatic repulsion and allow the strong force to kick in (remember, the strong force operates only over extremely short distances). In the Sun, high pressures and temperatures in the star's interior trigger the fusion reaction. The sunlight falling outside your window is generated by the conversion of 700 tons of hydrogen into helium each second. The helium nucleus has a mass about half a percent less than the four hydrogen nuclei. The "missing" mass is converted to the energy that eventually radiates out into space.

Since the 1950s there have been many attempts to harness nuclear fusion reactions to produce energy for human use. The problem has always been that it is very difficult to get protons to collide with enough energy to overcome the electrical repulsion between them and initiate the nuclear reaction.

One promising but technically difficult method is to confine protons in a very strong magnetic field while heating them with high-powered radio waves. Alternatively, powerful lasers can be used to heat pellets of liquid hydrogen. In this process, which goes by the name of *inertial confinement,* the rapid heating produces shock waves in the liquid, and these shock waves raise the temperature and pressure to the point where fusion is initiated.

A number of rather expensive programs are now underway in the United States, Europe, and Japan, where researchers are investigating ways to produce commercially feasible nuclear fusion reactors, perhaps sometime in the twenty-first century. Whether they will be successful is, at the moment, not known. If they are, however, the energy crisis will be over forever, since there is enough hydrogen in the oceans to power fusion reactions virtually forever.

Technology

Nuclear Weapons

In the years following World War II, nuclear weapons tests were conducted on land. This test took place at the Bikini Atoll on July 24, 1946.

Whenever humans have discovered a new source of energy, that source has quickly been turned into weapons. Nuclear energy, the most concentrated energy source known, is no exception. The earliest atomic bombs, developed in the United States as part of World War II's Manhattan Project, relied on fission reactions in concentrated uranium-235 or plutonium-239. The problem faced by the designers of the bomb was that there have to be a certain number of uranium-235 atoms to sustain a chain reaction to the point where large amounts of energy can be released. This quantity of uranium-235, called the *critical mass,* is about 10 kilograms—a chunk of uranium about the size of a grapefruit.

At the instant of an atom bomb explosion, a critical mass of the radioactive isotope is compressed into a small volume by a conventional chemical explosive. The normal chain reactions accelerate, as a flood of neutrons almost instantaneously splits all the available nuclear material and releases a destructive wave of heat, light, and other forms of radiation. The first atomic bombs carried the explosive power of many thousands of tons of chemical explosive.

Shortly after World War II, the United States and the Soviet Union

developed the hydrogen bomb, which relies on nuclear fusion. In a hydrogen bomb, a compact hydrogen-rich compound is surrounded by atom bomb triggers. When the atomic fission bombs are set off, they trigger much more intense fusion reactions in the hydrogen. Unlike atom bombs, which are limited by the critical mass of nuclear explosive, hydrogen bombs have no intrinsic limit to their size. The largest nuclear warheads, which could fit in a closet, carry payloads equivalent to many millions of tons *(megatons)* of conventional explosives. ■

Science in the Making

Cold Fusion

In 1989 two scientists at the University of Utah announced to the press that they had produced fusion reactions under very unusual circumstances. They claimed that they had experimental evidence that fusion had taken place in a simple glass apparatus at room temperature—an apparatus no more complicated than those normally used by high-school students in carrying out their experiments. This announcement caused a flurry of excitement. Could it be, people wondered, that the energy crisis was over and a new age of unlimited resources had come?

The experiment that attracted all this attention was simplicity itself. Two bars of the metal palladium were lowered into a bath that contained a special kind of water in which some of the hydrogen atoms, which normally have a lone proton in the nucleus, are replaced by deuterium, the isotope of hydrogen whose nucleus contains one proton and one neutron. The fusion of two hydrogen atoms into deuterium is the first step in the fusion process. The experimenters claimed that, in the presence of palladium, the deuterium in the water fused to form helium, releasing energy in the process. If their results turned out to be correct, the waters in the Earth's oceans could be "mined" for deuterium, and the human race would have acquired an energy source that was, for all practical purposes, inexhaustible.

Given such extraordinary claims, thousands of scientists rushed to reproduce the Utah experiment. Unfortunately, despite many scientist-years of research, no one was able to reproduce all the results originally claimed. Within about a month, the vast majority of scientists had concluded that the original experiment was flawed. It appears that measurements of both energy production and radioactivity (necessary to establish the presence of nuclear reactions) in the original experiment were simply incorrect.

Cold fusion is now cited as a classic example of how science is a self-correcting process. No result, no matter how compelling or satisfying, can stand without independent confirmation and testing. This incident also reveals why scientists are so reluctant to believe results that are announced directly to the press, without the standard (and exacting) process of peer review that goes into the publication of normal scientific results. ■

THINKING MORE ABOUT THE NUCLEUS

Nuclear Waste

When power is generated in a nuclear reactor, many more nuclear changes take place than those associated with the chain reaction itself. Fast-moving debris from the fission of uranium-235 strike other nuclei in the system—both the ordinary uranium-238 that makes up most of the fuel rods, and the nuclei in the concrete and metal that make up the reactor. In these collisions, the original nuclei may themselves undergo fission or absorb neutrons to become isotopes of other elements. Many of these newly produced isotopes are themselves radioactive. The result is that even when all of the uranium-235 has been used to generate energy, a lot of radioactive material remains in the reactor. This sort of material is called *high-level nuclear waste*. (The production of nuclear weapons is another source of this kind of waste.)

The half-lives of some of the materials in the waste can run to hundreds of thousands of years. What should we do with the waste? In particular, how can we dispose of it in a way that keeps it away from living things?

The management of nuclear waste begins with storage. Power companies usually store spent fuel rods at the reactor site for tens of years to allow the short-lived isotopes to decay. At the end of this period, long-lived isotopes that are left behind must be isolated from the environment. Some scientists have suggested that we incorporate these nuclei into stable minerals, either rocks or glass. The idea is that the electrons in these atoms will form the same kind of bonds that stable isotopes do, so that with judicious choice of a mineral, you can use the electrons to lock the radioactive nuclei into the mineral for long periods of time. The hope is that long-lived wastes can be sequestered from the environment until after they are no longer dangerous to human beings.

Plans now call for nuclear waste disposal by the incorporation of radioactive atoms into stable glasses, surrounding these glasses with successive layers of steel and concrete, and then burying them a kilometer or more under the surface of the Earth in stable rock formations. From our point of view, that seems to be the most reasonable available solution to this critical societal problem.

What responsibility do we have to future generations to insure that the waste we bury stays where we put it? Should the existence of nuclear waste restrain us in our development of nuclear energy? Should we, as some scientists argue, keep nuclear waste materials at the surface and use them for applications such as medical tracers and fuel for reactors?

Summary

The nucleus is a tiny collection of massive particles, including positively charged *protons* and electrically neutral *neutrons*. The nucleus plays an independent role from the orbiting electrons that control chemical reactions, and the energies associated with nuclear reactions are much greater. The number of protons—the *atomic number*—determines the nuclear charge, and therefore the type of element; each element in the periodic table has a different number of protons. The number of neutrons plus protons—*the mass number*—determines the mass of the *isotope*. Nuclear particles are held together by the *strong force,* which operates only over extremely short distances.

While most of the atoms in objects around us have stable, unchanging nuclei, many isotopes are unstable—they spontaneously change through *radioactive decay*. In *alpha decay,* a nucleus loses two protons and two neutrons. In *beta decay,* a neutron spontaneously transforms into a proton, an electron, and a neutrino. A third kind of radioactivity, involving the emission of energetic electromagnetic radiation, is called *gamma radiation*. The rate of radioactive decay is measured by the *half-life*—the time it takes for half of a collection of isotopes to decay into another element. Radioactive half-lives provide the key for *radiometric dating* techniques based on carbon-14 and other isotopes. Unstable isotopes are also used as radioactive tracers in medicine and other areas of science. Indoor radon pollution and nuclear waste are two societal problems that arise from the existence of radioactive decay.

We can harness nuclear energy in two ways. *Fission* reactions, as controlled in *nuclear reactors,* produce energy when heavy radioactive nuclei split apart into frag-

ments that together weigh less than the original isotopes. *Fusion* reactions, on the other hand, combine light elements to make heavier ones, as in the conversion of hydrogen into a smaller mass of helium in the Sun. In each case the lost nuclear mass is converted into energy.

Key Terms About the Nucleus

proton	mass number	beta decay	fission
neutron	strong force	gamma radiation	nuclear reactor
atomic number	radioactivity, or radioactive decay	half-life	fusion
isotope	alpha decay	radiometric dating	

Review Questions

1. Compare the size of a nucleus to the size of an atom.
2. Which is stronger—the binding of electrons into atoms, or the binding of protons and neutrons into nuclei?
3. Which are heavier—electrons or protons?
4. What fact about atomic nuclei suggests the existence of the neutron?
5. What is the atomic number?
6. What is an isotope?
7. How do you know that a strong force exists?
8. How is the strong force different from gravity and electromagnetism?
9. What is radioactive decay?
10. Define the term "half-life."
11. Explain how radioactivity can be used to date (*a*) a piece of wood, and (*b*) a rock.
12. Describe the major achievement of Marie Curie.
13. What is alpha decay? How does it change the nucleus?
14. What is beta decay? How does it change the nucleus?
15. What is gamma radiation? How does it change the nucleus?
16. What does it mean to say "Radioactivity is the philosopher's stone"?
17. How can we get energy from nuclear fission?
18. What is a chain reaction?
19. How does a nuclear reactor work?
20. How can we get energy from fusion?
21. What is a critical mass?
22. What is a radioactive tracer?
23. What is nuclear waste? Why is it a serious problem for society?

Discussion Questions

1. Would you say that Becquerel's discovery of radioactivity was good science or a lucky break? Explain your answer.
2. Why does radioactivity seem to be more common for heavier elements?
3. What was the hypothesis behind Rutherford's experiment on alpha decay? What did he prove?
4. What are the potential benefits and risks in using nuclear tracers in medical diagnosis?
5. "Critical mass" is a term that is widely used outside of nuclear science. What is its everyday meaning, and how does that relate to its scientific meaning?
6. Almost all of the atoms with which we come into daily contact have stable nuclei. Given that there are many more unstable isotopes than stable ones, how could this state of affairs have arisen? (*Hint:* What happens to unstable isotopes in nature?)
7. How is the principle of conservation of energy seen in (*a*) fission reactions, and (*b*) fusion reactions?
8. Discuss the pros and cons of nuclear power.
9. Can nuclear radiation escape from nuclear power plants? If so, how?

Problems

1. Use the periodic table to identify the element, atomic number, mass number, and electrical charge of the following combinations:
   a. 8 protons, 8 neutrons, 10 electrons
   b. 8 protons, 9 neutrons, 10 electrons
   c. 9 protons, 8 neutrons, 10 electrons
   d. 8 protons, 8 neutrons, 9 electrons

2. Use the periodic table to determine how many protons and neutrons are in each of the following atoms:
   a. ^{14}C
   b. ^{55}Fe
   c. ^{40}Ar
   d. ^{235}U

3. The average atomic weight of cobalt atoms (atomic number 27) is actually slightly greater than the average atomic weight of nickel atoms (atomic number 28). How could this situation arise?

4. Imagine that a collection of 1000 atoms of uranium-238 was sealed in a box at the formation of the Earth 4.5 billion years ago. Use the uranium-238 decay chain (Figure 11–4) to predict some of the things you would find if you opened the box today.

5. A particular isotope has a half-life of five days. A particular sample is known to have contained a million atoms when it was put together, but is now observed to have only about 100,000. Estimate how long ago the sample was assembled. Explain the relevance of this problem to the technique of radiometric dating.

Investigations

1. Read a historical account of the Manhattan Project. What was the principal technical problem in obtaining the nuclear fuel? Why did chemistry play a major role? What techniques are now used to obtain nuclear fuel?

2. What is the current status of U.S. progress toward developing a depository for nuclear waste? How do your representatives in Congress vote on matters relating to this issue?

3. What sorts of isotopes are used for diagnostics in your local hospital? Where are supplies of those

radioisotopes purchased? What are the half-lives of the isotopes, and how often are supplies replaced? What is the hospital's policy regarding the disposal of radioactive waste?

4. How much of the electricity in your area comes from nuclear reactors? What fuel do they use? Where are the used fuel rods taken when they are replaced? If the facility offers public tours, visit the reactor and observe the kinds of safety procedures that are used.

5. Obtain a radon test kit from your local hardware store and use it in the basement of two different buildings. How do the values compare? Is either at a dangerous level? If the values differ, what might be the reason?

6. Only about 90 elements occur naturally on Earth, but scientists are able to produce more elements in the laboratory. Investigate the discovery and characteristics of one of these human-made elements.

7. Read an account of the cold fusion episode. At what point in this history were coventional scien-tific procedures bypassed? Ultimately, do you think that the scientific method worked or failed?

8. Soon the United States government will take over responsibility for the nuclear wastes of the 50 states. What options do we have for waste storage? Do you think all the waste should be stored in one place? Should we try to separate and use the radioactive isotopes? What are the factors—social, political, and economic—that will help determine what happens to this nuclear waste?

Additional Reading

Alvarez, Luis. *Adventures of a Physicist*. New York: Basic Books, 1987.

Pflaum, Rosalynd. *Grand Obsession: Madame Curie and Her World*. New York: Doubleday, 1989.

Rhodes, Richard. *The Making of the Atomic Bomb*. New York: Simon and Schuster, 1986.

The movie *Fat Man and Little Boy*

The Ultimate Structure of Matter

All matter is made of quarks and leptons—the fundamental building blocks of the universe.

A Random Walk
The Library

The next time you head over to the library, wander through the stacks and think about what constitutes the fundamental building blocks of the library. Your first reaction would probably be to say that the fundamental building blocks are books. After all, the stacks are filled with row after row, shelf after shelf, of bound volumes. But a library is not just a collection of books—the volumes are arranged with an order to them. You could describe the set of rules that dictates how books are arranged in libraries—the Dewey decimal system, or the Library of Congress classification scheme, for example. Thus a complete description of a library at this most superficial level includes two things: books as the fundamental building blocks, and rules about how the books are organized.

But if you looked inside a book, you would quickly notice that various volumes are not as different from each other as they might seem at first. They are all made of an even more fundamental unit—the word. In this sense, you can say that the word is the fundamental building block of the library; and, as was the case for cataloged books, we require a set of rules, called grammar, that tells us how to put words together to make books. The discovery of words and grammar, then, takes you down to a more basic level in a probe of a library's reality.

You probably wouldn't be content very long with the notion of the word as the fundamental building block, because it soon becomes obvious that all of the thousands of words are different combinations of a small number of more fundamental things—letters. Only 26 letters provide the building blocks for all the thousands of words on all the pages in all the books of the library. Furthermore, we need a set of rules (spelling) that tells us how to put letters together into words. The discovery of letters and spelling would constitute perhaps the ultimate description of a library and its organization.

But, as we saw in Chapter 10, letters can be stored and manipulated in terms of a binary code. Call these entries in this code the "dots and dashes" of the library. We could argue that they, not the letters, are the most fundamental quantities in the library. And, just as we needed a rule to tell us how to use other "elementary" entities, we need a rule (call it a code) that tells us how to go from dots and dashes to letters.

The library, then, can be described in this way: We use the code to tell us how to put the dots and dashes together into letters, then spelling to tell us how to put the letters together into words, then grammar to tell us how to put the words together into books. Finally, we use the organizing rules to tell us how to put the books together into a library.

And this, as we shall see, is exactly how scientists have come to describe the entire physical universe.

What Is the Universe Made Of?

How many different kinds of material can you see when you look up from this book? You may see a wall made of cinder block, a window made of glass, a ceiling made of fiberglass panels. Outside the window you may see grass, trees, blue sky, and clouds. We encounter thousands of different kinds of materials every day. They all look different—what possible common ground could there be between a cinder block and a blade of grass? They all look different, but are they really?

In fact, for at least two millennia people who have thought about the physical universe have asked this question. Is the universe just what we see, or is there some underlying structure, some basic stuff, from which it's all made? You could even say that herein lies *the* scientific question.

Reductionism

The quest for the "ultimate building blocks" of the universe is referred to by philosophers as *reductionism*. Reductionism is an attempt to reduce the seeming complexity of nature by first looking for underlying simplicity, and then trying to understand how that simplicity gives rise to the observed complexity. This pursuit is a continuation of an old intellectual belief that the appearances of the world do not tell its true nature, but that its true nature can be discovered by the application of thought and, in the case of science, experiment and observation.

The Greek philosopher Thales (640?–546 B.C.) suggested that all materials are made of water. This supposition was based on the observation that in everyday experience water appears as a solid (ice), a liquid, and a gas (steam). Thus, alone among the common substances, water seemed to exhibit all the states of matter (see Chapter 9). To Thales, this observation suggested that water was in some sense fundamental. We no longer accept the idea that water is the fundamental constituent of matter, but we do believe that we can find other fundamental constituents.

The Building Blocks of Matter

To many people, the library analogy presents a profoundly satisfying way of describing complex systems. Some would even argue that everything you could possibly want to know about the library is contained in the dots and dashes and their organizing principles. In just the same way, scientists want to describe the complex universe by identifying the most fundamental building blocks and deduce the rules by which they are put together.

At first, you might say the most fundamental building block of the universe is the atom. All the myriad solids, liquids, and gases are made of just 100 or so different kinds of chemical elements. The complexity of materials that appears to the senses results from the many combinations of these relatively few kinds of atoms. The rules of chemistry tell us how atoms bind together to make all of the materials we see.

Early in this century scientists learned that atoms themselves are not really fundamental, but are made up of smaller, yet more fundamental, bits—nuclei and electrons. These particles arrange themselves according to their own set of rules, with massive neutrons and protons in the positively charged nucleus, and negatively charged electrons in orbit around the nucleus. A picture of the universe with only three fundamental building blocks—protons, neutrons, and electrons—is very simple and appealing. Protons and neutrons together form nuclei, while the electrons orbit the nucleus to form atoms. Electrons combine and interact with each other to form all the materials we know about.

But, just as letters and spelling gave us a false level of simplicity in the analogy of the library, this simple picture of the universe didn't stand up to more detailed experiments and observations. As we have hinted, the nucleus contains much more than just protons and neutrons, although this fact did not become clear to physicists until the post–World War II era. If we are going to follow the reductionist line in dealing with the universe, we have to start thinking about what makes up the nucleus. By common usage, the particles that make up the nucleus, together with particles such as the electron, were called *elementary particles,* to reflect the belief that they comprised the basic building blocks of the universe. The study of these particles and their properties is the domain of a field known as **high-energy physics,** or **elementary-particle physics.**

■ Discovering Elementary Particles

Nowhere in nature is the equivalence of mass and energy more obvious than in the interactions of elementary particles. Imagine that you have a mechanism for accelerating a proton to very high velocities; velocities approaching the speed of light. This mechanism might be astronomical in nature, or it might be a machine that accelerates particles. But once the proton has been accelerated, it has a very high kinetic energy. If this high-energy proton collides with a nucleus, the nucleus can be split apart. In this collision process, some of the kinetic energy of the original proton can be converted into mass according to the equation $E = mc^2$. When this happens, new kinds of particles that are neither protons nor neutrons can be created.

Cosmic Rays

During the 1930s and 1940s, physicists used a natural source of high-energy particles, the so-called cosmic rays, to study the structure of matter. **Cosmic rays** are particles (mostly protons) that rain down continuously on the atmosphere of the Earth after being emitted by stars in our galaxy and in others.

Space is full of cosmic rays. When they hit the atmosphere, they collide with molecules of oxygen or nitrogen and produce sprays of very fast-moving secondary particles. These secondary particles, in turn, can make further collisions and produce even more particles as a cascade builds up in the atmosphere. It is not uncommon for a single incoming particle to produce billions of secondary particles by the time the cascade reaches the surface of the Earth. Indeed, on average, several rays pass through your body every minute of your life.

Physicists in the 1930s and 1940s set up their apparatus on high mountaintops and observed what happened when the fast-moving primary cosmic rays or the slightly slower-moving secondary particles collided with nuclei. A typical apparatus incorporated a gas-filled chamber several centimeters across. Midway in the chamber was located a thin sheet of target material—lead, for example. Cosmic rays occasionally collided with one of the nuclei in the piece of lead, producing a spray of secondary particles. By studying particles in that spray, physicists hoped to understand what was going on inside the nucleus.

By the early 1940s, when the international effort in physics research shut down temporarily because of World War II, physicists working with these cosmic ray experiments had discovered particles in the universe in addition to the proton, neutron, and electron. And when the research effort started up again after the war, these discoveries multiplied as more and more particles were found in the debris of nuclear collisions, both by cosmic ray physicists and by those working at the new particle accelerators we will discuss shortly.

The net result of these discoveries was that the nucleus could no longer be considered as a simple bag of protons and neutrons. Instead, we had to think of the nucleus as a very dynamic place. All kinds of newly discovered elementary particles in addition to protons and neutrons were found there. These exotic particles are created in the interactions inside the nucleus, and they give up their energy (and, indeed, their very existence) in subsequent interactions to make other kinds of particles. This constant dance of the elementary particles inside the nucleus has been well documented since these early explorations.

Technology

Detecting Elementary Particles

If elementary particles are smaller even than an individual nucleus, how do we know they're there? Experimental physicists have raised detection of elementary particles to a fine art over the years. Nevertheless, the basic technique used in any detection process is the same: the particle in question interacts with matter in some way, and we measure the changes in matter that are the effects of that interaction.

If an elementary particle has an electrical charge, for example, it may tear electrons loose as it goes by an atom. Thus a charged elementary particle moving through material will leave a string of ions in its wake, much as a speedboat going across a lake leaves a trail of troubled

Ernest O. Lawrence posed in the 1930s with his invention, the cyclotron, which was the first particle accelerator.

water. Over the years many techniques have been used to measure the presence of those ions. For example, you can make a photographic emulsion in which the ions will show up as dark tracks when the emulsion is developed. Alternatively, an apparatus can be set up in such a way that the disturbed ions act as places where bubbles are likely to nucleate in a liquid kept under high pressure, or droplets of water will appear in a gas. In these detectors a string of bubbles or droplets will mark the paths of the particles.

A more modern method of detecting charged particles is to allow the particles to pass though a grid of thin conducting wires (usually made of gold). As a particle passes a wire, it exerts a force on the electrons in the metal, creating a small pulse of current. By measuring the time that this pulse arrives at the end of the wire, and by putting together the information from many wires, a computer can reconstruct with high precision the path taken by a particle.

Uncharged particles such as neutrons are much more difficult to detect because they do not leave a string of ions in their path. Typically, we do not detect the passage of an uncharged particle directly; instead we wait until it collides with something. If that collision produces charged particles, then we can detect them by the techniques outlined above and can work backward and deduce the property of the uncharged particle. ■

Particle Accelerators: The Essential Tool

For a time physicists had to sit around and wait for nature to supply high-energy particles (in the form of cosmic rays) so that they could study the fundamental structure of matter. The arrival of cosmic rays could not be controlled, and it could be very time-consuming waiting for one to hit. Physicists quickly realized that they had to build machines that could produce streams of "artificial cosmic rays"—**particle accelerators** that scientists could turn on and off at will, and that would take the place of the natural, but sporadic, cosmic rays in experiments. At the beginning of the 1930s at the University of California at Berkeley, Ernest O. Lawrence began producing a new kind of accelerator called a *cyclotron*, an invention for which he won the 1939 Nobel prize in physics.

The cyclotron works by applying an electrical force to groups of charged particles, usually electrons or protons, producing a large acceleration according to Newton's second law (force = mass × acceleration: see Chapter 2) by exerting that force over and over again. The particles are accelerated until they are moving at almost the speed of light. Once they have acquired this much kinetic energy, they are allowed to collide with other particles. Those collisions provide the interactions that scientists wish to study.

The 4-story-high Fermilab CDF particle detector. Particle detectors have gone from being small, desk-top pieces of apparatus to huge high-technology instruments like the one shown.

Lawrence's first cyclotron was no more than a dozen centimeters across and produced energies that were pretty puny by today's standards. A modern particle accelerator is a huge, high-tech structure capable of producing energies as high as all but the very highest cosmic rays. Called a **synchrotron,** its main working part is a large ring of magnets that keep the accelerated particles moving in a circular track.

One aspect of Maxwell's equations that we didn't explore in detail in Chapter 5 is that magnetic fields exert a force on moving charged particles. That force tends to make charged particles move in a circular track. As a particle in a synchrotron moves around the circle, the fields of the large electromagnets are adjusted to keep its track within a small chamber (typically several centimeters on a side) in which a near-perfect vacuum has been produced. This chamber, in turn, is bent into the large circle that marks the particle's orbit. Each time the particles come around to a certain point, an electric field boosts their energy. As the velocity increases, the field strength in the magnets is also increased to compensate, so that the particles continue around the circular track. Eventually, the particles reach the desired speed and they are brought out into an experimental area where they undergo collisions.

As the energy required to stay at the frontier of particle physics increases, so too does the size of accelerators. The highest-energy accelerator in the world today is at the Fermi National Accelerator Laboratory (Fermilab) outside of Chicago, Illinois. There, protons move around a ring almost 2 kilometers (about a mile) in diameter.

The **linear accelerator** provides an alternative strategy for making high-velocity particles. This device relies on a long, straight vacuum tube into which electrons are injected. The electronics are arranged so that an electromagnetic wave travels down the tube, and electrons ride this wave more or less the way a surfer rides a wave on the ocean. The largest linear accelerator in the world, at the Stanford Linear Accelerator Center in California, is about 3 kilometers (almost 2 miles) long.

The Elementary Particle Zoo

At the beginning of the 1960s, when the first generation of modern particle accelerators began to produce copious results, scientists discovered that the number of "elementary particles" that must reside inside the nucleus is very large. The last time either of the authors looked, the list had gotten past 200 and was still growing. These particles typically

(*a*) Fermilab in Illinois and (*b*) Stanford Linear Accelerator (SLAC) in California are two of the world's most powerful particle accelerators.

(*a*)

(*b*)

Table 12–1 • Summary of Elementary Particles

Type	Definition	Examples
leptons	do not take part in holding together the nucleus	electron, neutrino, mu, tau
hadrons	participate in holding nucleus together	proton, neutron, roughly 200 others
antiparticles	particles with same mass, but opposite charge and other properties	positron

are named by their discoverers, using Greek letters, and they come in a wide variety of types, summarized in Table 12–1. Following are some important classifications.

Leptons

Some elementary particles do not participate in the strong force that holds the nucleus together, and they are not part of the nuclear maelstrom. These non-nuclear particles are called **leptons,** for "weakly interacting ones." The two leptons that we have encountered so far are the *electron,* which is normally found in orbit around the nucleus rather than in the nucleus itself, and the *neutrino,* a light neutral particle that hardly interacts with matter at all. Since the 1940s, physicists have discovered a total of six kinds of leptons. Two of these particles are like the electron except that they are heavier (particles called the mu and the tau), and two others are kinds of neutrinos (one neutrino associated with each of the massive mu and tau particles). If you keep in mind that the electron and the neutrino are typical leptons, you will have a pretty good idea of what they're like.

Hadrons

All of the different kinds of particles that exist inside the nucleus are referred to collectively as **hadrons,** or "strongly interacting ones." The array of these particles is truly spectacular. Hadrons include particles that are stable like the proton, particles that undergo radioactive decay in a matter of minutes like the neutron (which undergoes beta decay), and still other particles that undergo radioactive decay in 10^{-24} seconds. The latter kind of particles do not live long enough even to travel across a single nucleus! Some hadrons carry an electrical charge, while others are neutral. But all of these particles are subject to the strong force, and all participate in some way in holding the nucleus together, and thus in making the physical universe possible.

Antimatter

For every particle that we see in the universe, it is possible to produce an antiparticle. Every **antimatter** particle has the same mass as its matter twin, but the particles have opposite charge and opposite magnetic characteristics. The antiparticle of the electron, for example, is a positively charged particle known as the **positron.** It has the same mass as the electron, but a positive electrical charge. Antinuclei, composed of antiprotons and antineutrons and orbited by positrons, could in principle form antiatoms. Indeed, some scientists have proposed the existence of entire galaxies, in every way like our own, but composed entirely of antimatter.

When a particle collides with its antiparticle, both masses are converted completely to energy in the most efficient and violent process known in the universe—a process called *annihilation*. The original particles disappear, and this means that energy appears as a spray of rapidly moving particles and electromagnetic radiation. This fact has long been adopted by science fiction writers in their descriptions of futuristic weapons and power sources. (The starship *Enterprise* on *Star Trek,* for example, has matter and antimatter pods as its power source.)

Although antimatter is fairly rare in the universe, it is routinely produced in particle accelerators. High-energy protons or electrons strike nuclear targets, and the energy of the particles is converted to equal numbers of other particles and antiparticles. Thus the existence of antimatter is verified daily in laboratories.

Science in the Making

The Discovery of Antimatter

In 1932, Carl Anderson, a young physicist at the California Institute of Technology, was performing a rather straightforward cosmic ray experiment of the type described in the text. Cosmic rays entered a type of detector called a "cloud chamber." In Anderson's cloud chamber, a cos-

Carl Anderson identified the positron (the antiparticle of the electron) from the distinctively curved path left in a bubble chamber. In Anderson's original photograph (*a*) the positron path curves upward and to the left. In a more recent photograph (*b*) an electron (e−) and a positron (e+) curve opposite directions in a magnetic field.

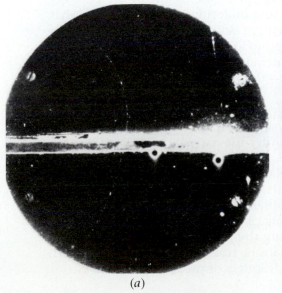

(*a*)

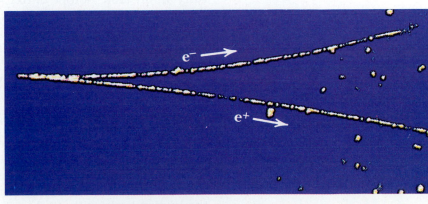

(*b*)

mic ray particle would move through a moisture-laden gas, leaving behind a string of ions. By pulling out a piston at the bottom of the chamber, the gas pressure was lowered, and the liquid (usually alcohol) that had been in gaseous form condensed out into droplets. The ions acted as nuclei for the condensation of these droplets, so that the path of the particle was marked by a string of droplets in the chamber.

The key innovation in Anderson's experiment was the positioning of the cloud chamber between the poles of a powerful magnet. These magnets caused electrically charged cosmic rays to move in curved tracks, with the amount of curving dependent on the particle's mass, speed, and charge. Furthermore, the tracks of positively and negatively charged particles curved in opposite directions under the influence of the magnetic field.

Soon after he switched on his apparatus, Anderson saw tracks of particles whose mass seemed to be identical to that of the electron, but whose tracks curved in the opposite direction from those of electrons being detected. This feature, he concluded, had to be the result of a "positive electron," a phrase he contracted to positron. Although no one realized it at the time, Anderson was, in fact, the first human being to see antimatter.

One amusing sequel to this discovery is that when Carl Anderson received a Nobel prize for his work in 1936, he still had not been promoted to full professor at Caltech. ■

The Human Body

Positron Emission Tomography

The study of elementary particles often seems quite abstract—the danger is always present of becoming lost in a kind of fantasy land of strange names, strange concepts, and strange ideas. Situations do arise, however, where elementary particles play a very important role in the real world. Positron emission tomography (PET) is one of these.

In this medical technique, glucose molecules (see Chapter 20) that carry an unstable isotope of oxygen (an isotope produced in nuclear reactors) are injected into a patient's bloodstream. These molecules are taken up by organs in the body, including the brain. They will go to the parts of the brain that need it—the parts that require energy at the time (see Figure 12–1).

This particular isotope of oxygen emits a positron, the antiparticle of an electron, which is relatively easy to detect from outside the body. A PET-scan works like this: After the material is injected into the bloodstream, the patient is asked to do something—talk, read, do mathematical problems, or just relax. Each of these activities uses a different region of the brain. Scientists watching the emission of positrons can see those regions of the brain quite literally "light up" as they are used. In this way, scientists can study the normal working of the human brain without disturbing the patient, as well as detect and study abnormalities that can perhaps be treated. Thus elementary particles have their role to play in the frontiers of medical research. ■

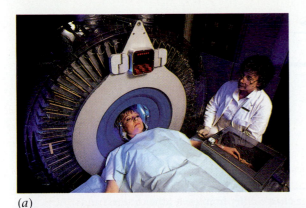

(a)

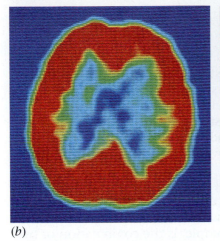

(b)

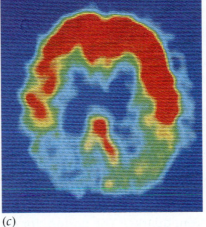

(c)

Figure 12–1
Positron emission tomography, commonly called the PET scan, reveals activity in the human brain (*a*) A patient undergoing a PET scan; (*b*) a scan of a normal brain. The bright spots are places where large amounts of glucose (a simple sugar used by most cells for energy) are being used by the brain. (*c*) A scan of a person suffering from Alzheimer's disease.

Quarks

When chemists understood that the chemical elements could be arranged in the periodic table, it wasn't long before they realized what caused this regularity. Different chemical elements were not "elementary," as Dalton had suggested, but were structures made up of things more elementary still. The same thing is true of the hundreds of elementary hadrons, or nuclear particles. They are not themselves elementary, but are made up of things more elementary still—things that are given the name **quark** (pronounced "quork"). First suggested in the late 1960s, quarks have come to be accepted by physicists as the truly fundamental building blocks of the hadrons. Even though they never have been (and probably cannot be) seen in the laboratory, the concept of quarks has brought order and predictability to the complex zoo of elementary particles.

Quarks are different from other elementary particles in a number of ways. Unlike any other known particle, they have fractional electrical charge—that is, they have electrical charges that are equal to $\pm$ ⅓ or $\pm$ ⅔ the charge on the electron or proton. We believe that quarks make up all the hadrons, but once they are locked into these particles, no amount of experimental machination will ever pry them loose. Quarks existed as free particles only briefly in the very first stages of the universe (see Chapter 18).

Table 12–2 • **Quark Properties**

Name of Quark	Symbol	Electrical Charge*
down	d	$-\frac{1}{3}$
up	u	$+\frac{2}{3}$
strange	s	$-\frac{1}{3}$
charm	c	$+\frac{2}{3}$
bottom	b	$-\frac{1}{3}$
top	t	$+\frac{2}{3}$

* Quarks with the same charge differ from each other in mass and other properties.

In spite of these strange properties, the quark picture of matter is a very appealing one. Why? Because instead of dealing with numerous hadrons, only six kinds of quarks (and six antiquarks) occur in the universe. The quarks, like many things in elementary-particle physics, have been given fanciful names: up, down, strange, charm, top, and bottom (see Table 12–2). Until recently, we had seen elementary particles that contain only five of these six. The discovery of the sixth and final quark, the top quark, was announced in May, 1994.

From these six simple particles, all of the hadrons that we know about—all those hundreds of particles that whiz around inside the nucleus—can be made. (We have to stress, however, that leptons are *not* made from quarks.) The proton, for example, is the combination of two up quarks and one down quark, while the neutron is the combination of two down quarks and one up quark. In this scheme, the charge on the proton, equal to the sum of the charges on its three quarks, is

$$\frac{2}{3} + \frac{2}{3} + (-\frac{1}{3}) = +1,$$

while the charge on the neutron is

$$\frac{2}{3} + (-\frac{1}{3}) + (-\frac{1}{3}) = 0.$$

In the more exotic particles, pairs of quarks circle each other in orbit, like some impossible star system.

Quarks and Leptons: The Dots and Dashes of the Universe

The quark model gives us a picture of the universe that restores the kind of simplicity that was brought by both Dalton's atoms and Rutherford's nucleus. All of the elementary particles in the nucleus are made from various combinations of six kinds of quarks. These elementary particles are then put together to make the nuclei of atoms. The leptons—primarily the electrons—are stationed outside the nucleus to make the complete atoms, and different atoms interact with each other to produce what we see in the universe. In this scheme, the quarks and leptons are the "dots and dashes" of the universe; they are the basic stuff from which everything else is made. The fact that there are six leptons and six quarks—that the universe contains as many dots as dashes—has not escaped the notice of physicists. This phenomenon is built into almost all theories of elementary particles. The question of why nature should be arranged this way is still very much open, however.

Quark Confinement

It would be nice to be able to find quarks and study them in the laboratory, and physicists have conducted extensive searches for them. Yet, at least at present, there has been no generally accepted experimental isolation of a quark. In fact, particle theorists now believe that quarks cannot be isolated in the laboratory—indeed, that they can never be pried loose from the particles in which they exist. In these theories, once a quark is taken up into a particle, it is "confined" in that particle forever.

Here's an analogy that may help you think about quark confinement. Suppose you cut a rubber band and you were asked to isolate just one end of it. (Think of the rubber band as a particle and the very end of it as being the quark.) You could grab hold of the rubber band and pull it, perhaps even break it. You would then have two shorter rubber bands, but you would never have the end of a rubber band by itself. No matter how many times you broke the rubber band apart, you would get the same result. There just is no such thing as an "end" not attached to something else.

Elementary particles seem to be just the same. You can hit them as hard as you like in an attempt to shake the quarks loose, but every time you start to pull out a quark, you've also supplied enough energy to the system to make more quarks and antiquarks, and those new particles will immediately be taken up into ordinary particles. If you hit one particle hard enough, you will wind up with lots of other elementary particles, the things that correspond to the short pieces of the rubber band in our analogy.

What does it mean that the fundamental building blocks of the universe are things we can never isolate and study? Does that mean they aren't real? Without question, elementary particles behave *as if* they were made up of quarks, and you might want to think about the question of the reality of atoms (see Chapter 7) for a historical precedent to this situation.

The Four Fundamental Forces

In our excursion into the library, finding the dots and dashes wasn't enough to explain what we saw. We had to know the rules by which they are converted into books. In the same way, if we are going to understand the fundamental nature of the universe, we have to understand not only the dots and dashes (quarks and leptons), but also the forces that arrange them and make them behave the way they do.

Table 12–3 • The Four Forces

Force	Relative Strength*	Range	Gauge Particle
gravity	10^{-39}	infinite	graviton
electromagnetic	1/137	infinite	photon
strong force	1	10^{-13} cm	gluon
weak force	10^{-5}	10^{-15} cm	W and Z

* Relative to the strong force.

One useful analogy is to think of the quarks and leptons as the bricks of the universe. The universe appears to be built of these two different kinds of bricks that are arranged in different ways to make everything we see. But you cannot build a house using bricks alone. There has to be something like mortar to hold the bricks together. The "mortar" of the universe—the things that hold the elementary particles together and organize the physical universe into the structures we know—are the forces. At the moment, we know of only four fundamental forces in nature. Two of these, gravity and electricity, were known to nineteenth-century physicists and are part of our everyday experience. They are forces with infinite range—that is, two objects such as stars and planets can exert these forces on each other even though they are very far apart.

The other two forces are less familiar to us, because they operate in the realm of the nucleus and the elementary particle. They have a range comparable to the size of the nucleus (or smaller), and hence play no role in our everyday experience. The strong force holds the nucleus together, while the weak force is responsible for processes such as beta decay (see Chapter 11) that tear nuclei and elementary particles apart.

Each of the the four fundamental forces is different from the others in strength and range (see Table 12–3). The important point about the four forces is that whenever anything happens in the universe, whenever an object changes its motion, it happens because one or more of these forces is acting.

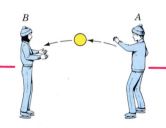

Figure 12–2

The exchange of a baseball between two skaters provides an analogy for the exchange of a gauge particle. The skater who throws the ball recoils, and the other recoils when the ball reaches her. Thus both skaters change direction, and, by Newton's first law, we say that a force acts between them.

Force as an Exchange

We know that forces cause matter to accelerate—nothing happens without a force. We have talked about the gravitational force, the electromagnetic force, the strong force, and the weak force. Each has its own distinctive effects on nature. We have not, however, asked how these forces work.

The modern understanding of forces may be thought of schematically as in Figure 12–2. Every force between two particles corresponds to the exchange of a third kind of particle, called **gauge particles** for historical reasons. That is, a first particle (an electron, for example) interacts with a second particle (say another electron) by the exchange of a gauge particle. The gauge particles produce the fundamental forces, like electricity, that hold everything together.

In Chapter 2 we used the analogy of someone standing on roller skates throwing baseballs to explain Newton's third law of motion. Suppose this same person was throwing baseballs, but another person standing on roller skates caught those baseballs some distance away. The person who threw the first baseball would recoil, as we discussed. The person who subsequently caught the baseball would also recoil. We could describe the situation this way: two people stand still before anything happens. After some time, the two people are moving away from each other. From Newton's first law, we conclude that a repulsive force had acted between those two people. Yet it's very clear in this analogy that that repulsive force is intimately connected with (a physicist would say "mediated by") the exchange of the baseballs.

In just the same way, we believe that every fundamental force is mediated by the exchange of some kind of gauge particle (see Figure 12–3). For example, the electrical force is mediated by the exchange of *photons*. That is, the magnet holding notes onto your refrigerator is exchanging huge numbers of photons with atoms inside the refrigerator metal to generate the magnetic force.

In the same way, the gravitational force is mediated by particles called *gravitons*. Right now you are exchanging large numbers of gravitons with the Earth, and it is this exchange that prevents you from floating up into space. The four fundamental forces and the gauge particles that are exchanged to generate each of them are listed in Table 12–3.

The two familiar forces of gravity and electromagnetism act over long distances, for example, because they are mediated by massless, uncharged particles (of which the familiar photon is one). The weak interaction, on the other hand, has a short range because it is mediated by the exchange of very massive particles (the W and Z particles have masses about 80 times that of the proton). Like the photon, the W and Z are particles that can be seen in the laboratory—they were first discovered in 1983, and are now routinely produced at accelerators around the world.

The situation with the strong force is a bit more complicated. The force that holds quarks together is mediated by particles called *gluons* (they "glue" the hadrons together). These particles are supposed to be massless, like the photon; but, like the quarks, they are confined to the interior of particles.

Unified Field Theories

Although a universe with six kinds of quarks, six kinds of leptons, and four kinds of forces may seem to be a relatively simple one, physicists have discovered an even greater underlying simplicity. The four fundamental forces turn out not to be as different from each other as their properties might at first suggest. The current thinking is that all four of these "fundamental" forces may simply be different aspects of a single underlying force.

Scientists believe that the four forces appear to us to be different because we are observing them at a time when the universe has been around for a long time and is at a relatively low temperature. The situation is somewhat analogous to freezing water. When water freezes it can adopt many apparently different forms—powdered white snow, solid ice blocks,

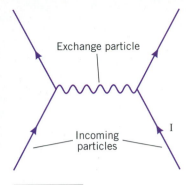

Figure 12–3
Exchange diagrams, introduced by physicist Richard Feynman, provide a model for particle interactions and the fundamental forces. Two incoming particles (such as two electrons) exchange a gauge particle (a photon) and thus are deflected by the force.

Richard Feynman (1918–1988)

delicate hoarfrost on tree branches, or a slippery layer on the sidewalk. You might interpret these forms of frozen H_2O as very different things, and in certain respects they are distinct. But heat them up and they're all simply water.

Similarly, the four forces only look different at the relatively low temperatures of our present existence. Heat matter up to trillions of degrees and the different forces are not really different at all. Theories in which fundamental forces are seen as different aspects of the same force go under the general name of **unified field theories.**

The first unified field theory in history was Isaac Newton's synthesis of earthly gravity and the circular motions observed in the heavens. To medieval scientists, earthly and heavenly motions seemed as different as the strong and electromagnetic forces do to us. Nevertheless, they were unified in Newton's theory of universal gravitation. In the same way, scientists today are working to unify the four fundamental forces.

The general idea of these theories is that if the temperature can be raised high enough—that is, if enough energy can be pumped into an elementary particle—the underlying unity of the forces will become clear. At a few laboratories around the world, it is possible to take protons and antiprotons (or electrons and positrons), accelerate them to extremely high energies, and let them collide. (As noted above, the proton-antiproton collisions involve the process of annihilation between particle and antiparticle.) When these collisions occur, for a brief moment the temperature in the volume of space about the size of a proton is raised to temperatures that have not been seen in the universe since it was less than a second old. In the resulting maelstrom, particles are produced that can be accounted for only if the electromagnetic and weak forces become unified.

In 1983, experiments at the European Center for Nuclear Research and the Stanford Linear Accelerator Center demonstrated conclusively that this kind of unification does occur. When protons and antiprotons (at the former laboratory) or electrons and positrons (at the latter) were accelerated and allowed to collide head-on, W and Z particles were seen in the debris of the collisions. Not only were the reactions seen, but the properties of the resulting particles and their rates of production were exactly those predicted by the first unified field theories.

The expectations of today's physicists regarding the unification of forces can be illustrated in a simple flow diagram (Figure 12–4). Scientists

Figure 12–4

The four forces become unified at extremely high temperatures, equivalent to those at the beginning of the universe. At 10^{-43} second after the moment of creation, the universe had already cooled sufficiently that gravity separated from the other three forces. The strong force separated at 10^{-33} second, while the weak and electromagnetic forces separated at 10^{-10} second.

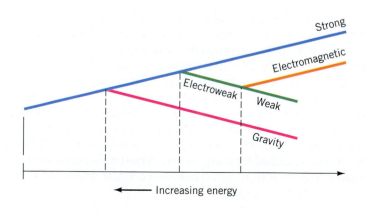

have already seen the unification of the electromagnetic and weak force in their laboratories. The resulting force, which physicists call the *electroweak force,* will be studied in great detail by the next generation of particle accelerators. At much higher energies, energies that will probably never be attained in Earth-based laboratories, we expect the strong force to unify with the electroweak. The theories that make this prediction constitute the *standard model*—the best model we have today of the elementary particles. Physicists have accumulated a fair amount of experimental evidence supporting the standard model.

Finally, at still higher energies, we hope to see the force of gravity unify with the strong-electroweak force. No theory yet describes this unification successfully, and attempts to develop a successful model are very much a frontier issue.

The unification of the four forces has no obvious effect on our everyday world, since it requires such exotic high-energy conditions to see it happen. Nevertheless, the unification of forces played a crucial role in the formation of the universe. Much of our understanding of what the universe is and how it came to be in its present state (Chapter 18) revolves around the various kinds of unification of the forces of nature.

Science in the Making

TOEs and TOOs

Unified theories that attempt to explain the existence of quarks, leptons, and the equivalence of all the forces in a single set of equations have been nicknamed *Theories Of Everything,* or TOEs. Some scientists, such as Cambridge University physicist Stephen W. Hawking, believe that the discovery of a valid TOE will be the supreme scientific discovery. In his best-selling book *A Brief History of Time* he writes:

> "If we do discover a complete theory, it should in time be understandable in broad principle by everyone, not just a few scientists. Then we shall all, philosophers, scientists, and just ordinary people, be able to take part in the discussion of the question of why it is that we and the universe exist. If we find the answer to that, it would be the ultimate triumph of human reason—for then we would know the mind of God."

Not all scientists agree with Hawking. Even the most elegant theory of everything is incomplete without additional knowledge of the organizing principles. It is doubtful, for example, that a physicist could start with a theory of everything and predict the existence of life. Thus we also need to find *Theories Of Organization,* or TOOs, that reveal how matter and energy are ordered into larger, complex systems. Such theories may or may not embody new general principles.

The relationship between TOEs and TOOs is something like that between computer hardware and software. A complete description of a computer system should include hardware diagrams that describe the machine down to every wire, screw, and microchip. But hardware by itself does nothing. Our description must also include listings and instructions for the computer program—the software that is not predictable from the hardware alone. ■

THINKING MORE ABOUT ELEMENTARY PARTICLES

The Superconducting Supercollider

Until 1993, physicists were building what would probably have been the last of the great particle accelerators. The Superconducting Supercollider (SSC), begun in the rolling Texas prairie south of Dallas, has been called the greatest high-tech engineering project in history. The SSC was designed basically as a synchrotron, but one of gigantic proportions. In a circular tunnel some 85 kilometers (about 50 miles) around, dug a hundred meters (about 300 feet) underneath the surface, particles were to have circulated through a series of electromagnets made from superconducting wire (see Chapter 10), eventually to achieve potentials of 80 trillion volts. While bunches of protons sped around through the system one way, antiprotons were to have circulated the other. When both beams achieved their highest energy, they would have been allowed to collide head-on, so that all of their energy would have been dumped into a volume about the size of a proton, raising its temperature to unprecedented levels on Earth. In the resulting maelstrom, physicists hoped to see the details of the unification of weak and electromagnetic forces and, perhaps, to learn why particles have mass.

That dream is dead, defeated by a budget-conscious Congress. You might think that all scientists would have been enthusiastic about building the SSC, but many weren't. The authors, for example, had completely different views of the desirability of building this machine. The reason for the debate was simple: over a decade-long construction period, the SSC would have cost over 10 billion dollars.

Proponents argued that maintaining the traditional U.S. leadership at the frontier of science required that the SSC be built here. They pointed out that the romance of finding the ultimate answers to questions about the nature of the universe would help to attract young people into careers in science, and that the payoff from understanding the phenomenon of mass would have been even greater than that which followed from the nineteenth-century research into the nature of electricity. Like their nineteenth-century counterparts, they could not predict what that payoff might be.

Opponents argued that this enormous expenditure would result in a lowering of support for other areas of science research and education, while the SSC would benefit only a few highly visible scientists. They supported more international cooperation with similar (though smaller) efforts in Europe. And they predicted that much more economic benefit would come from discoveries in small laboratories than from one more mega-experiment.

What is the best thing for governments to do when one field of science needs a large and expensive facility like the SSC? How much weight should be given to immediate payoffs, and how much to the long-term benefits of basic research? Is it essential for the United States to be the leader in all areas of science?

Summary

High-energy physics, or *elementary-particle physics,* deals with bits of matter that we cannot see, and forces and energies that we can barely imagine. Nevertheless, the study of the subatomic world holds the key to understanding the structure and organization of the universe.

All matter is made up of atoms, which are made up of even smaller particles, electrons and the nucleus; but these are not the most fundamental building blocks of the universe. Physicists originally studied the collisions between energetic *cosmic rays* and nuclei to study elementary particles. They now employ *particle accelerators,* including *synchrotrons* and *linear accelerators* to collide charged particles at near-light speeds. These scientists have discovered over 200 subatomic particles.

One class of particles, the *leptons* (including the electron and neutrino), are not subject to the strong force and thus do not participate in holding the nucleus together. Nuclear particles called *hadrons* (including the proton and neutron), according to present theories, are made from *quarks*—odd particles that have fractional electrical charge and cannot exist alone in nature. Together, leptons and quarks are the most fundamental building blocks of matter that we know. Each of these particle has an *antimatter* particle, such as the *positron,* the positively charged antiparticle of the electron.

The four known forces—gravity, electromagnetism, the strong force, and the weak force—cause particle interactions that lead to all of the organized structures we see in the universe. Particle interactions are mediated by the exchange of *gauge particles*—a different gauge particle for each of the different forces. Two masses, for example, will exchange gravitons (the gauge particle of gravity) as they are attracted to each other, and two charged particles will exchange photons, much the same way that two skaters will be "repelled" by each other if a mass is thrown from one to the other.

While the four known forces appear quite different to us, we believe that early in the universe, when temperatures were extremely high, the four forces were unified into a single force. At the forefront of modern physics research is the search for a *unified field theory* that describes this single force.

Key Terms About Particle Physics

high-energy physics, or elementary-particle physics

cosmic rays

particle accelerator

synchrotron

linear accelerator

leptons

hadrons

antimatter

positron

quarks

gauge particles

unified field theory

Review Questions

1. What is reductionism?
2. What are fundamental building blocks of a library? Why is there more than one correct answer?
3. What is a synchrotron?
4. How can we detect the presence of subatomic particles?
5. What is a lepton, and how do we know it exists?
6. What is a hadron, and how do we know it exists?
7. What is antimatter, and how do we know it exists?
8. What is a quark? Is there any way to prove that it exists?
9. How many different kinds of quarks are there?
10. Describe how quarks and leptons are put together to make all the matter we see.
11. Explain what it means for quarks to be confined.
12. What are the four fundamental forces? Which is the strongest? The weakest?
13. What role do gauge particles play in the universe?
14. What particle is exchanged to generate each of the four fundamental forces?
15. What is a unified field theory? Give an example.
16. In what sequence will the fundamental forces unify?
17. What is a PET scan? How does it employ elementary particles?
18. What was the SSC? Why was it discontinued?

Discussion Questions

1. Identify what might be considered the "fundamental units" and the rules of organization of (*a*) a grocery store, and (*b*) a parking garage. How many levels of organization can you identify? (Remember, not all questions have only one correct answer.)
2. Which particle-antiparticle interaction releases more energy: an electron-positron annihilation, or a proton-antiproton annihilation? How does the law of conservation of energy come into play?
3. What are the similarities in the modern argument over the reality of quarks and the nineteenth-century argument over the reality of atoms? What are the differences?

4. Why is it appealing for physicists to think that there are only six kinds of quarks, as opposed to more than a hundred hadrons?

5. What observations led Carl Anderson to conclude that he had discovered a particle with the same mass as the electron, but with a positive electrical charge?

6. What does it mean to say that all four fundamental forces were unified?

7. How might you detect the presence of a charged elementary particle?

8. How will we know when we have identified the truly fundamental building blocks?

Problems

1. What is the electric charge of an antiproton? An antineutron? Why?

2. A hadron called the lambda particle is made from two down quarks and one strange quark. What is the charge of the lambda particle?

3. A particle called the pi meson is made from an up quark and an anti-down quark. What is the charge of this particle?

Investigations

1. How did your congressional representative and senators vote on SSC funding in Congress? Why did they vote that way? Why do you agree or disagree with their vote?

2. Discuss the SSC controversy (*a*) in terms of the Big Science versus Little Science dichotomy discussed in Chapter 1; and (*b*) in terms of the basic research versus applied research dichotomy.

3. Locate the nearest PET-scan facility and arrange a visit. Where do the physicians obtain the special form of glucose used in the procedure? What kind of educational training would you need to operate such a facility?

4. Watch an episode of *Star Trek* and discuss the use of matter and antimatter in the propulsion system of the *Enterprise*. Can you find any other uses of antimatter in science fiction stories?

Additional Reading

Lederman, Leon. *The God Particle*. New York: Houghton-Mifflin, 1993.

Riordan, Michael. *The Hunting of the Quark*. New York: Simon and Schuster, 1987.

Trefil, James. *From Atoms to Quarks*. New York: Anchor, 1993.

Weinberg, Steven. *Dreams of a Final Theory*. New York: Pantheon, 1992.

Albert Einstein and the Theory of Relativity

All observers, no matter what their frame of reference, see the same laws of nature.

A Random Walk

The Airport

You walk down the ramp into the plane and get into your seat. It's been a long day—classes and maybe an exam or two. You doze off for a moment, then wake with a start. For a moment, it appears to you that the plane has started to move backward, away from the gate. Then, as you become more alert, you realize that your plane isn't moving at all, but that another plane has pulled into the gate next to yours.

During that brief moment between sleep and waking, you were seeing the world through eyes unaffected by years of experience. You were realizing that there is always more than one way to view any kind of uniform motion. One way, of course, is to say that you are stationary and the other plane is moving with respect to you. But you could also say that the other plane is stationary and you are moving with respect to it.

Which point of view is right?

One of the great scientific discoveries of the early twentieth century, the theory of relativity, grew out of thinking about this sort of question.

Frames of Reference

A **frame of reference** is the physical surroundings from which you observe and measure the world around you. If you read this book at your desk or in an easy chair, you experience the world from the frame of reference of your room, which seems firmly rooted to the solid Earth. If you read on a train or in a plane, your frame of reference is the vehicle that moves with respect to the Earth's surface. And you could imagine yourself in an accelerating spaceship in deep space, where your frame of reference would be different still. In each of these reference frames you are what scientists call an "observer." An observer looks at the world from a particular frame of reference with anything from casual interest to a full-fledged laboratory investigation of phenomena that leads to a determination of natural laws.

For human beings who grow up on the Earth's surface, it is natural to think of the ground as a fixed, immovable frame of reference, and to refer all motion to it. After all, train or plane passengers don't think of themselves as stationary while the countryside zooms by. But, as we saw in the opening example, there are indeed times when we lose this prejudice and see that the question of who is moving and who is standing still is largely one of definition.

From the point of view of an observer in a spaceship above the solar system, there is nothing "solid" about the ground you're standing on. The Earth is rotating on its axis and moving in an orbit around the Sun, while the Sun itself is performing a stately rotation around the Galaxy. Thus, even though a reference frame fixed in the Earth may seem "right" to us, there is nothing special about it.

Descriptions in Different Reference Frames

Different observers in different reference frames may provide very different accounts of the same event. To convince yourself of this idea, think about a simple experiment. While riding in a car, take a coin out of your pocket and flip it. You know what will happen—the coin will go up in the air and fall straight back into your hand, just as it would if you flipped it while sitting in a chair in your room (Figure 13–1a). But now ask yourself this question: how would a friend standing by the roadside, watching your vehicle go by, describe the flip of the coin?

To that person it would appear that the coin went up into the air, of course, but by the time it came down the car would have traveled some distance down the road. As far as your friend on the ground is concerned, the coin traveled in an arc (Figure 13–1b).

So you, sitting in the vehicle, say the coin went straight up and down, while someone on the ground says it traveled in an arc. You and the ground-based observer would describe the path of the coin quite differently, and you'd both be correct in your respective frames of reference. The universe we live in possesses this general feature—different observers will describe the same event in different terms, depending on their frames of reference.

Does this mean that we are doomed to live in a world where nothing

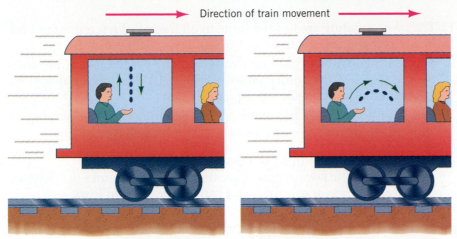

Direction of train movement

Apparent direction of coin fall

(*a*) Frame of reference: inside the train   (*b*) Frame of reference: outside the train

Figure 13–1
The path of a coin flipped in the air depends on the observer's frame of reference. (*a*) A rider in the car sees the coin go up and fall straight down. (*b*) An observer on the street sees the coin follow an arching path.

is fixed, where everything depends on the frame of reference of the observer? Not necessarily. The possibility exists that even though different observers give different descriptions of the same event, they will agree on the underlying laws that govern it. Even though the observers disagree on the path followed by the flipped coin, they may very well agree that motion in their frame is governed by Newton's laws of motion and the law of universal gravitation.

The Principle of Relativity

Albert Einstein came to his theories of relativity by thinking about a fundamental contradiction between Newton's laws and Maxwell's equations. You can see the problem by thinking about a simple example. Imagine you're on a moving railroad car and you throw a baseball. What speed will the baseball have according to an observer on the ground?

If you throw the ball forward at 40 kilometers per hour from a train traveling 100 kilometers per hour, the ball will appear to a ground-based observer to travel 140 km/hour—40 km/hour from the ball plus 100 km/hour from the train. If, on the other hand, you throw the ball backwards, the ground-based observer will see the ball moving at only 60 km/hour—the train's 100 km/hour minus the ball's 40 km/hour. In our everyday world, we just add the two speeds to get the answer, and this notion is reflected in Newton's laws.

Suppose, however, that instead of throwing a ball you turned on a flashlight and measured the speed of the light coming from it. In Chapter 6 we saw that the speed of light is built into Maxwell's equations. If every observer is to see the same laws of nature, they all have to see the same speed of light. In other words, the ground observer would have to see light from the flashlight moving at 300,000 km/sec, and *not* 300,000 km/sec plus 100 km/hr. In this case, velocities wouldn't add, as our intuition tells us they must.

Albert Einstein thought long and hard about this paradox, and he realized that it could be resolved in only three ways:

1. The laws of nature are not the same in all frames of reference (an idea Einstein was reluctant to accept on philosophical grounds), or;

2. Maxwell's equations could be wrong and the speed of light depends on the speed of the source emitting the light (in spite of abundant experimental support for the equations), or;

3. Our intuitions about the addition of velocities could be wrong, in which case the universe might be a very strange place indeed.

Einstein focused on the third of these possibilities.

The idea that the laws of nature are the same in all frames of reference is called the *principle of relativity,* and can be stated as follows:

Every observer must experience the same natural laws.

This statement is the central assumption of Einstein's **theory of relativity.** Hidden beneath this seemingly simple statement lies a view of the universe that is both strange and wonderful. The extraordinary theoretical effort required to understand the consequences of this one simple assumption occupied Einstein during much of the first decades of the twentieth century.

We can begin to understand his work by recalling what Isaac Newton had demonstrated three centuries earlier, that all motions fall into one of two categories: uniform motion or acceleration (Chapter 2). Einstein therefore divided his theory of relativity into two parts—one dealing with each of these kinds of motion. The easier part, first published by Einstein in 1905, is called **special relativity** and deals with all frames of reference in uniform motion relative to one another—reference frames that do not accelerate. It took Einstein another decade to complete his treatment of **general relativity,** mathematically a much more complex theory, which applies to any reference frame whether or not it is accelerating relative to another.

At first glance, the underlying principle of relativity seems obvious, perhaps almost too simple. Of course the laws of nature are the same everywhere—that's the only way that scientists can explain how the universe behaves in an ordered way. But once you accept that central assumption of relativity, be prepared for some surprises. Relativity forces us to accept the fact that nature doesn't always behave as our intuition says it must. You may find it disturbing that nature sometimes violates our sense of the "way things should be." But you'll have little problem with relativity if you just accept the idea that the universe is what it is, and not necessarily what we think it should be.

Another way of saying this is to note that our intuitions about how the world works are built up from experience with things that are moving at modest speeds—a few hundred, or at most a few thousand, miles per hour. None of us has any experience with things moving near the speed of light, so when we start examining phenomena in that range our intuitions won't necessarily apply. Strictly speaking, we shouldn't be surprised by anything we find.

Relativity and the Speed of Light

As the example of the train and the flashlight shows, one of the most disturbing aspects of the principle of relativity has to do with our everyday notions of speed. According to the principle, any observer, no matter what his or her reference frame, should be able to confirm Maxwell's description of electricity and magnetism. Since the speed of light is built into these equations, it follows that:

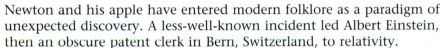

> **The speed of light, *c*, is the same in all reference frames.**

Strictly speaking, this statement is only one of many consequences of the principle of relativity. However, so many of the surprising results of relativity follow from it that it is often accorded special status and given special attention in discussions of those theories.

Albert Einstein (1879–1955)

Science in the Making

Einstein and the Streetcar

Newton and his apple have entered modern folklore as a paradigm of unexpected discovery. A less-well-known incident led Albert Einstein, then an obscure patent clerk in Bern, Switzerland, to relativity.

One day, while riding home in a streetcar, he happened to glance up at a clock on a church steeple (Figure 13–2). In his mind he imagined the streetcar speeding up, moving faster and faster, until it was go-

Figure 13–2

Albert Einstein, moving away from a clock tower, imagined how different observers might view the passage of time. If Einstein were traveling at the speed of light, for example, the clock would appear to him to have stopped, even though his own pocket watch would still be ticking.

ing at almost the speed of light. Einstein realized that if the streetcar were to travel at the speed of light, it would appear to someone on the streetcar that the clock had stopped. The passenger would be like a surfer on a light-wave crest—a crest that originated at 12 noon, for example—and the same image of the clock would stay with him.

On the other hand, a clock moving with him—his pocket watch, for example—would still tick away the seconds in its usual way. Perhaps, Einstein thought, *time as measured on a clock, just like motion, is relative to one's frame of reference.* ■

■ Special Relativity

Time Dilation

Think about how you measure time. The passage of time can be measured by any kind of regularly repeating phenomenon—a swinging pendulum, a beating heart, or an alternating electrical current. To get at the theory of relativity, though, it's easiest to think of a rather unusual kind of clock. Suppose, as in Figure 13–3, we had a flashbulb, a mirror, and a photon detector. A "tick-tock" of this clock would consist of the flashbulb going off, the light traveling to the mirror, bouncing down to the detector, and then triggering the next flash. By adjusting the distance between the light source and mirror, these pulses could correspond to any desired time interval. This unusual "light clock," therefore, serves the same function as any other clock—in fact, you could adjust it to be synchronized with anything from a grandfather's clock to a wristwatch.

Figure 13–3

A light clock incorporates a flashing light and a mirror. A light pulse bounces off the mirror and returns to trigger the next pulse. Two light clocks, one stationary (*a*) and one moving (*b*), illustrate the phenomenon of time dilation. Light from the moving clock must travel farther, and so it appears to a stationary observer to tick more slowly.

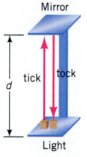

(a) Stationary light clock

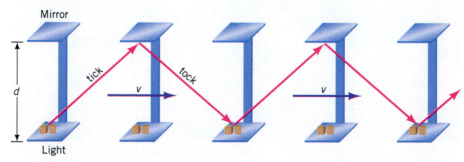

(b) Moving light clock

Now imagine two identical light clocks: one next to you on the ground (Figure 13–3a), and the other whizzing by in a spaceship (Figure 13–3b). Imagine further that the mirrors are adjusted so that both clocks would tick at the same rate if they were standing next to each other. How would the moving clock look to you?

Standing on the ground, you would see the ground-based clock ticking along as the light pulses bounce back and forth between the mirror and detector. When you looked at the moving clock, though, you would see the light following a longer, zig-zag path. If the speed of light is indeed the same in both frames of reference, it should appear to you that the light in the moving frame takes longer to travel the zig-zag path from light to detector than the light on the ground-based clock. Consequently, from your point of view on the ground, the moving clock must tick more slowly. The two clocks are identical, but *the moving clock runs slower*. This surprising phenomenon is known as **time dilation,** and it is an essential consequence of relativity.

Remember that each observer regards the clocks in his or her own reference frame as completely normal, while all other clocks appear to be running slower. Thus, paradoxically, while we observe the spaceship clock as slow, observers in the speeding spaceship see the Earth-based clock moving and, as above, believe that the Earth-based clock is running more slowly than theirs.

Relativity's prediction of time dilation can be tested in a number of ways. Scientists have actually documented relativistic time dilation by comparing two extremely accurate atomic clocks, one on the ground and one strapped into a jet aircraft. Even though jets travel at a paltry hundred-thousandth of the speed of light, the difference in the time recorded by the two clocks can be measured.

Time dilation can also be observed with high-energy particle accelerators that routinely produce unstable subatomic particles (see Chapter 12). The normal half-life of these particles is well known. When accelerated to near the speed of light, however, these particles last much longer because of the relativistic slowdown in their decay rates.

Thus, although the notion that moving clocks run slower than stationary ones violates our intuition, it seems to be well documented by experiment. Why, then, aren't we aware of this effect in everyday life? To answer that question, we have to ask how big an effect time dilation is. How much do moving clocks slow down?

The Size of Time Dilation

We have tried, in general, to talk about science in everyday terms and stay away from formulas in this book. But we have now run into a rather fundamental question that requires some simple mathematics to answer. In this section, you'll be able to follow the kind of thought process used by Einstein when he first formulated his revolutionary theory.

Consider the two identical light clocks in Figure 13–3, one moving at a velocity, *v*, and one stationary on the ground. Each clock has a light-to-mirror separation distance of *d*. (The various symbols we are using are summarized in Table 13–1.)

The notation for the time it takes for light to travel the distance *d* from the light to its opposite mirror—that is, one "tick" of the stationary

Table 13–1 ● Symbols for Deriving Time Dilation

Symbol	Description
v	Velocity of the moving light clock relative to the ground
d	Distance between the clock's light and mirror
t_{GG}	Time for one tick (ground clock, ground observer)
t_{MG}	Time for one tick (moving clock, ground observer)
t_{GM}	Time for one tick (ground clock, moving observer)
t_{MM}	Time for one tick (moving clock, moving observer)
c	Speed of light, a constant

clock—is a little trickier, because we have to keep track of which clock we're looking at *and* from which reference frame we're looking. We will use two subscripts—the first subscript to tell us if the clock is on the ground (*G*) or moving (*M*), and the second subscript to indicate if the observer is on the ground or moving. Thus, t_{GG} is the time for one tick of the ground-based clock as observed by an observer on the ground. On the other hand, t_{MG} is the time for one tick of the moving clock from the point of view of this ground-based observer. According to the principle of relativity, all observers see clocks in their own reference frames as normal. Or, in equation form,

$$t_{GG} = t_{MM}.$$

As ground-based observers, we are interested in determining the relative values of t_{GG} and t_{MG}—what we see as "ticks" of the stationary versus the moving clocks. In the stationary ground-based frame of reference, one tick is simply the time it takes light to travel the distance d.

$$\text{time} = \frac{\text{distance}}{\text{speed}}.$$

Substituting values for the light clock into this equation,

$$\text{time for one tick} = \frac{\text{light-to-mirror distance}}{\text{speed of light}},$$

or,

$$t_{GG} = \frac{d}{c}$$

where c is the standard symbol for the speed of light.

We argued that to the observer on the ground, it appears that the light beam in the moving clock travels on a zig-zag path as shown in Figure 13–3, and that this made the moving clock appear to run more slowly. In what follows, we will show how to take an intuitive statement like this one and convert it into a precise mathematical equation. We begin by labeling the dimensions of our two clocks.

The moving clock travels a horizontal distance of $v \times t_{MG}$ during each of its ticks. In order to determine the value of t_{MG}, we must first determine how far light must travel in the moving clock as seen by the observer on the ground. As illustrated in Figure 13–4, we know the lengths of the two shortest sides of a right triangle. One side has length d, representing the vertical distance between light and mirror (a distance, remember, that is the same in both frames of reference). The other side is $v \times t_{MG}$, which corresponds to the distance traveled by the moving clock as observed in the stationary frame of reference. The distance traveled by the moving light beam in one tick is represented by the hypotenuse of this right triangle, and is given by the Pythagorean theorem:

▶ **In words:**

The square of the length of a right triangle's long side equals the sum of the squares of the lengths of the other two sides.

▶ **In words (applied to our light clock):**

The square of the distance light travels during one tick equals the sum of the squares of the light-to-mirror distance and the horizontal distance the clock moves during one tick.

▶ **In symbols:**

(distance light travels)$^2 = d^2 + (vt_{MG})^2$.

We can begin to simplify this equation by taking the square roots of both sides.

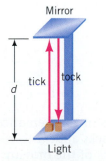

(a) Stationary light clock

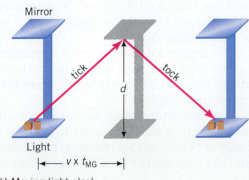

(b) Moving light clock

Figure 13–4

Light clocks with dimensions labeled. Both the stationary clock (a) and the moving clock (b) have light-to-mirror distance, d. During one tick the moving clock must travel a horizontal distance $v \times T_{MG}$.

$$\text{distance light travels} = \sqrt{d^2 + (vt_{MG})^2}$$

Remember, time equals distance divided by velocity. So the time it takes light to travel this distance, t_{MG}, is given by the distance, $\sqrt{d^2 + (vt_{MG})^2}$, divided by the velocity of light, c.

$$t_{MG} = \frac{\sqrt{(d^2 + v^2 t_{MG}^2)}}{c}$$

We now must engage in a bit of algebraic manipulation. First, square both sides of this equation.

$$t_{MG}^2 = \frac{d^2}{c^2} + \frac{v^2 t_{MG}^2}{c^2}$$

But we saw previously that $t_{GG} = d/c$, so, substituting,

$$t_{MG}^2 = t_{GG}^2 + \frac{v^2 t_{MG}^2}{c^2}.$$

Dividing both sides by t_{MG}^2 gives

$$\frac{t_{MG}^2}{t_{MG}^2} = \frac{t_{GG}^2}{t_{MG}^2} + \frac{v^2 t_{MG}^2/c^2}{t_{MG}^2}$$

or,

$$1 = \left(\frac{t_{GG}}{t_{MG}}\right)^2 + \left(\frac{v}{c}\right)^2.$$

Finally, regrouping yields

$$t_{MG} = \frac{t_{GG}}{\sqrt{1 - (v/c)^2}}.$$

This equation expresses in mathematical form what we said earlier in words—that moving clocks appear to run slower. It tells us that t_{MG}, the time it takes for one tick of the moving clock as seen by an observer on the ground, is equal to the time it takes for one tick of an identical clock on the ground *divided by a number less than one.* Thus the time required for a tick of the moving clock will always be greater than that for a stationary one.

The expression $\sqrt{1 - (v/c)^2}$ is a number, called the *Lorentz factor,* that appears over and over again in relativistic calculations. In the case of time dilation, the Lorentz factor arises from an application of the Pythagorean theorem.

One important point to notice is that if the velocity of the moving clock is very small compared to the speed of light, the quantity $(v/c)^2$ becomes very small and the Lorentz factor is almost equal to one. In this case, the time on the moving clock is equal to the time on the stationary one, as our intuition demands that it should be. Only when speeds get very high do the effects of relativity become important.

Science by the Numbers

How Important Is Relativity?

To understand why we aren't aware of relativity in our everyday life, let's calculate the size of the time dilation for a clock in a car moving at 70 km/hr (about 50 miles per hour).

The first problem is to convert the familiar speed in km/hr to a speed in meters per second so we can compare it to the speed of light. There are $60 \times 60 = 3600$ seconds in an hour, so a car traveling 70 km/hr is moving at a speed of

$$70 \text{ km/hr} = \frac{70,000 \text{ m}}{3600 \text{ s}}$$
$$= 19.4 \text{ m/s}$$

For this speed, the Lorentz factor is:

$$\sqrt{1 - (19.4/300,000,000)^2} = 0.9999999999999999$$

Thus the passage of time for a stationary and speeding car differs by only one part in the sixteenth decimal place.

To get an idea of how small the difference is between the ground clock and the moving one in this case, we can note that if you watched the moving car for a time equal to the age of the universe, you would observe it running *10 seconds slow* compared to your ground clock.

For an object traveling at 99% of the speed of light, however, the Lorentz factor is

$$\sqrt{1 - (v/c)^2} = \sqrt{1 - (0.99)^2}$$
$$= \sqrt{0.0199}$$
$$= 0.1411$$

In this case, you would observe the stationary clock to be ticking about seven times as fast as the moving one—that is, the ground clock would tick about seven times while the moving clock ticked just once.

This numerical example illustrates a very important point about relativity. Our intuition and experience tell us that the exterior clock on our local bank doesn't suddenly slow down when we view it from a moving car. Consequently, we find the prediction of time dilation to be strange and paradoxical. But all of our intuition is built up from experiences at very low velocities—none of us has ever moved at an appreciable fraction of the speed of light. For the everyday world, the predictions of relativity coincide precisely with our experience. It is only when we get into regions near the speed of light, where that experience isn't relevant, that the "paradoxes" arise. ■

The Human Body

Space Travel and Aging

While humans presently do not experience the direct effects of time dilation in their day-to-day lives, at some future time they might. If we ever develop interstellar space travel with near-light speed, then time dilation may wreak havoc with family lives (and genealogists' records).

Imagine a spaceship that accelerates to 99% of the speed of light and goes on a long journey. While 15 years seem to pass for the crew of the ship, more than a century goes by on Earth. The space explorers return almost 15 years older than when they left, but biologically younger than their great-grandchildren! Friends and family all would be long-since dead.

If we ever enter an era of extensive high-speed interstellar travel, people may drift in and out of other people's lives in ways we can't easily imagine. Parents and children could repeatedly leapfrog each other in age, and the notion of relatedness could take on complex twists in a society with widespread relativistic travel. ■

Distance and Relativity

A long list of results from relativity run counter to our intuition. They can be derived by procedures similar to (but more complicated than) the one we just gave for working out time dilation. In fact, using arguments like those above, Einstein showed that *moving yardsticks must appear to be shorter than stationary ones in the direction of motion* (see Figure 13–5).

This series of four computer-generated images shows the changing appearance of a network of balls and rods as it moves toward you at different speeds. (*a*) At rest—the normal view; (*b*) at 50% of light speed that array appears to contract; (*c*) at 95% of light speed the lattice has curved rods; and (*d*) at 99% of light speed the network is severely distorted.

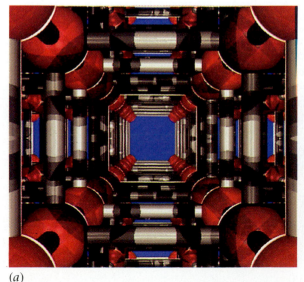

(*a*)

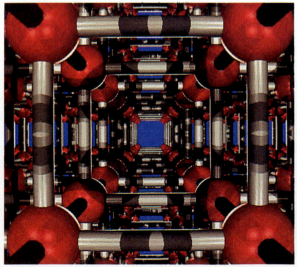

(*b*)

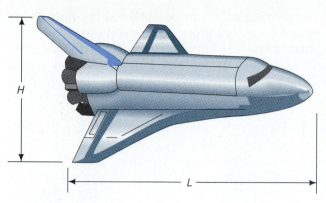

(*a*) Spaceship at rest

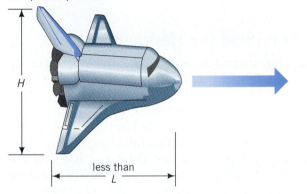

(*b*) Spaceship at high speed

The equation that relates the ground-based observer's measurement of a stationary object's length, L_{GG}, to that observer's measurement of the length of an identical moving object, L_{MG}, is

$$L_{MG} = L_{GG} \times \sqrt{1 - (v/c)^2}.$$

The term on the right-hand side of this equation is the familiar Lorentz factor that we derived from our study of light clocks. The equation tells

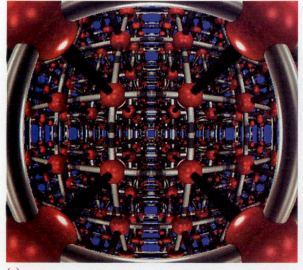

(*c*)

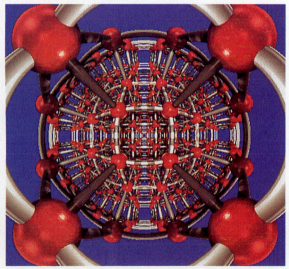

(*d*)

us that the length of the moving ruler can be obtained by multiplying the length of the stationary ruler by a number less than one, and thus must appear shorter. This phenomenon is known as **length contraction.**

Note that the height and width of the moving object do not appear to change—only the length along the direction of motion. A moving basketball, then, takes on the appearance of a pancake at speeds near those of light.

Length contraction is not just an optical illusion. While relativistic shortening doesn't affect most of our daily lives, the effect is real. Physicists who work at particle accelerators inject "bunches" of particles into their machines. As these particles approach light speed, the bunches are observed to contract according to the Lorentz factor, an effect that must be compensated for.

So What About the Train and the Flashlight?

Now that we understand a little about how relativity works, we can go back and unravel the paradox we discussed earlier in this chapter—the problem of how both an observer on the ground and an observer on the train could see light from a flashlight moving at the same speed.

Velocity is defined to be distance traveled divided by the time required for the travel to take place. Since both length and time appear to be different for different observers, it should come as no surprise that the rule that tells us how to add velocities (such as the velocity of the light and the train) might be more complicated than we would expect. The simple intuition that tells us that we should add the velocity of the train to the velocity of the ball, like our notions of time and space, is valid at small velocities but breaks down for objects moving near the speed of light. For those objects, a more complex addition has to be done, and when it is, we find that both observers see the light moving at a velocity of c.

Mass and Relativity

Perhaps the most far-reaching consequence of Einstein's theory of relativity was the discovery that mass, like time and distance, is relative to one's frame of reference. So far we have been faced with two strange ideas:

1. Clocks run fastest for stationary objects; moving clocks slow down. As the speed of light is approached, time slows down and approaches zero.
2. Distances are greatest for stationary objects; moving objects shrink in the direction of motion. As the speed of light is approached, distances shrink and approach zero.

Einstein showed that a third consequence followed from his principle:

3. Mass is lowest for stationary objects; moving objects become more massive. As an object's velocity approaches the speed of light, its mass approaches infinity.

Einstein showed that if the speed of light is a constant in all reference frames—which must follow from the central assumption of the theory of relativity—then an object's mass depends on its velocity. The faster an object travels, the greater its mass and the harder it is to deflect from its course. If a ground-based observer measures an object's stationary or "rest" mass, m_{GG}, then the apparent mass, m_{MG}, of that object moving at velocity v is

$$m_{MG} = \frac{m_{GG}}{\sqrt{1 - (v/c)^2}}.$$

Once again the Lorentz factor comes into play. As we observe an object approach the speed of light, its mass appears to us to approach infinity.

This property of mass leads to the common misperception that relativity predicts that nothing can travel faster than the speed of light. In fact, the only thing we can conclude is that nothing that is now moving at less than the speed of light can be accelerated to or past that speed. It also says that, should there exist objects already moving faster than light, they could not be decelerated to speeds of less than c, and that the only objects that travel at the speed of light (such as photons) are those that have zero rest mass.

Mass and Energy

Time, distance, and mass—all quantities that we can easily measure in our homes or laboratories—actually depend on our frame of reference. But not everything in nature is so variable. The central tenet of relativity is that natural laws must apply to *every* frame of reference. Light speed is constant in all reference frames in accord with Maxwell's equations. Similarly, the first law of thermodynamics—the idea that total amount of energy in any closed system is constant—must hold, no matter what the frame of reference. Yet here, Einstein's description of the universe seems to run into a problem. He claims that the observed mass depends on your frame of reference. But, in that case, kinetic energy—defined as mass times velocity squared—could not follow the conservation of energy law. In Einstein's treatment, faster frames of reference seem to possess more energy than slower ones. Where does the extra energy come from?

Conservation of energy appeared to be violated because we missed one key form of energy in our equations: mass itself. In fact, Einstein was able to show that the amount of energy contained in any mass turns out to be the mass times a constant.

▶ **In words:**

All objects contain a rest energy (in addition to any kinetic or potential energy), which is equal to the object's rest mass times the speed of light squared.

▶ **In equation form:**

rest energy = rest mass $\times$ (speed of light)^2.

▶ **In symbols:**

$$E = mc^2.$$

This familiar equation has become an icon of our nuclear age, for it defines a new form of energy. It says that mass can be converted to energy, and vice versa. Furthermore, the amounts of energy involved are prodigious (because the constant, the speed of light squared, is so large). A handful of nuclear fuel can power a city; a fist-sized chunk of nuclear explosive can destroy it.

Until Einstein traced the implications of special relativity, the nature of mass and its vast potential for producing energy was hidden from us. Now nearly a quarter of all electrical power in the United States is produced in nuclear reactors that confirm the predictions of Einstein's theory every day of our lives (see Chapter 11).

General Relativity

Special relativity is a fascinating and fairly accessible intellectual exercise, requiring little more than an open mind and a lot of basic algebra. General relativity, which deals with all reference frames including accelerating ones, is much more challenging in its full rigor. While the details are tricky, you can get a pretty good feeling for Einstein's general theory by thinking about the nature of forces.

The Nature of Forces

Begin by imagining yourself in a completely sealed room that is accelerating at exactly one "*g*"—the Earth's gravitational acceleration. Could you devise any experiment that would reveal one way or the other if you were on Earth or accelerating in deep space?

If you dropped this book on Earth, the force of gravity would cause it to fall to your feet. If you dropped the book in the accelerating spaceship, however, Newton's first law tells us that it will keep moving with whatever speed it had when it was released. The floor of the ship, still accelerating, will therefore come up to meet it. To you, standing in the ship, it appears that the book falls, just as it does if you are standing on Earth.

From an external frame of reference, of course, these two situations would involve very different descriptions. In the one case the book falls due to the force of gravity; in the other the spaceship accelerates up to meet the free-floating book. But no experiment you could devise *in your reference frame* could distinguish between acceleration in deep space and the force of the Earth's gravitational field.

In some deep and profound way, therefore, gravitational forces and acceleration are equivalent. Newton saw this connection in a way, for his laws of motion equate force with an accelerating mass. But Einstein went a step further by recognizing that what we call gravity versus what we call acceleration is a purely arbitrary decision, based on our choice of reference frame. Whether we think of ourselves as stationary on a planet with gravity, or accelerating on spaceship Earth, makes no difference in the passage of events.

Although this connection between gravity and acceleration may seem a bit abstract, you already have had experiences that should tell you it is true. Have you ever been in an elevator and felt momentarily heavier when it starts up or momentarily lighter when it starts down? If so, you know that the feeling we call "weight" can indeed be affected by acceleration.

The actual working out of the consequences of this notion of the equivalence of acceleration and gravity is complicated, but a simple analogy can help you visualize the difference between Einstein's and Newton's views of the universe. In the Newtonian universe, forces and motions can be described by a ball rolling on a perfectly flat surface with neatly inscribed grid lines. The ball rolls on and on, following a line exactly, unless an external force is applied. If, for example, a large mass rests on the surface, the rolling ball will change its direction and speed—it will accelerate in response to the force of gravity. Thus, for Newton, motion occurs along curving paths in a flat universe.

The description of that same event in general relativity is very different. In this case, as shown in Figure 13–6, we would say that the heavy

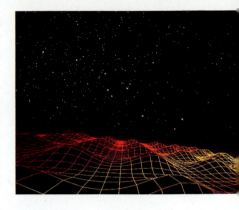

A computer-generated image of a gravity field reveals masses as gravity wells on otherwise flat grid.

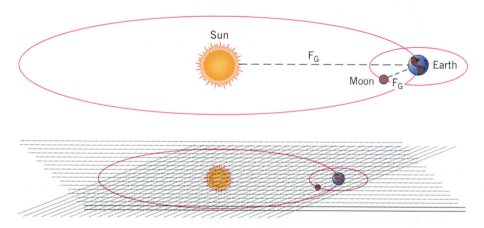

(a) Newtonian universe: gravitational forces in a "flat" universe

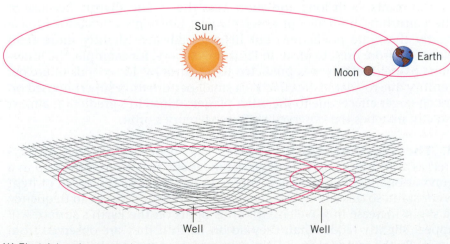

(b) Einstein's universe: motion in a curved universe.

Figure 13–6

Newtonian and Einsteinian universes treat the motion of rolling balls in different ways. In the Newtonian scheme, a ball travels in uniform motion unless acted upon by a force; motion occurs along curved paths in a flat universe. In the Einsteinian universe, a ball's mass distorts the universe; it moves in a straight line across a curved surface.

object distorts the surface. Peaks and depressions on the surface influence the ball's path, deflecting it as it rolls across the surface. For Einstein, the ball moves in a straight line across a curved universe.

Given these differing views, Newton and Einstein would give very different descriptions of physical events. Newton would say, for example, that the Moon orbits the Earth because of an attractive gravitational force between the two bodies. Einstein, on the other hand, would say that space has been warped in the vicinity of the Earth-Moon system, and this warping of space governs the Moon's motion. In the relativistic view, space deforms around the Sun, and planets follow the curvature of space like marbles rolling around in the bottom of a curved bowl.

We now have two very different ways of thinking about the universe. In the Newtonian universe, forces cause objects to accelerate. Space and time are separate dimensions that are experienced in very different ways. This view more closely matches our everyday experience of how the world seems to be. In Einstein's universe, objects move according to distortions in space, while the distinction between space and time depends on your frame of reference.

Predictions of General Relativity

The mathematical models of Newton and Einstein are not just two equivalent descriptions of the universe. They lead to slightly different quantitative predictions of events. In three specific instances, the predictions of general relativity have been confirmed.

1. The Gravitational Bending of Light   One consequence of Einstein's theory is that light can be bent as it travels along the warped space near strong gravitational centers such as the Sun. Einstein predicted the exact amount of deflection that would occur near the Sun, and his prediction was confirmed by precise measurements of star positions during a solar eclipse in 1919. Today these measurements are made with much more precision by measuring the deflection of radio waves emitted by distant galaxies called quasars (see Chapter 18).

2. Planetary Orbits   In Newton's solar system, the planets adopt elliptical orbits, with long and short axes that rotate slightly because of the perturbing influence of other planets. Einstein's calculations make nearly the same prediction, but his axes advance slightly more than Newton's from orbit to orbit. In Einstein's theory, for example, the innermost planet Mercury was predicted to advance by 43 seconds of arc per century due to relativistic effects—a small perturbation superimposed on much larger effects due to the other planets. Einstein's prediction almost exactly matches the observed shift in Mercury's orbit.

3. The Gravitational Red Shift   The theory of relativity predicts that as a photon (a particle of electromagnetic radiation) moves up in a gravitational field, it must lose energy in the process. The speed of light is constant, so this energy loss is manifest as a slight decrease in frequency (a slight increase in wavelength). Thus lights on the Earth's surface will appear slightly redder than they do on Earth if they are observed from space. By the same token, a light shining from space to the Earth will be slightly shifted to the blue end of the spectrum. Careful measurements of laser light frequencies have amply confirmed this prediction of relativity.

Recent advances in highly sensitive electronics are now providing more opportunities for researchers to measure the predictions of general relativity. One of the most intriguing tests will be conducted as part of a satellite mission in the mid-1990s. Meticulously machined quartz spheres will be set into rotation and carefully measured. According to general relativity, these spheres should develop a small wobble as they rotate in the Earth's gravitational field. Sensitive electronics will detect any perturbations of this sort.

As scientists get better and better at making precision measurements, more and more tests of the extremely small differences between Newtonian and relativistic predictions of physical events can be made. A few years ago, for example, a group of scientists at institutions in the Washington, D.C., area proposed an experiment in which light from a laser would be sent out over the city from the University of Maryland, reflected from a mirror on top of the National Cathedral, and received by detectors at the Naval Research Laboratory. Because of the rotation of the Earth, there should be a tiny difference in travel time between light traveling east and light traveling west. The theory of relativity predicts an additional, tinier difference as well. With good enough clocks and short enough laser pulses, these sorts of differences can be measured, and they provide just as good a check on general relativity as instruments in a satellite.

Who Can Understand Relativity?

Einstein's theory of relativity was extraordinary, but when first introduced was difficult to grasp in part because it relied on some complex mathematics that were unfamiliar to many scientists at the time. Furthermore, while the theory made specific predictions about the physical world, most of those predictions were exceedingly difficult to test. Soon after the theory's publication it became conventional wisdom that only a handful of geniuses in the world could understand it.

Einstein did make one very specific prediction, however, that could be tested. His proposal that the strong gravitational field of the Sun would bend the light coming from a distant star was different from other theories. The total eclipse of the Sun in 1919 gave scientists the chance to test Einstein's prediction. Sure enough, the apparent position of stars near the Sun's disk was shifted by exactly the predicted amount.

Around the world, front-page newspaper headlines trumpeted Einstein's success. He became an instant international celebrity and his theory of relativity became a part of scientific folklore. Attempts to explain the revolutionary theory to a wide audience began almost immediately.

Few scientists may have grasped the main ideas of general relativity in 1915, when the full theory was first unveiled, but that certainly is not true today. The basics of special relativity are taught to tens of thousands of college freshmen every year, while hundreds of students in astronomy and physics explore general relativity in its full mathematical splendor.

If this subject intrigues you, you might want to read some more, watch TV specials or videos about relativity, or even sign up for one of those courses!

THINKING MORE ABOUT RELATIVITY

Was Newton Wrong?

The theory of relativity describes a universe about which Isaac Newton never dreamed. Time dilation, contraction of moving objects, and mass as energy play no role in his laws of motion. Curved space-time is alien to the Newtonian view. Does that mean that Newton was wrong? Not at all.

In fact, all of Einstein's equations reduce *exactly* to Newton's laws of motion, *at speeds significantly less than the speed of light*. This feature was shown specifically for time dilation in the Science by the Numbers section in this chapter. Newton's laws, which have worked so well in describing our everyday world, fail only when dealing with extremely high velocities or extremely large masses. Thus Newton's laws represent an extremely important special case of Einstein's more general theory.

Science often progresses in this way, with one theory encompassing previous valid ideas. Newton, for example, merged discoveries by Galileo of Earth-based motions and Kepler's laws of planetary motion into his unified theory of gravity. And some day Einstein's theory of relativity may be incorporated into an even grander view of the universe.

Summary

Every observer sees the world from a different *frame of reference*. Descriptions of actual physical events are different for different observers, but the *theory of relativity* states that all observers must see the universe operating according to the same laws. Since the speed of light is built into Maxwell's equations, this principle requires that all observers must see the same speed of light in their frames of reference.

Special relativity deals with observers who are not accelerating with respect to each other, while *general relativity* deals with observers in any frame of reference whatsoever. In special relativity, simple arguments lead to the conclusion that moving clocks appear to tick more slowly than stationary ones—a phenomenon known as *time dilation*. Furthermore, moving objects appear to get shorter in the direction of motion—the phenomenon of *length contraction*. Finally, moving objects become more massive than stationary ones, and an equivalence exists between mass and energy, as expressed by the famous equation, $E = mc^2$.

General relativity begins with the observation that the force of gravity is connected to acceleration, and describes a universe in which heavy masses warp the fabric of space-time and affect the motion of other objects. There are three classic tests of general relativity—the bending of light rays passing near the Sun, the changing orientation of the orbit of Mercury, and the redshift of light passing through a gravitational field.

Key Terms About Relativity

frame of reference	special relativity	time dilation
theory of relativity	general relativity	length contraction

Key Equations

Time dilation:	$t_{MG} = \dfrac{t_{GG}}{\sqrt{1 - (v/c)^2}}$
Length contraction:	$L_{MG} = L_{GG} \times \sqrt{1 - (v/c)^2}$
Mass effect:	$m_{MG} = \dfrac{m_{GG}}{\sqrt{1 - (v/c)^2}}$
Rest mass:	$E = mc^2$

Review Questions

1. What is a frame of reference? Give examples of frames of reference you have been in today.

2. What is the central idea of Einstein's theory of relativity?

3. What is the difference between special and general relativity?

4. What is time dilation? How fast does something have to be moving for time dilation to be appreciable?

5. What is the Lorentz factor?

6. According to an observer on the ground, how does the length of a moving object compare to the length of an identical object on the ground? How does the mass compare?

7. What is the relation between the mass of an object and its energy?

8. How can we say that gravitational forces and acceleration are equivalent?

9. Explain three tests of general relativity. Are these the only tests possible?

Discussion Questions

1. Imagine arriving by spaceship at the solar system for the first time. Identify three different frames of reference that you might choose to describe the Earth.

2. Did Einstein disprove Newton's laws of motion? Explain your answer.

3. In Chapter 2 we talked about the idea that Newton's work was profoundly in tune with the time in which he lived. In what sense might you say that relativity is in tune with the twentieth century?

4. The twentieth century has been called the age of relativism, where each person has his or her own ethical system and no set of values are absolute. Do you agree? Does the theory of relativity imply that no values are absolute?

Problems

1. You are traveling 80 km/hour in your car when you throw a ball 50 km/hour. What is the ball's apparent speed to a person standing by the road when the ball is thrown (a) straight ahead, (b) sideways, and (c) backward?

2. Calculate the Lorentz factor for objects traveling at 1%, 50%, and 99.9% of the speed of light.

3. What is the apparent mass of a 1-kilogram object that has been accelerated to 99% of light speed?

4. If a moving clock appears to be ticking half as fast as normal, at what percent of light speed is it traveling?

5. Draw a picture illustrating how a spaceship passing the Earth might look at 1%, 90%, and 99.9% of light speed.

Investigations

1. Read a biography of Albert Einstein. What were his major scientific contributions? For what work did he receive a Nobel prize?

2. Take a bathroom scale into an elevator in a tall building, stand on it, and record your weight under acceleration and deceleration. Why does the scale reading change?

3. Read the novel *Einstein's Dream* by Alan Lightman. Each of the chapters explores different time-space relationships. Which chapters teach you something about Einstein's theory of relativity?

4. Investigate the influence of Einstein's theory of relativity on twentieth-century art and philosophy.

Additional Reading

Lightman, Alan. *Einstein's Dream*. New York, Pantheon, 1993.

The Earth and Other Planets

The Earth, one of nine planets that orbit the Sun, formed 4.5 billion years ago from a great cloud of dust.

A Random Walk
Looking at the Night Sky

Walk outside and look at the sky tonight just after sunset. Chances are you will find two or three particularly bright objects that stand out among the stars, even in the haze and illumination of a city. They don't seem to twinkle like stars, but shine steadily. If you looked at them through binoculars, they would appear to be small disks.

If you look at the same bright objects on successive nights, you'll notice that over a period of weeks or months, they seem to wander among the stars, never appearing in exactly the same place two nights in a row. The Greeks called them "wanderers," or planets, and assigned them the names of the gods. In the evening and morning, for example, you are likely to see Venus, the goddess of love, and swift-moving Mercury, the messenger of the gods; and the night sky is often dominated by Jupiter, the Roman name for the king of the gods.

Today we know that those disks of light in the sky are objects similar in many ways to our own planet, the Earth. They show us that we are part of a system that includes not only the Earth, but the Sun and the other planets as well. Our probes have visited all of them and landed on two—Mars and Venus. Visionaries talk of the day when science fiction will become reality and human beings will work and live on these, our nearest neighbors in the cosmos. In every sense of the word, the planets are the next frontier.

The Sun and Its Planets

The Copernican revolution radically altered human perceptions of our place in the universe (see Chapter 2). Rather than occupying what was assumed by many to be the center of creation, the Earth became just one of a number of planets orbiting the Sun. The **solar system,** which includes the Sun, the planets and their moons, and all other objects gravitationally bound to the Sun, displays several distinctive characteristics. Describing these features and explaining in detail how they came to be remains one of the main challenges faced by scientists today.

Features of the Solar System

How can we deduce the origin and present state of the solar system? Until recently, all of our observations of the Sun and planets have been made from the surface of the Earth. We see points of light moving in

The solar system

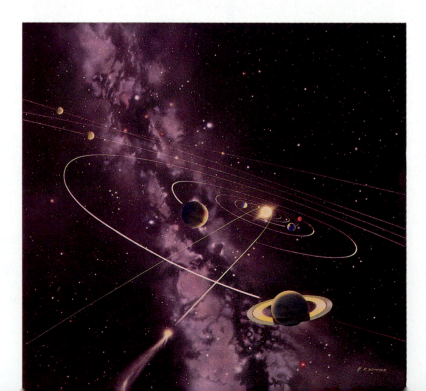

the sky, but how can that information be translated into a vivid picture of a dynamic system?

Humans have been studying the solar system for thousands of years, making observations and proposing models. Ancient scholars recorded, among other things, the changing positions of the brightest planets, such as Venus and Jupiter. The development of the telescope by Galileo (see Chapter 2) led to the discovery of numerous new, faint objects, including dozens of moons and other small bodies. Our present understanding of the solar system, therefore, represents the cumulative effect of centuries of observation.

As astronomers gathered data on the solar system, they noticed several striking regularities regarding the orbits of planets and the distribution of mass, patterns that provide the key to understanding the evolution of our home.

Planetary Orbits   Think about what Newton's laws tell us regarding satellites orbiting a central body. A satellite can go in any direction: east to west or west to east, around the equator or over the poles. There are no constraints regarding the orientation of the orbit, and planets could orbit any which way around the Sun. Yet in our solar system we see two very curious features:

- All of the planets move in orbits in the same direction around the Sun, and this direction is the same as that of the rotation of the Sun.

- All the orbits of the planets are in more or less the same plane. The solar system resembles a bunch of marbles rolling around on a single flat dish.

Why is the solar system arranged in this orderly way? What do these features tell us about its history?

The Distribution of Mass   You could imagine a solar system in which mass was evenly distributed, with all planets more or less the same size and same chemical composition. But our solar system is not that way at all. Instead:

- Virtually all of the material of the solar system is contained within the Sun, with only a small fraction in the planets and other objects in orbit.

- There are two distinct kinds of planets. Near the Sun, in what we call the "inner" solar system, we find planets like the Earth—relatively small, rocky, high-density worlds. These are called the **terrestrial planets,** and include **Mercury, Venus,** the Earth, **Mars,** and (although it isn't really a planet) the **Earth's Moon.** Farther out from the Sun, in the "outer" solar system, we find huge worlds made primarily of liquids and gases. We call them "gas giants," or **Jovian planets,** and they are **Jupiter, Saturn, Uranus,** and **Neptune. Pluto,** the outermost planet, is something of an anomaly, being small and rocky. The planets of the solar system and some of their characteristics are listed in Table 14–1.

Table 14–1 • The Planets and Their Characteristics

	Mercury	Venus	Earth	Mars	Jupiter	Saturn	Uranus	Neptune	Pluto
Diameter (km)	4880	12,104	12,756	6787	142,800	120,000	51,800	49,500	6000
Mass (Earth = 1)	0.055	0.815	1	0.108	317.8	95.2	14.4	17.2	0.003
Density, g/cm³ (water = 1)	5.44	5.2	5.52	3.93	1.3	0.69	1.28	1.64	2.06
Number of moons	0	0	1	2	16	18	15	8	1
Length of day (in Earth hours)	1416	5832	24	24.6	9.8	10.2	17.2	16.1	154
Period of one revolution around Sun (in Earth years)	0.24	0.62	1.00	1.88	11.86	29.5	84.0	164.9	247.7
Average distance from sun (millions of kilometers)	58	108	150	228	778	1427	2870	4497	5900

The planet Saturn is distinguished by its extensive system of rings. Rapid rotation of this gaseous body causes the bulge around the equator.

- Interspersed with the planets are a large number of other kinds of objects. All the planets except the innermost Mercury and Venus are orbited by one or more **moons.** While some moons are little more than boulders a few kilometers across, others are much larger, and Saturn's largest moon Titan is about the same size as Mercury. Saturn also has dramatic *rings* composed of millions of tiny moons. Small, rocky **asteroids** that circle the Sun like miniature planets are found primarily in orbits between Mars and Jupiter, in what is called the **asteroid belt,** although some have orbits that cross the Earth's. Far beyond Pluto, we find a swarm of icy **comets** with compositions something like a "dirty snowball." Occasionally, one is jostled loose from its orbit and becomes part of the realm of the planets, creating a spectacular display in the sky.

These regularities in the distribution of the solar system's mass give us important hints as to how the system was formed—new theories that will lead to even more observations of the heavens.

Science in the Making

The Discovery of Pluto

Five of the planets—Mercury, Venus, Mars, Jupiter, and Saturn—are visible to the naked eye and have been known from ancient times. The other three, most distant planets—Uranus, Neptune, and Pluto—were discovered after the invention of the telescope. The most recent discovery of a planet occurred on February 18, 1930, when Clyde Tombaugh,

The discovery photos for Pluto reveal the shift of one point of light between January 23, 1930 (left) and January 29, 1930 (right). The white arrows point to Pluto.

a Kansas farm boy employed as a technician at Lowell Observatory in Flagstaff, Arizona, uncovered convincing evidence for the existence of a new planet.

As a teenager, Tombaugh had built a small telescope, using parts from an old cream separator to make its stand. He drew sketches of the surface of Mars and sent them to Lowell Observatory, which was then engaged in observations of the red planet. The sketches, made with a small amateur's instrument, corresponded so well to what astronomers at Lowell were seeing through their state-of-the-art telescope that Tombaugh received a job offer by return mail.

At Lowell, he was given the task of doing a systematic search of the skies for what was then called Planet X. The founder of the observatory, American astronomer Percival Lowell, had predicted the existence of such a planet on the basis of some rather questionable data on variations in the orbit of Neptune. Tombaugh's task was straightforward, if tiring. As the Earth swept around in its orbit, he took photographs of each section of the sky, then took another photograph a few days later. The two photographs were then put into a machine that would show first one photograph, then the other, in an eyepiece. As the photographs were "blinked," any object that had moved between the time of the two photographs would appear to jump back and forth, while stars would remain stationary.

The main problem is that the plane of the solar system is littered with asteroids, each of which could show up as a moving light on such photos. The key point wasn't that Tombaugh found something out there that moved—there were plenty of such objects. The point was to find a point of light that moved as much as Kepler's laws tell us that a planet out beyond Neptune would move in a few days. That is exactly what Tombaugh found on that day in February, some ten months into his search.

After becoming one of only three human beings to have discovered a new planet, Tombaugh went back to college. Much to his surprise, he was not allowed to take introductory astronomy. "They cheated me out of four hours!" he says.

And his old telescope? When one of the authors asked him a few years ago whether he was going to donate it to the Smithsonian Institution, the 83-year-old astronomer replied "They want it, but they can't have it. I'm not through using it yet!" ■

The Nebular Hypothesis

The modern theory of the solar system's formation was first put forward by the French mathematician and physicist Pierre Simon Laplace (1749–1827). His was a simple idea, and one that takes into account many of the distinctive characteristics of the solar system—the rotation of the Sun, the orbits of the planets, and the distribution of mass into several large objects and lots of smaller ones. According to the model, called the *nebular hypothesis,* long ago (we now estimate about 4.5 billion years ago) a large cloud of dust and gas floated in space in the region now occupied by the solar system. Such dust and gas clouds, called **nebulae,** are common throughout our galaxy, the Milky Way. They typically contain more than 99% hydrogen and helium, with lesser amounts of all the other naturally occuring elements.

Under the influence of gravity, the nebula slowly, inexorably, started to collapse on itself. As it did so, it must have begun to rotate faster and faster. In a sense, the collapsing, rotating cloud was analogous to an ice skater beginning a spin. Going into the spin she will have her arms extended out, but as the spin progresses she pulls her arms in and spins faster and faster—sometimes so fast that you can barely make out her features. In the same way, as the nebula from which the solar system was formed began to collapse, it began to spin faster and faster.

This spin had several consequences. For one thing, it meant that some of the material in the outer parts of the cloud began to spin out into a flat disk. This common consequence of fast rotation is familiar to anyone who has watched a pizza maker create a flat disk of dough by spinning a mass overhead. If you imagine the solar system at this stage of its formation as a large pancake with a big lump in the middle, you won't be far wrong. The big lump represents the material that will eventually become the Sun, and the material in the thin flattened disk will eventually become the planets and the rest of the solar system (see Figure 14–1).

The flattening of the nebula into a disk explains another feature of the solar system. The planets had to form in this rotating disk of material, and hence their eventual orbits had to lie close to the disk's plane. The fact that all planetary orbits lie near the same plane, then, is a simple consequence of the solar system's rapid rotation as the nebular cloud began to contract.

The formation of stars like the Sun is discussed in more detail in Chapter 17, so here we can confine our attention to what happened in the outer, thin regions of the disk. In any conglomeration of matter like the spinning disk, purely by chance, matter is more densely collected in some regions than elsewhere. These regions exert a stronger gravitational force than their neighbors, so that nearby matter tends to gravitate to them. Once the nearby matter has come in, the concentration of matter at that point is even greater, and it will pull even more material into it. As more material accumulates, solid grains start to stick together.

This ultimate consequence of gravitational force leads to the rapid breakup of the disk into small objects called *planetesimals,* which range in size from boulders to masses several kilometers across. Once this has happened, the process of gravitational attraction goes on at a grander scale—larger objects capture smaller ones and continue growing.

An ice skater in a spin twirls faster as she pulls her arms into her body. Similarly, the solar nebula began to spin faster and faster as mass was pulled toward the central region.

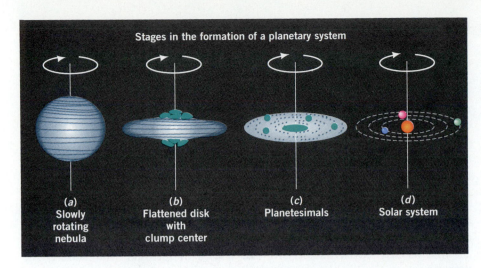

Stages in the formation of a planetary system

(a) Slowly rotating nebula

(b) Flattened disk with clump center

(c) Planetesimals

(d) Solar system

Figure 14–1
As the nebula that formed the solar system collapsed, it began to rotate and flatten into a disk. The stages in solar system formation include (a) a slowly rotating nebula, (b) a flattened disk with massive center, (c) planets in the process of birth represented as mass concentrations in the nebula, and (d) the solar system.

About the time that this consolidation was going on in the early solar system, the material at the center—more than 99% of the nebula's original mass—began to turn into a star. Light energy began to radiate out from the Sun, and temperature differences began to develop in the disk. Those parts nearest the sun warmed quite a bit, while those farther out warmed only a little. As a result, the inner and outer solar systems developed differently. In the warm inner system, materials such as water, hydrogen, and helium were in gaseous form, while farther out they were frozen into solids.

Thus everyday physical processes having to do with phases of matter and the response to temperature—processes as familiar as boiling water and making ice—explain one of the crucial facts about the solar system. The terrestrial planets—Mercury, Venus, Earth, and Mars—were formed from materials that remain solid at high temperatures. Consequently, they are small, rocky, high-density worlds.

Farther out in the solar system, we find the Jovian planets: Jupiter, Saturn, Uranus, and Neptune. The compositions of those planets are essentially the same as the material concentrated in the original nebula—that is, they contain large amounts of hydrogen and helium. These planets formed from material that remained solid (or at least liquid) because of

The surface of Venus is shrouded in clouds, but the *Magellan* spacecraft produced radar images of the surface. In this computer-generated view of a Venusian volcano the vertical relief has been greatly exaggerated.

The largest planet, Jupiter, displays colorful atmospheric bands with swirls and eddies indicating turbulence. Jupiter's moon Io is visible against the planet in this photograph taken by NASA's *Voyager 1* in 1979.

their lower temperature so far from the Sun. Consequently, they have a markedly different chemical composition from the planets in the inner solar system.

In passing, we should note that the Jovian planets probably had their own complement of high-density materials. Scientists suspect that beneath the thousands of kilometers of helium, hydrogen, and other condensed gases on these planets is concealed a core like a small terrestrial planet, whose composition is very much like that of the Earth and its neighbors. But this rocky matter represents only a small fraction of these planets' total mass.

Some astronomers argue that the largest planets—Jupiter and Saturn—formed by a process more like that of a small star than through the accretion of planetesimals. The details of the structure and formation of the Jovian planets remains a rich ground for debate in the sciences.

Just as any construction site has a pile of leftover materials lying around when the building is finished, so too does the solar system have its "scrap pile." These leftovers take the form of the rocky asteroids and icy comets that still orbit the Sun. They represent the matter that never got taken up into planets.

The Human Body

Gravity and Bones

The planets of our solar system vary greatly in mass, from tiny Pluto—only 1/500th the mass of Earth—to Jupiter, which is 317 times more massive than our own planet. Consequently, the force of gravity at the surface of each planet is different—you would weigh only a small fraction of what you do now if you were on the surface of Pluto, but many times your present weight on Jupiter. (See Chapter 2 for a review of the connection between mass and weight.)

In humans, bones in the skeleton support our weight. Bone is not a static material, like concrete, but changes in response to its environment. It constantly rebuilds itself, replacing the calcium-rich minerals that form the solid structure. In fact, you can think of what goes on in bones as being analogous to the remodeling of a house. First, the old material is removed (a process called resorption), then new material is added. If a bone is subject to unusual stress (or lack of it), it will change gradually in response, adding or subtracting mass.

If you were to go to the surface of the Moon, for example, you would weigh one-sixth of your Earth weight. Your bones would start to remodel themselves in response to the lower force of gravity, unless you kept them under stress by an extended program of vigorous exercise. Astronauts spending a few weeks or months in the weightless environment of space have experienced significant bone loss—which, fortunately, is quickly made up after the return to Earth. Scientists are less certain, however, of the extremely long-term effects of low gravity on bone loss. What might happen during the long months of a space flight to Mars, for example? Studies on astronauts now may well determine whether such ambitious interplanetary flights take place in the foreseeable future. ■

Evolution of the Earth and Moon

The collapse of the solar nebula into the Sun and planets was the first step in the evolution of the solar system. Following the initial accretion of planets each object evolved in its own distinctive way.

The Formation and Early History of the Earth

Once the countless thousands of planetesimals were formed, the formation of planets followed quickly. As planetesimals moved through their orbits, they collected smaller planetesimals through the process of gravitational attraction. Then these larger planetesimals collided and coalesced into the beginnings of a planet. As the process of accumulation went on, the growing planet gradually swept up all the debris that lay near its orbit.

If you had been standing on the surface of the newly forming Earth during this stage, you would have seen a spectacular display. A constant rain of debris left over from the initial period of planetary formation fell to the surface, constantly adding mass to the Earth. During this period, called the **great bombardment,** the large amounts of kinetic energy carried by the shower of stones were converted into heat, which was added to the newly forming planet. By some accounts, much of the Earth's surface would have glowed bright red from this accumulating heat, and each large impact would have been accompanied by a spectacular splash of molten rock. Although the addition of material to the Earth has slowed considerably since the beginning, it has not stopped. Every time you see a **meteor** (often called a shooting star) for example, you are seeing an object roughly the size of a grain of sand being added to our planet. Scientists estimate that the mass of the Earth grows by about 20 metric tons (20,000 kilograms, or 2×10^7 gm) per day by accretion of material falling from space.

When the nebular hypothesis was first proposed in the eighteenth century, there seemed little chance that any direct observational evidence could be found to support it. In 1992, however, astronomers using the Hubble Space Telescope (see Chapter 17) were able to detect thick masses of dust encircling newborn stars in a region of space called the Orion Nebula. It appears that in these cases we are seeing distant solar systems in the process of being born—observations that give us a measure of confidence in our model of how planets come into existence.

The Orion Nebula is an active region of star formation.

Science by the Numbers

The Growth of the Earth

The Earth's mass is approximately 6×10^{27} gm. Adding 2×10^7 grams per day, how many years would it take to double the Earth's mass?

▶ **Reasoning:** First, we have to calculate how many grams are added to the Earth each year by multiplying the daily added mass by the number of days in a year. The Earth's total mass, divided by the mass added each year, gives us the time it would take to double the present mass.

▶ **Solution:** First we determine the yearly mass added to the Earth.

$$\text{mass added per year} = (365 \text{ days/year}) \times (2 \times 10^7 \text{ gm/day})$$
$$= 730 \times 10^7 \text{ gm/year}$$
$$= 7.3 \times 10^9 \text{ gm/year}$$

Then, divide the Earth's total mass by the mass added every year.

$$\text{number of years} = \frac{6 \times 10^{27} \text{ gm}}{7.3 \times 10^9 \text{ gm/year}}$$
$$= 0.82 \times 10^{18} \text{ years}$$
$$= 8.2 \times 10^{17} \text{ years}$$

This number is the time (in years) that would be required to double the mass of the Earth at its present rate of growth. This immense time—nearly a billion billion years—is vastly greater than the lifetime of our planet, which is a "paltry" 4.5 billion years. From this calculation we see that the total amount of mass now being added to the Earth is trivial. ■

The Layered Structure of the Earth

Each time another planetesimal hit the early Earth, all of its kinetic and potential energy was converted into heat. That heat diffused rapidly through the planet, which glowed red hot and reached temperatures of thousands of degrees in its deep interior. Eventually, scientists postulate, the Earth either melted completely or else was heated to high enough temperatures so that it was very soft all the way through. Heavy, dense materials (like iron and nickel) sank under the force of gravity toward the center of the planet, while lighter, less-dense materials floated to the top. The result of this process, called **differentiation,** is that the present-day Earth has a very definite layered structure (as shown in Figure 14–2).

In a sense, what happened to the Earth long ago isn't too different from what happens to a mixture of oil and water that is shaken up and then allowed to stand. Eventually, the lighter oil will float to the top and the heavier water will sink to the bottom under the influence of gravity. In just the same way, the Earth separated into layers of different density when it underwent differentiation.

At the center of the Earth, with a radius of about 3400 kilometers (2000 miles), we find the **core,** made primarily of iron and nickel metal. Temperatures at the center of the Earth are believed to exceed 5000°C but pressures are so high—about 3.5 billion grams per square centimeter (almost 50 million pounds per square inch)—that the iron-nickel inner core is solid. A little farther out the pressures are somewhat lower, so that the outer region of the iron-nickel core is actually a liquid.

The metal core is overlain by a thick layer, the **mantle,** that is rich in the elements oxygen, silicon, magnesium, and iron. Metallic bonding predominates in the core, but the mantle features minerals with primarily ionic bonds between negatively charged oxygen ions and positively charged silicon, magnesium, and other ions. Mantle rocks are similar in composition to some familiar surface rocks, but the atoms in these high-pressure materials are packed together in much denser forms.

At the very outer layer of the Earth is the **crust,** which is made up of the lightest materials. The crust's thickness ranges from less than 10

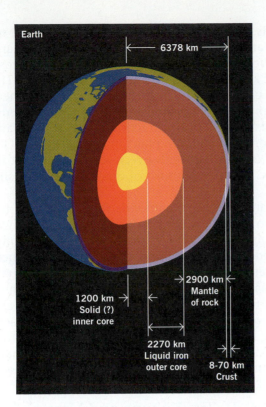

Figure 14–2
The layered Earth. The principal layers, which differ in chemical composition and physical properties, are the core, the mantle, the crust, and the atmosphere. When looked at in detail, each of these layers is itself composed of smaller layers.

- Earth
- 6378 km
- 1200 km Solid (?) inner core
- 2900 km Mantle of rock
- 2270 km Liquid iron outer core
- 8-70 km Crust

kilometers (about 6 miles) in parts of the oceans, to as much as 70 kilometers (about 45 miles) beneath parts of the continents. The crust is the only layer of the solid Earth with which human beings have had contact, and it remains the source of almost all the rocks and minerals that we use in our lives.

You might be wondering how we can speak with such confidence about parts of the Earth that no human being has ever seen. In the next chapter, we will talk about seismology, a branch of science that has provided (among other things) our present picture of the Earth's interior.

Technology

Producing High Pressure

The force of gravity, pulling inward on all of the Earth's layers, results in immense internal pressures—pressures at the planet's center exceeding three million times the atmospheric pressure. What changes affect rocks and minerals at these extreme conditions? High-pressure researchers, who have learned to sustain laboratory pressures greater than those at the center of the Earth, are providing surprising answers.

Of all the materials that form deep within the Earth, none holds more fascination than diamond, the high-pressure form of carbon. Not only do diamonds make magnificent gemstones, but as the hardest known substance they also serve as the most efficient abrasive for machining the tough metal parts of modern industry. Until the mid-

Researchers attain high pressures, equivalent to those deep inside the Earth and other planets, using the diamond anvil cell. Looking through such diamond cells you can observe pressurized samples such as this high-pressure ice crystal that formed at room temperature by squeezing water.

1950s diamonds were available only from a few natural sources, but in 1954 scientists at General Electric discovered how to manufacture diamonds by duplicating the extreme temperatures and pressures that exist hundreds of kilometers beneath the Earth's surface. The researchers squeezed black carbon between the jaws of a massive metal vise and heated their sample with a powerful electrical current. Although the first experiments yielded only a fraction of a carat of diamond, large factories (including a major GE plant in Worthington, Ohio) now produce dozens of *tons* of diamonds every year—an output that has greatly exceeded the total amount of diamonds mined since Biblical times.

The Earth taught us how diamonds were made, but now scientists use diamonds to learn how the Earth was made. The highest sustained laboratory pressures available today are obtained by clamping together two tiny pointed anvils of diamond. Samples squeezed between the diamond-anvil faces have been subjected to pressures of several million kilograms per square centimeter—significantly greater than the pressure at the center of the Earth. At such extreme conditions, crustal rocks and minerals compress to new, dense forms with compact atomic structures occupying less than half their original volumes. Dramatic changes in chemical bonding are also observed, with many ionically bonded compounds transforming to metals at high pressure. ■

EXAMPLE 14–1: The Volume and Mass of the Earth's Core

What fraction of the Earth's volume is taken up by the core? What fraction of the Earth's mass is in the core?

▶ **Reasoning:** We must solve this problem in two parts. First, we need to compare the volume of the Earth's core, a sphere with a 3500-km radius, to the volume of the entire Earth with approximately 6400-km radius. The volume of a sphere is given by the formula

sphere volume = $\frac{4}{3} \times \pi \times$ radius^3,

where the constant π is approximately 3.14.

▶ **Solution:** The ratio of the core's volume to the entire Earth's volume is thus

$$\frac{\text{vol}_{core}}{\text{vol}_{Earth}} = \frac{\frac{4}{3} \times 3.14 \times 3500^3}{\frac{4}{3} \times 3.14 \times 6400^3}.$$

Note that the numerical terms $\frac{4}{3}$ and 3.14 cancel out on the right side of this equation, so

$$\frac{\text{vol}_{core}}{\text{vol}_{Earth}} = \frac{3500^3}{6400^3}$$
$$= \frac{4.3 \times 10^{10}}{2.6 \times 10^{11}}$$
$$= 0.17, \text{ or } 17\%.$$

Thus the core accounts for about 17% of the Earth's volume, while the outer layers—the crust and mantle—account for the other 83%.

▶ **Reasoning:** The second part of this problem considers the mass of the core compared to the mass of the entire Earth. The average density of the highly compressed iron core is about 10 grams per cubic centimeter—slightly less than the density of lead. This value is almost twice as great as the 5.5-gram-per-cubic-centimeter average density of the entire Earth, which is composed mostly of rocks that are much less dense than iron.

▶ **Solution:** The mass of an object equals its volume times its density. Thus the mass of the Earth is given by

$$mass_{Earth} = vol_{Earth} \times density_{Earth}.$$

Similarly,

$$mass_{core} = vol_{core} \times density_{core}.$$

The relative mass of the core, $mass_{core}/mass_{Earth}$, is thus given by

$$\frac{mass_{core}}{mass_{Earth}} = \frac{vol_{core} \times density_{core}}{vol_{Earth} \times density_{Earth}}.$$

But we calculated above that $vol_{core}/vol_{Earth} = 0.17$, and we know the average values of core and Earth densities are 10 and 5.5 grams per cubic centimeter, respectively. Thus

$$\frac{mass_{Core}}{mass_{Earth}} = \frac{0.17 \times (10\,gm/cm^3)}{(5.5\,gm/cm^3)}$$
$$= 0.31, \text{ or } 31\%.$$

The core thus represents slightly less than a third of our planet's mass. ▲

The Formation of the Moon

The origin of the Moon, the only large body in orbit around the Earth, poses one of the oldest puzzles in planetary science. The Moon is one of the largest bodies in the solar system—larger than the planets Mercury and Pluto, and almost as large as Mars. The Moon's density and chemical composition, subjects of intense study by the astronauts of the Apollo lunar missions, are quite different from the Earth as a whole, though they are remarkably similar to those of the Earth's mantle. The problem of the Moon's origin can be stated simply: How could it have arisen in the same region of space as the Earth, when its composition is so different?

Various attempts have been made to solve this problem. Some astronomers argued that the Moon was formed elsewhere in the solar system in a region of low-density, mantle-like material and was later captured by the Earth. Other scientists suggested that the Moon was somehow thrown out of a rapidly rotating Earth after differentiation had taken place. Indeed, in the early twentieth century, astronomers often pointed to the Pacific Basin as the likely "birth scar" of the Moon. Neither of these theories can meet rigorous testing, however. It turns out to be very difficult for a planet to capture a large body in a nearly circular orbit, and even harder for a planet to throw off a Moon-sized chunk.

The current theory (and one that seems to the authors to have a reasonable chance of being right) is called the "big splash." In this theory, the Earth underwent differentiation as described earlier, but while it was still in a formative state, it was struck by an object about the size of Mars. This collision blew a huge quantity of mantle rocks out into orbit around the Earth. The Moon then formed from this material, which explains the Moon's unusual composition, density, and large size. Meanwhile, the great crater created by this epic impact was filled and eventually weathered away, so no evidence of it appears today on Earth.

Formation of the Other Planets

The natural processes that occurred during the Earth's formation affected the other planets as well. Mercury, Mars, and the Moon, for example, display surface cratering that suggests that large chunks of rock bombarded all of these planets late in their formation. The Earth undoubtedly looked like this over 4 billion years ago, but all evidence of those early craters has been weathered away. On Mercury and the Moon, which have no atmosphere, there has been nothing to affect the craters and so they are still there.

The early bombardment may also have affected other characteristics of the terrestrial planets. The direction of rotation of Venus, for example, is opposite that of the Earth. (Planets revolve around the Sun, but rotate about their axes.) The Earth's axis of rotation, furthermore, is tilted at 23 degrees to the plane of its orbit, while Uranus has its axis of rotation close to the plane of its orbit—a full 90 degrees from an upright orientation. Current thinking is that these differences resulted from the more or less random collisions with large objects, perhaps hundreds of kilometers in diameter, that marked the end of the main phase of planetary formation. You might expect that the details of these late-stage collisions were different for each planet. Thus the nebular hypothesis not only explains how it is that the planets all have their orbits in the same plane and move in the same direction around the Sun, but also allows us to explain why the rotations of individual planets can be so different.

An artist's impression of the surface of the Earth during the great bombardment.

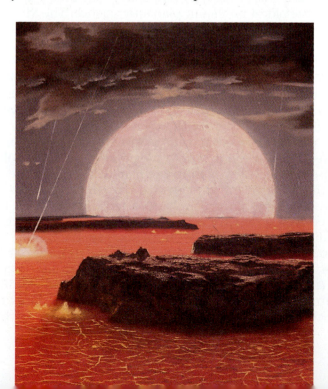

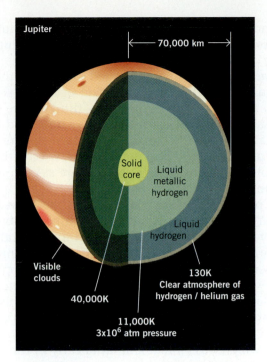

Figure 14–3
A theoretical view of the interior of Jupiter, one of the Jovian planets. Most of the planet's volume is highly compressed hydrogen and helium.

This scenario for the formation of the Earth also has important implications for the origins of life on our planet. During the period of the great bombardment, the Earth was constantly being hit by huge objects—chunks of rock the size of a state or even a country. At the very least, the tremendous energy released by such collisions would be enough to vaporize any oceans that had formed. Each collision would, quite literally, sterilize the planet. Even if life had come into existence during this period, it would have been wiped out by the impacts. Thus the ancestors of modern living things could not have gotten their start until the end of the bombardment, which occurred between 4.2 and 3.8 billion years ago. This fact will become very important when we discuss the origins of life in Chapter 23.

The structure of the outer giant planets is layered, like that of the terrestrial planets, but they do not have a well-defined solid surface like the Earth and Moon. Moving down from space into the body of Jupiter or Saturn would be a strange experience. You would move through progressively denser and denser layers of clouds and then pass imperceptibly into a layer where the gases change into liquids because of the high pressure. In fact, landing on Jupiter would be more like landing on a giant ice cream sundae than landing on the planet Earth or the Moon. In Figure 14–3, we show a typical structure for one of the Jovian planets.

The Evolution of Planetary Atmospheres

The Earth didn't always have the kind of atmosphere it has today. In fact, scientists now suggest that originally it had no atmosphere at all, and that once an atmosphere formed, its chemical composition evolved gradually to that of the present. The question of how the Earth's atmosphere arose and changed is extremely important, because this history is inseparable from understanding the origin and evolution of life on our planet.

The early Earth may well have collected some gases through gravitational attraction. During the early formation of the Sun, large amounts of material and radiation were thrown off—think of the start-up of the Sun as being analogous to a cranky car engine on a cold morning. This flood of materials would have blown off any atmosphere the Earth had accumulated. Thus, for all intents and purposes, we can think of the first stage in the evolution of the Earth's atmosphere as corresponding to an airless ball of hot (or even molten) rock.

During the period of cooling that followed the great bombardment and melting, large amounts of water vapor, carbon dioxide, and other gases would have been released from deep within the solid Earth. Countless volcanoes and fissures, all belching steam and other materials, would have blown their gases into the newly forming atmosphere. In other words, the gases that formed the ancestor of today's atmosphere were probably originally locked into the rocks near the Earth's surface when the original atmosphere was swept away. Subsequently, a completely new atmosphere was released by a process called *outgassing*.

Outgassing, which was violent and rapid early in the Earth's history, has not ended today. We tend to think of volcanic eruptions as involving the flow of red-hot lava, but if you remember pictures of eruptions, you probably recall large clouds of smoke and steam that accompany the glowing lava flows. Even today, more than 4.5 billion years after the planet's formation, volcanoes release large amounts of gases from the interior of the Earth.

The conventional estimate at the moment is that the principal result of outgassing in the early Earth was the production of an atmosphere composed primarily of methane (CH_4), ammonia (NH_3), carbon dioxide (CO_2), hydrogen (H_2), and water (H_2O), though scientists are engaged in a spirited debate on the subject. Presumably, the same sort of process was occurring on the other terrestrial planets. For a time, the atmosphere was probably too hot for water to condense from a gas to a liquid, but eventually atmospheric temperature dropped and torrential rains began to fill the ocean basins.

Once a planet has acquired an atmosphere by outgassing, there are several ways that its atmosphere can evolve and change. The simplest is *gravitational escape*. The molecules in an atmosphere heated by the Sun may move sufficiently fast so that appreciable fractions of them can actually escape the gravitational pull of their planet. The Moon, Mercury, and Mars are examples of bodies that had denser atmospheres early in their history, but lost much of these gases through gravitational escape long ago. Most of the light elements such as hydrogen and helium were presumably lost in the same way from the Earth, but the heavier gases such as carbon dioxide and water vapor remained because they were too heavy to escape the Earth's gravitational force.

A second cause of atmospheric change—one that operates only on the Earth—is the effect of living things. To the best of our knowledge, no life exists anywhere else in the solar system (although some scientists argue that there may have been life at one time on Mars). By the time the Earth was 1 billion years old, photosynthetic organisms had evolved to use the Sun's energy to power the chemical reactions essential for life. In photosynthesis, carbon dioxide and water are taken into the structures of living things, and oxygen is given off as a waste product. As the number of living things on the planet increased, the amount of free oxygen increased as well, until today it comprises about 20% of the atmosphere.

We tend to think of oxygen as a benign and beautiful substance, but from a chemical point of view it's really rather nasty stuff. As we saw in Chapter 7, oxygen reacts violently with many materials (think of fire burning or the explosion of hydrogen or gasoline). In fact, the production of oxygen by living things in the early Earth can be thought of as the first global pollution event. Thus life forms both affect and are affected by the atmosphere of the planet on which they reside. In fact, most scientists now argue that you can tell whether a planet has life on it simply by looking at its atmosphere.

The Outer Solar System

Space probes sent out from Earth in the 1970s and 1980s have visited most of the outer planets, and as a result we have a pretty good idea of what the outer solar system is like. The distances to the giant outer planets are immense. The closest gas giant, Jupiter, orbits at five times the Earth-Sun distance—over 800,000,000 kilometers away. Saturn is twice that far, while Uranus and Neptune are several billion kilometers away. This far out in the system, the Sun looks like a small marble in the sky, and its warming effects are feeble indeed. Compounds that are normally gases on the Earth—carbon dioxide, nitrogen, methane, and so on—are found in liquid or even solid form as we move to the frigid outer solar system. The Jovian planets are fairly simple in structure, although we don't know much about many of their details. Astronomers are much more interested in the amazingly varied moons that encircle Jupiter, Saturn, Uranus, and Neptune. Astronomers have found dozens of such moons (in fact, it is a matter of semantics whether something can be reasonably called a "moon" as opposed to just a large orbiting rock).

Each of these moons can be thought of as a small laboratory that sheds some light on the formation of terrestrial planets. The moon Io, which circles close to Jupiter, has active volcanoes on it—the only moon in the solar system known to have them. Scientists think that flexing and twisting of the moon's body due to the powerful gravitational forces of Jupiter produces the energy to drive those volcanoes.

Saturn's moon Titan, one of the largest moons in the solar system (it is about the same size as the planet Mercury), may well be something of a laboratory for the early Earth. It may have frigid lakes of methane (CH_4) and ethane (C_2H_6) on its surface and, from the beginning of the solar system, simple chemical reactions in its atmosphere have probably produced a layer of organic materials there. Thus Titan may have something to tell us about how organic materials accumulate, and about how cellular life developed on our own planet.

The outermost planet, Pluto, remains something of a mystery. It can be studied through telescopes, of course, but has yet to be visited by a space probe. Pluto is a small planet—only about 0.2% of the mass of the Earth—and is circled by a moon called Charon that is only slightly smaller than the planet. Its orbit is highly elliptical and, for a period of its orbital "year" (about 250 Earth years), brings Pluto inside the orbit of Neptune. Because it is a small rocky planet, unlike the neighboring gas giants, some planetary scientists argue that Pluto actually formed as a captured comet or asteroid, rather than a planet in the usual sense of the word. Indeed, the origin of Pluto is very much an open question in planetary astronomy.

Science in the Making

The Voyager Satellites

On August 20, 1977, a rocket blasted off from Cape Canaveral in Florida, to set a small space probe on its course. Sixteen days later, another rocket did the same. These two probes, called *Voyager 1* and *2*, respectively, spent the next 15 years moving past the planets of the outer solar system, providing scientists with their first close-up look at the Jovian planets and their moons.

Each spacecraft had ten scientific instruments on board, each designed to measure a different aspect of the deep space environment. The ones that had the greatest public impact were the cameras that took pictures of moons, rings, and planets, but measurements were also made of magnetic fields, cosmic ray abundances, and infrared and ultraviolet radiation. Taken together, the Voyager probes produced a good deal of the detailed information about the outer solar system contained in this chapter.

One interesting feature of spaceflights like this is that by the time a spacecraft has been in flight for a while, all of its instrumentation has become obsolete. Between its encounters with Jupiter and Neptune, for example, the computers on *Voyager 2* were reprogrammed to make significant changes in the way they analyzed and transmitted data. As a result, the rate of transmission during the Neptune flyby in 1989 was not significantly different from that of the Jupiter encounter in 1979, despite the greater distance and lower light levels at Neptune.

Voyager 2 passed within 4,800 kilometers of Neptune in August, 1989.

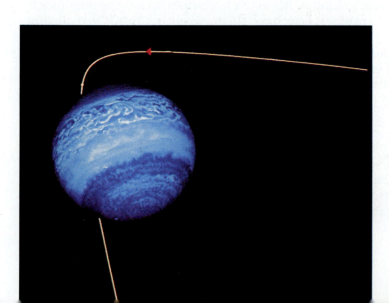

Among their discoveries, the *Voyagers* found new rings around all the Jovian planets, recorded a volcanic eruption on Io, tripled the number of known moons around Uranus, and clocked record winds on the surface of Neptune. Today, both *Voyagers,* together with a couple of earlier space probes called *Pioneers,* have moved out of the solar system and into interstellar space. *Voyager 2* is expected to keep returning data until its plutonium power supply runs down, sometime around 2020. By that time, it may have reached the place where the Sun's magnetic field blends into the magnetic field of our galaxy, thereby becoming the first human-made object to have broken free from all the influences of the Sun. ■

Asteroids, Comets, and Meteors

As the solar system formed, not all the material in the planetary disk was taken up into the bodies of the planets and moons. Even after hundreds of millions of years of accumulation and bombardment, a lot of debris was still floating around out there and remains even today. This debris comes in two main forms—asteroids and comets—which in a sense mimic the compositional differences of the inner and outer planets.

Asteroids   Like the inner terrestrial planets, asteroids are small rocky bodies in orbit around the Sun. Most asteroids are found in a broad, circular asteroid belt between Mars and Jupiter—material thought to be a collection of planetesimals that never managed to collect into a single planet. The most likely explanation is that the nearby planet Jupiter had a disrupting gravitational effect. In addition, many asteroids possess orbits that cross the Earth's orbit, and they produce occasional large impacts on our planet.

Comets   Comets can best be thought of as "dirty snowballs." Unlike asteroids, they consist of chunks, sometimes many miles in diameter, of material such as water ice and methane ice in which is embedded a certain amount of solid, rocky material or dirt. Also unlike asteroids, most of the comets in the solar system circle the Sun far outside the orbit of Pluto. Billions of them are to be found in a region called the *Oort cloud* (named

Comets are distinguished by their bright gaseous tails, which stream away from the Sun.

after Dutch astronomer Jan Oort, 1900–1992, who first postulated its existence). Occasionally, when the orbit of one of these comets is perturbed, a comet will be deflected so that it falls toward the Sun. When this happens, the increasing temperature of the inner solar system begins to boil off materials, and we see a large "tail," blown away from the Sun by the solar wind, that reflects light to us.

Sometimes a comet will be captured and fall into a regular orbit around the Sun. The most famous of these periodic comets is Halley's Comet (see Chapter 2), which returns to the vicinity of the Earth about every 76 years. Halley's return in 1910 was quite spectacular, because the comet passed near the Earth when it was at its highest temperature and therefore had its largest and most spectacular tail. The return in 1986 was much less spectacular because the comet was on the far side of the Sun when it was at its brightest. The next predicted return in 2061, unfortunately, will probably be just as unspectacular.

In addition to providing us with one of the first historical tests of the law of universal gravitation, comets may have had an important effect on the evolution of life on Earth. Many scientists suspect (see Chapter 23) that the impacts of comets or asteroids may have drastically altered the earth's climate and produced mass extinctions, or killings, at various times in the Earth's history. Most Earth scientists, for example, now believe that the dinosaurs and other life forms that thrived 65 million years ago were driven to extinction following the impact of a large comet or asteroid that hit the Earth at a site near Mexico's Yucatan Peninsula.

Meteoroids, Meteors, and Meteorites   Meteoroids are small pieces of ancient space debris in orbit around the Sun. Occasionally one of these bits, perhaps the size of a sand grain, will fall into the Earth's atmosphere where it becomes briefly visible as a meteor. Most meteors burn up completely to microscopic particles of ash that slowly, imperceptibly, rain down on Earth. The meteors' bright streaks of light record the path of this burning. Occasionally, if the object is big enough so that only the outer surface burns, a piece of rock may actually reach the Earth's surface. Any such rock that has fallen to Earth from space is called a **meteorite.**

Meteor showers are a set of spectacular, regularly occurring events in the night sky. During a shower, you can see bright streaks in the sky every minute or so. They are caused by the collision of the Earth with clouds of small debris that travel around the orbits of comets. Some of these clouds may be comets that have been broken up by the gravitational

The 1,200 meter-wide Meteor Crater in Arizona formed from a collision about 20,000 years ago.

Table 14–2 • Major Meteor Showers

Name	Date for Maximum	Extreme Limits	Hourly Rate of Meteors
Quadrantid	Jan 3	Jan 1–4	30
Perseid	Aug 12	July 29–Aug 17	40
Aquarid	May 4	May 2–6	5
Orionids	Oct 22	Oct 18–26	13
Taurid	Nov 1	Sep 15–Dec 15	5
Leonid	Nov 17	Nov 14–20	6
Geminid	Dec 14	Dec 7–15	55

Note: The name identifies the constellation in which the radiant point is found. The best time to observe most showers is in the early morning, although not all the showers peak then.

pull of one of the planets. Table 14–2 lists some of the most spectacular meteor showers.

Meteorites are extremely important in the study of the solar system because they represent the material from which the system was originally made. They are analyzed intensely by scientists, both to get a notion of how and when the Earth was made, and to learn what kinds of materials human beings will find when they leave the Earth to explore the rest of the solar system.

THINKING MORE ABOUT PLANETS

Human Space Exploration

Since astronaut Neil Armstrong became the first human to walk on the surface of the Moon in 1969, the scientific community has debated the question of how the exploration of the solar system should be carried out. The question is this: Should future missions to the planets carry people, or should they carry only machines?

Those who advocate exploration by machines point to the enormous technical difficulties involved in providing a safe habitat for human beings in the harsh environment of space. Why, they ask, should we make the enormous, expensive effort to put a human being on the surface of Mars, for example, when we can learn just as much by sending instrument packages and robots controlled from the Earth?

On the other side of the issue, scientists advocating space exploration by astronauts argue that no machine has the flexibility and ingenuity of a human being. They note that no matter how well designed a machine might be, when it is millions of miles from Earth things can go wrong, and only a trained astronaut can salvage the mission. They point out that even a mammoth project like the Hubble Space Telescope needed astronauts to replace the optical systems after the main mirror was built incorrectly. Besides, they argue, if one goal of the space program is to establish human colonies on other bodies in the solar system, you can't do that with machines.

What do we hope to learn from our studies of the solar system? Is colonization of the rest of the solar system the real long-term goal of the space program? How much extra effort (and taxpayer's dollars) is it worth expending to put people instead of machines on the surface of Mars?

Summary

The Earth formed along with the Sun and other planets in our *solar system* from a *nebula*—a large dust cloud rich in hydrogen and helium—approximately 4.5 billion years ago. As that dust cloud began to contract from gravitational forces, it also began to rotate and flatten out into the disk that now defines the planetary orbits. More than 99% of the original nebula's mass concentrated at the center, which became the Sun.

Gradually, the matter in the flat disk began to form clumps under its own local gravitational forces. The largest of these masses swept up more and more debris as they orbited the early Sun, and they began to define a string of planets. *Terrestrial planets,* those nearest the Sun, were subjected to high temperatures and strong solar winds, so that most gases such as hydrogen, helium, and water vapors were swept out into space. Thus the inner four planets—*Mercury, Venus, Earth,* and *Mars*—are dense, rocky places with relatively low content of gaseous elements.

The Earth's formation was probably typical of these planets. After the principal mass of the Earth had been collected together, additional rocks and boulders showered down in the *great bombardment,* adding matter and heat energy to the planet. Dense iron and nickel separated from lighter materials by the process of *differentiation* and sunk to the center to form a metallic *core.* Most of the Earth's mass concentrated in the thick *mantle,* while the lightest elements formed a thin *crust.* The *Moon,* Earth's only large satellite, may have formed when a planet-sized body hit the Earth early in its history.

The solar system's outer *Jovian planets—Jupiter, Saturn, Uranus,* and *Neptune*—are quite different from the inner planets. Lying beyond the strong effects of solar heat and wind, they accumulated large amounts of gases such as hydrogen, helium, ammonia, and water. These outer planets are thus giant balls of ices, with thick atmospheres and great frigid oceans of nitrogen, methane, and other compounds that are gases on Earth. Beyond the Jovian planets lies *Pluto,* a rocky body that is the smallest planet. All of the planets except Mercury and Venus, the two closest to the Sun, have *moons* in orbit.

Interspersed with the planets and their moons are many other kinds of objects. Small, rocky *asteroids,* most of which are concentrated in an *asteroid belt* between Mars and Jupiter, circle the Sun like miniature planets. Far outside the solar system, swarms of "dirty snowballs" called *comets* are concentrated in the Oort cloud. If a comet's distant orbit is disturbed, it may fall toward the Sun and create a spectacular display in the night sky. When a piece of interplanetary debris hits the Earth's atmosphere, it creates a *meteor,* or shooting star, which burns up with a fiery trail. Occasionally, a meteor fragment will hit the Earth and become a *meteorite.*

Key Terms About the Solar System

solar system	moons	meteor
terrestrial planets	asteroid	differentiation
(Mercury, Venus, Earth, Earth's Moon, Mars)	asteroid belt	core
Jovian planets	comet	mantle
(Jupiter, Saturn, Uranus, Neptune)	nebula	crust
Pluto	great bombardment	meteorite

Review Questions

1. What is a terrestrial planet?
2. What is a Jovian planet?
3. State the nebular hypothesis. How does it explain the orbits of the planets?
4. What are planetesimals and what role do they play in the formation of planets?
5. Why are the terrestrial planets different in composition from the Jovian planets?
6. What is differentiation? How has this process affected the Earth?
7. What is the Earth's core? What is it made of?
8. How does the high pressure in the core affect chemical bonds?
9. What is the Earth's mantle? What is it made of?
10. Describe the "Big Splash."
11. What is outgassing? Is it still going on today?
12. Describe two ways that a planet's atmosphere can evolve.
13. What is an asteroid? Where are most asteroids found?

14. What is the difference between a meteor and a meteorite?

15. What is a comet? Where are most comets found?

16. How could asteroids and comets affect life on Earth?

17. Which planets have moons?

Discussion Questions

1. What distinctive characteristics of the Earth make it suitable for life? How has life altered the chemistry of the Earth's atmosphere?

2. Why are meteorites important to our study of the solar system?

3. Review the history of studies of the solar system. What new technologies have enhanced our ability to make observations on objects in the solar system?

4. What factors might influence the development of life on other planets?

5. What sources of data might help us determine more about how the Earth's Moon formed?

6. How does Clyde Tombaugh's work fit into the scientific method?

7. Do you think that we should continue to explore space? Why or why not?

Problems

1. Given the diameters of the planets in Table 14–1, what are the relative volumes of Earth, Mercury, and Jupiter? From the same table, what are the relative masses?

2. From the values of mass in Problem 1, calculate the densities (mass divided by volume) for Earth, Mercury, and Jupiter. Why do you think they are different?

3. How many asteroids 10 kilometers in diameter would be required to make a planet about the size of Mars? Neglect the effects of compression in the planet's interior.

4. If the average thickness of the Earth's crust is 10 kilometers, what fraction of the solid Earth's total volume is in the crust?

Investigations

1. Investigate the history of unmanned planetary probes. What are the names, dates, and target planets of these probes? What countries sponsored them? What kinds of data did they return? Are any planetary missions now underway?

2. Read a history of the Apollo missions to the Moon. What theories about the Moon's origins prevailed before these missions? What new data changed theories about the origin of the Moon?

3. How would you respond to an argument that went as follows: No one was present when the Earth was formed, so how can scientists talk about the details of the formation process?

4. Listen to *The Planets*, a suite for orchestra by the British composer Gustav Holst. In what ways do the musical descriptions of each planet reflect the physical characteristics of that planet? What other sources of inspiration, beside scientific data, did Holst use in creating these pieces? Which planets did he omit and why?

5. There are many more meteor showers than the ones listed in Table 14–2. Find out which ones may be coming in the next month or two and plan a meteor-watching party.

6. Scientists are attempting to document the paths of asteroids with Earth-crossing orbits. Investigate this research and comment on the probability that a large asteroid might hit the Earth. Should we increase funding for asteroid monitoring?

7. Research NASA's plans for upcoming space probes. Do you think your representatives and senators will vote to approve funding?

Additional Reading

Comins, Neil F. *What if the Moon Didn't Exist.* New York: Harper-Collins, 1993.

Greeley, Ronald. *Planetary Landscapes,* 2nd ed. New York: Chapman and Hall, 1993.

Hartmann, William K. *Moons and Planets,* 3rd ed. Belmont, California: Wasdsworth, 1993.

Sagan, Carl. *Comet,* New York: Random House, 1985.

Voyage Through the Universe series, Alexandria, Virginia: Time-Life Books.

Plate Tectonics

The entire Earth is still changing, due to the slow convection of soft, hot rocks deep within the planet.

A Random Walk
A Construction Site

Think about the last time you passed a new construction site after a rain shower. You could probably see little valleys freshly carved by water running over the bare earth, and shallow pools where fine-grained material collected in layers—features that are small-scale examples of the erosion of soil by rain. If the earth-movers had dug a deep pit you might have noticed different layers of earth and rock freshly exposed—layers that represent sediments deposited by water long ago.

Or think about the last time you were at the beach and the wind was blowing. Did you feel a lot of sand and grit in the air? The wind is constantly moving soil at the Earth's surface, shifting and changing what is there.

These sorts of small-scale changes in the Earth's surface are mirrored by much more dramatic large-scale changes. When Mount St. Helens, a volcano in Washington state, erupted in 1980, the entire side of a mountain was blown away and many square kilometers of forest were flattened. Large earthquakes in our lifetime will change the course of rivers and destroy villages and towns.

Perhaps the most extraordinary changes of all are ones that we cannot easily perceive in human time scales without sophisticated instruments. The most permanent features of the Earth that we know—the continents and ocean crust—are moving. The continent of North America is about a meter farther from Europe than it was when you were born, the result of forces deep within the Earth.

The Earth's surface is constantly changing. No feature of the Earth—no desert or broad plain, no mountain or ocean—is permanent. Every feature is constantly changing, constantly evolving into something different.

The Dynamic Earth

How do we know that the Earth's surface is changing? We rarely observe significant changes in the landscape, but you can make a simple estimate that will convince you that the Earth must be a very dynamic planet. Mountains appear to be as permanent as anything could be, yet, as we see in the following section, it is easy to convince yourself that mountains must wear away in times much shorter than the 4.5-billion-year age of the Earth. We will also see that continents are not as permanent as they may appear to be.

Science by the Numbers

How Long Could a Mountain Last?

While there is no such thing as a "typical" mountain, for the purposes of this rough estimate let's think of a mountain as a rectangular mass about 2 kilometers high, about 4 kilometers long, and 4 kilometers wide (that's about 1.2 miles by 2.5 miles by 2.5 miles). The volume of this mid-sized mountain is

$$\text{volume} = \text{length} \times \text{height} \times \text{width}$$
$$= 2 \text{ km} \times 4 \text{ km} \times 4 \text{ km}$$
$$= 32 \text{ km}^3.$$

Expressed in cubic meters, that is

$$2000 \text{ m} \times 4000 \text{ m} \times 4000 \text{ m} = 3.2 \times 10^{10} \text{ m}^3.$$

Think about a stream running down a mountainside. You know that such a stream carries a certain amount of sand, silt, and dirt with it. You can see this because the stream has a sandy bottom, and you can watch it depositing sand in little eddies and still water along its side. You might also see gravel and boulders in the stream—evidence that, from time to time, heavy rains cause much more violent movement of material down the mountainside. Where did the rock and sand come from? Obviously, they had to be worn off of materials higher up on the mountain. Thus the existence of the stream means that the mountain is constantly being eroded away.

You can estimate how long a mountain might survive against erosion by a stream. Suppose, for example, that four principal streams run off the sides of the mountain, and that each stream carries an average of one-tenth of a cubic meter of earth per day off that mountain (the actual amount would vary from day to day depending on the kind of rock, the amount of water flowing downhill, and other factors). One-tenth of a cubic meter per day is not very much material. It corresponds to a pile of sand, dirt, and gravel about 50 centimeters (a foot and a half) on a side, a pile that could fit under an ordinary kitchen chair. If you think about the amount of material you might collect if you put your hand down into a stream for a while, you'll see that the number is reasonable. Over a period of a year, the four streams might thus remove

4 streams $\times$ 0.1 meter^3/stream-day $\times$ 365 days/year = 146 meter^3/year

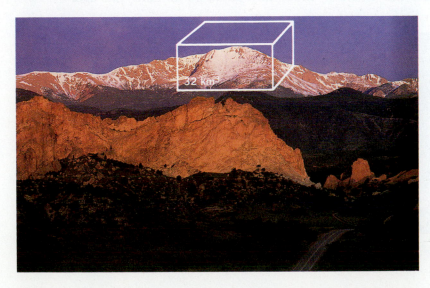

Pike's Peak near Colorado Springs, Colorado, may be approximated as 2 $\times$ 4 $\times$ 4 km rectangular block of rock.

Every year, therefore, close to 150 cubic meters of material—about six dump trucks full—could be removed from a mountain by normal erosional processes of streams.

If the mountain streams remove about 150 cubic meters each year, then the lifetime of the mountain can't be much longer than the volume of the mountain, divided by the volume lost each year:

$$\frac{3.2 \times 10^{10} \text{ m}^3}{150 \text{ m}^3/\text{year}} = 0.0213 \times 10^{10} \text{ years}$$
$$= 213{,}000{,}000 \text{ years.}$$

This estimate, while very rough and not directly applicable to any specific mountain, tells us that under normal circumstances mountains can't last more than a few hundred million years. All mountains must disappear and be eroded away to low, rounded hills in times much shorter than the 4.5-billion-year age of the Earth. ■

The Case of the Disappearing Mountains

If all mountains weather away, why are there any mountains at all left on Earth? We must conclude that if the erosion of mountains takes a few hundred million years, then any mountain that existed when the Earth first formed 4.5 billion years ago would have been worn away long ago, and the Earth's surface by now should be smooth and featureless. The only way that mountains could still exist on the surface of the Earth is for mountains to be continuously formed.

Thus everyday observations of erosion and some very simple arithmetic leads us to a startling conclusion: tremendous forces must be acting on the Earth, creating new mountain chains as the old ones are worn down. The surface of the Earth cannot be static. Although mountains seem to human beings to symbolize eternal solidity, they are transitory. Geologists who map the distribution and ages of rocks have shown that the Appalachian Mountains were formed just a few hundred million years ago, and the Rocky Mountains about 60 million years ago.

Steep slopes and angular peaks characterize young mountains such as the 65-million-year old Rocky Mountains in Alberta, Canada (*a*) photographed from Mt. Rae. The 400-million-year old Great Smokey Mountains in North Carolina (*b*) display the rounded character of older mountains.

(*a*)

(*b*)

Newton's laws of motion (Chapter 2) tell us that nothing happens unless a force acts. What forces in the Earth could create entire mountain ranges? Until recently this question remained one of the greatest puzzles in geology.

Volcanoes and Earthquakes—Evidence of Forces in the Earth

Most geological processes such as mountain building and erosion are slow by human standards and take thousands or millions of years to effect noticeable change. But volcanoes and earthquakes may transform a landscape in an instant, thus revealing the tremendous energy stored in our dynamic planet.

Volcanic eruptions provide the most spectacular process by which new mountains are formed. In a typical **volcano,** subsurface molten rock called *magma,* concentrated in the Earth's upper mantle or lower crust, breaks through to the surface, as shown in Figure 15–1. This breakthrough may be sudden, giving rise to the kind of dramatic events we saw when Mount St. Helens exploded in 1980, or it may be relatively slow, with a stately surface flow of molten rock, known as *lava.* In both cases, however, magma eventually breaches the surface and hardens into new rock.

Earthquakes occur when rock suddenly breaks along a more or less flat surface, called a *fault.* Have you ever stretched a thick, strong elastic

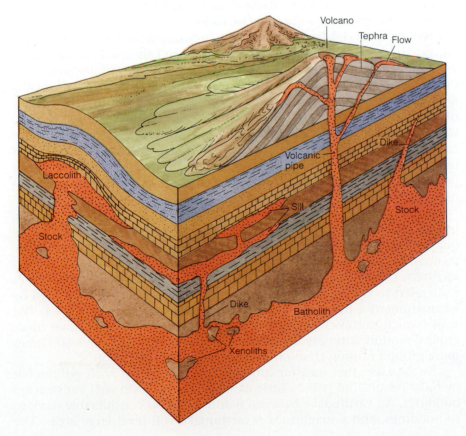

Figure 15–1

A cross-section of a volcano reveals a magma chamber, which stores molten rock, and a system of pipes, cracks, and vents that lead to the surface. The terms in orange refer to the kinds of rock formed from cooled magma. Xenoliths are the original rocks encased in this cooled magma.

Recent volcanic eruptions. (*a*) Mount St. Helens in Washington state erupted in a violent explosion on July 22, 1980. (*b*) Puu O'O on the flanks of Hawaii's most active volcano, Kilaeua, erupted periodically throughout most of the 1980s. This photograph was taken in 1983. (*c*) Mount Pinatubo in the Philippines erupted in 1990, spreading a thick blanket of ash over thousands of square kilometers.

band, only to have it snap back painfully against your hand? You gradually added elastic potential energy to the band, and that energy was suddenly released—converted to violent kinetic energy. The same thing happens in the Earth when stressed rock suddenly snaps. This sudden release of energy is called an **earthquake.**

When a rock suddenly breaks, tremendous amounts of potential energy are released. The two sides of the fault can't fly apart like an elastic material can, so the energy is transmitted in the form of a sound wave or *seismic wave* (see Chapter 6). These waves, traveling at speeds of several kilometers per second, cause the ground to rise and fall like the surface of the ocean. Normally "solid" ground sways and pitches in a motion that can cause severe damage to buildings and other structures. If an earthquake occurs under or near a large body of water, these violent motions of the ground can transfer energy into great waves that can devastate low-lying coastal areas. Such waves are called *tsunamis*, a Japanese term for harbor or bay waves.

Great earthquakes, such as the one that killed 20,000 people in India in September, 1993, are fortunately relatively rare. But smaller earthquakes, barely noticeable to the average person, occur every day by the thousands. Earthquakes are usually rated on the *Richter scale,* after Charles Richter, a United States geologist who devised it. Technically, the Richter scale refers to the amount of ground motion that would be measured by an instrument a fixed distance from the center of the earthquake. The scale is such that each increase of 1 unit corresponds to 10 times more ground motion. Thus an earthquake that measures 7 will have 100 times more ground motion than one that measures 5, and so on.

Earthquakes that measure around 5 on the Richter scale will be felt by most people, but will do little damage in areas with well-constructed buildings. An earthquake between 6 and 7 will do considerable damage to buildings, and a magnitude 8 earthquake will level large areas. The earthquake that occurred in the San Francisco Bay area in October, 1989, measured 7.1, and was dubbed the "Pretty Big One" by Californians who are waiting for what they call the "Big One." The January 1993 earthquake that caused so much damage to the freeway system of Los Angeles measured 6.7 on the Richter scale. No earthquakes greater than 9 on the Richter scale have ever been recorded, probably because no rocks can store that much energy before they let go.

The puzzle remains, however: Where does all the energy that powers volcanoes and earthquakes come from?

The Movement of the Continents

Geologists have faced other mysteries as well. Think about a large map of the world. In your mind, move North and South America eastward toward Eurasia and Africa. Have you ever noticed how the two seem to fit together (Figure 15–2)?

This jigsaw-puzzle pattern was one of the first indications that the Earth's surface is not static, but in a state of constant flux. English statesman and natural philosopher Francis Bacon (1561–1626) pointed out this fact in 1620. It wasn't until the beginning of the twentieth century, however, that anyone took the parallel coastlines seriously enough to question the origin of this pattern. In 1912 a German meteorologist named Alfred Wegener (1880–1930) proposed that the Earth's continents are in motion. The reason that the Americas fit so well into the coastline of Europe and Africa, he suggested, is that they were once joined and have since been torn apart.

Wegener's theory, called *continental drift,* was dismissed by most earth scientists. He amassed some geologic evidence to back it up, such as the matching locations of distinctive rock formations on opposite sides of the Atlantic Ocean. But most of this evidence was fragmentary and uncon-

Figure 15–2

A map of the world's continents reveals the similar shapes of coastlines on the two sides of the Atlantic Ocean.

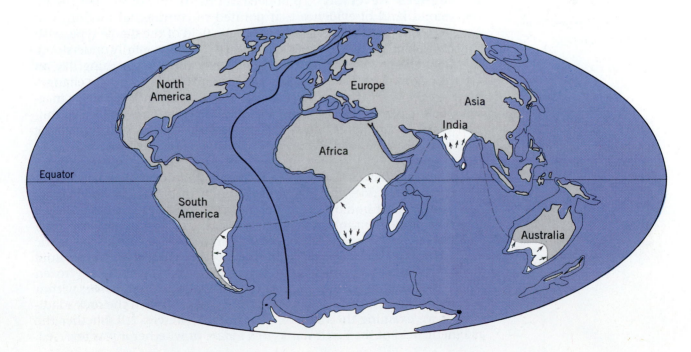

vincing to other scientists—indeed, some of Wegener's arguments later turned out to be wrong. Furthermore, he failed to provide a reasonable mechanism by which continents could move. For most of the twentieth century, continental drift was regarded as a far-fetched exercise in theory, and few geologists paid much attention to it. Beginning about 1960, however, geologists and oceanographers obtained new evidence to support one aspect of Wegener's notion—the idea that the continents are not fixed.

The discovery that continents do indeed move required the merging of very different kinds of newly acquired geological evidence—topographic profiles of the oceans' floors, maps of rock magnetism, and data on rock ages.

1. Ocean Floors When the contours of ocean floors were mapped in the years following World War II, oceanographers discovered remarkable, unsuspected features. Most scientists thought that the deep ocean bottoms were simply flat plains, passively collecting the sediments that gradually eroded off the ancient continents. Instead they found steep-walled canyons and lofty mountains, indicating that the seafloor is as dynamic and changing as the continents themselves (Figure 15–3). The longest mountain range on Earth, for example, is not on a continent, but in the middle of the Atlantic Ocean. Called the Mid-Atlantic Ridge, this feature extends from Iceland in the North Atlantic to the South Sandwich Islands in the South Atlantic. These ridges are sites of continuous geological activity, including numerous earthquakes, volcanoes, and lava flows. Similar ridges, also called rises, were found on the floors of all the Earth's oceans. In fact, oceanographers have now mapped more than 85,000 km of ocean ridges.

2. Magnetic Reversals To understand the nature of magnetic data, the second kind of evidence that pointed to continental motion, you have to recall that the Earth has a magnetic field of the dipole type, with north and south magnetic poles. For reasons that are not fully understood, this field changes direction sporadically over time—something like an electromagnet in which the direction of current in the coils changes occasionally (see Chapter 5). Well over 300 reversals of the Earth's magnetic field have been recorded in ancient rocks spanning about 200 million years. During recent episodes of reverse fields, the north magnetic pole of the Earth was located somewhere in what is now Antarctica, and the south magnetic pole somewhere in the Canada-Greenland region.

When lava flows out of a fissure in the Earth, it contains small crystals of natural iron oxides, including the naturally magnetic mineral magnetite. These bits of iron ore act as tiny magnets, and, since the rock is still in a fluid state, their magnetic dipoles are free to turn around and align themselves in a north-south direction parallel to the Earth's magnetic field. Think of them as small compass needles embedded in the fluid rock. Once the rock hardens, however, the bits of magnetite are frozen in place—they can no longer move. Thus the volcanic rock carries within it a memory of where the magnetic north pole was when the rock solidified. If we examine the tiny compass needles, we can tell whether the magnetic field of the Earth was as it is today, or whether it was reversed.

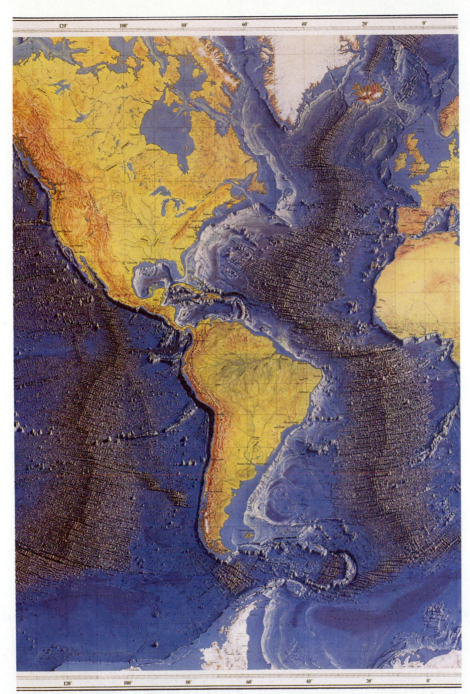

Figure 15–3
A topographic map of ocean floors reveals dramatic mountain chains, deep flat planes, canyons, and trenches. These features suggest that ocean basins are active geological regions.

The field devoted to the study of this sort of effect is called *paleomagnetism*, and it came into maturity in the early 1960s.

In the mid-1950s, an oceanic exploration ship named *Pioneer* began taking magnetic measurements near the ocean floor off the coast of Washington state. At first the data made little sense, but as more and more profiles were obtained, a puzzling pattern of stripes on the ocean floor began to emerge. It seemed that there were parallel strips of rock in which the magnetic direction in neighboring strips alternated. We show some of the original data in Figure 15–4. Through the mid-1960s, Earth scientists collected more and more data of this type, and it soon became clear that the Mid-Atlantic Ridge, the East Pacific Rise, and many other places on ocean floors show the same pattern of alternating magnetic stripes.

How could this configuration of magnetic stripes be explained? In an environment in which the orientations of north and south magnetic poles were switching back and forth over time, the only way to get the observed striping is shown in Figure 15–5. The seafloor must be getting wider, or *spreading*, as magma—new molten rock—comes from deep within the Earth and erupts through fissures on the seafloor. Over time,

Figure 15–4

Measurements in the late 1950s and early 1960s revealed magnetic stripes running nearly parallel to the Vancouver province and Washington state coastlines.

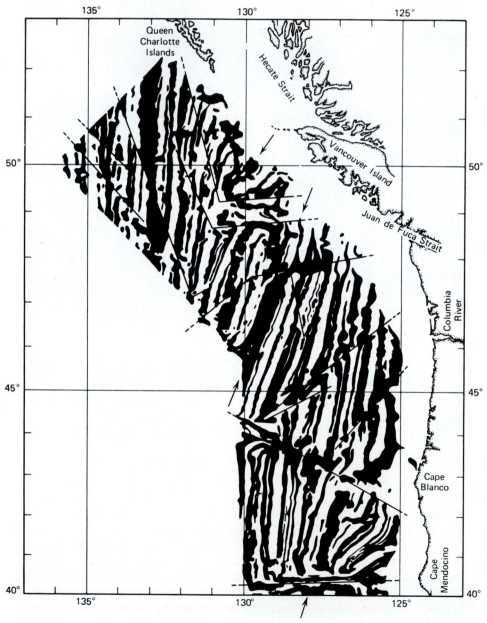

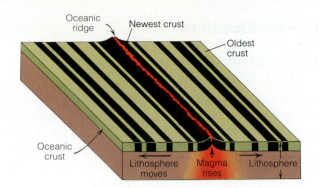

Oceanic ridge · Newest crust · Oldest crust · Oceanic crust · Lithosphere moves · Magma rises · Lithosphere

Figure 15–5
Magnetic stripes that parallel ocean ridges must form as new magma wells up from the fissure and pushes out to the sides. In this cross-sectional view, older rocks lie farther from the ridge. (The "lithosphere" includes the uppermost mantle and all of the crust.)

the rock at the surface today will be pushed aside as the continents move apart and as more magma comes up to take its place. The compass needles in this newer rock will point to the location of the magnetic north pole when they reach the surface. Each new batch of magma will solidify and lock in the current magnetic orientation, despite any further reversals. Each time the Earth's magnetic field reverses, the dipole direction of the natural magnets changes and gets locked into the rock. Thus, over long periods of time, we would expect to see alternating bands of magnetic orientation—exactly the kind of zebra-striped pattern of alternating north and south compass orientations shown in Figure 15–4.

3. Rock Dating The conclusions from magnetic data were reinforced by studies of the age of volcanic rocks in the oceanic crust. Volcanic rocks contain radioactive isotopes that can be used for dating (see Chapter 11). Rocks near the Mid-Atlantic Ridge and other similar features were found to be quite young, a few million years old or less. Rocks collected successively farther away from the ridge proved to be successively older.

New Support for the Theory

The new data on the topography of ocean floors, as well as the magnetic properties and ages of its rocks, suggested to many scientists that the width of the Atlantic Ocean must be increasing yearly as the seafloor spreads. Eurasia and North America, it appears, are moving apart. Could the hypothesis that this distance is increasing be tested by an experimental measurement?

Up until the late 1980s no one was able to measure the motion of continents directly. The breakthrough came from an unexpected source—radio astronomy. Astronomers in North America and Europe trained radio telescopes on distant quasars (see Chapter 18). By measuring the times of arrival of the crests of the same radio wave, astronomers were able to get very accurate measurements of the distance between their telescopes. By repeating these sorts of experiments over a period of a few years, they were able to measure exactly the separation of the continents. Their work shows that North America is separating from Europe at the rate of about 5 centimeters (about 2 inches) a year.

Science by the Numbers

The Age of the Atlantic Ocean

If continents are constantly moving at speeds up to several centimeters per year, what would a map of the Earth's surface have looked like millions of years ago? We can make an educated guess by using present rates of movement and "reversing the tape," so to speak. We know, for example, that the floor of the Atlantic Ocean is spreading at about 5 cm per year. Assuming that the rate of spreading has remained more or less constant, how old is the Atlantic Ocean?

To calculate the answer, we need to apply the familiar equation for travel time:

$$\text{time of travel} = \frac{\text{distance}}{\text{speed}}$$

(You use the same relationship every time you estimate the driving time for a vacation.) The speed in this equation is the spreading rate, 5 cm per year, and the present width of the Atlantic is approximately 7000 km, or 7×10^8 cm. The Atlantic Ocean began to open, therefore, at

$$\text{time} = \frac{7 \times 10^8 \text{ cm}}{5 \text{ cm/year}}$$
$$= 1.4 \times 10^8 \text{ years}$$
$$= 140 \text{ million years.}$$

Think about what the Earth must have looked like 140 million years ago, during the age of the dinosaurs. Europe and the Americas would be joined together and there would be no Atlantic Ocean. In fact, by retracing the wandering paths of continents, earth scientists have discovered that 200 million years ago what we now call the continents of North America, South America, Eurasia, and Africa were locked together in one giant continent called Pangaea (Figure 15–6). A globe of

Figure 15–6

A map of the ancient continent of Pangaea reveals how the positions of present-day continents appeared 200 million years ago.

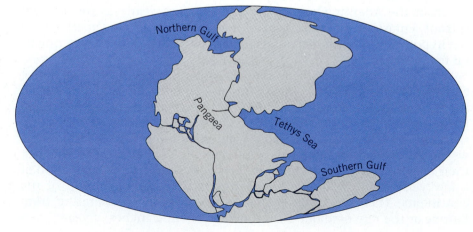

that ancient Earth would be all but unrecognizable, with a huge land mass on one side, and a giant ocean on the other.

By the same token, 100 million years from now the Earth's continents will have moved thousands of kilometers and appear completely different from today's arrangement. ▪

Plate Tectonics—A Unifying View of the Earth ▬

The compelling model of the dynamic Earth that has emerged from studies of paleomagnetism, rock dating, and much other data is called **plate tectonics.** "Tectonics" is related to the word "architect" and carries the connotation of building or putting things together. Plate tectonics develops a picture of the world that explains many of the Earth's large-scale surface features and related phenomena.

The central idea of the plate tectonics theory is that the surface of the Earth is broken up into about a dozen large pieces (as well as a number of smaller ones) called **tectonic plates** (see Figure 15–7). Each plate is a rigid, moving sheet of rock up to 100 km (60 miles) thick, composed of the crust and part of the upper mantle. Oceanic plates have an average 8- to 10-km thickness of dense rock known as *basalt* on top of the mantle rock. Continental plates have an average 35-km thickness of lower-density rock, such as *granite* capping the basalt.

The tectonic plate boundaries are not the same as those of the continents and oceans. Some plates have continents on all or part of their surface, while some are covered only by oceanic crust. Most of the North American continent, for example, rests on the 8000-km-wide North Amer-

Figure 15–7
The major plates of the Earth with their directions of motion. The length of the arrows indicates the relative speed of plate motions.

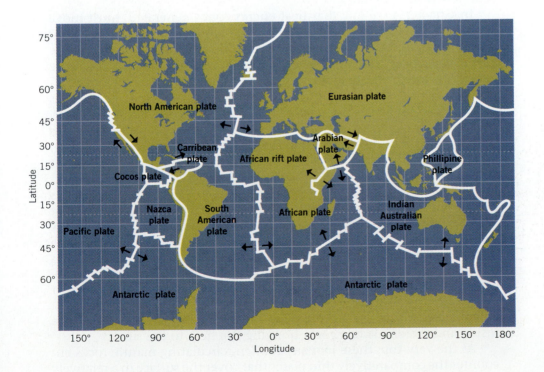

ican Plate, which extends from the middle of the Atlantic Ocean to the edge of the Pacific Plate on the West Coast.

About one quarter of the Earth's surface is covered by continent; the rest is ocean. On time scales of millions of years, the plates shift about on the planet's surface, carrying the continents with them like passengers on a raft. Thus it is not the motion of the continents themselves that is fundamental to understanding the dynamics of the Earth, but the constant motion of the underlying plates. Continental motion (what Wegener called continental drift) is just one manifestation of that plate motion.

The Convecting Mantle

Nothing happens without a force, but what force could possibly be large enough to move not only the continents, but also the plates of which they are a part? According to current theory, continents move as a result of the force generated by **mantle convection** deep within the Earth—motions driven by our planet's internal heat energy.

Two sources of energy contribute to the Earth's interior heat. Some of this heat energy is left over from the gravitational potential energy released during the great bombardment and differentiation of the mantle and core as the Earth formed (see Chapter 14). The decay of uranium and other radioactive elements (see Chapter 11), which is fairly common throughout the Earth's core and mantle, provides a second important source of heat energy. These elements decay over time and produce energetic, fast-moving decay products, particles that collide with atoms and molecules in surrounding rocks and give up their energy as heat. Deep in the Earth all this heat energy softens rocks to the point that they can flow slowly like hot taffy.

In Chapter 4 we saw that heat energy, once generated, must move spontaneously toward cooler regions. The heat generated in the interior of the Earth, for example, must eventually be radiated into space. If the Earth were somewhat smaller, or if only small amounts of heat were generated, the energy could be carried to the surface entirely by conduction—the movement of heat by atomic collisions. This process operates in the Moon and Mars, for example. In the case of the Earth, however, there is too much internal heat to be carried by conduction alone. Rocks in the mantle have been heated to the point where they are able to flow, and deep within the mantle of the Earth immense convection cells are set up. In a sense, the rocks of the mantle behave like water in a boiling pot.

There are, of course, very important differences between the Earth's mantle and a boiling pot—in particular, the time scales. Solid rock cannot flow significantly on time scales comparable to human lifetimes. But give soft, hot rocks a few hundred million years and they can move large distances. Earth scientists now estimate that the Earth's convection cells go through a full cycle—with hot rocks rising from the lower mantle, cooling near the surface, and falling back to be replaced by other warmer rocks—on a time scale of about 200 million years. In effect, the Earth behaves something like a giant spherical stove, with the burners on the inside and the circulating rocks bringing the heat to the surface.

At the very top, these partially molten, circulating mantle rocks encounter the comparatively thin plates that cover the surface in a relatively cool, brittle layer. Along the oceanic ridges the brittle plates fracture and basaltic lava erupts, initiating seafloor spreading and plate motion. The plates move along with the convection cells underneath them, more or less as a film of oil on boiling water would move along on top of the water. The plates shift around, bash into each other, join each other, and are split apart, in a constant dance. And on top of these plates, floating along like a thin scum, with no control whatsoever over their destiny, are the continents—the places that we call the "solid Earth."

The theory of plate tectonics caused a revolution in the earth sciences. For the first time, scientists had a single, coherent picture of our planet that includes the oceans, the land masses, the planet's deep interior, and the interconnections between all these systems. There was a time when geologists of different specialties—women and men who studied the ocean currents or ancient fossils, ore deposits or the planet's deep interior—had little to say to each other. That situation is no longer true. All of these different disciplines now share a common way of looking at the Earth, and thus are able to give each other ideas and gain breadth and depth from interactions with each other. It may be that your college or university no longer has a department of geology but instead has a department called Earth Sciences or Environmental Sciences. These changes in name are more than just bureaucratic shufflings. They represent a very real change in the way scientists view and study the Earth.

Science in the Making

Reactions to Plate Tectonics

At first, geologists confronted by the observations of magnetic stripes on the ocean floors, rock ages, and other seemingly odd data tried to find an explanation that did not require new seafloor to be created. All geologists had been taught a "fixed Earth" model in their college courses, and that was the standard picture of our planet they used in their work. Most of these scientists were not prepared to accept a radical new idea—that the continents might actually move—without a fight.

Some experts questioned the statistical significance of the magnetic data, and they wondered whether the distinctive striped patterns weren't the result of some as-yet unknown effect (such as small electric currents running around the ocean floor). Others simply tried to ignore the whole thing. But as geologists and oceanographers collected more data—as new measurements of magnetic patterns and more precise rock ages were obtained—the evidence for seafloor spreading simply became overwhelming.

In a very short time—less than a decade from the time the first puzzling data from the *Pioneer* expedition came in—geologists had for the most part accepted a theory that radically changed many of the central principles of their discipline. This dramatic change in perspective brings us back to a point we made in Chapter 1. Good scientists will eventually accept the implications of their observations, whether those implications violate preconceived concepts or not. A scientist can't look at data without some preconceived notions. Few scientists, for example, took the original continental drift arguments of Wegener seriously because, in part, no obvious mechanism could cause entire continents to move. But as more and more data supported the more-convincing plate tectonics model, the majority of earth scientists readily changed their notions as the data demanded it.

The fact that the fixed Earth—one of the most revered and widely accepted geological theories—could be abandoned in the space of a few years when confronted by powerful contrary evidence indicates that the scientific method works. It also shows that many of the arguments one hears from the proponents of pseudosciences such as UFOs, astrology, and the like—arguments to the effect that the scientific community routinely closes its mind to new ideas and will not accept them—are simply wrong. Scientific theories, unlike the claims of pseudoscience, can be falsified. When confronted with overwhelming evidence, scientists are, indeed, prepared to accept new ideas and reject the old "conventional wisdom." ■

Plate Boundaries

The boundaries between the Earth's tectonic plates are active sites that determine much of the geological character of the surface. Three main types of boundaries separate the Earth's tectonic plates: divergent, convergent, and transform.

1. Divergent Plate Boundaries We saw one aspect of plate motion when we talked about seafloor spreading at the Mid-Atlantic Ridge, where new plate material is formed. We can now understand how such a spreading feature arises. In Figure 15–8, we show what happens when plates happen to lie above a zone where magma comes to the surface. Not only does the volcanic action form a chain of mountains, but the motion of the magma also pushes the two adjoining plates farther and farther away from each other. The newly erupted molten material cools to rock and becomes new plate material. As the brittle tectonic plates crack and separate, shallow earthquakes of relatively low energy occur. This mechanism drives the seafloor spreading that gave us our first indication of the nature of continental motion. Such a spreading zone of crustal formation is called a **divergent plate boundary.**

When a divergent plate boundary occurs on the ocean floor, the

Figure 15–8

A divergent plate boundary defines a line along which new plate material is formed from volcanic rock.

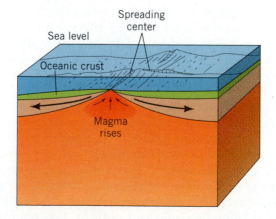

seafloor spreads, basalt lava erupts from the newly created fissures, the two plates are pushed apart, and any continents that might be located on other portions of those plates are pushed apart as well. Eurasia and North America, for example, are separating right now at the rate of about 5 cm per year; consequently, the Atlantic Ocean is getting wider. Note that old spreading centers, such as the Mid-Atlantic Ridge, are always located in the middle of an ocean.

New spreading centers, on the other hand, may begin anywhere, even in the middle of a continent. If a continent happens to be sitting above what will eventually become a divergent plate boundary, then the continent itself will literally be torn apart. The Great East African Rift Valley, which extends south from Ethiopia, along the east coastal interior of central Africa to the coast of Mozambique, and north into Israel and Syria, is a modern-day example of this motion. Millions of years in the future, the western part of Africa may be separated from the eastern part by an ocean. In fact, at the point where the Great Rift Valley crosses the African coast, this sea is already beginning to form. Look at the Dead Sea, Red Sea, and Gulf of Aqaba and you will see this rift in progress.

2. Convergent Plate Boundaries The Earth is not getting any larger. If new material pushes tectonic plates apart at places like the oceanic ridges, then old plate material must be pushed together and taken into the Earth somewhere else. A place where two plates are coming together is called a **convergent plate boundary.**

At most convergent plate boundaries, one plate sinks beneath another to form a **subduction zone.** The plate that is subducted, or "taken beneath," in this way sinks down to rejoin the mantle material from which it came.

Earth scientists observe three broadly different kinds of surface features associated with convergent plate boundaries, all shown in Figure 15–9. First, if no continents are on the leading edge of either of the two

A satellite photograph of a portion of the Great Rift Valley. The narrow body of water defines a divergent plate boundary at which new plate material is being created and plates are moving out to either side.

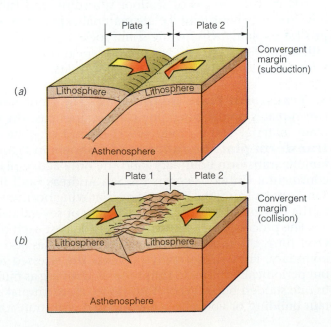

Figure 15–9
Convergent plate boundaries may display a variety of surface features, depending on the distribution of continental material. (*a*) When neither tectonic plate carries continental material at the convergent boundary, then a subduction zone is formed. A deep ocean trench and chain of island volcanoes are often created in this case. (*b*) When both tectonic plates carry continental material at the boundary, the continental materials buckle and fold to form nonvolcanic mountains. (The "asthenosphere" is part of the upper mantle.)

converging plates, the result will be a *deep ocean trench*. As one plate penetrates deep into the Earth, it can bend and buckle the adjacent plate to produce a deep furrow in the ocean floor. Melting of the subducting slab at depth can generate magma. The hot magma slowly rises toward the surface and, if the magma penetrates the overlying ocean crust, continued eruptions of lava will build a chain of volcanic islands adjacent to the trench. The Mariana Trench, the deepest point in the world's oceans (11 km or about 7 miles deep—2 km deeper than Mt. Everest is high) near the volcanic coastline of the Philippines, is an example of just such a subduction zone and volcanic terrain.

If continents ride on top of both converging plates, they will collide. Continental material will be compressed together like crumpled cloth and will be pushed up to form a high, jagged mountain chain. The Himalayas, for example, which began to form about 30 million years ago, are still growing taller at about 1 centimeter per year as the once-separate Indian subcontinent collides with Asia. Similarly, the Ural Mountains mark the point at which Europe and Asia were welded together, and the Alps mark the point at which the Italian peninsula was joined to Europe. All of these geological processes involve the production of midcontinent mountain chains.

Finally, if a continent rests at the leading edge of only one of the two colliding plates, the denser oceanic plate will subduct beneath the continent. Just as in the case where two oceanic plates converge, a deep trench may form a short distance off shore, while the continental material may be crumpled into a coastal mountain range. In addition, the subducting tectonic slab may partially melt and thus provide magma that rises to form a chain of volcanic mountains parallel to the coast. The Andes Mountains of South America and the Cascade Mountains of the northwestern United States are spectacular examples of this phenomenon.

Roughly speaking, oceanic plate material on the Earth renews itself about every 200 million years. The ocean crust is constantly being replaced by the process of seafloor spreading and subduction, while the lower-density continental plate material experiences no subduction. This process explains why, while some rocks on ancient continents formed billions of years ago and are still preserved there, no rocks on the ocean floor are older than about 200 million years.

3. Transform Plate Boundaries   The third kind of boundary between plates occurs when one plate scrapes past the other, with no new plate material being produced. This kind of plate contact is called a **transform plate boundary** and is shown in Figure 15–10. The most famous transform boundary (and the only active plate boundary in the continental United States) is the San Andreas Fault in California. At the San Andreas Fault, the Pacific Plate is moving northwestward with respect to the North American Plate at the rate of several centimeters a year.

The process by which two plates slide past each other is not smooth. Over time, the motion of the plates compresses and strains rocks at the boundary. Friction normally prevents the stressed rocks from moving, but periodically the rocks simply break, moving as much as several meters in one sudden burst. When they do so, an earthquake occurs. No mountain building or volcanism is associated with transform boundaries.

The San Andreas fault in California marks the transform boundary between the North American plate and the Pacific plate.

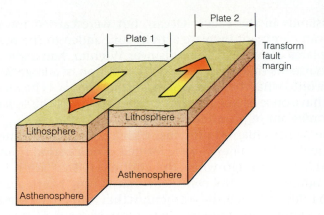

Figure 15–10
A transform plate boundary, showing the relative motions of the adjacent plates.

The Human Body

Upright Posture

Have you ever wondered why human beings are among the very few animals on the Earth that walk upright? Most scientists today suspect that it was the adaptation of upright walking, which freed our hands, that led eventually to the use of tools and development of increased brain size in humans (see Chapter 24). Richard Leakey, a well-known paleoanthropologist, has suggested that our upright posture might have resulted indirectly from the movement of the Earth's tectonic plates.

His argument goes like this: 30 million years ago, most of eastern Africa was covered by a lush jungle. No fewer than 20 different species of apes, including our ancestral species, flourished in that environment and were especially well adapted to living in trees. When a divergent boundary started to pull the continent apart along the East African Rift Valley, the environment started to change. The forest began to disappear, to be replaced first by open plains dotted with stands of trees, and finally, 3 million years ago, by the savannah that exists there today. Most of the apes became extinct long before our 3-million-year-old ancestors, but Leakey argues that walking upright and being able to get from one forest "island" to another would have been a distinct advantage in that sort of environment. The result, according to Leakey: Today there are only three kinds of descendants from those apes in Africa—gorillas, chimpanzees, and human beings. ■

The Geological History of North America

The epic movements of tectonic plates provided geologists with a new way of thinking about the history of the Earth's surface. Earth scientists now use our understanding of plate tectonics to tell us something about the formation of our own continent. The oldest parts of the North American continent are in northeastern Canada. Here we find large geological formations of rocks, several billion years old, that form the core of the continent. Over long periods of time, land was added to this continent by tectonic activity. Most of the western part of the United States, for example, is made up of small chunks of land called *terranes*—masses of rock several hundred kilometers across. Originally, these terranes were large islands in the Pacific Ocean, but were carried toward the North American continent by plate activity and added to the mainland as tectonic plates converged. The hills near Wichita, Kansas, for example, are old mountains that once marked the addition of a large South American terrane onto what then comprised North America. (The idea that Wichita might have once had oceanfront property is one of the strange discoveries that comes out of plate tectonics.)

The Appalachian Mountains, which may at one time have rivaled the Himalayas in majesty, were formed over a period from about 450 to 300 million years ago when the continents that are now Eurasia and Africa rammed into the continent that is now North America. A series of long folds and fractures—structures that formed the present-day Appalachian Mountains—appeared in the surface rocks. This process explains, for example, why roads in the mountainous regions of the eastern part of the United States tend to run from southwest to northeast—they follow the mountain valleys that were created by erosion of these folded rocks. Thick wedges of sediments eroded off the mountains, forming the Coastal Plains of eastern North America and contributing to the sediments of the Great Plains.

The dramatic geological features of the western United States record a great variety of mountain-forming events. The Rocky Mountains rose approximately 60 million years ago from a broad warping, and subsequent folding and fracturing, of continental material. The Colorado Plateau, comprising parts of the states of Colorado, Arizona, and New Mexico, experienced a more gentle uplift, as rivers incised features such as the Grand Canyon. The Sierra Nevada range formed more recently when molten rock pushed up a huge block of sediments. These processes of uplift and erosion continue to this day in many places around the world.

Another Look at Volcanoes and Earthquakes

Plate tectonics provides us with a dynamic picture of the Earth. Plates are continually moving over the hot, partially molten rocks in the mantle. They crash together, rip apart, and scrape by each other. In the process, tall mountain chains are uplifted and worn down, and wide ocean basins are opened and closed as continents come together and split apart. Nothing on the Earth is permanent, because heat continuously flows from the hot interior to the cooler surface, and mantle convection provides the primary mechanism for that heat transfer. The scale of these ongoing processes is vast—nearly outside our ability to comprehend, but we are occasionally reminded of the power of geological processes.

For thousands of years humans have realized that the Earth's most violent events—volcanoes and earthquakes—do not occur randomly. Earthquakes are common in California and Alaska, but extremely rare in Kansas or Florida. Volcanoes are commonplace in Hawaii and the Pacific Northwest, but never appear in New York or Texas. Why should this be? Plate tectonics provides an answer.

Plates and Volcanism

The global distribution of volcanoes may be understood in terms of the principles of plate tectonics. Volcanoes are common in three geological situations: along divergent plate boundaries, near convergent plate boundaries, or above places called "hot spots."

1. Divergent Plate Boundaries The formation of new crust along volcanic spreading ridges of divergent plate boundaries is the principal way that new crustal rocks are formed. New basalt plate material forms at the rate of a few centimeters per year along about 85,000 kilometers of oceanic ridges around the world.

2. Convergent Plate Boundaries Volcanoes are also ubiquitous near subduction zones, except where two continental tectonic plates collide. As water-rich crustal material plunges into the mantle it becomes hotter and may partially melt. This magma—highly mobile fluid rock— rises to the surface to form chains of volcanoes, typically about 200 km inland from the line of subduction. The "ring of fire," a dramatic string of volcanoes that borders much of the Pacific Ocean, is a direct consequence of plate subduction (see Figure 15–11).

The volcanoes that form the Cascade Mountain chain along the northwest coast of the United States (including Mount Rainier and Mount St. Helens) are striking examples of the processes associated with subduction of an oceanic plate beneath a continent. Frequent dramatic eruptions of similar volcanoes in Central America, Japan, and the Philippines point to other places where subduction and volcanism occur in tandem.

3. Hot Spots Finally, **hot spots** are a dramatic type of volcanism indirectly associated with plate tectonics. Earth scientists recognize dozens of hot spots around the world—Hawaii, Yellowstone Park, Iceland, and others where large isolated chimneylike columns of rising hot rock, also known as mantle plumes, rise to the surface more or less like bubbles coming to the surface in water being heated on a stove. These plumes may originate in the lower mantle or even at the core-mantle boundary. On a geological time scale, sources of hot spots are relatively stationary, so if a tectonic plate slowly moves over the fixed hot spot, the result will

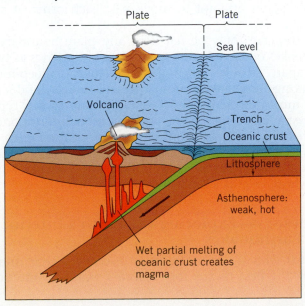

Figure 15–11
Volcanoes form above a subduction zone when heated plate material partially melts. The hot magma rises through the overlying crust to form a chain of volcanic islands.

Figure 15–12
The Hawaiian Islands stretch along a northwest-southeast line that reveals the northwesterly motion of the Pacific Plate over a fixed hot spot. As the Pacific Plate moves, new volcanic islands are created to the southeast, while older islands (ages in millions of years) erode away.

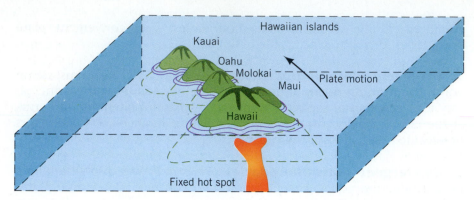

be a chain of volcanoes like the Hawaiian Islands that are built through a series of basaltic lava eruptions (see Figure 15–12). These islands were created one at a time as the Pacific Plate moved above the localized hot spot. The present-day volcano Kilauea on the "big island" of Hawaii, the site of most of the island chain's active volcanism, is directly over a hot spot. The volcanic islands to the northwest are progressively older, as well as smaller owing to erosion, revealing that the motion of the Pacific Plate is also toward the northwest. In fact, a series of eroded submarine peaks that were islands millions of years ago stretches hundreds of kilometers farther to the northwest. In several million years, the most northwesterly of the present Hawaiian islands will have been eroded beneath the waves, but new volcanic activity has already begun on the ocean floor southeast of Kilauea, promising new islands that will replace the old.

Earthquakes

Stress builds up in brittle rock for several reasons. Heated rock expands and cooling rock contracts, changes that cause a solid formation to warp and distort. Rock may also become stressed in response to changes in pressure, as overlying mountains wear away or new layers of sediment weigh down. And, of course, stress builds up to extreme levels as two tectonic plates attempt to move past each other at a transform plate boundary.

Earthquakes may be felt near any plate boundary. Minor shallow earthquakes occur near divergent plate boundaries as two oceanic plates move apart. Stronger earthquakes, including "deep-focus" earthquakes originating more than 100 km down, occur near subduction zones. Many of the most destructive shocks in Japan are of this type. In the United States, earthquakes at the transform plate boundary along the San Andreas Fault receive the most attention because of the fault's unusual activity, length, and proximity to major population centers. There are, however, occasional earthquakes in the middle of plates—in Missouri, for example—whose origins are not fully understood.

Seismology—Exploring the Earth's Interior with Earthquakes

Scientists who study earthquakes discovered that these violent events provide the best means for exploring the deep interior of our planet.

Geologists can't get their hands on much more than the outer 10 km or so of rock layers. Everything we know about the interior of the Earth deeper than a few kilometers has to be obtained by indirect means. The science of **seismology**—the study and measurement of vibrations within the Earth—is dedicated to deducing our planet's inner structure.

The basic idea of seismology is simple. When an earthquake or explosion occurs, waves of vibrational energy (see Chapter 5) move out through the rocks. Some of these waves travel through the center of the Earth, some move along the surface, and still others bounce off layers deep within the planet.

So-called seismic waves come in two principal types. *Compressional* or *longitudinal waves* are like sound: the molecules in the rock move back and forth in the same direction as the wave. *Shear waves,* on the other hand, are transverse waves, like water waves in which the molecules move up and down perpendicular to the direction of wave motion. These kinds of waves travel through rock at different velocities depending, among other things, on the rock type, its temperature, and the pressure. After a major earthquake, scientists at laboratories around the world record the intensity and time of arrival of the various kinds of waves—both those that pass along the surface and those that pass through the interior of the Earth (see Figure 15–13). By comparing the arrival behavior of waves

A geologist uses a seismograph to detect sound waves passing through the Earth.

Figure 15–13

Seismic waves passing through the Earth can take a variety of paths. The speeds of *P* (compressional) and *S* (shear) waves will be different depending on the type of rock, its temperature, and the pressure.

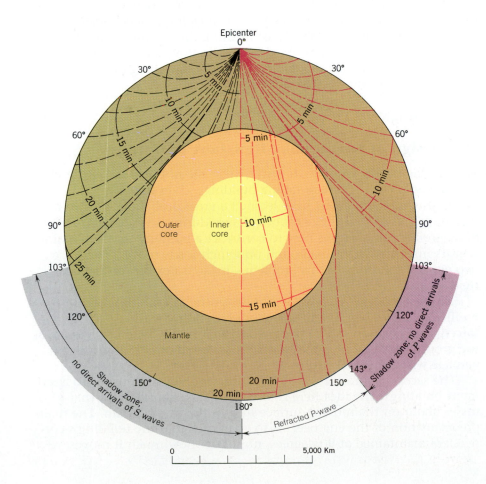

Figure 15–14

A seismic tomograph provides a picture of the Earth's deep interior, based on millions of individual measurements of seismic wave velocities. This image reveals hot rocks (red) around the Pacific Ocean rim, and cooler rocks (blue) under the continents.

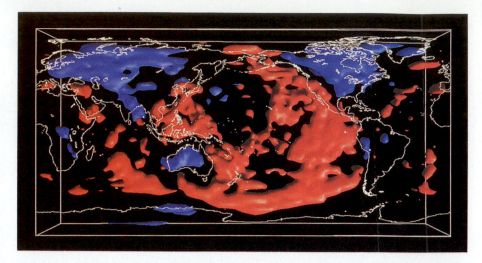

from the same earthquake at many different sites on the Earth, a computer can construct a picture of the material through which those waves passed.

The picture of the Earth we gave in Chapter 14, in which we talk about the solid and liquid core, the layered mantle, and the thin brittle crust, came from studies of seismic waves. Now, as more and more seismic data are collected, and ever-faster computers permit new ways to process those data, a new branch of Earth science called *seismic tomography* is enabling geophysicists to obtain astonishing three-dimensional pictures of the Earth's interior (Figure 15–14). We are now able to document the basic movements of cold subducting slabs, hot upwelling mantle plumes, and the convection cells that drive plate tectonics.

 Technology

The Design of Earthquake-Resistant Buildings

Most of the people who die in earthquakes die not because of the violent shaking, but because of falling buildings. Over the past few decades, structural engineers have learned a great deal about how to design buildings so that they do not collapse during an earthquake. The basic problem they face is how to construct a building that will maintain its integrity when the ground on which it sits moves. There are two general solutions to this problem—make it flexible, or make it rigid.

The first strategy, widely used in designing tall buildings, is best typified by a tree bending in the wind. A building with a specially reinforced steel skeleton can be designed so that it will bend and vibrate as the ground shakes, but come back to its original orientation without damage when the quake stops.

The second approach is widely used in individual houses and apartment buildings. The idea here is to construct the building so that it tosses about on the moving surface like a ship on the ocean. The main preoccupation of the engineers is to guarantee that the building's corners are maintained at 90 degrees, no matter how much it moves around, by reinforcement of corners and rigid connections to the foundation and roof. Sometimes this rigidity can be obtained simply by covering all the walls of the building with plywood sheets before the outer siding is installed. ■

THINKING MORE ABOUT PLATE TECTONICS

Earthquake Prediction

One of the most important roles scientists can play in modern society is to provide warnings of natural disasters such as earthquakes, volcanic eruptions, and large storms. In the summer of 1992, for example, early warnings allowed residents of southern Florida to escape the path of Hurricane Andrew. The storm caused billions of dollars of property damage, but surprisingly little loss of life. Similarly, volcanic eruptions, which are usually preceded by numerous small earthquakes, also can be forecast with some certainty. The successful prediction of Mount Pinatubo's 1991 eruption in the Philippines gave thousands of people living on the flanks of the long-dormant volcano time to flee.

But earthquake prediction is another story. Earthquakes occur because of the predictable and measurable gradual buildup of stresses in the Earth's crust. But the sudden failure of rock is not so predictable. You are familiar with this situation if you've ever gradually bent a pencil to the breaking point. It's easy to predict that a bent pencil will eventually break, but it's very hard to say exactly at what point the failure will occur. Much of the scientific effort related to earthquake prediction centers on the careful documentation of stress buildup in earthquake-prone areas. But these measurements alone tell us little about the exact time when failure will occur.

In their efforts to predict earthquakes, many scientists search for "precursor events"—measurable phenomena that precede the quake. There is much anecdotal evidence that such events may indeed occur. Folklore has it that some domesticated animals become highly agitated before a strong earthquake, or that changes are seen in the flow of well water or hot springs. Similarly, scientists have noticed changes in the regularity of geysers and have recorded swarms of minor quakes just prior to some big earthquakes. Nevertheless, no reliable methods for predicting earthquakes have yet been devised. At best, we can reliably predict the probability that a large earthquake might occur in a given area during a period of a few decades.

Ironically, if we do develop a way to predict the timing of earthquakes precisely, a whole new series of problems could arise. Suppose you could predict with 80% certainty that there would be a major earthquake in the Los Angeles basin sometime in the next 30 days. Would you then announce it? What would happen in the Los Angeles basin if you made such an announcement? What would happen if you made the announcement and the earthquake never occurred? What would happen if you didn't make the announcement and the earthquake did occur? These are not easy subjects to deal with.

Summary

The surface of the Earth constantly changes. Mountains are created and worn away, while entire continents slowly shift, opening up oceans and closing them again.

Plate tectonics, a relatively new theory that explains how a few thin, rigid *tectonic plates* of crustal and upper mantle material are moved across the Earth's surface by *mantle convection,* provides a global context for these changes. According to this theory, plates move over a partially molten, underlying section of the Earth's mantle, like rafts on the ocean, in response to the convection of pliable mantle rocks.

Three different kinds of observations—the geological features of ocean floors; parallel stripes of magnetic rocks, situated symmetrically about volcanic ridges on the ocean floor; and the ages of these rocks—provided direct evidence that new crust is being created at *divergent plate boundaries.* Meanwhile, old crust returns to the mantle in *subduction zones,* where plates converge. *Convergent plate boundaries* in the ocean create deep trenches and associated volcanic islands. When an oceanic plate subducts beneath a plate carrying a continent, an offshore trench and a chain of continental volcanoes parallel to shore result. The collision of two plates that carry continents at their margins produces mountain ranges of crumpled continental material. *Transform plate boundaries* occur where two plates scrape by each other.

Most *volcanoes* form near plate boundaries, either along the volcanic ridges of diverging plates, or above subducting plates. Other volcanoes such as the Hawaiian Islands form above *hot spots* originating in the Earth's mantle. *Earthquakes* occur when stressed rock ruptures. Earthquakes may be felt at all plate boundaries. The only plate boundary in the United States, California's San Andreas Fault, is a transform boundary in which one block of crust moves horizontally past the opposing block. The science of *seismology,* which documents the passage of earthquake-generated sound waves through the Earth, is providing new insights into the dynamic processes that drive plate tectonics.

Key Terms About Plate Tectonics

volcano	convergent plate boundary
earthquake	
plate tectonics	subduction zone
tectonic plates	transform plate boundary
mantle convection	
divergent plate boundary	hot spots
	seismology

Review Questions

1. What evidence suggests that mountain ranges are not permanent features on the Earth's surface?

2. What evidence suggests that Europe, Africa, and North and South America were once joined?

3. What pieces of evidence pointed to the process of seafloor spreading, or diverging plates?

4. Hot mantle rocks are sometimes compared to a pot of soup. Why?

5. Identify three kinds of plate boundaries.

6. Explain how the Alps, Himalayas, and Ural Mountains formed as a result of plate motions.

7. How do mountains form at divergent plate boundaries?

8. How do mountains form at convergent plate boundaries?

9. Why does a ring of volcanoes and earthquakes surround much of the Pacific Ocean?

10. What North American mountain range may have been the tallest in the world approximately 300 million years ago?

11. State some direct evidence for the movement of tectonic plates.

12. What is the Richter scale and where might you read about it?

13. How do the ages of rocks on the ocean floor help support the theory of seafloor spreading?

Discussion Questions

1. Identify two kinds of volcanic mountain chains that bear a close relationship to plate boundaries.

2. What plate do you live on? How many adjacent plates are there? What kinds of boundaries do you find to the north, south, east, and west?

3. The continent of Antarctica has rocks with plant and animal fossils that suggest the Antarctic climate was once temperate. Explain at least two different ways in which these warm-climate fossils might have ended up in what is now a polar region. How might you test your hypotheses?

4. Volcanic islands, including the Azores, Canaries, and Iceland, lay scattered across the Atlantic

Ocean. If you were to date the rocks on these and other Atlantic islands, what pattern do you predict you would find?

5. Geology has been called an integrated science, because it calls on several scientific disciplines to help explain features and processes of the Earth. Explain how geologists have used other sciences to answer the following questions:
   a. How old is a piece of rock?
   b. How is heat transfered from the Earth's deep interior to the surface?
   c. How does the Earth's magnetic field change over time?
   d. What is the structure of the Earth's interior?
   e. What is the topography of the seafloor?

6. At some convergent plate boundaries, deep ocean trenches lie a short distance from tall mountains. Why are these two contrasting features related to each other?

7. Popular media sometimes describe the California coast as poised to "slide into the ocean." Based on your understanding of plate tectonics, is this a plausible occurrence? Why or why not?

Problems

1. If the African Rift Valley opens up at the rate of 5 cm per year, how long will it be before a body of water 1000 km wide divides the African continent?

2. Estimate the probable lifetime of your favorite mountain. (Hint: Get its dimensions from a map, then estimate its volume.)

Investigations

1. Examine original sources related to Wegener's continental drift theory. Why was this theory rejected by the majority of earth scientists in the 1920s? Compare and contrast the major features of the continental drift theory with plate tectonics theory.

2. How did ancient civilizations explain the occurrence of earthquakes and volcanoes?

3. Does plate tectonics operate on any other terrestrial planet in our solar system? Why or why not?

4. In your library, examine newspaper reports of major volcanoes and earthquakes during the past 20 years (each student could take one year). Plot these events on a world map. Do you see any obvious geographic patterns? How do the locations of these events relate to the plate boundaries shown in Figure 15–7?

5. A few years ago someone predicted an earthquake on the New Madrid fault in Missouri, the site of a destructive shock in 1812. He was widely believed and schoolchildren were trained in what to do in case of an earthquake. The earthquake did not take place and still has not occurred. What evidence did the amateur scientist use to make his case? How would you analyze that evidence?

6. How did Wegener's theory of continental drift differ from modern plate tectonics theory?

Additional Reading

Glen, William. *The Road to Jamarillo*. Stanford, California: Stanford University Press, 1982.

McPhee, John. *Basin and Range*. New York: Farrar, Strauss, and Giroux, 1981.

———. *In Suspect Terrain*. New York: Farrar, Strauss, and Giroux, 1983.

———. *Rising from the Plains*. New York: Farrar, Strauss, and Giroux, 1986.

Parker, Ronald. *Inscrutable Earth*, New York: Scribner, 1984.

Cycles of the Earth

All matter above and beneath the surface of the Earth moves in cycles.

A Random Walk
Through the Day

Every place we go, every day of our lives, we experience Earth's cycles. A day may start out clear and sunny, with mild temperatures and gentle breezes. But, suddenly, the wind picks up as a line of dark clouds presses in from the west. With the wind comes the first threatening rumble of thunder, flashes of distant lightning, and spurts of rain. A heavy downpour and powerful gusts of wind follow, as thunder booms and lightning illuminates the grey sky around us. Trees sway, windows rattle, and limbs come crashing down at the height of the storm.

Within an hour skies are sunny again, though the temperature is noticeably cooler. The air has a fresh, clean smell, and the sky seems more deeply blue than before. In the span of a few hours we have experienced one of Earth's many cycles—the atmospheric cycle called weather.

Day and night, summer and winter, life and death—these and many other cyclical changes characterize our dynamic planet. As the author of the Book of Ecclesiastes wrote 3000 years ago:

One generation passeth away, and another generation cometh:
  but the earth abideth forever;
The sun also ariseth, and the sun goeth down, and hasteth
  to the place where he arose;
The wind goeth toward the south, and turneth about
  unto the north; it whirleth about continually,
  and the wind returneth again according to his circuits;
All the rivers run into the sea; yet the sea is not full; unto the
  place from whence the rivers come, thither they return again. . .
The thing that hath been, it is that which shall be;
  and that which is done is that which shall be done:
  and there is no new thing under the sun.

Cycles Small and Large

Recycling

Think about the last time you drank a can of soda. What did you do with the aluminum can when you were through? You may have taken the time to place the can in a recycling bin. Alternatively, you may have tossed it into a trash container, or even by the side of the road. What difference does it make? Where do the aluminum atoms end up?

The atoms that make up the Earth, with the exception of a few radioactive isotopes (see Chapter 11), will last virtually forever. A single aluminum atom, for example, will appear in many different guises during its lifetime. It may form part of a swirling lava flow in which it is tightly bonded to oxygen atoms. It may then be incorporated with those oxygen atoms into a solid rock. As the rock weathers away, the atom may be concentrated in soil with other aluminum atoms, where it is mined. Giant smelters separate the aluminum atom from its oxygen neighbors— a process that consumes prodigious amounts of energy—to produce the aluminum metal that made your soda can. Once discarded, the atom may be recycled into new cans, or it may go back to the soil where it once again bonds to oxygen.

The Earth has a vast, though finite, number of aluminum atoms. The advantage to recycling aluminum is that we save all the energy that went into finding concentrated aluminum sources and breaking aluminum-to-oxygen bonds. But no matter what you do with your can, the aluminum atoms are still part of the Earth.

The Earth also has a finite number of carbon atoms, and every time you breathe you take part in the grand cycle by which they move through the environment. You take in carbon atoms with your food, in which chemical energy is stored in the bonds between atoms. When you release that energy by "burning" the carbon-bearing molecules in your cells, carbon dioxide is produced as a by-product, carried to your lungs by your blood, and returned to the environment when you breathe out. Some of those carbon atoms may be taken into plants and incorporated into plant tissue through the process of photosynthesis. Someday, you may take in that same carbon atom when you eat an apple and begin the whole process again.

The Three Great Cycles of Earth Materials

The histories of the aluminum and carbon atoms are just two examples of the many paths within the three great cycles of materials in the Earth's atmosphere, oceans, and interior. Two central ideas frame our understanding of our constantly changing planet:

> **Earth materials, above and beneath the surface, move in cycles**
> **and**
> **a change in one cycle affects the others.**

Some of Earth's cycles—weather, for example—can vary on a scale of hours, while the cycles that form and alter rocks may take hundreds of millions of years. But all of these processes illustrate a central theme in the earth sciences: our planet's atoms are constantly moving, constantly recycling. Above the Earth, the gases of the atmosphere flow in the great cycles of weather, the seasons, and global climate. Water moves from rivers to oceans to glaciers and to clouds as it takes part in the Earth's dynamic hydrological cycle. And the solid surface of the Earth itself slowly forms, alters, erodes away, and forms again in the stately rock cycle.

Many of the Earth's cycles are driven by the tendency of heat to spread out—to flow from hot to cold in what we described as the second law of thermodynamics (see Chapter 4). The Earth has two primary sources of heat energy: the Sun, and its own geothermal processes, each of which drives its own thermal cycle. More heat energy from the Sun falls at the equator than at the poles, and heat transfer by convection moves gases in the atmosphere and water in the oceans in the great cycles that control weather and climate. Similarly, heat energy stored in the Earth's core and mantle drives the convection cycles that move the tectonic plates.

Thus Earth's cycles reflect the most basic properties of matter and energy. These cycles may be studied at many levels, from an atomic scale for individual elements such as aluminum or carbon, to the global cycles involved in plate tectonics. We find it especially useful to consider the Earth in terms of the three most familiar cycles around us: the cycles of air, water, and rock.

The Atmospheric Cycle

The Earth's atmosphere and oceans play the most important role in redistributing heat across the surface of the planet. The atmosphere also has chemical cycles involving oxygen and carbon, but those cycles are intimately bound up with the presence of living things on the Earth, and we will wait to discuss them until Chapter 25, when we look at the Earth's ecosystems. The circulation of gases near the Earth's surface—both the short-term variations of weather, and the longer patterns of climate—is called the **atmospheric cycle.**

Figure 16–1

If the Earth did not rotate, the circulation of the atmosphere would take place in convection cells that create high-altitude winds from equator to pole, and low-altitude winds from pole to equator.

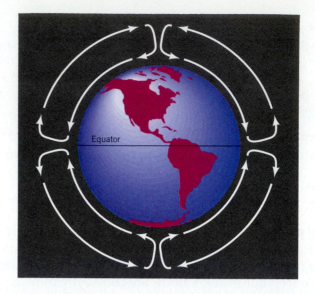

The General Circulation of the Atmosphere

The atmosphere circulates in vast rivers of air that cover the globe from the equator to the poles. This circulation is powered by the energy of the Sun. Air in the tropics is heated and rises. If the Earth did not rotate, we would expect to have a situation like the one shown in Figure 16–1. Warm air would rise at the equator, cool off, and sink at the poles. This pattern of flow is the familiar convection cell we saw in Chapter 4. Such a pattern arises whenever a fluid is heated nonuniformly in a gravitational field.

If the Earth did not rotate, then, in the Northern Hemisphere prevailing winds would flow from north to south. In fact, they do nothing of the kind. The weather patterns in North America move, in general, from west to east—we live in a region of what meteorologists call *prevailing westerlies*. This behavior of the Earth's atmosphere results from the fact that the Earth rotates. The rotation breaks the north-south atmospheric convection cell that would exist in the absence of rotation into three cells in each hemisphere as shown in Figure 16–2. In addition, the rotation "stretches out" the shape of the air circulation pattern in each cell. In the cell nearest the equator, the winds at the surface tend to blow from

Figure 16–2

Atmospheric convection on the rotating Earth, showing the principal air circulation cells. Compare this diagram with the opening photograph in Chapter 14.

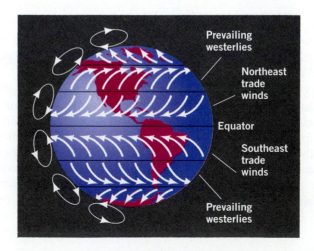

east to west—the so-called trade winds that drove sailing ships from Europe to North America. In temperate zones, the effect is to cause the winds to blow from west to east, creating regions in which weather patterns also usually move from west to east. Finally, in the Arctic and Antarctic, the winds blow once again from east to west.

Similar patterns of atmospheric motion can be seen on all the planets in the solar system that have atmospheres. In some cases, like the planet Jupiter, the rapid rotation of the planet and the atmospheric dynamics cause more than three convection cells. Jupiter, in fact, has no fewer than 11.

Weather and Climate

The word **weather** refers to daily changes in rainfall, temperature, amount of sunshine, and other variables that are part of our everyday lives. The term **climate,** on the other hand, refers to trends in the weather longer than the cycle of seasons. Thus a city such as Seattle may have hot sunny weather on a particular day, but will have a generally moderate, rainy climate. The weather in a given region results partly from the general circulation in the atmosphere, and partly from local disturbances and variations.

The most important variations underlying the daily weather are differences in atmospheric pressure—in the amount of air actually present at a given location. Air, like any other fluid, can slosh around. There will be regions of higher pressure, where more air than average accumulates, and regions of lower pressure as well. When low-pressure areas develop in the general flow, air from the surrounding regions tends to rush in to fill up those lows. In the Northern Hemisphere, the rotation of the Earth forces the air to flow in a generally clockwise direction around an area of high pressure and in a counterclockwise direction around an area of low pressure. (It is just the reverse in the Southern Hemisphere.) This flow gives rise to the circular patterns we often see in satellite weather photographs.

Air in low-pressure areas rises, cooling as it does so. At some altitude, the temperature falls to the point where water droplets condense out. Low-pressure areas are thus often characterized by extensive cloud cover and precipitation. In high-pressure areas, air is moving downward and becoming warmer and relatively drier, which is why high-pressure areas are associated with sunny and clear weather. A typical weather map will show two or three high- or low-pressure areas across a continent such as North America at any given time, and it is the constant interplay and change among these patterns as they move across the continent that produces the daily changes of weather.

Another important determinant of the weather in North America is the position of the **jet stream.** This high-altitude stream of fast-moving winds marks the boundary between the northern polar cold air mass and the warmer air of the temperate zone. It generally shifts north and south on an annual basis, but small kinks in the jet stream can prevent low-pressure areas from forming (a process that can produce drought) and can allow cold air to spill down into the temperate zones (a process nicknamed the Alberta Clipper or Siberian Express), creating cold waves

Jupiter's atmosphere displays complex banding and turbulence that result from high-velocity winds.

Figure 16–3

The jet stream is a fast-moving, high-altitude air current above North America. (*a*) The jet stream often follows a relatively straight path, with minor undulations. (*b*) Strongly developed undulations may pull a mass of cold arctic air to the south.

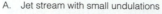

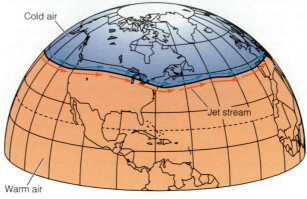

A. Jet stream with small undulations

Cold air

Jet stream

Warm air

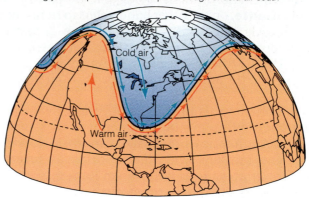

B. Strongly-developed undulations pull a trough of cold air south

Cold air

Warm air

(see Figure 16–3). You sometimes see the jet stream referred to in weather reports as an explanation for unusual weather patterns.

Common Storms and Weather Patterns

Many kinds of severe weather conditions affect our world. Because of our long acquaintance with these conditions, we have given them familiar names. Here are a few:

Tropical storms are severe storms that start as low-pressure areas over warm ocean water. They draw energy from the warm water, growing and rotating in great *cyclonic* patterns hundreds of kilometers in diameter. Tropical storms can cause serious damage when they strike land because of the high winds they generate and the fact that the low-pressure area at their center causes an increase in sea level. Once over land, these storms lose their source of energy and eventually die out.

Tropical storms that begin in the Atlantic Ocean off the coast of Africa and affect North America are called *hurricanes;* those that begin in the North Pacific are called *typhoons.* Until recently these violent weather systems often hit unprotected coastlines with little warning. Today, weather satellites spot and track tropical storms long before they approach land (see Figure 16–4).

Tornadoes, much smaller-scale phenomena than hurricanes, are rotating air funnels some tens to hundreds of meters across. Tornadoes descend from storm clouds to the ground, causing intense damage along the path where the funnel touches the ground. They are the most violent weather phenomenon known, with air speeds in excess of 500 km (about 300 miles) per hour.

Monsoons include any wind systems on a continental scale that seasonally reverse their direction. These winds arise primarily from seasonal variations in relative temperatures over land and sea. Land cools off and warms up faster than the ocean when the seasons change; and, since atmospheric pressure increases over colder areas, these winds tend to blow from cold (higher pressure) to warm (lower pressure) regions. Thus in summer monsoons come from the ocean and bring rain, and in winter from the land and bring drought. Monsoon systems occur across eastern Africa and southern Asia, northern Australia, and the Gulf Coast of the United States.

El Niño is a weather cycle in the Pacific basin that recurs every four to seven years. The name means Christ Child and comes from the fact that the phenomenon, when it happens, usually begins around Christmas time. El Niño can cause severe storms and flooding all along the western coast of the Americas, and drought from Australia to India.

El Niño is an example of a coupling between two of the Earth's cycles—in this case, the atmospheric and water cycles. It requires both winds and ocean currents to work. Here's what happens: Normally, the winds off the coast of Peru blow westward. They move the warm ocean water westward also, like water sloshing in a bathtub, to the western Pacific. As the surface water moves west, the colder, nutrient-rich water from the deep ocean wells up and supports marine life and fish-eating birds. Because the eastern Pacific is cooler, air cools and descends there, forming a zone of high pressure and dry weather off the coast of South America. Meanwhile, the western Pacific is warmer, so air warms and rises above it in a rainy, low-pressure zone.

Every four to seven years, however, this pattern changes, signifying the beginning of an El Niño event. Warm surface water sloshes to the east, the water temperature in the eastern Pacific increases by a few degrees, and

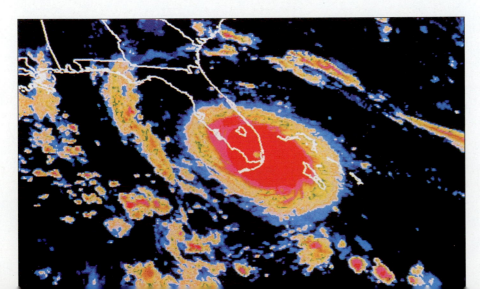

Figure 16–4

A satellite weather photo reveals Hurricane Andrew approaching the Florida coast on August 24, 1992.

the normal atmospheric patterns switch places. Westerly winds replace the normally easterly flowing trade winds. This wind reversal reinforces the movement of warm water eastward to the coast of South America, where air warms and rises, creating rainy conditions. The marine life and birds are no longer supported by the nutrients of the cold, deep water. The western Pacific is now cold and dry. Eventually, the water sloshes back and the whole cycle repeats. Historical records tell us that this four-to-seven-year cycle has been repeating in the Pacific basin since the 1600s, and may have been going on since the last ice age.

 Technology

Doppler Radar

Radar has been a staple tool for weather forecasters for decades, and you may see radar maps of local weather conditions on TV every night. The way radar works is simple. Microwaves are sent out from a central antenna. When they encounter objects such as raindrops, snowflakes, or ice in the air, the waves are reflected back. Each kind of material produces a distinct pattern of reflected waves because its density is different from that of the surrounding air. Using these reflection patterns, a map of local storm activity can be assembled.

Ordinary radar, however, cannot detect winds, even winds of high velocity. The reason is that the densities of moving air and stationary air are usually not very different, and the two produce the same reflection pattern. But in many situations it is important to detect air currents. Near airports, for example, sudden downdrafts create violent air turbulence called *wind shear,* an extremely dangerous condition that we need to be able to detect.

Doppler radar is designed to detect motions of the air itself. It works like this: Reflected waves are analyzed not only for their intensity, as in ordinary radar, but for their frequency as well. From the difference between the emitted and reflected frequencies, a simple analysis of the Doppler effect yields the velocity of the object from which the wave was reflected. In this way high winds and atmospheric turbulence can be detected at a safe distance. ■

Understanding Climate

While the daily weather is often dominated by the position of the jet stream and the creation of high- and low-pressure zones, the long-term climate depends on more lasting features of the Earth's surface. These factors include the distribution of heat due to the stabilizing temperature of oceans, and the presence of mountains, which force air masses up over them. The climate is also extremely sensitive to the amount of sunlight that falls on the atmosphere and the amount of heat that is radiated back to space.

At the moment, our best attempts at predicting long-term climate depend on complex computer models of the atmosphere called *global circulation models* (GCM). In a typical GCM, the computer splits the world's surface into squares several hundred kilometers on a side, and slices the atmosphere into about 10 vertical compartments. In each of these little boxes the laws of motion and thermodynamics are used to calculate the amount of heat that flows in and out, how much water vapor comes out of the air, and so on. The computer balances the inflow and outflow from all of the boxes in the atmosphere and projects forward in time to try to predict long-term climate trends. Our current models are still rather crude (they have a great deal of difficulty accounting for the effects of clouds or the ocean, for example), but they represent the best attempts to date to understand what affects the Earth's climate. These models also play a critical role in discussing various types of ecological changes such as global warming (see Chapter 25).

The Hydrological Cycle ▬

Water plays a vital role in the unique chemistry of the Earth's outer layers. Water saturates the air, falls to the ground as precipitation, moves through a complex system of rivers and streams, and is stored for long periods in underground reservoirs, oceans, and ice. Water shapes the surface of our planet, and it provided the medium in which life began. The combination of processes by which water moves from repository to repository above, below, and on the Earth's surface is called the **hydrological cycle.**

The total amount of water on the Earth's surface has stayed roughly the same since the water formed. Water first reached the surface during the outgassing of the young, volcano-covered Earth (see Chapter 14). When the planet's surface temperature finally fell below 100°C, this water condensed into liquid form and began to fill the ocean basins. Relatively minor processes still add and remove water from the Earth. High in the atmosphere, ultraviolet rays from the Sun break up water molecules, freeing hydrogen atoms, which, because of their low mass, may escape into space. At the same time, at converging and diverging plate boundaries and other sites of volcanism, small amounts of new water are emitted from the Earth's deep interior. These losses and gains are in rough equilibrium, and in any case both are rather small—by one estimate, no more than one or two Olympic-sized swimming pools of water per year. Thus, for all intents and purposes, we can treat the Earth as if it has had a fixed amount of water at its surface for billions of years. The water that we have now is all there is.

Reservoirs of Water

The Earth boasts several major water repositories. In addition to oceans, lakes, and rivers, significant amounts of water are locked into the Earth's polar ice caps and glaciers, bodies of ice that form in regions where snowfall exceeds melting. **Ice caps** are layers of ice that form at the north and south polar regions of the Earth. **Glaciers** are large bodies of ice that slowly flow down a slope or valley under the influence of gravity. Approximately 96% of glaciers (by volume) occur in Antarctica and

Figure 16–5

An aquifer is a rock layer that stores water and can be tapped by a well. The High Plains aquifer provides much of the water used for irrigation in this region. (*a*) A map of the aquifer with contour lines showing the depth to the water level (in meters). (*b*) A cross-section of the aquifer reveals the slope of the water-bearing rock formations.

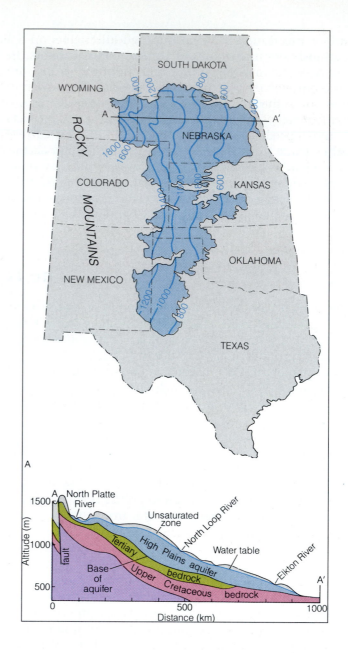

Greenland, while the rest are widely scattered in mountainous areas.

All of these places where water occurs simply tap into the same central supply. During its lifetime, a given molecule of water will cycle through many different kinds of bodies, over and over again.

The hydrological cycle with which most of us are familiar involves the short-term back-and-forth transfer of water molecules between the oceans and the land. Water evaporates off the surface of the ocean, forms into clouds, falls as rain on the land, and then returns to the oceans via rivers and streams. Most terrestrial life depends upon this simple cycle.

A vast part of this cycle remains unseen, however. Some of the water that falls on the continents does not immediately return to the ocean;

rather it seeps into the Earth to become **groundwater.** There, it goes into large *aquifers*—bodies that are, in effect, underground storage tanks of water. By some estimates, more than 98% of the world's fresh water is stored as groundwater. Water typically percolates into the ground and fills the tiny spaces between grains of sandstone and other porous rock layers. These layers of water-saturated rock are often bounded by impermeable materials such as clay, which keeps the water from seeping away (see Figure 16–5). Humans tap into these aquifers when they drill wells to supply water for cities and agriculture.

One problem with using aquifers as a water supply is that it may take many thousands of years to fill them, but only a few years to drain them. For example, when farmers in the central United States take water from the High Plains aquifer, one of the great underground reservoirs, they are, in effect, mining the water. Underground water is not a renewable resource except over very long time scales.

The fixed amount of water on the Earth has had many important geological consequences. From time to time, much of the Earth's water supply becomes locked into glaciers that advance across land from the poles—a period called an **ice age.** When this happens, the amount of water available to fill the ocean basins decreases, and the sea level drops. During the most recent advance of glaciers some 20,000 years ago, for example, the eastern coast of the United States was about 250 km farther east than it is today, and a land bridge made it possible to walk from Alaska to Siberia. This land bridge provided a route that was taken by the ancestors of many Native Americans when they moved into the Americas from Siberia.

Ice Ages

The periodic onset of massive glaciers is one of the most intriguing features of the Earth's climate. We are now in the middle of an *ice age*—a period of several million years during which glaciers have repeatedly advanced and retreated. In fact, geologists say that we are living in an *interglacial period,* between two major advances of glaciers. About 20,000 years ago, massive glaciers spread down from eastern and central Canada, covering a good deal of northern North America, and then receded to Greenland about 10,000 years ago, as illustrated in Figure 16–6. Glaciers have come and gone many times, and it appears that the periods of cyclic glaciation like the one in which we now live have occurred relatively often during the past 2 million years of Earth history.

The occurrence of the present ice age, and the resulting cycles of glaciation, may have had a profound influence on human evolution. In fact, some scholars have argued that the constant radical changes in climate during recent human history is responsible for the development of technology. They point out, for example, that the emergence of modern human beings and the disappearance of Neanderthals in Europe about 35,000 years ago coincided with one of these glacial periods. Thus you might say that the fact that you are sitting in a heated or air-conditioned room, reading a book produced on a modern printing press, is ultimately due to the fact that ice ages are part of the Earth's climate.

Figure 16–6
The extent of the most recent North American glaciation, approximately 20,000
years ago.

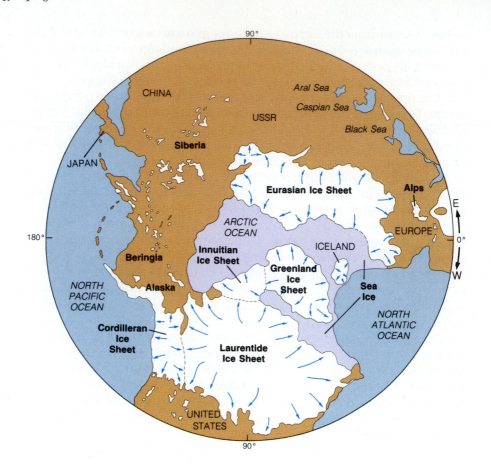

Milankovitch Cycles

The main causes of periodic glaciations were first explained by Milutin
Milankovitch, a Serbian civil engineer, in the early part of the twentieth
century. His theory was simple: The relation between the Earth and the
Sun is affected by a number of variations in the Earth's rotation and orbit.
These variations cause slight changes in the amount of solar radiation
absorbed by the Earth.

The easiest of the orbital effects to understand is the precession of
the Earth's axis of rotation. If you have ever watched a child's top spinning, you know that sometimes the top spins rapidly around its axis, but
the axis itself describes a lazy circle in space. This circular motion is called
a *precession*. In the same way that a top may precess every second or two,
the Earth's axis, which is tilted at an angle of 23 degrees, precesses once
every 23,000 years. While this is going on, the axis of the Earth's elliptical
orbit also moves around the Sun due to the gravitational effects of the
other planets. The net effect of these two motions is that when the Earth
is farthest from the sun, during the summer, the axis of the Earth tilts
the Northern Hemisphere toward the Sun for half of a 23,000-year cycle,
but gradually shifts so that the Southern Hemisphere is tilted toward the
Sun for the other half. Today the Earth's axis is such that the Northern
Hemisphere is tilted away from the Sun during the winter months (the
period in which the Earth is actually closest to the Sun). About 11,500

years from now, however, the situation will be reversed, and the Northern Hemisphere will be tilted toward the Sun during this same period.

Other orbital effects involve a slow change in the angle of the axis of rotation (it rocks back and forth by about a degree every 41,000 years), and a small change induced in the shape of the Earth's orbit by other planets (a cycle that takes about 100,000 years).

Milankovitch proposed that the global climate changes in cycles (now called Milankovitch cycles) when all three of these effects reinforce each other. If we find ourselves in a period of decreasing solar energy absorption and increased precipitation, more snow will fall in the winter and it will stay on the ground longer in the summer. Snow and ice reflect sunlight, so this extra snow and ice further cool the Earth and more snow falls and stays on the ground even longer. Thus a decline in absorption of sunlight may trigger a sequence of events that can lead eventually to glacial advance. By the same token, a period of increased absorption of sunlight will result in warmer periods during which glaciers will tend to retreat.

Other factors important in controlling the extent and distribution of glaciers include the distribution of land masses (which changes because of tectonic motion) and the location of mountain chains (which alters wind and precipitation patterns). Recent evidence also suggests that the amount of volcanic dust and gases in the atmosphere can cause short-term changes in global temperatures. Finally, some scientists suspect that the energy output of the Sun is cyclical, and that variations in the Sun's energy output impose a cyclic variation in the Earth's temperature.

For the record, most scientists think that we are in the process of heading into a new phase of glacial advance within the next 10,000 years. This cooling trend may be offset for a time, however, by global warming potentially resulting from an enhanced greenhouse effect (see Chapter 25).

Science in the Making

Milutin Milankovitch Decides on His Life's Work

Milutin Milankovitch didn't seem to be headed toward a career in the sciences. Trained in Vienna as a civil engineer, he designed reinforced concrete structures in central Europe for a few years before becoming a professor of mathematics in his native Belgrade just before World War I. Swept up in the nationalistic movements that were then (as now) prominent in the Balkans, he became friends with some poets who specialized in writing patriotic verse. One evening, drinking coffee to celebrate a new book of verse, Milankovitch and his friends came to the attention of a banker at a neighboring table. The banker was so taken with the poems that he bought ten copies on the spot.

With the new money in their pockets, the friends started to celebrate with wine. After the first bottle, Milankovitch says in his journal, "I looked back on my earlier achievements and found them narrow and limited." By the third bottle, he had decided to "grasp the entire universe and spread light to its farthest corners."

He then methodically set apart a few hours each day to study and interpret climate records. Even during World War I, when he served as an engineer on the Serbian general staff and became a prisoner of war, he kept at it. A member of the Hungarian Academy of Sciences arranged for him to have a desk at the academy after he gave his word of honor that he would not try to escape, and a good deal of the work described in the text was done under those conditions. The final work, published in 1920, was quickly recognized and accepted by the scientific community. ■

Ocean Currents: The Physical Circulation of the Oceans

The Earth's oceans are far from static. It would be very misleading to think of them as a series of gigantic bathtubs or passive sinks into which the Earth's water flows. In fact, the Earth's oceans, even more than the atmosphere, are important in redistributing heat across the surface of the planet, and thus in determining climate.

Each ocean basin, such as the North Atlantic, has surface currents flowing in it (see Figure 16–7). **Currents** are like rivers of moving water within the larger ocean. Some surface currents carry warm water from the equator, where a large amount of heat energy from the Sun is absorbed, toward the cooler poles. At the same time, other surface currents carry cold water from the poles back to the equator to be heated and cycled again. These great *gyres,* as they are called, have a profound effect on the weather of the land they flow past. In the North Atlantic, for example, the Gulf Stream current starts in the Caribbean and flows past the eastern coast of the United States. It comes near England, making

Figure 16–7

Ocean currents play a major role in redistributing heat.

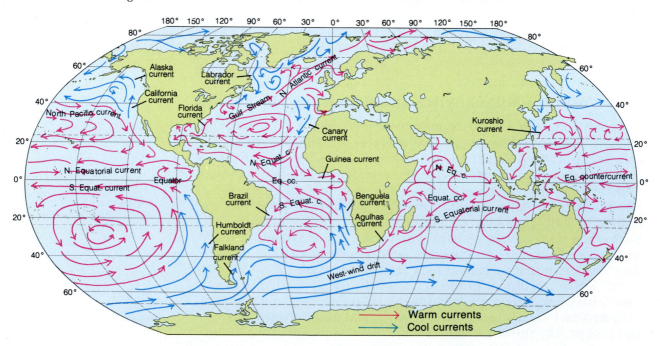

the British Isles much warmer than you might expect them to be based on their latitude, which is farther north than Maine. Indeed, the city of London possesses a mild climate even though it is on about the same latitude as Canada's Hudson Bay.

A much colder current, on the other hand, flows along the western coast of Europe back to the tropics to complete the cycle. In each of the Earth's great ocean basins, a similar set of currents can be seen. In the North Pacific, the Kuroshio current carries warm waters from the tropics past the east coast of Japan, while the California current carries cold water back down toward the equator along the west coast of North America.

In addition to this rather rapid circulation of water at the surface, we find deeper three-dimensional circulation of water in the ocean. When the effects of surface currents and wind along a coast act in such a way as to push surface water away from the land, colder water from the depths comes up to create an area of upwelling. The waters along the coast of California display this phenomenon, which does much to explain why the ocean there is so cold. On a larger scale, water from the Arctic and Antarctic, which is both very cold and salty because so much fresh water is removed to form ice, sinks to the bottom of the sea and rolls sluggishly toward the equator, producing the slow, deep movement that characterizes much of the bottom currents in the world's ocean basins.

Chemical Cycles in the Oceans

Just as the water in the oceans is in constant physical motion, so too are the chemicals that make it salty. It used to be thought that the oceans were simply passive receptacles for materials washed into them from the land. The old argument went that rivers flowing into the ocean carried dissolved minerals and salts with them, and when water evaporated, these minerals were left behind so that the oceans were becoming saltier and saltier with time. In this view, the extremely salty Dead Sea in Israel was always cited as a place where salts had accumulated over the longest periods of time.

In fact, we now know that the saltiness of the oceans has not changed appreciably over several hundred million years, and probably has been roughly constant since soon after the oceans formed. Instead of thinking of the oceans as passive receptacles, it is more accurate to think of them as large test tubes in which a constant round of chemical reactions goes on that affects and is affected by the input of minerals from the world's rivers.

The saltiness of the oceans comes primarily from the presence of sodium and chlorine, but many other dissolved minerals occur in seawater as well. Each kind of element follows a different set of reactions once it comes into the seas, and it will remain in solution for different periods of time. Calcium, for example, may be dissolved or *leached* out of limestone by water and carried into the ocean. Once there, a given calcium ion can be expected to float around for about 8 million years, but eventually it will be taken up into the skeleton of some sea creature, incorporated, and sink to the bottom when that creature dies. There it may once more

Table 16–1 ● Some Typical Residence Times for Elements in the Ocean

Element	Concentration (parts per million)	Residence Time (million years)
Sodium	10,800.	260.
Calcium	413.	8.
Chlorine	19,400.	infinite
Gold	0.00005	0.042
Potassium	387.	11.
Copper	0.003	0.05

be formed into limestone, lifted up by tectonic forces, and eventually carried back to the sea through erosion or dissolution.

The average length of time that any given atom will stay in the ocean water before it is removed by some chemical reaction is called the *residence time* (see Table 16–1). The element sodium (one of the atoms that gives seawater its distinctive taste) enters the ocean after having been leached out of various kinds of rocks. A sodium ion will stay in suspension on average for about 260 million years before it is incorporated again into various kinds of clays and muds on the ocean bottom. Once so incorporated, it can go through the same cycle of uplift and erosion as limestone. Chlorine ions, on the other hand, tend to stay in the ocean almost forever, though these atoms may leave the ocean for short periods of time.

Have you ever gone to the beach and noticed that your skin was salty at the end of the day, even if you never went in the water? Bubbles of sea foam burst, evaporate, and spew little bits of salt (sodium chloride, or NaCl) into the air. This salt is picked up by the wind and may be suspended for some time before it sticks to your skin. Sailors know that a rain in the open ocean is often salty, and rains that fall near coasts also contain a fair amount of sodium and chlorine. These salty rainfalls often give crops raised in coastal areas a distinctive taste. The artichoke fields south of San Francisco are one example of such a phenomenon, and the distinctive taste of the mutton from sheep that graze on the coastal meadows of Brittany in France is another. The chlorine that leaves the ocean in this way is quickly returned in rain or by coastal streams.

Only when a body of salt water is evaporated, forming dry salt deposits like those surrounding the Great Salt Lake in Utah, is chlorine removed from its water environment for any appreciable length of time. But even these deposits, slowly buried and compressed, could eventually (in perhaps a few hundred million years) rise to the surface as salt domes, where they would weather away, returning the salt once again to the sea.

The chemical cycles in the ocean, then, can be pictured very simply. The Earth's rivers continuously transport all sorts of elements into the sea. Each of these elements resides for a certain amount of time in the ocean and then is removed by one sort of reaction or another. The supply of every kind of atom is being constantly renewed, and the oceans may never be any less salty than they are today.

The Human Body

Element Residence Times

Your body, like the oceans, constantly recycles atoms. Some of your body's cells, such as the lining of the intestines, are replaced every few days. You need fresh supplies of carbon, oxygen, and nitrogen every day to help replace these cells. Red blood cells last much longer—on average 120 days—but you need a regular intake of iron to produce these important cells. Failure to digest enough iron can lead to anemia, a condition characterized by fatigue due to insufficient red blood cells.

Atoms in your bones and tendons last much longer—a decade or more, on average—but even those atoms are constantly being replaced. Gradual loss of calcium atoms in bone, for example, is of special concern in older people who may not consume enough replacement calcium. Osteoporosis, a disease in which bones become weak and brittle, may result from this calcium deficiency. A similar mysterious loss of bone calcium affects astronauts who spend more than a few days in the weightless environment of space.

Some harmful elements, including lead, mercury, and other so-called heavy metals, do not easily recycle once taken into the body, because we have no effective biological process to remove them. Their average residence times, in other words, are much longer than a human lifetime. Concentrations of these atoms can thus build up over time and can result in sickness or even death. You may hear about efforts in your community to test drinking water for high concentrations of lead that are found in old plumbing systems. ■

Science by the Numbers

The Ocean's Gold

We've just said that every element can be found in seawater. How much gold is there in a cubic kilometer of seawater?

According to Table 16–1, gold is present in the ocean at a concentration of 0.00005 parts per million. A concentration of 1 part per million corresponds to 1 milligram (mg) of a solid dissolved in a liter (l) of water. A liter is a volume measurement equal to a thousandth of a cubic meter (m³). The total amount of gold in a cubic meter of seawater, therefore, is 0.00005 milligrams per liter times 1000 liters per meter cubed:

$$(0.00005 \text{ mg/l}) \times (1000 \text{ l/m}^3) = 0.05 \text{ mg/m}^3.$$

A gram contains 1,000 milligrams, so

$$(0.05 \text{ mg/m}^3) \times (1 \text{ g/1,000 mg}) = 0.00005 \text{ g/m}^3.$$

A cubic kilometer contains 1000 m × 1000 m × 1000 m = 10^9 m³, so the total amount of gold in a cubic kilometer of seawater is

$$(10^9 \text{ m}^3) \times (0.00005 \text{ g/m}^3) = 5 \times 10^4 \text{ gm}$$

Every cubic kilometer of ocean water holds about 50,000 grams (about 100 pounds) of gold. At $300 an ounce, that much gold is worth about half a million dollars.

The total amount of gold dissolved in the world's oceans is vast, but there is no known economical way to extract and concentrate these riches. The equipment and energy required to process that much water would cost far more than the value of any gold recovered. ■

The Rock Cycle

When the Earth first formed there were no rocks. Four and a half billion years ago the great bombardment, the process that built the Earth from the solar nebula, released prodigious amounts of energy as swarms of meteorites crashed into the growing planet, converting gravitational potential energy into heat. That heat produced a molten ball orbiting the Sun. There was no land, no oceans, and no atmosphere. Only when the bombardment subsided and the Earth began to cool did rocks appear. First, as the temperature dropped below the melting point of the surface rocks, the outer crust of the Earth must have gradually solidified, like the first layer of ice on a pond in winter. Then, when surface temperatures dropped below the boiling point of water, the first rains must have fallen. Together, these two events began the **rock cycle,** a cycle of internal and external Earth processes by which rock is created, destroyed, and altered.

Igneous Rocks

Igneous rocks, which solidify from a hot liquid and thus were the first to appear on the Earth, come in two principal types. **Volcanic** or **extrusive rocks** solidify on the Earth's surface in what are by far the most spectacular of all rock-forming events, volcanic eruptions. Red-hot fountains and flows of lava ooze down the slopes of the growing volcanic cone. The most common variety of volcanic rock is *basalt,* a dark, even-textured rock rich in oxides of silicon, magnesium, iron, calcium, and aluminum. Basalt makes up most of the rock in Hawaii, as well as most of the new material formed at mid-ocean ridges. Other volcanoes feature rocks richer in silicon; if these magmas mix with a significant amount of water or other volatile (easily boiled) substance, the volcanic rock can become the frothy rock *pumice.*

Igneous rocks that harden underground are called **intrusive rocks.** Dark-colored basalt often exploits underground cracks near volcanoes to form layers or sheets of igneous rock. The Palisades on the Hudson River near New York City formed in this way. Lighter in color and density, *granite* is perhaps the most common intrusive rock in the Earth's crust. Hard, durable granite, with its attractive pale grayish or pinkish color and speckled array of light and dark minerals, makes an ideal ornamental building stone. New England is particularly famous for its many fine granite quarries.

(a)

(b)

Igneous rocks are still being formed on Earth—for example, when new plate material is formed at diverging boundaries (see Chapter 15) or in active volcanoes. In other places, such as the Yellowstone Park region, hot springs and geysers reveal hidden sources of underground heat and may indicate places where intrusive igneous rocks are forming today.

Igneous rocks. (a) Devil's Tower in Wyoming represents the neck of a would-be volcano that may have never quite reached the surface. The surrounding sediments have subsequently eroded away. (b) Granite from a quarry in Barre, Vermont, solidified from a magma deep underground.

Sedimentary Rocks

When the first rains began to fall on the first igneous rocks, the process of weathering began. Small grains washed off the recently hardened volcanic rocks, flowed down through streams and rivers into the seas, and were deposited on the seafloors when the fast-moving waters of the rivers met the slower currents of the oceans. Over time, layers of sediment accumulated, especially at the mouths of rivers near the shores of the Earth's new oceans. As more and more sediment collected, these layers became thicker and thicker. In many places on the Earth right now—the Mississippi River delta that extends into the Gulf of Mexico, for example—layers of sediment may reach several kilometers in thickness.

As the first sediments were buried deeper and deeper, their temperature and the pressure on them increased. In addition, water flowed through the layers of sediments, dissolving and redepositing gluelike chemicals—something like the crusty deposits that can build up on an ordinary faucet when water drips continuously. The net result of all of these processes—pressure, heat, and the effects of mineral-laden water flowing between the grains—was to weld the bits of sediment together into new layered rocks. This kind of rock, appropriately called **sedimentary rock,** is made up of grains of material worn off previous rocks. Other common sedimentary rocks, including salt deposits, may form from layers of chemical precipitates.

While uniform sedimentary rocks can form at the base of a single mountain or cliff, the collection of grains often comes together from

Figure 16–8

A spectacular example of sedimentary rocks in Utah. The different colored bands correspond to layers of different kinds of materials being deposited on the floor of a long-vanished ocean. The surrounding sediments have been eroded away to provide material for new sedimentary rocks downstream.

many different places. The grains in a single fragment of sedimentary rock being formed in the Mississippi delta, for example, may have come from a cliff in Minnesota, a valley in Pennsylvania, and a mountain in Texas. Similarly, sediments deposited near the mouth of the Colorado River carry bits of history from much of the North American West. Deltas inevitably contain particles from all the rocks in their rivers' drainage area.

As you travel across the United States you will encounter many common varieties of sedimentary rock. They are easy to spot in road cuts and outcrops because of their characteristic layered appearance, like the pages of a book or a many-layered cake (Figure 16–8). *Sandstone* forms mostly from sand-sized grains of quartz (silicon dioxide or SiO_2), the commonest mineral at the beach, and from other hard mineral and rock fragments. Many sandstones represent ancient beaches, deserts, or stream beds—places where concentrations of sand are found today. Sandstones usually feel rough to your touch, and you can just barely see the individual grains that have been cemented together.

Shales and *mudstones* form from sediments that are much finer-grained than sand. These rocks commonly accumulate beneath the calm waters of lakes or in the deep ocean basins, places often teeming with life. There organisms, both large and small, die and are buried in muddy ooze, where they may eventually form into fossils (see Chapter 24) that provide us with much information about the evolution of life on Earth.

Limestone, another distinctive type of sedimentary rock, forms from the calcium carbonate ($CaCO_3$) skeletons of sea animals. Some limestones grow from a gradual rain of microscopic debris or broken shells, while others represent a coral reef that spread across the floor of a shallow sea. Like shales and mudstones, limestones commonly bear fossils.

Given what we know about plate tectonics, and about the constant movement of materials around the surface of the Earth, it should come as no surprise that just because sedimentary rocks originally formed at the bottom of the ocean, they have not necessarily stayed there since their formation. Indeed, it is not at all unusual to see sedimentary rocks in mountain passes thousands of meters above the ocean or in the middle of continents thousands of kilometers from the nearest open water. Some of the best exposures of limestone that the authors have ever seen are in downtown Nashville, Tennessee, and on the cliffs of Lookout Mountain high above Chattanooga, Tennessee, nearly as far from the ocean as it is possible to get in modern North America.

One of the best places to get an appreciation of sedimentary rocks is at the beach. If you pick up a handful of sand, you will notice that each grain is different. Some are dark colored, some are light. Some have sharp, angular edges, some are smooth and worn down. Each of these grains of sand once was part of a rock in a drainage of the rivers that feed into the ocean. As the rock weathered away, each grain was chipped off and carried to the sea by wind and water. Eventually, the grains of sand you hold in your hand will be formed into solid rock and subjected to the forces of plate tectonics. That sandstone may some day be uplifted to an altitude well above sea level, where the grain may be weathered again and start the whole cycle all over. Each grain of sand in your hand, then, may have made the trip from rock to beach to sandstone many times in its life.

The story of the grain of sand provides a good model for the way materials move about the surface of the Earth. The atoms of rocks, just like those of the air, water, or your body, are always shifting around, but it is always the same matter—the same atoms—that are recycling.

Coral Reefs: A Special Kind of Rock ▉

Coral reefs are the ocean homes of countless billions of creatures that build their homes, bit by bit, from calcium carbonate. Reefs thrive in shallow, clear ocean water with temperatures above 18°C (about 36°F). *The Pelican Island* by British poet James Montgomery (1771–1854) captures the extraordinary phenomenon of entire oceanic islands rising from this biological process:

The Great Barrier Reef in Australia is a location of active limestone formations.

I saw the living pile ascend,
The mausoleum of its architects,
Still dying upwards as their labours closed;
Slime the material, but the slime was turn'd
To adamant, by their petrific touch;
Frail were their frames, ephemeral their lives,
Their masonry imperishable. . . .
  Atom by atom, thus their burthen grew,
Even like an infant in the womb, till Time
Deliver'd ocean of that monstrous birth,
—A coral island, stretching east and west.

Hundreds of millions of years ago, massive limestone reefs like the one in Montgomery's poem thrived in shallow seas—in New York, Montana, Texas, and other places where limestone mountains now stand.

Metamorphic Rocks

It may happen that sedimentary rocks are buried deep within the Earth and thus subjected to intense pressure and heat. There they will be turned into yet another kind of rock, transformed by the Earth's extreme conditions into **metamorphic rock.** If shales or mudstones are buried like this they may eventually turn into brittle, hard *slates,* the kind of rock that school blackboards used to be made from. Even higher temperatures and pressure can transform slates into spectacularly banded rocks, called *schists* and *gneisses,* which often boast fine crystals of garnets and other high-pressure minerals. Roadcuts and outcrops of these metamorphic rocks can look like an intensely folded cloth or a giant cross-section of swirled marble cake. Sandstones, when exposed to high temperature and pressure also metamorphose, recrystallizing to a durable rock in which the original sand grains fuse into a solid mass known as *quartzite.*

The Green Mountains in New Hampshire are made of metamorphic rocks— rocks that have been altered by intense pressure and temperature deep within the Earth, and then uplifted and eroded into mountains.

The Story of Marble

Of all the metamorphic rocks, none tells a more astonishing tale than *marble,* a rock of extraordinary beauty. If you ever travel the roads of Vermont, chances are you will pass an outcrop or roadcut of distinctive greenish-white cast—a rock with intricate bands and swirls. These marbles take a high polish and have been prized for centuries by sculptors and architects. But no works of humans can match the epic process that formed the stone.

Most marbles began as limestone, rocks that originate primarily from the skeletal remains of sea life. Over the ages these limestones in the area we now call Vermont were buried deeper and deeper, crushed under the weight of many kilometers' thickness of sands and shales and more limestones in an ancient sea. But no ocean or sea can last forever in our dynamic planet. An ancient collision of the Eurasian and North American plates compressed and deformed this ocean basin, crumpling the layered rock into tight folds and subjecting the sedimentary pile to intense temperatures and pressures. The buckled and contorted formations were uplifted to high elevations when the Appalachian Mountains formed. During the intense pressures and high temperatures associated with the converging tectonic plates, the limestones were metamorphosed to the marble that we use today. Many millions of years of erosion and uplift have exposed these ancient rocks, gradually to weather away and begin the cycle again. And humans, in a futile quest for immortality, quarry the marble for their monuments and tombstones and other transient reminders of Earth's incessant change.

Igneous, sedimentary, and metamorphic rocks all participate in the rock cycle. Igneous rocks, once formed, can be weathered to form sedimentary rocks, and can themselves undergo metamorphism. Layers of sedimentary rocks also can be transformed into metamorphic rocks. All three kinds of rocks can be subducted into the Earth, partially melted, and reformed as new igneous rocks. Thus the rock cycle never ceases.

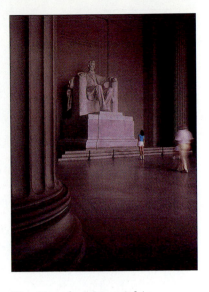

The Lincoln Memorial is one of many famous monuments carved from marble—a metamorphosed limestone.

Science in the Making

James Hutton and the Discovery of ''Deep Time''

Near the town of Jedburgh in Scotland is a curious cliff that reveals vertical layers of rock overlain by horizontal layers. How could such a sequence have occurred? In the last decades of the eighteenth century, Scottish scientist James Hutton (1726–1797), a man who is often called "the father of modern geology," studied this remarkable cliff and realized that he was seeing the result of an incredibly long period of geological turmoil.

Knowing what you now know about sedimentary rocks, you will realize when you look at this cliff that you are seeing the end product of a long chain of events (Figure 16–9). First, a series of sedimentary rocks was laid down in the usual horizontal fashion, one flat layer on top of another. Then some tectonic activity disrupted those layers, breaking and folding them until they were tilted nearly vertically. Then, as the result of still more tectonic activity, the rocks found them-

Figure 16–9

This series of diagrams shows several stages in the history of the Jedburgh outcrop. (*a*) Layers of sediment were gradually deposited in water. (*b*) Those sediments, deeply buried, were compressed and tilted during tectonic activity. (*c*) Uplift brought those tilted sediments to the surface, where they were partly eroded. (*d*) The rocks subsided, and a new cycle of sedimentation began.

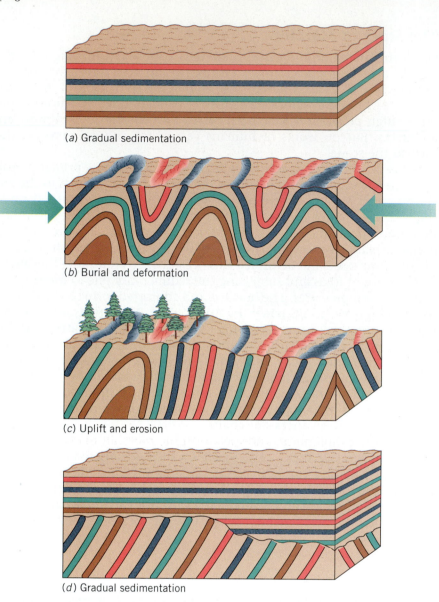

(*a*) Gradual sedimentation

(*b*) Burial and deformation

(*c*) Uplift and erosion

(*d*) Gradual sedimentation

selves at the bottom of an ocean and another layer of sedimentary rocks formed on top of them. Finally, an episode of uplift and erosion has brought the rocks to our view.

Hutton realized that geological forces must have been operating for very long times indeed. Each step of the formation process—gradual sedimentation, burial, folding, uplift, more sedimentation, and so on—would require countless generations, based on observations of ongoing geological processes. In the words of nature writer John McPhee, Hutton had discovered "deep time." In order for a formation like the one at Jedburgh to exist, the Earth had to exist not for thousands of years or even hundreds of thousands of years, but for many millions of years.

Today, we know that the age of the Earth is calculated in billions of years, and the existence of structures like this is not surprising. At

James Hutton recognized the immense spans of time required to form this spectacular outcrop at Siccar Point, Berwickshire, Scotland.

the time of its interpretation by James Hutton, however, the rocks at Jedburgh provided a totally new insight into the inconceivable antiquity of our planet.

In the words of Hutton himself, the testimony of the rocks offered

"No vestige of a beginning, no prospect of an end." ■

The Interdependence of Earth Cycles

We have described the atmospheric, hydrological, and rock cycles as if they were completely independent of each other; as if they operated alone in splendid isolation. In fact, each cycle affects and is affected by the others.

The amount of rainfall in a given location will affect the rate of erosion and thus the amount of sediment being deposited in deltas—and therefore the amount of sedimentary rock being formed. In this way the atmospheric and water cycles affect the rock cycle. In the same way, the breakdown of rock is essential to the formation of soils in which plants grow. The presence of plants, in turn, affects the absorption of sunlight at the Earth's surface and thus the energy balance that controls the movement of the winds and ocean currents. Thus, although each of the three cycles operates on a different time scale, they constantly influence each other.

And, finally, over the span of hundreds of millions of years, the global cycle of plate tectonics, which controls the distribution of the Earth's mountains and oceans, influences all other cycles.

417

THINKING MORE ABOUT CYCLES

Beach Erosion

Something about shorelines appeals to people. Beachfront property is considered to be highly desirable, and over the past half century America's prosperity has resulted in large-scale development of the Atlantic and Gulf coasts. As a result, Americans are becoming very aware of the effects of natural cycles on property values.

Beaches, like any other system in nature, are not static but change in response to environmental forces. Waves lift sand grains up, carry them around, and deposit them somewhere else. Large waves tend to move sand away from a beach and deposit it in offshore bars, while smaller waves tend to move the sand back toward the beach. Thus a seasonal movement of sand occurs on many beaches—offshore in the winter (when storms send in large waves), and onshore during the summer. In addition, waves normally strike a beach at a slight angle, a phenomenon that moves sand along the beachfront. Large storms may completely destroy beaches and dunes, which are rebuilt farther inland over time. Thus every beach is a dynamic, shifting system. Left to itself, a beach will move around in response to the forces of wave and storm.

If the beach were left to itself, this would cause no problems. If, however, waterfront properties worth many millions of dollars have been developed on or near the beach, such movements have enormous economic consequences for homeowners. Should governments use public funds to try to protect such homes? Should the government issue low-cost insurance to indemnify the owners for loss? Should insurance be issued to allow people to rebuild beachfront homes after a storm?

Summary

Matter that forms the outer layers of the Earth follows many cycles, driven by the energy of the Sun and the Earth's inner heat energy. The *atmospheric cycle* of the *weather* redistributes solar energy from the warmer equatorial regions to higher latitudes through the development of global convection cells of air. The prevailing westerly flow of weather across North America marks one of these large cells, while the *jet stream* delineates the boundary between this flow and the contrary cell to our north. The *climate*, in contrast to weather, varies much more slowly in response to ocean circulation, the Sun's energy output, the positions of continents and mountain ranges, and other relatively fixed conditions.

The *hydrological cycle* traces the path of water as it evaporates from the oceans, falls back to Earth as rain, and forms lakes, rivers, *ice caps, glaciers,* and *groundwater* reservoirs. During unusually cold climatic periods more water falls as snow, creating a white reflective blanket that further reduces the amount of absorbed solar radiation. This situation, if prolonged, can lead to an *ice age,* during which ocean levels drop significantly and great sheets of ice cover the land at high and middle latitudes. Temperatures in these latitudes may be moderated, however, by ocean *currents* that are important in redistributing temperatures at the Earth's surface.

The solid materials that make up the Earth's crust are subject to the *rock cycle*. The first rocks to form on the cooling planet were *igneous rocks,* which are formed from hot, molten material. *Volcanic* or *extrusive* igneous rocks solidify on the surface, while *intrusive* igneous rocks cool underground. The first igneous rocks were subjected to weathering by wind and rain, which eventually produced layers of sediment and the first *sedimentary rocks*. Sandstones, shales, limestones, and other sedimentary rocks have been deposited in ocean basins, layer upon layer, in sequences often many kilometers thick. Igneous and sedimentary rocks were subsequently buried and transformed by the Earth's internal temperature and pressure to form *metamorphic rocks*. Each of the three major rock types—igneous, sedimentary, and metamorphic—can be converted into the others by the ongoing processes of the rock cycle.

Key Terms About Earth Cycles

atmospheric cycle	ice age
weather	current
climate	rock cycle
jet stream	igneous rock
hydrological cycle	volcanic or extrusive rock
ice caps	intrusive rock
glaciers	sedimentary rock
groundwater	metamorphic rock

Review Questions

1. What happens to matter during a cyclical process?
2. What is the difference between weather and climate? Describe the weather and climate of your area.
3. What is the prevailing direction of the jet stream? How does it influence your weather?
4. What are the principal repositories of water in the Earth?
5. How does the atmosphere distribute heat across the Earth's surface?
6. What are the differences between ice caps and glaciers?
7. What factors might cause glaciers to advance from polar areas to more temperate zones?
8. How do ocean currents affect local climate?
9. Why do we call groundwater in most areas a "nonrenewable resource"?
10. Why were igneous rocks the Earth's first rocks?
11. What are the three principal kinds of rocks, and how do they form?
12. If you were driving past a large roadcut through rock, what features might you observe to tell you its origin?

Discussion Questions

1. Describe three places where you might find volcanic rocks forming today. Describe three places where you could watch sedimentary rocks forming today. Where would you have to go to watch metamorphic rocks form?
2. The thickness of polar ice caps depends critically on the location of continents; much thicker ice can accumulate on land than in water. Where is the only polar continent now?
3. What kind of rock underlies your campus? Are there any rock outcrops nearby? If so, was your area ever under a large body of water?

4. How do oceans redistribute the Earth's heat? How does the atmosphere accomplish this? Do rocks redistribute heat? Which global cycle do you think is most efficient in transferring heat? Why?
5. You can often distinguish surface ocean currents because their color is different from the surrounding ocean. Why do you think this is so?

Problem

1. How much copper is in a cubic kilometer of seawater? (*Hint:* Refer to Table 16–1.)

Investigations

1. Look at some weather maps in your local newspaper over a period of several weeks and see if observing weather patterns to the west of your location is a good predictor of your weather. Why should this be so?
2. Where does your water come from at your college? Is the water processed or treated in any way? How long is that source of water expected to last? What alternatives exist if that supply is totally depleted?
3. The three kinds of rocks—igneous, sedimentary, and metamorphic—are described in this chapter as being quite distinct, yet some earth scientists have engaged in an intense debate about the origins of certain rocks, such as granites, that formed at high temperature deep within the Earth. Some scientists claim that these rocks are igneous, while others say they are metamorphic. How could such a debate arise, and how could it be resolved? (*Hint:* Think about making taffy.)
4. Investigate the biological cycle of calcium in your body. Where in your body are calcium atoms used? How often are they replaced? How much calcium do you need to consume each day? What are the best food sources of this element?
5. What are the "doldrums" and how do they form? What role have they played in poetry and literature?

Additional Reading

Bascomb, Willard. *Waves and Beaches.* New York: Doubleday, 1964.

Trefil, James. *A Scientist at the Seashore.* New York: Scribners, 1984.

———. *Meditations at Sunset.* New York: Scribners, 1987.

van Andel, Tjered. *Tales of an Old Ocean.* New York: Norton, 1974.

The Stars

The Sun and other stars use nuclear-fusion reactions to convert mass into energy. Eventually, when a star's nuclear fuel is depleted, it must burn out.

A Random Walk
A Shopping Trip

Think about the remarkable range of elements you can buy at your local shopping center. The hardware store stocks aluminum siding, copper wire, and iron nails. The drugstore sells iodine for cuts, zinc and calcium compounds as dietary supplements, and perhaps bottles of oxygen for patients with breathing difficulties. The jeweler displays rings and necklaces of silver, gold, and platinum set with diamonds, a pure form of carbon. And your local electronics dealer offers an amazing assortment of audio and video equipment made possible by integrated circuits of silicon, perhaps doped with small amounts of aluminum or phosphorus.

Where did all these elements come from? Were they always part of the universe, or was there a time in the history of the universe when most of them weren't around? It may come as a surprise to you to learn that the answers to questions like these lie in studies of the stars.

■ Energy and the Stars

Astronomy, the study of objects in the heavens, is perhaps the oldest science. When you examine the night sky, the most striking feature is the thousands of visible stars. Each **star** is an immense fusion reactor in space. The Sun, the nearest star to Earth, is just one of countless trillions of stars in our universe.

You have already learned enough about the way the universe works to understand some of the implications of your nighttime view of the stars. The fact that you can see the stars, for example, means that they are emitting electromagnetic radiation (see Chapter 6). Every star you see sends radiant energy into space. Your eye intercepts a tiny fraction of that energy and converts it into the image you see, but similar amounts of energy radiate out in every direction from the star.

The laws of nature, including the laws of thermodynamics that describe the behavior of energy, apply everywhere in the universe. If a star sends energy out into space, then the star must have a source of that energy. Furthermore, since every star is a finite object, it must have a finite store of energy. This simple observation leads to perhaps one of the most profound insights about the universe.

> **All stars have a beginning and an ending.**

In other words, the magnificent display that we see in the nighttime sky is a temporary phenomenon. It has lasted from the time that stars first formed until the present, a span of perhaps 15 billion years. However, the view we see in the heavens will not last forever. Each star will eventually run out of energy.

Science by the Numbers

The Sun's Total Energy Budget

To understand the energy budget of stars, we must first determine how much energy a star releases. If you go outside on a warm day, you can feel the heat of the Sun on your body. In fact, if you measured the sunlight you would find that a square meter on the Earth's surface receives an average of about 750 watts (750 joules/second), just about enough energy to light your desk and heat your room on a cool evening. Above the Earth's atmosphere, away from the absorbing effects of air and dust, a square meter receives about twice that much, or 1.5 kilowatts.

Imagine an immense sphere centered on the Sun with the surface passing through the Earth (see Figure 17–1). The Earth's orbit would be a line on the sphere's circumference. None of the Sun's radiant energy can disappear during its transit between the Sun and this imaginary sphere, so the total amount of energy going through the sphere is exactly the same as the total amount of energy generated by the Sun. Every single square meter on this sphere will receive the same amount of energy as a square meter located above the Earth's atmosphere.

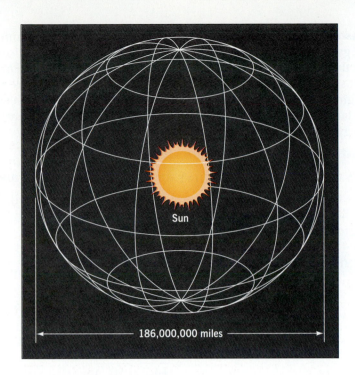

Figure 17–1
An imaginary sphere centered about the Sun with a radius equal to the Sun–Earth distance. Each square meter of this sphere would receive the same amount of energy each second as a one-meter-square panel above the Earth's atmosphere. The total energy passing through the sphere is the same as the total number of square meters in the sphere (its area) multiplied by the energy passing through each square meter.

First, we must calculate the total area of this immense sphere with a radius equal to the Sun-to-Earth distance—about 150 million kilometers (1.5×10^{11} m), or 93 million miles. The surface area of a sphere is given by

$$\text{area} = 4 \times \pi \times \text{radius}^2,$$

where π is the constant 3.14. We can calculate how many square meters are in this imaginary sphere:

$$\begin{aligned}
\text{area} &= 4 \times 3.14 \times (1.5 \times 10^{11}\ \text{m})^2 \\
&= 4 \times 3.14 \times 2.25 \times 10^{22}\ \text{m}^2 \\
&= 28.3 \times 10^{22}\ \text{m}^2 \\
&= 2.83 \times 10^{23}\ \text{m}^2
\end{aligned}$$

The total energy of the Sun is simply this area times the energy per square meter at the distance of the sphere.

$$\text{Sun's total energy} = (\text{energy per meter}^2) \times (\text{sphere's total area})$$
or
$$\begin{aligned}
\text{energy} &= (1.5\ \text{kilowatts/meter}^2) \times (2.83 \times 10^{23}\ \text{meter}^2) \\
&= 4.24 \times 10^{23}\ \text{kilowatts.}
\end{aligned}$$

This number is astronomical in every sense of the word. ◼

EXAMPLE 17–1: Moving Closer: The Sun's Energy and Venus

What is the energy per square meter from the Sun on the planet Venus, which is about 108 million kilometers from the Sun (the next nearest planet from Earth to the Sun)?

▶ **Reasoning:** The key idea to remember is that the total amount of energy passing the orbit of Venus at a radius of 108 million km is exactly the same as the energy passing the orbit of the Earth at a radius of 150 million km—4.24×10^{23} kilowatts. The area of the imaginary sphere at Venus's orbital radius is much smaller than a similar sphere at the Earth's orbital radius, so Venus receives more energy per square meter.

▶ **Solution:** From the equation above,

Sun's total energy = (energy per meter²) × (sphere's total area),

where, as before, the area of a sphere at the orbit of Venus will be $4 \times \pi \times r^2$, where r is the distance between Venus and the Sun.
Therefore:

$$
\begin{aligned}
\text{(energy per meter}^2 \text{ at Venus)} &= \frac{\text{Sun's total energy}}{\text{sphere's total area at Venus}} \\
&= \frac{4.24 \times 10^{23} \text{ kilowatts}}{4 \times 3.14 \times (1.08 \times 10^{11} \text{ m})^2} \\
&= \frac{4.24 \times 10^{23} \text{ kw}}{4 \times 3.14 \times (1.17 \times 10^{22} \text{ m}^2)} \\
&= \frac{4.24 \times 10^{23} \text{ kw}}{14.7 \times 10^{22} \text{ m}^2} \\
&= 2.89 \text{ kilowatts per m}^2
\end{aligned}
$$

Thus Venus, which is about three-fourths the Earth's distance to the Sun, receives twice the energy per square meter. ▲

The Sun's Energy Source: Fusion

Once people understood conservation of energy—that energy has to come from somewhere, and that the total amount of energy in a closed system is constant—two key questions about the stars arose naturally: What is their energy source? How can they continue to burn, emitting huge amounts of energy into space, yet remain seemingly unchanged for such long periods of time?

In the nineteenth century, several scholars attempted to explain the Sun's energy source. One astronomer, for example, calculated how long the Sun could burn if it were composed entirely of the best fuel available at that time—anthracite coal. (The answer turns out to be about 10,000 years.) The Sun's energy source also figured in the famous debate at the end of the nineteenth century regarding the age of the Earth (see Chapter 3).

Today, we understand that the Sun is indeed "burning" a fuel, but that fuel is hydrogen, which is consumed through nuclear fusion (see Chapter 11). The stars are born in the depths of space, in clouds of gases and other debris. In such a cloud, matter happens to be more concentrated in some regions than in others. These regions of higher mass concentration attract their neighbors by the ordinary force of gravity. Under the influence of gravity, clumps of matter come together, some eventually building up to very large masses.

As a new star begins to form, and as more and more mass pours into it from the surrounding region, the pressure and temperature at the center begin to climb. The new star will be made primarily of hydrogen gas, since that is the most common material in the universe. As a star forms and heats up, the electrons are torn from the hydrogen and other atoms, creating a *plasma* made up primarily of protons (the nucleus of the hydrogen atom) and electrons.

Normally, protons would repel each other (see Chapter 5). As matter accumulates in the new star, however, the protons move faster as they are pressed closer and closer together under tremendous gravitational pressure. Eventually, they acquire enough energy to overcome the electrical repulsion between them. They start to fuse.

The fusion process in the Sun's core does not take place all at once, with four particles suddenly coming together to make a helium-4 nucleus. Instead, it takes place in three steps.

STEP 1: $P + P \rightarrow D + e^+ + neutrino$
> Two protons come together to form a deuterium nucleus (an isotope of hydrogen made up of one proton and one neutron), a positron (e^+—the anti-particle of the electron as described in Chapter 12), and a neutrino.

STEP 2: $D + P \rightarrow {}^3He + photon$
> Another proton collides with the deuterium produced in the first step to form ${}^3He$, the isotope of helium that has two protons and one neutron in its nucleus. A photon in the form of an energetic gamma ray is also produced.

STEP 3: ${}^3He + {}^3He \rightarrow {}^4He + 2\ protons + photon$
> Two ${}^3He$ nuclei collide to form ${}^4He$, two protons, and a photon (another gamma ray).

The net effect of this three-step process, called *hydrogen burning*, is that four protons are converted into a ${}^4He$ nucleus with a few extra particles thrown in. As we saw in Chapter 11, the sum of the masses of all the particles produced in this reaction amounts to *less* than the mass of the original four protons. The lost mass has been converted into energy—the nuclear energy that powers the Sun and eventually radiates out into space.

How long could the Sun burn hydrogen at its present rate? If you simply add up all the hydrogen in the Sun and ask how long it could last, the answer turns out to be something like 75 billion years. Actually, no star ever consumes all of its hydrogen in this way. The hydrogen burning is generally confined to a small region in the center of the star called the *core*. The best current estimates of the total lifetime of our Sun are about 11 billion years—that is, our star is almost halfway through its hydrogen-burning phase.

Science in the Making

The Solar Neutrino Problem

Our explanation of the Sun's energy source is a theory—a very plausible one—that is subject to experimental verification. In fact, a great deal of observational evidence supports the notion that the hydrogen-burning reactions just outlined account for the Sun's energy. One crucial piece of evidence, however, seems to indicate that we might not know as much about the interior of the Sun as we'd like to think. Beginning in the early 1970s, a large experiment located a mile underground in a gold mine in Lead, South Dakota, has been returning results that are puzzling, to say the least. The purpose of this experiment is to detect the neutrinos that should be given off in the Sun's nuclear reactions. Almost all of these neutrinos are expected to pass right through the outer layers of the Sun, and most of the time they pass right through the Earth as well. Occasionally, however, one will interact with an atom in the apparatus of this experiment, so that it can be detected. Thus the experiment provides us with a "telescope" that can "see" right down to the center of the Sun. If we believe we know what reactions are going on there, we should be able to predict how many neutrinos should be seen.

When these measurements are taken, scientists see only about one-third to one-half of the number of neutrinos they had expected to see. In the late 1980s, other experiments of this type were set up in Europe and in the former Soviet Union, and these experiments seem to bear out the original results. The missing one-half to two-thirds of the expected neutrinos are known as the "solar neutrino problem."

One possible explanation of the solar neutrino problem is tied to the nature of elementary particles such as neutrinos, and the fact that there are three different kinds of neutrinos (see Chapter 12). According to this idea, the predicted number of ordinary neutrinos is produced in the Sun, but while they're in transit to the Earth, some of them are converted into the two other kinds of neutrinos (the ones associated with the mu and tau particles). Such transitions between neutrino types are possible in some versions of modern unified field theories. In this way, by the time the neutrino stream reaches the Earth, only a fraction of the original particles will be ordinary neutrinos capable of interacting with ordinary atoms in an experimental apparatus. Such a solution to the problem would be enormously satisfying to astrophysicists, but only time will tell if it is the correct one. ■

Technology

Dyson Spheres, an Ultimate Technology

Sometimes it's fun to speculate about how an extremely advanced technological civilization would use the energy from a star. One approach to the question is to note that most of the energy generated by stars is radiated into space and, from the point of view of planet dwellers, wasted. Physicist Freeman Dyson has suggested that one way to avoid this energy loss would be to dismantle a large planet and convert its material into a sphere surrounding a star—a hypothetical structure now called a Dyson sphere. A Dyson sphere would not only intercept and utilize all of a star's energy, it also could support a tremendous population. Such an immense structure would have an internal surface area comparable to more than 500 million Earthlike planets, and it would be uniformly heated in eternal daylight.

A star enclosed in a Dyson sphere would radiate its energy in the visible and higher energies, as usual, but all that energy would be absorbed by the sphere, which would then be warmed and radiate the energy into space as infrared radiation. Scientists have suggested that one way to search for advanced extraterrestrial civilizations would be to look for the characteristic infrared signal. ■

The Anatomy of Stars

Stars are much more than uniform balls of gas. They have a complex and dynamic interior structure that is constantly changing and evolving. The Sun is the only star that is near the Earth. It is, in fact, the only star for which we have the kind of detailed knowledge that allows us to talk about how the various parts function. In this sense, the Sun is not only the giver of life on our planet, but also the giver of knowledge. Without such a star close by to study, we would probably not know much about the billions of stars that are farther away.

The Structure of the Sun

We believe that the Sun is a rather ordinary star. At its center is the stellar core, comprising about 10% of the Sun's total volume. This core is the Sun's furnace, where nuclear fires rage, and energy generated in the core streams out from the center. Deep within the Sun, this energy transfer takes place largely through the collisions of the high-energy particles— protons and positrons, for example—that are generated by the core's nuclear reactions.

About four-fifths of the way out, however, the energy-transfer mechanism changes, and the hydrogen-rich material in the Sun begins to undergo large-scale convection. This outer region, comprising the upper 200,000 kilometers (about 125,000 miles) of the Sun, is called the *convection zone*. Thus energy is brought from the core to the surface in a stepwise process—first by collisions, then by convection.

The only part of the Sun that we actually see is a thin outer layer. We can peer perhaps 150 kilometers (about 100 miles) into the Sun; any deeper and the stellar material becomes too dense to be transparent. The outer part of the Sun—the part that actually emits most of the light we see—is called the *photosphere*. The Sun does not have a sharp outer boundary, but gradually becomes thinner and thinner farther away from the surface. These gaseous layers are not usually visible from the Earth. But during a total eclipse of the Sun, when the Moon passes in front of the Sun, the Sun's spectacular halo—called the *chromosphere* and the *corona*—may become visible for a few minutes (see Figure 17–2).

The Sun constantly emits a stream of particles—mainly ions (electrically charged atoms) of hydrogen and helium—into space around it. This stream of particles, called the **solar wind**, blows by the Earth all the time. Because the particles are charged, they affect the magnetic fields of the planets, compressing the fields on the "upstream" side and dragging them out on the "downstream" side (see Figure 17–3). The interaction of the solar wind with the outer reaches of the Earth's atmosphere also gives rise to the aurora borealis, or northern lights.

Figure 17–2
The Sun's chromosphere and corona become visible during a total eclipse. This halo of incandescent plasma is normally blotted out by light from the Sun's main disk.

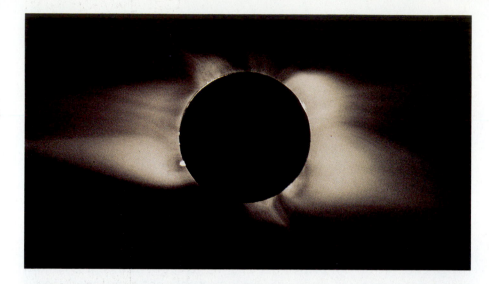

Figure 17–3
The magnetic field of the Earth is swept out into a long tail by the solar wind.

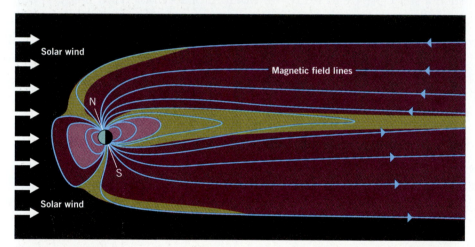

Northern lights.

The flow of energy from the Sun, then, is a complex affair. Beginning with the conversion of mass in fusion reactions, the energy slowly percolates outward, first in collisions and later in great convection cells under the solar surface. It takes a few tens of thousands of years for the energy to work its way to the photosphere, but only eight minutes to cover the distance between the Sun and the Earth.

Once the sunlight reaches our planet, a tiny fraction of it is converted by the process of photosynthesis in plants (see Chapter 21) into chemical energy stored in large molecules. This energy is the primary source of motive power for all living things on the planet.

The Human Body

Why Is the Visible Spectrum Visible?

In Chapter 5 we saw that, of all the possible waves in the electromagnetic spectrum, the human eye can detect only the small interval of wavelengths between red and violet. One reason why the human eye is made this way was discussed in Chapter 5—the Earth's atmosphere is transparent to these wavelengths, so it is possible for the waves to travel long distances through the air. But there is another aspect to the development of the human eye that has to do not with the Earth, but with the Sun. Because of the fusion reactions at its core, the temperature of the outer part of the Sun is quite high—about 5500°C, (for reference, the melting point of gold is about 1065°C). Every object above absolute zero radiates electromagnetic waves, and both the total amount of energy and the wavelengths of the radiation depend on the body's temperature.

In Figure 17–4, we show the amount of energy that the Sun radiates at each wavelength, with the visible spectrum represented by the

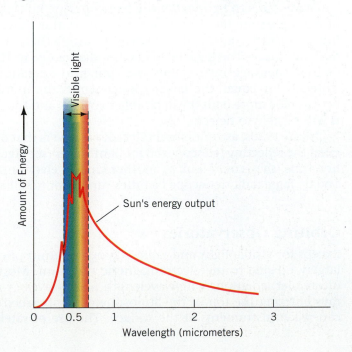

Figure 17–4
The Sun's peak output of energy is in the middle of the visible spectrum, with lesser amounts of energy emission at different wavelengths.

vertical lines. We see that the Sun's peak output of energy is in the middle of the visible spectrum. Thus the wavelengths between red and violet are visible for two reasons—the air is transparent to them, and the Sun, our main source of light, emits the greatest proportion of its energy in that form.

Science fiction writers often use this fact when they portray imaginary beings from other planets. Those from planets around cooler stars than the Sun may be given large eyes so they can absorb more of the scarcer photons. For the reasons cited above, however, humans don't require such large optical collectors. ■

Measuring the Stars with Telescopes and Satellites

Our only source of data on distant stars is electromagnetic radiation—energy streaming through space at 3×10^8 meters per second. To collect and analyze radio waves, microwaves, light, and other radiation, astronomers have devised a variety of **telescopes,** devices to focus and concentrate radiation from distant objects.

Earth-Based Telescopes

The first telescopes could examine only visible light, and when most people use the word telescope today, they mean an instrument for gathering and concentrating this form of radiation. The classical reflecting telescope (see Figure 17–5a) has a large mirror that reflects and focuses light to produce an image of the object being studied. (See Chapter 6 for a review of the reflection and refraction of light.)

Many modern light-gathering telescopes are built differently. Instead of having a solid block of glass for a mirror, they have a series of small, independently controlled, lightweight mirrors that, taken together, produce an image. The Keck telescope on Mauna Kea in Hawaii, the biggest of these new-age instruments, was put into operation in 1992. Other things being equal, the larger the telescope, the more useful it will be because the more light it will be able to collect and the fainter the objects it will be able to detect.

In the 1930s, astronomers built radio receivers that did for radio waves what the reflecting telescope did for light waves (see Figure 17–5b). For the first time, they could look at another kind of electromagnetic radiation. Today, large radio telescope facilities can be found all around the world.

Orbiting Observatories

Except for visible light and radio waves, the atmosphere of the Earth is largely opaque to the electromagnetic spectrum. Most infrared and all ultraviolet radiation, long-wavelength microwaves, X-rays, and gamma rays entering the top of the atmosphere are absorbed long before they can reach instruments at the surface. The use of satellite observatories

Several telescopes take advantage of the excellent conditions at the summit of Hawaii's Mauna Kea.

The Very Large Array in Socorro, New Mexico, features a Y-shaped arrangement of linked radio telescopes. The effect of many small telescopes is equivalent to one very large telescope.

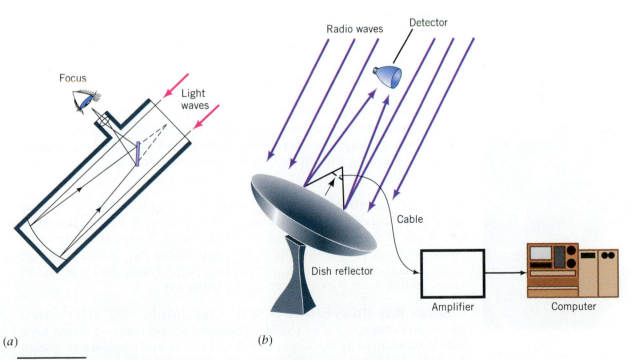

(a)   (b)

Figure 17–5

Schematic diagrams of telescopes. In an optical telescope (a) light strikes a curved mirror and is focused on a light-sensitive detector such as the eye or a piece of film. In a radio telescope (b) radio waves from space strike a curved metal dish that focuses the waves onto an antenna. Signals are amplified and processed by computer.

during the last half of the twentieth century has ushered in a golden age of astronomy. In this period, we have been able to put instruments into orbit far above the atmosphere and, for the first time, record and analyze the entire flood of electromagnetic radiation coming to us from the cosmos. Following are descriptions of some particularly important orbiting observatories, both current and historical.

Cosmic Background Explorer (COBE)   The Cosmic Background Explorer measures microwave radiation that is present as background noise in every direction of the sky. COBE's data have played a significant role in our understanding of the early history of the universe (see Chapter 18).

Infrared Astronomical Satellite (IRAS)   Launched in 1983 by the United States, United Kingdom, and Netherlands, IRAS functioned for nine months and gave us our first view of infrared radiation in the universe. Since every object in the universe, including the infrared telescope itself, emits some infrared radiation, parts of IRAS had to be cooled to within a few degrees of absolute zero with liquid helium in order to prevent the telescope from "seeing itself." IRAS photos gave us new views of the dust clouds that clog the central disks of galaxies and of regions in the Milky Way where new stars are forming. IRAS stopped functioning only when it ran out of liquid helium—as one commentator remarked, it was "good to the last drop."

Hubble Space Telescope (HST)   Launched in 1990, HST is a reflecting telescope with a 2.4-meter mirror designed to give unparalleled resolution in the visible and ultraviolet wavelengths. Despite the fact that flaws in the manufacturing of the main mirror prevented some planned observations and garnered a great deal of bad publicity, HST made an impressive series of discoveries even before a new optical system was installed by astronauts in late 1993. Among its observations detailed photographs of storm systems on Saturn, and discoveries of details of how stars form in dust clouds. Perhaps the most striking Hubble photographs were those of Jupiter, showing the effects of the impacts of comet Shoemaker–Levy in July, 1994.

Roentgen Satellite (ROSAT)   An X-ray satellite launched in 1990 by the United States, United Kingdom, and Germany, ROSAT is the latest in a series of satellites equipped to detect X-rays. X-rays are emitted by energetic processes; hence the X-ray map of the sky indicates where violent, high-energy events are occurring. ROSAT will, over a period of years, complete an X-ray survey of the entire sky.

Gamma Ray Observatory (GRO)   Launched in 1991, GRO detects the highest energy end of the electromagnetic spectrum—gamma rays, which are emitted by black holes, exploding stars, and some active distant galaxies called quasars (see Chapter 18). GRO and HST are the first two telescopes in the National Aeronautics and Space Administration's (NASA's) Great Observatories Program, whose ultimate goal is to have permanent orbiting observatories monitoring all parts of the electromagnetic spectrum. Two such observatories now in the planning stage at NASA are SIRTF (Space Infrared Telescope Facility) and AXAF (Advanced X-Ray Astronomy Facility).

Astronaut F. Story Musgrave helped to repair the Hubble Space Telescope (HST) during a space shuttle mission in December, 1993. The quality of HST images improved dramatically following the repairs.

The Zoology of Stars

Knowing how one star, our own Sun, works should help us understand the other stars we see. When you look at the stars in the night sky, one of the first things you notice is that they don't all look the same. Some are very bright; some are barely visible. Some seem to have a reddish color; some are almost blue. Part of these differences in appearance arises because the stars are different distances from the Earth. An unusually bright star located far away will appear dim to us, while an average star located nearby might appear very bright. The brightness of a star is also related to the amount of energy the star is producing. Astronomers often refer to the total energy emitted by a star as its *luminosity*.

Differences in the appearances of stars due to distance are taken into account by distinguishing between a star's **apparent magnitude,** which is the brightness it appears to have when viewed from the Earth, and its **absolute magnitude,** which is the brightness it would have if viewed from a standard distance. The brightest stars are said to be first magnitude, the next brightest second magnitude, and so on. The human eye can see down to sixth magnitude, and telescopes routinely detect stars down to magnitude 23.

Much of the differences in the stars' appearance, though, arises from the many varieties of stars themselves. Some stars shine a thousand times brighter than the Sun, while others are a thousand times dimmer. Some stars contain 40 times more mass than the Sun, while others have much less. As we shall see, we can bring order to this tremendous diversity of stars by recognizing that the behavior of every star depends primarily on just two factors: its total mass and its age.

The Astronomical Distance Scale

When we look at the sky, we see a two-dimensional display—all stars look equally distant. To add the third dimension to this picture, we must find the distance to the stars. Astronomers customarily measure these great distances in **light-years**—the distance light travels in one year, or about 10 trillion kilometers (about 6.2 trillion miles). Alternatively, they may use a unit called the *parsec*, a distance of about 3.3 light-years. This convenient measure roughly corresponds to the average distance between nearest neighbor stars in our galaxy.

In practice, no single method can be used to find the distance to every star. Just as you might use a ruler, a tape measure, and a surveyor's tape to measure successively larger distances, astronomers measure distances to stars with a series of "yardsticks," each appropriate to a particular distance scale.

For short distances (up to a few hundred light-years), several different methods involving simple geometry can be used. For nearby stars, for example, the angle of sight to the star measured at opposite ends of the Earth's orbit (see Figure 17–6) can be used to work out the distance. Navigators on the Earth's surface use a similar method, called *triangulation*, to determine the position of ships.

For greater distances, a standard type of star called a *Cepheid variable* is used. These stars, the first of which was discovered in the constellation of Cepheus, show a regular behavior of steady brightening and dimming

Figure 17–6

The triangulation of stellar distances. By measuring the angle of sight to a given star from two points of known separation, we can determine the star's distance from us.

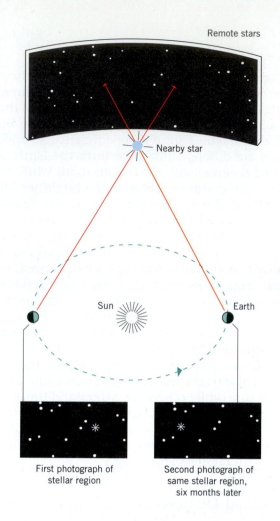

Remote stars

Nearby star

Sun

Earth

First photograph of stellar region

Second photograph of same stellar region, six months later

over a period of weeks or months. Henrietta Leavitt (1868–1921) of Harvard College Observatory showed that the absolute magnitude of these stars is related to the time it takes for them to go through the dimming-brightening-dimming sequence. Thus we can watch a Cepheid variable for a while and deduce how much energy it is pouring into space. This measurement, together with a knowledge of how much energy we actually receive, tells us how far away it is.

The method of Cepheid variables can be used to distances of many millions of light-years and was, as we shall see in Chapter 18, a crucial ingredient in the birth of modern cosmology.

The Hertzsprung–Russell Diagram

Early in this century two astronomers, Ejnar Hertzsprung of Denmark and Henry N. Russell of the United States, independently discovered a way to find order among the diversity of stars. The product of their work, called the *Hertzsprung–Russell (H–R) diagram*, is a simple graphical technique widely used in astronomy. It works like this: On a graph's vertical axis, astronomers plot the amount of energy given off by a star, as measured by estimating the star's distance and brightness. On the graph's horizontal axis, they plot the star's temperature, as determined by its spectrum (see Chapter 6). Each star has its own characteristic combination of energy and temperature, and so it appears as a single point on the Hertzsprung–Russell diagram. The Sun, for example, is one of the points highlighted in Figure 17–7.

When stars are plotted this way, the majority (the ones that are like the Sun) fall on a band that stretches from the upper left to the lower right in the diagram. That is, most stars conform to a trend from very hot stars emitting lots of energy, down to relatively cool stars emitting less energy. Objects in this grouping are called **main-sequence stars**. All of these stars are in the hydrogen-burning phase of their lives, and their energy is derived from the fusion reactions that we described earlier in this chapter.

Two additional clumpings of stars appear in the H–R diagram. One clumping, in the upper right corner, corresponds to stars that emit a lot of energy but whose surfaces are very cool. These stars must be very large so that the low temperature (and low-energy emission of each square foot of surface area) is compensated by the large surface area. Stars of this type are called **red giants,** and they often do appear somewhat reddish in the sky.

Another grouping of stars appears in the lower left corner of the H–R diagram. These stars, called **white dwarfs**, have very low emission of energy but very high surface temperatures; they are very small and very hot.

Both red giants and white dwarfs play crucial roles in the life cycles of stars, as we shall see shortly.

Figure 17–7

A Hertzsprung–Russell diagram plots a star's temperature versus its energy output. Stars in the hydrogen-burning stage, including the Sun, lie along the main sequence, while red giants and white dwarfs represent subsequent stages of stellar life.

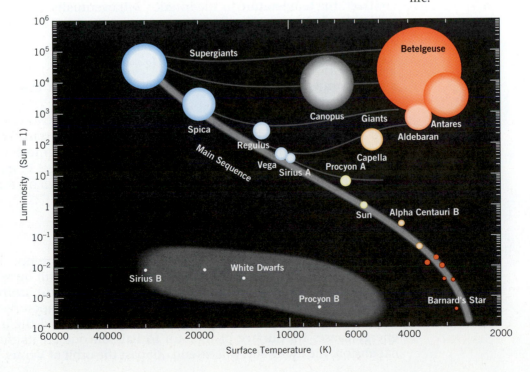

The Death of Stars

Sometime in the future, the core of the Sun will run out of hydrogen, as indeed every star must eventually deplete its hydrogen fuel. In order to understand the spectacular events that occur when a star dies, we have to understand a little more about the lives of stars.

One way to look at the life of a star like the Sun is to think of it as a continual battle against the force of gravity. From the moment when the Sun's original gas cloud started to contract, the force of gravity acted on every particle, forcing it inward and trying to make the entire structure collapse on itself. When the nuclear fires ignited in the core of the Sun 4.5 billion years ago, gravity was held at bay. The increase in temperature in the center raised the pressure in the star's interior and balanced the inward pull of gravity. But, in the long view of things, this balance can only be a temporary state of affairs. The Sun can stave off the inward tug of gravity only as long as it has hydrogen to burn. When hydrogen fuel in the core is depleted, the amount of energy generated in the core will decrease, and gravity will begin to take over. The Sun will begin to contract.

This dramatic situation will have two effects. First, the temperature in the region immediately surrounding the core will begin to rise. Any remaining hydrogen in that region, which had not burned because it had been at too low a temperature, will begin to burn. Thus a hydrogen-burning shell will begin to form around the extinguished core. The second effect is that helium, the "ash" of hydrogen burning, will have become concentrated in the Sun's core. The Sun's structure then begins to look like an onion, with a core of helium surrounded by a hydrogen-burning layer. As the pressure continues to increase, and the Sun continues to contract, the temperature in the interior will eventually get hot enough so that the helium itself will begin to undergo nuclear fusion reactions. The net reaction will be

$$^4\text{He} + {}^4\text{He} + {}^4\text{He} \rightarrow {}^{12}\text{C},$$

a process called *helium burning*, in which the helium in the core burns to make carbon.

This notion, that the ashes of one nuclear fire serve as fuel for the next, is central to an understanding of what goes on in stars. With stars like the Sun, the temperature never gets high enough to ignite the carbon, so the helium burning is the final energy-producing stage. In more massive stars, this process of successive burning cycles can go on for quite a while, as we shall see.

The Death of the Sun

The Sun will burn at more or less its present size and temperature for billions of years more. But, in its final stages, our star will undergo dramatic changes. When the core burns out, the hydrogen-burning shells surrounding the central region are pulled in. This temporary collapse increases the amount of energy generated by fusion, and the increased energy causes the surface of the Sun to balloon out. At its maximum expansion, the dying Sun will extend out past the orbit of Venus. Because the solar wind also increases during this period, however, the Sun's mass drops and the planets move outward. In the end, only Mercury is actually swallowed. During this phase of its life, the Sun will emit its energy through a much larger surface than it does now, and that surface will appear to be very cool—red hot to our eyes. In fact, our Sun will become a red giant, and the helium in the Sun's core will burn to produce an inner core primarily of carbon.

As carbon accumulates in the core, a slow collapse will ensue until some other force intervenes. In the case of the Sun, that force will come from the Pauli principle—the principle (see Chapter 7) that tells us that no two electrons can occupy the same state. As the core starts to collapse, its electrons are compressed into a smaller and smaller volume. They reach the point (what we called the "full parking lot") where they can no longer be pushed together. At this point, the Pauli principle takes over and the electrons' intrinsic need for "elbow room" controls the Sun's size. The collapse stops for the simple reason that the electrons can't be pushed together any closer than they already are.

Astronomers call this countervailing force exerted by electrons the *degeneracy pressure*. They think of it as a permanent outward force on every element in the star—an outward force that cancels the inward force of gravity.

When the Sun reaches this stage, it will be rather small—probably about the size of the Earth—and it will no longer be generating energy through nuclear reactions. It will be very hot and will take a long time to cool off. During this phase, the temperature of each part of the Sun's surface will be very high, but, because the Sun will be so small, the total amount of radiation coming from it will not be very large. It will be, in other words, a white dwarf. The carbon that is the end product of helium burning will remain locked in the white dwarf, and will not be returned to the cosmos.

One way to think about the life cycle of the Sun is to follow the path that it traces out in the H–R diagram. As a nebular cloud, the Sun had a low temperature and low energy. It entered the H–R diagram from the lower right corner. By the time it had become a full-fledged star, it had moved to the main sequence, where it sits now. Six billion years from now, when the Sun runs out of hydrogen, it will move into the upper right corner of the H–R diagram, as a red giant. It will then move quickly down into the lower left corner to become a white dwarf. Any star like the Sun will trace out a similar path on the H–R diagram (see Figure 17–8).

Now that we understand the life story of the Sun, from its birth in a contracting dust cloud to its death as a cooling white dwarf, there is one point that should be noted. During the billions of years that the Sun is on the main sequence, the total amount of energy it gives off stays roughly constant. Since it first entered the main sequence 4.5 billion years ago, for example, the amount of energy generated by the Sun has increased by only about 30%. This long-term stability has important implications for the development of life on this planet.

In Chapter 23, for example, we will see that the existence of oceans early in Earth's history was crucial—without them, there would be no life today. But had the Sun flared up in the past, the oceans could easily have evaporated, turning the Earth into a copy of Venus. Had the Sun

Figure 17–8

The life cycle of the Sun on a Hertzsprung–Russell diagram. The Sun started hydrogen burning in its core more than 4.5 billion years ago on the main sequence (at point 1) and it will remain near that point on the diagram for several billion years more. As the hydrogen in the core is consumed, however, a short period of helium burning (point 2) will move the Sun's position on the diagram rapidly upward toward the red giant stage (point 3). Once the helium is consumed, the nuclear fusion reactions will cease and gravitational contraction will cause the Sun to heat up (point 4). Eventually the Sun will cool to a white dwarf (point 5).

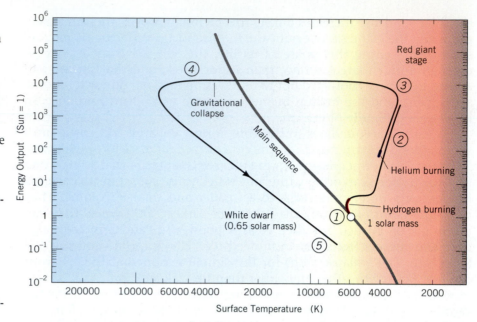

dimmed down, the oceans could have frozen solid, destroying whatever life they held. Thus, although the Sun's birth and death may be spectacular events, life depends critically on the fact that, in between, the Sun has been steady.

The Death of Very Large Stars

Stars up to about eight times the mass of the Sun will carry out the life cycle of main sequence to red giant to white dwarf much as the Sun does. Such stars will have different lifetimes, depending on their size, but will have essentially the same life history.

One of the paradoxes of astronomy is that larger stars—those with the most hydrogen fuel—have the shortest lifetimes. This paradox arises because the largest stars have to burn hydrogen at a prodigious rate in order to overcome the intense force of gravity. Thus a star 10 times as massive as the Sun may live for a very short time—perhaps only 30 million years—compared to the Sun's 11-billion-year span. At the same time, stars much less massive than the Sun may live for hundreds of billions of years, rather than the Sun's 11 billion.

Large stars die differently from the Sun. For these stars, the pressure exerted by gravity is high enough so that the helium in the core not only burns to carbon, but the carbon can also undergo fusion reactions to produce oxygen, magnesium, silicon, and other larger nuclei. For such a star, the successive collapses and burnings will produce a layered onion-like structure such as that shown in Figure 17–9.

In fact, this chain of nuclear burning goes on until iron, the element with 26 protons, is produced. As we noted in Chapter 11, iron is the most tightly bound nucleus. Energy is required to break the iron nucleus apart (nuclear fission) and to add more protons and neutrons to it (nuclear fusion). Thus, it is impossible to extract energy from iron by any kind of nuclear reaction.

The cores of large stars will eventually fill up with iron "ash," and, no matter how high the pressure and temperature get, iron simply will not burn to produce a countervailing force to gravity. In fact, the iron core builds up until the force of gravity becomes so great that even the degeneracy pressure of the electrons cannot prevent collapse. At the incredible pressures and temperatures at the center of the star, the electrons actually combine with protons inside the iron nuclei, forming neutrons, a process that is the exact opposite of radioactive beta decay (see Chapter 11). Within a second or so all of the protons in the iron nuclei are turned into neutrons, and all of the electrons disappear. At this point, the core of the star begins a catastrophic collapse. The collapse will go on until another force appears on the scene to counteract gravity. In this case, the force is provided by the degeneracy pressure of the neutrons which, like electrons, are subject to the Pauli exclusion principle.

The core collapses so fast that it falls inward beyond the point where the degeneracy pressure of the neutrons can balance gravity. Like an acrobat jumping on a trampoline, the star's falling matter first bounces inward, and then rebounds as the neutrons exert a counterpressure. Meanwhile, the outer gaseous envelope of the star has suddenly lost its support and begins a free-fall toward the interior of the star. When the collapsing envelope of dense gas meets the rebounding core of neutrons, intense shock waves are set up in the star, and the entire outer part of the star literally explodes. Someone watching this event from a distance will see a sudden brightening of the star in the sky, usually in a matter of a day or so. We call this dramatic event a **supernova**. Supernovas probably happen about every 30 years in the Milky Way galaxy. We don't normally see all of these events because of intervening dust, but we do see them in neighboring galaxies.

During the explosion, intense shock waves tear back and forth across the exploding star, raising the temperature enough to form all of the chemical elements in the periodic table. In a complex set of collisions, some of the nuclei up to iron that have been created by the successive fusion reactions soak up neutrons and undergo beta decay (see Chapter

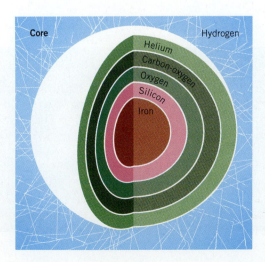

Figure 17–9
The interior of a large star displays concentric shells of fusion reactions, yielding progressively heavier elements toward the core.

11) to form nuclei up to uranium and beyond. All elements beyond iron are created in the short-lived maelstrom of the supernova explosion.

For a while, a supernova is surrounded by a cloud of ejected material. This expanding cloud dissipates into interstellar space, leaving behind the core of neutrons that was created in the collapse. A star that is being held up by degeneracy pressure of neutrons is called a **neutron star**.

Neutron Stars and Pulsars

A neutron star is, in essence, a giant nucleus—incredibly dense and very small. A typical neutron star might be 10 miles across, small enough to fit within the city limits of even a moderate metropolis. Several significant things happen when a large star shrinks down into something the size of a city. For one thing, the rate of rotation of the star goes up substantially. Just as an ice skater increases her spin when she pulls in her arms, a star rotates faster and faster as it contracts. In fact, some neutron stars in our galaxy rotate 1000 times a second. You can compare this to the sedate motion of the Sun, which rotates once every 26 days.

Neutron stars do not give off much light, and they would probably have gone undetected if some of them didn't exhibit unusual behavior in the radio part of the spectrum. The reason for this behavior can be understood if you follow the collapse that leads to the neutron star. As the star collapses, the strength of its magnetic field increases. If a normal star has a dipole field (see Chapter 5), for example, then during the collapse the field lines are dragged in with the material of the star so the field becomes much more concentrated and intense. Some neutron stars in our galaxy possess fields as much as a trillion times that of the magnetic field at the surface of the Earth.

These two effects—a strong magnetic field and rapid rotation—may combine to produce a special kind of neutron star, which astronomers call a **pulsar**. Fast-moving particles speed out along the intense magnetic field lines of the rotating neutron star, and these accelerating particles give off electromagnetic radiation, as shown in Figure 17–10. Most of this radiation is in the radio range, so the neutron star's signal is seen primarily with radio telescopes.

Figure 17–10

A schematic diagram of a pulsar reveals its two key attributes—rapid rotation and an intense magnetic field. This combination of traits produces a pulsing light-house-like pattern of energetic radiation.

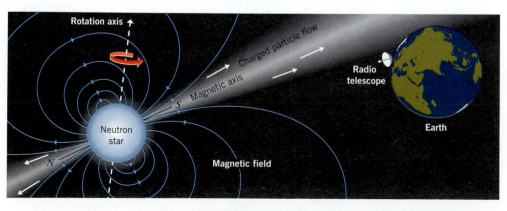

One way of thinking about a pulsar is to imagine it as being somewhat like a searchlight in the sky. Radio waves are continuously emitted along an axis that goes between the north and south magnetic poles of the neutron star, and this line describes a circle in space as the neutron star rotates. If you are standing in the line you will see a burst of radio waves every time the north or south pole of the pulsar is pointing toward you, and nothing when it's not. You will, in other words, see a series of pulses of radio waves. The signature of a pulsar in the sky is a series of regularly spaced pulses, typically some tens to thousands per second. The pulsar represents one possible end state of a supernova. All pulsars are neutron stars, although all neutron stars are probably not seen as pulsars by earthbound astronomers.

We know of several pulsars that are the remnants of previous supernovas. The Crab Nebula (see Figure 17–11), a supernova seen from Earth in A.D. 1054, contains at its core one of the first pulsars discovered. Likewise, Supernova 1987A (see the following section) is also expected to reveal a pulsar when all the dust clears.

Our current theories of stellar evolution say that stars more than 10 times as massive as the Sun will go through the supernova process we've just described, and eject large amounts of heavy elements into space.

Figure 17–11

In A.D. 1054, a supernova was observed from Earth, forming what is now called the Crab Nebula. This photograph shows the nebula today, with material from the supernova being returned to interstellar space. At the center of the nebula is a pulsar.

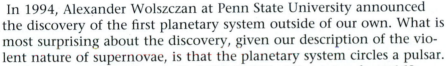

Science in the Making

The Discovery of a Planetary System

In 1994, Alexander Wolszczan at Penn State University announced the discovery of the first planetary system outside of our own. What is most surprising about the discovery, given our description of the violent nature of supernovae, is that the planetary system circles a pulsar.

This particular pulsar (called PSR B1257+12) rotates about 160 times per second. If a planet is orbiting the pulsar, then the gravitational force exerted by the planet will pull the pulsar toward it, and there will be a slight shift in the arrival times of the radio pulses. If the pulsar is moving away from us, the crests will arrive farther apart than average, and the opposite will be true if it is moving toward us, simply because the pulsar itself has moved in the 0.06 seconds between pulses. (This effect is similar to the Doppler effect discussed in Chapter 6.)

After monitoring the pulsar for several years, Wolszczan was able to gather enough data to unravel the effects of three objects in orbit around the pulsar—two planets about three times the size of the Earth, and one about the size of the Moon. If this claim holds up, it will be the first successful detection of another planetary system.

How could a pulsar, the product of a supernova, have planets? The best guess is that the star that exploded to form PSR B1257+12 was once part of a double star system, and that the disk from which the planets were formed (see Chapter 14) came from the breakup of the companion after the supernova. In any case, these planets would not receive sufficient energy from the pulsar to develop life (see Chapter 23). ■

This pair of photos was taken (*a*) on February 23rd before, and (*b*) on February 26th after the explosion of Supernova 1987a.

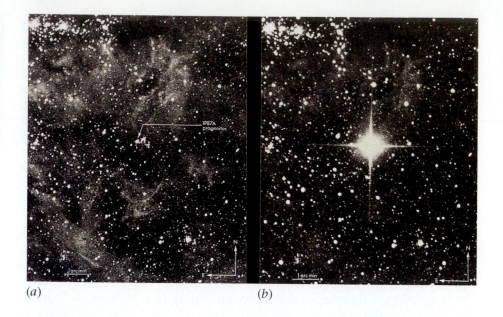

(*a*)   (*b*)

Supernova 1987A

On February 23, 1987, a supernova was seen in the Large Magellanic Cloud, a small galaxy-like structure near the Milky Way galaxy. Although the supernova was 170,000 light years from the Earth, it caused a great stir in science because it was the first supernova to be observed with modern observatories, including satellites. It was seen by large neutrino detectors and many ordinary telescopes on the Earth, and by X-ray and gamma-ray observatories above the atmosphere. Since it was the first supernova observed in 1987, it was given the name "1987A."

Supernova 1987A was a classic example of a supernova, and perhaps the biggest surprise to come out of the experience was that there were so few surprises. The intricate theories of nuclear reactions that take place in those incredibly complex few hours when a star explodes were largely confirmed. In the years following the appearance of 1987A, astronomers have observed a sequence of particles characteristic of the decay of different nuclei that we believe are created in supernovas. In each case, the predictions of theory seem to have been matched by what actually happened.

Black Holes

Occasionally, a large star may die in a way that does not lead to the formation of a pulsar. If a star is large enough—perhaps 30 times as massive as the Sun—there may be processes, as yet only imperfectly understood, by which even the degeneracy pressure of neutrons is overcome and the star collapses. The result is the ultimate triumph of gravity, a **black hole**. A black hole is an object so dense, a mass so concentrated, that nothing—not even light—can escape from its surface.

We do not know how often black holes are formed. Indeed, it comes as a surprise to most people to learn that no widely accepted astronomical black holes have been detected in our galaxy. Black holes are difficult to detect because you can't see them. Several promising candidates have been spotted, but none you could point to and say, "Yes, this is beyond doubt an example of a black hole."

The search for black holes concentrates on double star systems in which one star has evolved into a black hole. The idea is that even though we can't see the black hole itself, we can see its effect on its partner. The most striking thing to look for is material falling into the black hole. The enormous gravitational energy released in the process is partially converted into X-rays and gamma rays, which can be detected by orbiting observatories.

THINKING MORE ABOUT STARS

Generation of the Chemical Elements

Almost everything you see about you was made in a supernova. Your body, for example, is made primarily from elements that formed in some distant exploding star more than 4.5 billion years ago. We say this because, as we shall see in Chapter 18, the universe began its life with only light elements—hydrogen, helium, and small amounts of lithium. These elements formed the first stars and were processed in the first stellar nuclear fires. In stars like the Sun, elements heavier than helium may be made, but they remain in that star and never return to the cosmos.

In large stars, however, all the elements up to uranium (the element with 92 protons) and beyond are made and spewed back into the interstellar medium in the titanic explosions we call supernovas. These heavy elements enrich the surrounding galaxy, and when new stars are formed these elements are incorporated into them. The Sun, which formed fairly late in the history of our own galaxy, thus incorporated many heavy elements that had been made in previous supernovas.

In fact, you can think of the history of our galaxy as one of successive and cumulative enrichment by nuclear processing in large, short-lived stars. These stars, with lifetimes as short as tens of millions of years, take the original hydrogen in the galaxy and convert it into heavier elements. Thus we expect that older, smaller stars that have been shining since the early history of the universe will have fewer heavy elements than relatively young stars like the Sun—a prediction that is borne out by astronomical observations.

Think about what this means as you look around you. All the objects in your life—this book, your clothes, even your skin and bones—are made of atoms that formed in the hearts of giant stars long ago.

Summary

Astronomy is the study of objects in the heavens. Astronomers have discovered much about the nature and origins of *stars*. Stars such as our own Sun form from giant clouds of interstellar dust—clouds that gradually collapse under the force of gravity. This collapse subjects the star's atoms, primarily hydrogen, to tremendous temperatures and pressures. The life of a star is a continuous struggle against this gravitational force.

The extreme temperature and pressure conditions deep inside a star cause its hydrogen core to undergo nuclear fusion reactions, burning to create helium and heat energy. Ignition of these nuclear fires creates an outward flow of particles, called the *solar wind*. The fusion reaction proceeds in three steps, in which (1) two protons come together to form deuterium, (2) a proton and a deuterium nucleus come together to form helium-3, and (3) two helium-3 nuclei fuse to make helium-4. This energy creates the pressure that balances the force of gravity that pulls the star inward.

Stars that are burning hydrogen to produce energy

are said to be *main-sequence* stars. Larger stars burn hotter and emit more energy, while smaller stars are cooler and radiate less energy. Main-sequence stars are found in a simple bandlike pattern on a Hertzsprung-Russell diagram, which graphs a star's energy output versus its temperature.

We study stars with *telescopes*, instruments that gather and focus electromagnetic radiation. Earth-based telescopes detect visible and radio waves, while orbiting observatories detect all other regions of the electromagnetic spectrum. A star's brightness is measured in magnitude. The brightness of the star as seen from Earth is its *apparent magnitude,* which depends on both the amount of energy the star produces and its distance from us. The *absolute magnitude* is a star's brightness when viewed from a fixed distance and depends only on its energy output. The most powerful telescopes can detect stars that are hundreds of millions of *light-years* away.

When a star like the Sun consumes most of its core hydrogen, a helium-rich central region remains. The star once again begins to collapse under gravity, and internal temperatures rise again. Hydrogen burning begins in shells outside the core, while the core's helium may also combine in nuclear fusion reactions to form carbon. These new nuclear processes may cause a star like the Sun to expand briefly and become a *red giant*, a star whose outer layers glow red hot. Eventually, however, nuclear fuel must be exhausted. Gravity will dominate and the carbon-rich star will collapse to a very small, very hot *white dwarf*.

Stars at least eight times larger than the Sun may evolve beyond hydrogen and helium burning. If temperatures and pressures are high enough, carbon can undergo additional nuclear reactions to form elements as heavy as iron, the ultimate nuclear ash. Once iron is formed, however, there can be no more energy produced by these reactions and burning will cease. The sudden extinguishing of a star causes a catastrophic gravitational collapse and rebound—a *supernova*—in which the star literally explodes and spews all the chemical elements into the heavens. A dense, spinning *neutron star* or *pulsar* may be the only remnant of the original star. The largest stars may collapse into a *black hole*, an object so massive that not even light can escape its gravitational pull.

Key Terms About Stars

astronomy	main-sequence star
star	red giant
solar wind	white dwarf
telescope	supernova
apparent magnitude	neutron star
absolute magnitude	pulsar
light-year	black hole

Review Questions

1. What is a star?

2. Why do stars emit energy in the form of radiation?

3. What are the major layers of the Sun?

4. What two properties of stars do scientists plot on a Hertzsprung-Russell diagram? Why do they choose these properties?

5. In what ways is the Sun a typical star?

6. What is the solar wind?

7. Describe two kinds of telescopes. In what ways are they similar?

8. What are the advantages of placing a telescope in orbit?

9. Describe two ways to determine the distance to another star.

10. Describe hydrogen burning. Where does it take place?

11. Why must the Sun eventually die? What changes will the Sun undergo before it dies?

12. Why do large stars have shorter lives than smaller stars?

13. Why won't the Sun become a supernova or a black hole?

14. How are supernovas and neutron stars related to each other?

15. How are neutron stars and pulsars related to each other?

16. If iron is the ultimate nuclear ash, where do elements heavier than iron come from?

17. Why is it difficult to detect a black hole from Earth?

Discussion Questions

1. Why do we see stars only at night? Do they shine during the day?

2. How might you determine the age of a star from an Earth-based telescope? What measurements might you make?

3. In the science fiction movie classic *Star Wars*, Han Solo speaks the following line: "Fast? This is the ship that made the Kessel run in under 12 parsecs. She's fast enough for you, old man."

What grade would Captain Solo receive in Astronomy 101? Why?

4. What was happening on Earth when the energy in the sunlight falling outside your window today was first produced in fusion reactions at the Sun's core? Were the Pyramids built yet? Did people live in cities? Was Neanderthal still around?

5. Given the connection between the Sun and the human eye, speculate about the kinds of stars that might be found in solar systems from which different fictional extraterrestrials are supposed to have come. Would the Klingon and Vulcan stars in *Star Trek* have been much different from our own? Why or why not?

6. How can we talk about the evolution of stars over billions of years when human beings have been observing stars for only a few thousand years?

7. How does the principle of conservation of energy apply to a supernova?

8. Would a star surrounded by a Dyson sphere be visible at night? How would you go about searching for such a star? What kinds of instruments would you need to detect it?

9. Most stars we see are on the main sequence. Stars spend most of their lives consuming their initial stock of hydrogen. Is there a connection between these two statements? If so, what is it?

Problems

1. Mercury, the closest planet to the Sun, lies at an average distance of 93 million kilometers (about 58 million miles) from the Sun. How much energy falls on a square meter of Mercury's sunlit surface? What does this number imply about the possibility of life surviving on the planet?

2. If the Sun shines for 11 billion years at about its present energy output, how much total energy will the star send into space during its lifetime?

3. How much energy would fall on a one-square-meter detector above the atmosphere of Mars (2.3×10^{11} meters from the Sun)? Of Jupiter (7.8×10^{11} meters)? Of Pluto (5.9×10^{12} meters)?

4. How far away is Alpha Centauri, the nearest star? How long would it take to get there at a speed of 2000 miles per hour (the speed of a fast jet plane?)

Investigations

1. Locate some stars in the sky and find out their apparent magnitude. (You might want to start with

some familiar stars such as those in the Big Dipper.)

2. The Crab Nebula is the remains of a supernova event that was sighted on Earth almost 1000 years ago. It must have been visible as a brilliant object for several days. What cultures left a record of this astronomical event? How did they explain what they saw?

3. You can set up an analog to the astronomical distance scale by using two "yardsticks"—a ruler and a tape measure, for example—to measure distances. Measure the dimensions of your classroom this way. How would you make sure that distances on each yardstick were the same? Does this exercise suggest a way for astronomers to check the consistency of their distance scale?

4. The nineteenth-century American poet Walt Whitman (1819–1892) wrote the following poem about astronomy:

When I Heard the Learn'd Astronomer
When I heard the learn'd astronomer,
When the proofs, the figures,
were ranged in columns before me,
When I was shown the charts and diagrams,
to add, divide, and measure them,
When I sitting heard the astronomer where he lectured
with much applause in the lecture-room,
How soon unaccountable I became tired and sick,
Till rising and gliding out I wander'd off by myself,
In the mystical moist night air, and from time to time,
Look'd up in perfect silence at the stars.

Although Whitman was unimpressed by the facts and figures of the "learn'd astronomer," astronomers of the past century have changed the way we think about our place in the universe. In this respect, how does science complement poetry? How do poetry and astronomy differ as ways of understanding why we are here? How would you answer the poet today?

Additional Reading

Ferris, Timothy. *Coming of Age in the Milky Way*. New York: William Morrow, 1988.

Goldsmith, Donald. *Supernova! The Exploding Star of 1987*. New York: St. Martin's Press, 1989.

———. *The Astronomers*. New York: St. Martin's Press, 1991.

Sullivan, Walter. *Black Holes*. New York: Doubleday, 1979.

Trefil, James. *Space Time Infinity*. Washington DC: Smithsonian, 1983.

Cosmology

The universe began billions of years ago in the big bang, and it has been expanding ever since.

A Random Walk

The Campfire

One of life's most pleasant experiences is to sit around a campfire as dusk falls in the outdoors, watching the flames leap and dance. If you observe closely, you may notice that the color of the coals in the fire changes, depending on how hot the fire is. They are ordinarily red, but in a roaring blaze they can actually be white. Then, as the fire starts to go out and people start to unroll their sleeping bags and get ready for bed, the coals glow a dull red and, eventually, stop glowing altogether.

But even when the coals aren't glowing, they are giving off energy in the form of infrared radiation, which you can feel if you put your hand out to the fire. Even the next day, you can still feel the radiation given off by the cooling embers.

Would you believe that a phenomenon like this campfire experience led twentieth-century scientists to a completely new understanding of the structure and history of the universe in which we live?

The Nature of the Cosmos

On any given night you can see several thousand points of light in the sky, almost all of which lie within a collection of about 100 billion stars we call the **Milky Way** galaxy. A **galaxy** is a large assembly of stars (between millions and hundreds of billions of them), together with gas, dust, and other materials, that is held together by the forces of mutual gravitational attraction.

Copernicus, Kepler, and Newton had shown that the Earth is not at the center of the universe, and we knew that the Earth was only one of several planets circling the Sun. But until the first part of the twentieth century, we could still believe that our own Milky Way was the only large collection of stars in the universe. In the last 70 years or so, this comfortable picture of the universe has disappeared. We now know that the collection of stars we call home turns out to be only one of many billions of similar collections in the sky. In many ways, the more we learn about our universe, the less unique our own home planet Earth seems to be.

In 1924 a young American astronomer named Edwin Hubble established beyond a shadow of a doubt that the Milky Way is just one of a countless number of galaxies in the universe. In doing so, he set the tone for a century of progress in the new branch of science called **cosmology,** which is devoted to the study of the structure and history of the entire universe.

Science in the Making

Edwin Hubble and the Birth of Modern Cosmology

In 1919, Edwin Hubble (1889–1953) went to work at Mount Wilson Observatory near Los Angeles. Mount Wilson was then a lonely observatory far from any major city. Today it has been engulfed by the Los Angeles metropolitan area, but in those days it afforded astronomers a chance to look at the sky through clear, unpolluted air. It had what was then the world's largest telescope, whose mirror measured 100 inches (nearly 3 meters) across.

Edwin Hubble led an extraordinary life and excelled at an astonishing variety of endeavors. Born in Missouri, he became an honor student at the University of Chicago, where he lettered in track and played on a championship basketball team. He was also a talented amateur boxer, and at one point had to choose between continuing his education and becoming a professional prize fighter. He opted to take a Rhodes Scholarship, and studied law at Oxford. He practiced law for a while in Kentucky, decided he didn't like it, and returned to Chicago to study for a Ph.D. in astronomy. He got his degree at about the time the United States entered World War I. Hubble enlisted as a private in the infantry, and by the end of the war, he had risen to the rank of major. At age 30, he was ready to begin the career that was to change forever our view of humanity's place in the universe.

By the time Hubble was ensconced at Mount Wilson, astronomers had been debating for almost a century about the nature of the universe we live in. They had known for a long time that the Sun is one of a large collection of stars in the Milky Way (see Investigation 1 at the end of this chapter). Is this collection just one of many "island universes" in the vastness of space, they asked, or is the Milky Way all there is? Throughout the latter part of the nineteenth century and the early part of the twentieth, the debate raged among astronomers. The controversy continued for one simple reason: While it was evident that stars in the Milky Way are no more than about 100,000 light-years away, no one could provide an accurate measurement of the distance to any object that might lie outside the Milky Way.

The debate centered around things called "nebulae"—fuzzy structures in the sky that couldn't be seen clearly with the telescopes then available. They were long thought to be dust clouds or large collections of stars, but no one was sure whether they were in the Milky Way or outside of it. Because his new telescope allowed him to see individual Cepheid variable stars in some nebulae (something no one had been able to do before), Hubble was able to measure the distance to them. It turned out that the distance to the nearest one, the Andromeda nebula, was some 2 million light-years—far outside the bounds of the Milky Way. Thus, with a single observation, Hubble established one of the most important facts about the universe we live in: it is made up of countless galaxies, of which the Milky Way is but one. ■

Kinds of Galaxies

The Milky Way is a rather typical galaxy. As shown in Figure 18–1, it is a flattened disk about 100,000 light-years across, with a central bulge known as the nucleus. Bright regions in the disk, known as spiral arms,

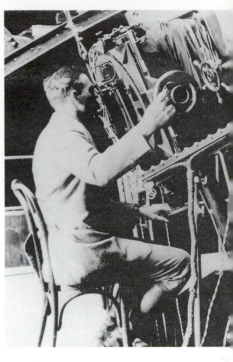

Edwin Hubble (1889–1953) at the 100-inch telescope of California's Mount Wilson observatory.

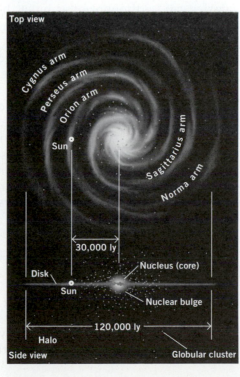

Top view

Cygnus arm
Perseus arm
Orion arm
Sun
Sagittarius arm
Norma arm

30,000 ly

Disk
Sun
Nucleus (core)
Nuclear bulge
120,000 ly
Halo
Side view
Globular cluster

Figure 18–1
A detailed portion of the Milky Way reveals a small fraction of the billions of stars that comprise our home galaxy.

Figure 18–2
A typical spiral galaxy, with a bright core and spiral arms where new stars are forming.

mark areas where new stars are being formed. About 75% of the brighter galaxies in the sky are of this spiral type (see Figures 18–1 and 18–2).

Another class of galaxies, known as ellipticals, resemble nothing so much as a cosmic football. The brightest elliptical galaxies tend to have more stars than spiral galaxies do, and comprise about 20% of bright galaxies. In addition to the relatively large and bright elliptical and spiral galaxies, the universe is littered with small collections of stars known as irregular and dwarf galaxies. Most astronomers think that these are probably the most common galaxies in the universe, given that they are faint and therefore difficult to detect.

Larger elliptical and spiral galaxies and smaller irregular and dwarf galaxies can be thought of as quiet, homey galaxies, where the process of star formation and death goes on in a stately, orderly way. But a small number of galaxies—perhaps 10,000 among the billions known—are quite different and are referred to collectively as active galaxies. The most spectacular of these unusual objects are the *quasars* (for quasi-stellar radio sources). Quasars are wild, explosive, violent objects, where as-yet unknown processes pour vast amounts of energy into space each second from an active center no larger than the solar system. Astronomers suggest that the only way to generate this kind of energy is for the center of a quasar to be occupied by an enormous black hole (with masses, in some cases, millions of times greater than that of the Sun), and for the energy to be generated by huge amounts of mass falling into this center.

Because they are so bright, quasars are the most distant objects we can see in the universe.

The Red Shift and Hubble's Law

Hubble's recognition of galaxies other than our own Milky Way wasn't the end of his discoveries. When he looked at the light from nearby galaxies, he noticed that the distinctive colors emitted by different elements seemed to be shifted toward the red (long-wavelength) end of the spectrum, compared to light emitted by atoms on Earth. Hubble interpreted this **redshift** as an example of the Doppler effect (see Chapter 6), the same phenomenon that causes the sound of a car whizzing past to change its pitch. Hubble's observation meant that distant galaxies are moving away from the Earth. Furthermore, Hubble noticed that the more distant a galaxy, the faster it moves away from us (Figures 18–3 and 18–4).

On the basis of measurements of a few dozen nearby galaxies, Hubble suggested that a simple relationship exists between the distance of an object from the Earth, and that object's speed away from the Earth.

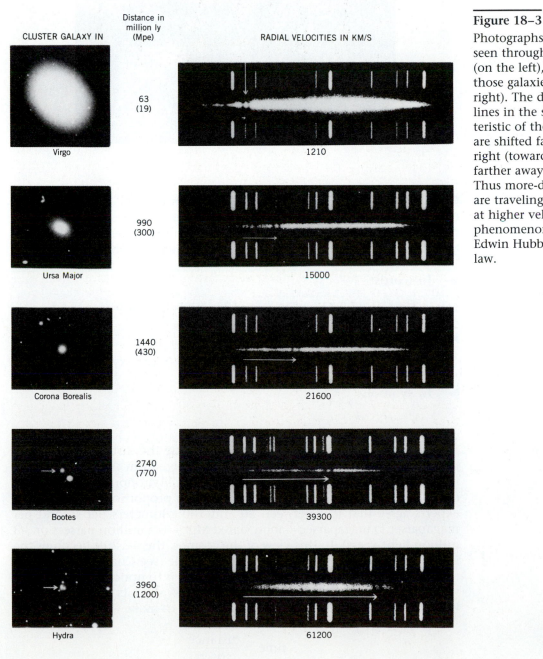

CLUSTER GALAXY IN — Distance in million ly (Mpe) — RADIAL VELOCITIES IN KM/S

Virgo — 63 (19) — 1210

Ursa Major — 990 (300) — 15000

Corona Borealis — 1440 (430) — 21600

Bootes — 2740 (770) — 39300

Hydra — 3960 (1200) — 61200

Figure 18–3

Photographs of galaxies as seen through a telescope (on the left), with spectra of those galaxies (on the right). The double dark lines in the spectra, characteristic of the calcium atom, are shifted farther to the right (toward the red) the farther away the object is. Thus more-distant galaxies are traveling away from us at higher velocities. This phenomenon was used by Edwin Hubble to derive his law.

Figure 18–4

Illustration of Hubble expansion. The more distant a galaxy is from the Earth, the faster it moves away from us.

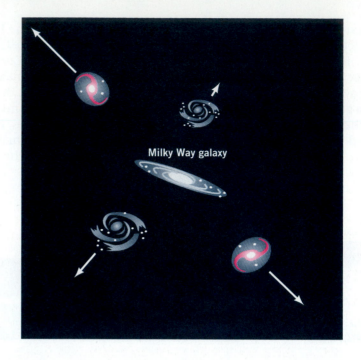

Comparing two galaxies—one twice as far away from the Earth as the other—the farther galaxy moves away from us twice as fast. This statement, which has been amply confirmed by measurements in the subsequent half-century, is now called **Hubble's law**. Hubble's law says:

▶ **In words:**

The farther away a galaxy is, the faster it recedes.

▶ **In equation form:**

galaxy's velocity = (Hubble's constant) × (galaxy's distance)

▶ **In symbols:**

$v = H \times d$

Hubble's law tells us that we can determine the distance to galaxies by measuring the redshift of the light we receive from them, whether or not we can make out individual stars in them. Astronomers continue to debate the exact value of Hubble's constant of proportionality, but most experts agree that it is between 50 and 100 kilometers per second per megaparsec (a megaparsec—abbreviated Mpc—is a million parsecs or 3.3 million light-years). In this view of the cosmos, the redshift becomes the final "ruler" in the astronomical distance scale (see Chapter 17).

One way of interpreting Hubble's constant is to notice that if a galaxy were to travel from the location of the Milky Way to its present position with a velocity v, then the time it would take to make the trip would be

$$\text{time} = \frac{\text{distance}}{\text{speed}}$$

$$t = \frac{d}{v}.$$

Substituting for v from Hubble's law,

$$t = \frac{d}{H \times d}$$
$$= \frac{1}{H}$$

Thus the Hubble constant provides a rough estimate of the time that the expansion has been going on and, hence, of the age of the universe. A Hubble constant of 50 km/sec/Mpc corresponds to an age of the universe of about 16 billion years, while a constant of 100 km/sec/Mpc corresponds to an age of about 8 billion years.

EXAMPLE 18–1: The Distance to a Receding Galaxy

Astronomers discover a new galaxy and determine from its redshift that it is moving away from us at approximately 100,000 kilometers per second (about one-third the speed of light). Approximately how far away is this galaxy? Assume an intermediate value of 75 km/sec/Mpc for the Hubble constant.

▶ **Reasoning:**   According to Hubble's law, a galaxy's distance equals its velocity divided by the Hubble constant.

▶ **Solution:**

$$\text{Distance (in Mpc)} = \frac{\text{velocity (in km/sec)}}{\text{Hubble's constant (in km/sec/Mpc)}}$$
$$= \frac{100,000 \text{ km/sec}}{75 \text{ km/sec/Mpc}}$$
$$= \frac{100,000}{75} \text{ Mpc}$$
$$= 1,333 \text{ megaparsecs}$$

Remember, a parsec equals about 3.3 light-years, so this galaxy is more than 4 billion light-years away. The light that we observe from such a distant galaxy began its trip about the time that our solar system was born. ▲

Science by the Numbers

Analyzing Hubble's Data

In his original sample, Hubble observed 46 galaxies, but was able to determine distances to only 24. Some of his data are given in Table 18–1.

How does one go about analyzing data like these? One common way is to make a graph. In this case, the vertical axis is the velocity of recession of the galaxy, and the horizontal axis is the distance to the galaxy. In Figure 18–5 we show the data as originally plotted by Hubble.

Table 18–1 • Some of Hubble's Data

Distance to Galaxy (in megaparsecs)	Velocity (in km/sec)
1.0	620
1.4	500
1.7	960
2.0	850
2.0	1090

Figure 18–5

Hubble's distance-versus-velocity data for five galaxies.

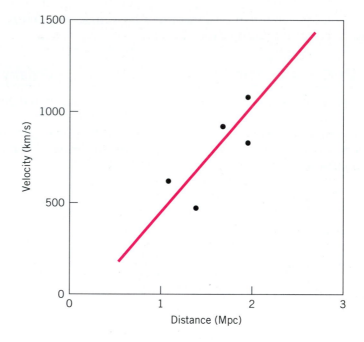

Looking at the data, the general trend of Hubble's law is obvious—the farther you go to the right (i.e., the farther away the galaxies are), the higher the points (i.e., the faster the galaxies are moving away). You also notice, however, that the points do not fall on a straight line, but are scattered. Confronted with this sort of situation, you can do one of two things. You can assume that the scattering is due to experimental error, and that more precise experiments will verify that the points fall on a straight line; or you can assume that the scatter is a real phenomenon and try to explain it. Hubble took the first alternative, so the only problem left was to find the line about which experimental error was scattering his data.

The way this is usually done is to find the line for which the sum of the distances between the line and each data point are smaller than for any other line. In effect, you find the line that comes closest to all the data points. The slope of this line, which measures how fast the velocity increases for a given change in distance, is the best estimate of Hubble's constant. ■

The Big Bang

Hubble's law reveals an extraordinary aspect of our universe: it is expanding. Nearby galaxies are moving away from us, and faraway galaxies are moving away even faster. The whole thing is blowing up like a balloon. This startling fact leads us, in turn, to perhaps the most amazing discovery of all. If you look at our universe expanding today and imagine moving backward in time (think of running a videotape in reverse), you can see that at some point in the past the universe must have started out as a very small object. In other words:

> **The universe began at a specific point in the past, and it has been expanding ever since.**

This picture of the universe—that it began from a small, dense collection of matter and has been expanding ever since—is called the **big bang theory**. This theory constitutes our best idea of what the early universe was like.

Think how different the big bang theory of the universe is from the theories of the Greeks or the medieval scholars, or even the great scientists of the nineteenth century whose work we have studied. To them, the Earth went in stately orbit around the Sun, and the Sun moved among the stars, but the collection of stars you can see at night with your naked eye or with a telescope was all that there was. Suddenly, with Hubble's work, the universe grew immeasurably. Our own collection of stars, our own galaxy, is just one of perhaps 100 billion known galaxies in a universe in which galaxies are flying away from each other at incredible speeds. It is a vision of a universe that began at some time in the distant past and will, presumably, end at some time in the future.

The Large-Scale Structure of the Universe

The Milky Way is part of a group of galaxies known as the Local Group, made up of ourselves, the Andromeda galaxy, and perhaps a dozen small "suburban" galaxies. The Andromeda galaxy is visible with the naked eye from the Earth—you can see it from a dark spot on a clear summer night as a fuzzy patch of light in the northeast. The Local Group, in turn, is part of the Local Supercluster, a collection of galaxies about 100 million light-years across.

We now know that literally billions of galaxies populate the universe, each a collection of billions of stars. Most galaxies seem to be clumped together into *groups* and *clusters*, many of which are, in turn, grouped into larger collections called *superclusters* of thousands of galaxies.

In the 1980s, astronomers began to make "redshift surveys" of the sky. In these surveys they not only look at distant galaxies, but they measure their redshifts so we know how far the galaxies are. In this way, it is possible to construct a full, three-dimensional picture of the distribution of matter in the cosmos. With the results from these observations, led primarily by the team of Margaret Geller and John Huchra at the Harvard-Smithsonian Astrophysical Observatory, astronomers have

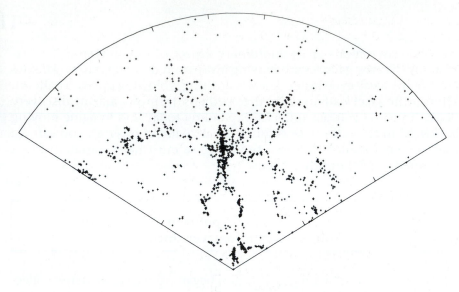

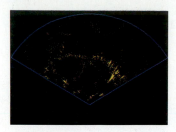

Figure 18–6
The large-scale structure of the universe. This figure shows a thin slice of the universe, with each galaxy represented as a point (the Milky Way lies at the lower point in this figure). Matter is concentrated in superclusters, with large, relatively empty areas called voids between them.

put together a picture of the universe that is very different from what you might expect. Instead of finding galaxies scattered more or less at random through space, they find that galaxies are collected into large structures that run for billions of light-years across the sky. In fact, you can get an excellent picture of the structure of the universe by imagining that you are taking a knife and slicing through a big pile of soapsuds. The result will give you a structure in which large empty spaces are surrounded by soap film. In exactly the same way, matter in the universe seems to be concentrated in superclusters on the surfaces of large empty areas called *voids* (see Figure 18–6). Attempting to understand the reason for this very complex structure in the universe remains one of the major tasks of modern cosmology.

Some Useful Analogies

The big bang picture of the universe is so important that we should spend some time thinking about it. Many analogies can be used to help us picture what the expanding universe is like, and we'll look at two. Be forewarned, however: none of these analogies is perfect. If you pursue any of them far enough they fail, because none of them captures the entirety and complexity of the universe in which we live. And yet each of the analogies can help us understand aspects of that universe.

1. The Raisin-Bread Dough Analogy One standard way of thinking about the big bang is to imagine the universe as being analogous to a huge vat of rising bread dough in a bakery (see Figure 18–7). If raisins scattered through the dough represent galaxies, and if you're standing on one of those raisins, then you would look around you and see other raisins moving away from you. You could watch as a nearby raisin moves away because the dough between you and it is expanding. A nearby raisin wouldn't be moving very fast, because there isn't much expanding dough between you and that raisin. A raisin three times as far away, however,

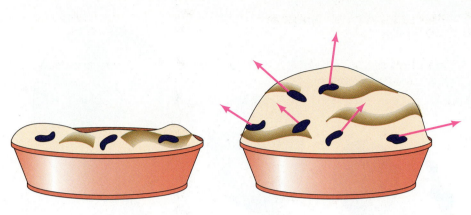

Figure 18–7
The raisin-bread dough analogy of the expanding universe. As the dough expands, all raisins move apart from each other—the farther apart the raisins, the faster the distance increases.

would move away faster—three times faster, in fact, because three times as much dough lies between you and the raisin.

The raisin-bread dough analogy is very useful because it makes it easy to visualize how everything could seem to be moving away from us, with objects that are farther away moving faster. If you stand on any raisin in the dough, all the other raisins look as though they're moving away from you. This analogy thus explains why the Earth seems to be the center of the universe. It also explains why this fact isn't significant—*every* point appears to be at the center of the universe.

But the expanding dough analogy fails to address one of the most commonly asked questions about the Hubble expansion: What is outside the expansion? A mass of bread dough, after all, has a middle and an outer surface; some raisins are nearer the center than others. But we believe the universe has no surface, no outside and inside, and no unique central position. In this regard, the surface of an expanding balloon provides a better analogy.

2. The Expanding-Balloon Analogy   Imagine that you live on the surface of a balloon in a two-dimensional universe. You would be absolutely flat, living on a flat-surface universe (similar to the way we are three dimensional, living in a three-dimensional universe). Evenly spaced points cover the balloon's surface, and one of these points is your home. As the inflating balloon expands, you observe that every other point moves away from you—the farther away the point, the faster away it moves (see Figure 18–8).

Where is the edge of the balloon? What are the "inside" and "outside" of the balloon in two dimensions? The answers, at least from the perspective of a two-dimensional being on the balloon's surface, are that every point appears to be at the center, and the universe has no edges, no inside, and no outside. The two-dimensional being experiences one continuous, never-ending surface. We live in a universe of higher dimensionality, but the principle is the same: our universe has no center and no inside versus outside.

The balloon analogy is also useful because it can help us visualize another question that is often asked about the expanding universe: What is it expanding *into*? If you think about being on the balloon, you realize

Figure 18–8
The expanding-balloon analogy of the universe. All points on the surface of the expanding balloon move away from each other—the farther apart the points initially, the faster they move apart.

that you could start out in any direction and keep traveling. You might come back to where you started, but you would never come to an end. There would never be an "into." The surface of a balloon is an example of a system that is bounded (in two dimensions), but that has no boundaries.

Evidence for the Big Bang

In Chapter 1 we pointed out that every scientific theory must be tested and have experimental or observational evidence backing it up. The big bang theory provides a comprehensive picture of what our universe might be like, but is there sufficient observational evidence to support it? In fact, three important pieces of observational evidence make the big bang idea extremely compelling to scientists.

1. The Universal Expansion   Edwin Hubble's observation of universal expansion provided the first strong evidence for the big bang theory. If the universe began from a compact source and has been expanding, then you would expect to see that expansion going on today. The fact that you do see such an expansion must be taken as evidence for a big bang event in the past. It is not, however, conclusive evidence. Many other theories of the universe have incorporated an expansion, but not a specific beginning in time. During the 1940s, for example, scientists proposed a theory called the *steady-state universe*. In this universe, galaxies move away from each other, but new galaxies are constantly being formed in the spaces that are being vacated. Thus the steady-state model describes a universe that is constantly expanding, constantly forming new galaxies, but with no trace of a beginning.

Because of the possibility of this kind of theory, the universal expansion, in and of itself, does not compel us to accept the big bang theory.

2. The Cosmic Microwave Background   In 1964, Arno Penzias and Robert W. Wilson, two scientists working at Bell Laboratories in New Jersey, used a primitive radio receiver to scan the skies for radio signals. Their motivation was a simple one. They worked during the early days of satellite broadcasting, and they were measuring microwave radiation to document the kinds of background signals that might interfere with radio transmission. They found that whichever way they pointed their receiver, they heard a faint hiss in their apparatus. There seemed to be microwave radiation falling on the Earth from all directions. We now call this radiation the **cosmic microwave background radiation.**

At first they suspected that this background noise might be an arti-fact—a fault in their electronics, or even interference caused by droppings from a pair of pigeons that had nested inside their funnel-shaped micro-wave antenna. However, a thorough testing and cleaning made no differ-ence in the odd results. A constant influx of microwave radiation of wavelength 7.35 centimeters flooded the Earth from every direction in space. And so the scientists asked a very simple question: Where is this radiation coming from?

In order to understand the answer to their question, you need to remember that every object in the universe that is above the temperature of absolute zero emits some sort of radiation (see Chapter 6). As we saw in the Random Walk that opened this chapter, a coal on a fire may glow white hot and emit the complete spectrum of visible electromagnetic radiation. As the fire cools it will give out light that is first concentrated in the yellow, then orange, and eventually dull red range. Even after it no longer glows with visible light, you can tell that the coal is giving off radiation by holding out your hand to it and sensing the infrared or heat radiation that still pours from the dying embers. As the coal cools still more, it will give off wavelengths of longer and longer radiation.

One way to think about the cosmic microwave background, then, is to imagine that you are inside a cooling coal on a fire. No matter which way you look, you will see radiation coming toward you, and that radia-tion will move progressively from white to orange to red light and eventu-ally all the way down to microwaves as the coal cools.

In 1964, a group of theorists at Princeton University (not far from Bell Laboratories) pointed out that if the universe had indeed begun at some time in the past, then today it would still be giving off electromag-netic radiation in the microwave range. In fact, the best calculations at the time indicated that the radiation would be characteristic of an object at a few degrees above absolute zero. When Penzias and Wilson got in contact with these theorists, the reason they couldn't get rid of the microwave signal became obvious. Not only was it a real signal, it was evidence for the big bang itself. For their discovery, Penzias and Wilson shared the Nobel prize in physics in 1978—not a bad outcome for a measurement designed to do something else entirely!

We said before that it is possible to imagine theories, such as the steady-state theory, in which the universe is expanding but has no begin-ning. However, it is impossible to imagine a universe that does not have a beginning but that produces the kind of microwave background we're talking about. Thus, Penzias and Wilson's discovery put an end to the steady-state theory.

In 1989, the satellite observatory called the Cosmic Background Ex-plorer (COBE) measured the microwave background to extreme levels of accuracy. The purpose of this measurement was to see, in great detail,

The Cosmic Background Explorer (COBE) produced this map of microwave radiation from the entire sky. Blue indicates regions that are 0.01 percent cooler than average, whereas red indicates 0.01 percent warmer regions. This map indicates that the early universe was not perfectly uniform—a situation that led to the present "clumpiness" of the universe.

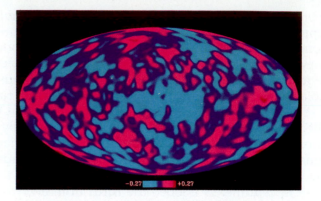

whether the predictions of the big bang theory about the nature of the cosmic microwave background radiation were correct.

These data established beyond any doubt that we live in a universe where the average temperature is 2.7 Kelvins (K). This finding reaffirmed the validity of the big bang theory in the minds of scientists.

3. The Abundance of Light Elements The third important piece of evidence for the big bang theory comes from studies of the abundances of light nuclei in the universe. For a short period in the early history of the universe, as we'll see at the end of this chapter, atomic nuclei could form from elementary particles. Cosmologists believe that the only nuclei that could have formed in the big bang are isotopes of hydrogen, helium, and lithium (the first three elements, with one, two, and three protons in their nuclei, respectively). All elements heavier than lithium were formed later in stars, as discussed in Chapter 17.

The conditions necessary for the formation of light elements were twofold. First, matter had to be packed together densely enough to allow enough collisions to produce a fusion reaction. Second, the temperature had to be high enough for those reactions to happen, but not so high that nuclei created by fusion would be broken up in subsequent collisions. In an expanding universe, the density of matter will decrease rapidly because of the expansion. Thus nuclei form in a very narrow window of opportunity. Calculations based on density and collision frequency, together with known nuclear reaction rates, make rather specific predictions about how much of each isotope could have been made before matter spread too thinly. Thus the cosmic abundances of elements such as deuterium (the isotope of hydrogen with one proton and one neutron in its nucleus), helium-3 (the isotope of helium with two protons and one neutron) and helium-4 (with two protons and two neutrons) comprise another test of our theories about the origins of the universe.

In fact, studies of the abundances of these isotopes find that they agree quite well with the predictions made in this way. The prediction for the primordial abundance of helium-4 in the universe, for example, is that it cannot have exceeded 25%. Observations of helium abundance are quite close to this prediction. If the abundance of helium were significantly higher than this or fell below 20% or so, the theory would be in serious trouble.

The Evolution of the Universe ▮

Our vision of an expanding universe leads us to peer back in time, to the early history of matter and energy. What can we say about the changes that must have taken place during the past 16 billion years?

Some General Characteristics of an Expanding Universe

Have you ever pumped up a bicycle tire with a hand pump? If you have, you may have noticed that after you've run the pump for awhile, the barrel gets very hot. All matter heats up when it is compressed.

The universe is no exception to this rule. A universe that is more compressed and denser than the one in which we live would also be hotter on average. In such a universe, the cosmic radiation background would correspond to a temperature much higher than 2.7 K (where it is today), and the wavelength of the background radiation would be shorter than 7.35 cm.

When the universe was younger, it must have been much hotter and denser than it is today. This cardinal principle guides our understanding of how the universe evolved. In fact, the big bang theory we have been discussing is often called the "hot big bang" to emphasize the fact that the universe began in a very hot, dense state and has been expanding and cooling every since.

In Chapter 9 we saw that changes of temperature may correspond to changes of state in matter. If you cool water, for example, it eventually turns into ice at the freezing point. In just the same way, modern theories claim, as the universe cooled from its hot origins it went through changes of state very much like the freezing of water. We will refer to these dramatic changes in the fabric of the universe as *freezings*, even though they are not actually changes from a liquid to a solid state.

A succession of freezings dominates the history of the universe. Six distinct episodes occurred, and each had its own unique effect on the universe that we live in. Between each pair of freezings was a relatively long period of steady and rather uneventful expansion. Thus, once we understand these crucial transitions in the history of the universe, we will have come a long way toward understanding why the universe is the way it is.

Let's look at the transitions in order, from the earliest to the most recent, as summarized in Figure 18–9. The first set of freezings involves the unification of forces we discussed in Chapter 12, while the last set involves the coming together of particles to form more complex structures such as nuclei and atoms.

10^{-43} Second: The Freezing of All Forces

The freezings that marked the very early stages of the universe do not involve particles at all, but rather the unification of forces we described in Chapter 12. Cosmologists calculate that the first freezing after the beginning of the universe took place at about 10^{-43} second. (This is really a small number—0.0000000000000000000000000000000000000000001 second!) Before this time, there was only a single unified force. At 10^{-43}

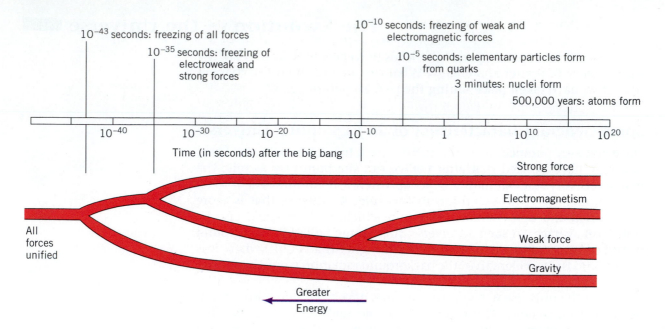

10^{-43} seconds: freezing of all forces

10^{-35} seconds: freezing of electroweak and strong forces

10^{-10} seconds: freezing of weak and electromagnetic forces

10^{-5} seconds: elementary particles form from quarks

3 minutes: nuclei form

500,000 years: atoms form

10^{-40} 10^{-30} 10^{-20} 10^{-10} 1 10^{10} 10^{20}

Time (in seconds) after the big bang

Strong force

Electromagnetism

All forces unified

Weak force

Gravity

Greater Energy

Figure 18–9

The sequence of "freezings" in the universe since the big bang. The earliest freezings involve the splitting of forces, while later freezings involve forms of matter.

second, gravity split off from the strong-electroweak force, so there were two fundamental forces acting in nature.

We cannot reproduce the unimaginably high temperatures that existed at this freezing in our laboratories, and we do not have successful theories that describe the unification of gravity with the other forces. Thus this earliest freezing remains both the theoretical frontier and the limit of our knowledge about the universe at the present time.

10^{-35} Second: The Freezing of the Electroweak and Strong Forces

Unified field theories that describe the behavior of all matter and forces (see Chapter 12) tell us that before the universe was 10^{-35} second old, the strong force was unified with the electroweak force. At 10^{-35} second, the strong force split off from the electroweak force. That is, before this time there were only two fundamental forces acting in the universe (the strong-electroweak force and the force of gravity), but after this time, there were three.

Two important events are associated with the "freezing" at 10^{-35} seconds. They are:

1. The Elimination of Antimatter Antimatter (see Chapter 12) is fairly rare in the universe we live in. We had to wait until the twentieth century before scientists were able to identify antimatter, and we have compelling evidence that no large collections of antimatter exist anywhere in the universe. We have already landed spaceships on the Moon, Mars, and Venus, for example. If any of those bodies had been made of antimatter, those spaceships would have been annihilated in a massive burst of gamma rays. As they did not, we conclude that none of those planets are made of antimatter.

By the same token, the solar wind, composed of ordinary matter, is constantly streaming outward from the Sun to the farthest reaches of the solar system. If any objects in the solar system were made of antimatter, the protons in the solar wind would be annihilating with materials in that body, and we would see evidence of it. The entire solar system, therefore, is made of ordinary matter. By the same type of argument, scientists have been able to show that our entire galaxy is made of ordinary matter, and that no clusters of galaxies anywhere in the observable universe are made of antimatter.

The question, then, is this: If antimatter is indeed simply a mirror image of ordinary matter, and if antimatter appears in our theories on an equal footing with ordinary matter, as it does, then why is there so little antimatter in the universe?

The unified field theories give us an explanation of this rather striking feature of the cosmos. In our laboratories we find one instance of a particle that decays preferentially into matter over antimatter—that is, a particle whose decay products more often contain more matter than antimatter. This particle is called the K_L^0 ("K-zero-long"), one of the many heavy particles such as protons and neutrons that were discovered in the latter part of the twentieth century.

If you take the theories that are successful in explaining this laboratory phenomenon and extrapolate them to the very early universe, you find that they predict that, during the freezing at 10^{-35} second, there were about 100,000,001 protons made for every 100,000,000 antiprotons. In the maelstrom that followed, the 100,000,000 antiprotons were annihilated with 100,000,000 protons, leaving only a sea of intense radiation to mark their presence. From the collection of the leftover protons, all the matter in the universe (including the Earth and its environs) was made. This discovery allowed high energy physicists to explain the very puzzling absence of antimatter in the universe, and led to the enormous burst of interest in the evolution of the early universe in the 1980s.

2. Inflation   According to some versions of the unified field theories, the freezing at 10^{-35} second was accompanied by an incredibly rapid (but short-lived) increase in the rate of expansion of the universe. This short period of rapid expansion is called *inflation*, and theories that incorporate this phenomenon are called *inflationary theories*.

One way to think about inflation is to remember that changes in volume are often associated with changes of state. Water, for example, expands when it freezes, which explains why water pipes may burst open when the water freezes in very cold weather. In the same way, scientists argue, the universe underwent a period of very rapid expansion during the period when the strong force froze out from the electroweak. Roughly speaking, at this time the universe went from being much smaller than a single proton to being about the size of a grapefruit.

Inflation explains another puzzling feature of the universe. We have repeatedly observed that the cosmic microwave background (which is an index of the temperature of the universe) is remarkably uniform. The temperatures associated with microwaves coming from one region of the sky differ from those coming from another region by no more than 1 part in 1000. But calculations based on a uniform rate of expansion say that different parts of the universe would not have been close enough together to have established a common temperature.

In the inflationary theory, the resolution of this problem is very simple. Before 10^{-35} second, all parts of the universe were in contact with each other because the universe was much smaller than you would have guessed based on a uniform rate of expansion. There was time to establish equilibrium before inflation took over and increased the size of the universe. The temperature equilibrium, established early, was preserved through the inflationary era and is seen today in the uniformity of the microwave background.

Thus, the coming together of the theories of elementary particle physics and the study of cosmology has produced solutions to long-standing problems and questions about the universe.

10^{-10} Second: The Freezing of the Weak and Electromagnetic Forces

At 10^{-10} second (that's one ten-billionth of a second) the weak and the electromagnetic forces were unified. In other words, before 10^{-10} second, there were only three fundamental forces operating in the universe. These were the strong, gravitational, and electroweak forces. After 10^{-10} second, the full complement of four fundamental forces was present.

The time of 10^{-10} second also marks another milestone in our discussion of the evolution of the universe. The modern particle accelerators of high energy physics can just barely reproduce the incredible concentration of energy associated with that event. This means that from this point forward it is possible to have direct experimental checks of the theories that describe the evolution of the universe.

10^{-5} Second: The Freezing of Elementary Particles

Up to this point, the matter in the universe had been in its most fundamental form—quarks and leptons (see Chapter 12). The remaining freezings involve the coming together of those basic particles to create the matter we see around us. At 10^{-5} second (10 microseconds), the first of these events occurred, when elementary particles were formed out of quarks.

Before 10^{-5} second, in other words, matter existed in the form of independent quarks and leptons—what we called the "dots and dashes" of the universe in Chapter 12. After this time, matter existed in the form of hadrons and leptons—that is, the ordinary elementary particles we see in our laboratories today. The universe composed of these particles kept expanding and cooling until nuclei and atoms formed.

Three Minutes: The Freezing of Nuclei

Three minutes marks the age at which nuclei, once formed, could remain stable in the universe. Before this time, if a proton and a neutron came together to form deuterium, the simplest nucleus, then the subsequent collisions of that nucleus with other particles in the universe would have been sufficient to knock the nucleus apart. Before three minutes, matter existed only in the form of elementary particles, which could not come together to form nuclei. Before that time, the universe consisted of a sea of high-energy radiation whizzing around between all the various species of elementary particles we discussed in Chapter 12.

At three minutes, a short burst of nucleus formation occurred, as we discussed earlier. Thus, from three minutes on, the universe was littered with nuclei, which formed part of the plasma that was the material of the early universe.

Before One Million Years: The Freezing of Atoms

The most recent transition occurred gradually between the time the universe was a few hundred thousand and a million years old. At this time, the background temperature of the hot, dense universe was so great that electrons could not bond to the atomic orbits to form atoms. Even if an atom formed by chance, its subsequent collisions were sufficiently violent that the atom could not stay together. Thus all of the universe's matter was in the form of a plasma, a hot fluid mixture of electrons and simple nuclei.

This freezing marks an extremely important point in the history of the universe, because it is a point at which radiation such as light was no longer locked into the material of the universe. You know from your own experience that light can travel long distances through the atmosphere (which is made of atoms). But light cannot travel freely through a plasma, which quickly absorbs light and other forms of radiation. Thus, when atoms formed, the universe became transparent and radiation was released. It is this radiation, cooled and stretched out, that we now see as the cosmic microwave background.

The formation of atoms marks an important milestone for another reason. Before this event, if clumps of matter happened to begin forming (under the influence of gravity, for example), they would absorb radiation and be blown apart. This means that there must have been a "window of opportunity" for the formation of galaxies. If galaxies are made of ordinary matter, they couldn't have started to come together out of the primordial gas cloud until atoms had formed—about 500,000 years after the big bang. By that time, however, the Hubble expansion had spread matter out so thinly that the ordinary workings of the force of gravity would not have been able to make a universe of galaxies, clusters, and superclusters. Known as the "galaxy problem," this puzzle remains the great riddle that must be answered by cosmologists.

Another way of stating this problem is to compare the clumpiness of matter in the universe to the smoothness of the cosmic background radiation. The background radiation seems to be the same no matter which way you look in the universe. This uniformity argues that the universe had a smooth, regular beginning. How can this statement be reconciled with the lumpy structure we see when we look at the distribution of matter?

Dark Matter and Ripples at the Beginning of Time

As complicated and complete as this history of the early universe may seem, significant gaps in our understanding of the evolution of the universe remain. Some of these gaps were closed with the development of unified field theories and the inflationary scheme of the universe. The problem of explaining the existence of galaxies, clusters, and superclusters, however, remains.

It now appears that, as impressive as the luminous objects in the sky are, they constitute less than 10% of the matter in the universe—perhaps a good deal less than 10%. The rest of the matter exists in forms that we cannot see, but whose effects we can measure. This mysterious new kind of material is called **dark matter**.

The easiest place to see evidence for dark matter is in galaxies such as our own Milky Way. Far out from the stars and spiral arms that we normally associate with galaxies, we can still see a diffuse cloud of hydrogen gas. This gas gives off radio waves, so we can detect its presence and its motion. In particular, we can tell how fast it is rotating. When we do these sorts of measurements, a rather startling fact emerges. In Chapter 2 we saw that Kepler's laws implied that any object orbiting around a central body under the influence of gravity will travel slower the farther out it is. The distant planet Jupiter, for example, moves more slowly in orbit around the Sun than does the Earth. Similarly, you would expect that when hydrogen molecules are far enough away from the center of the galaxy, these more distant atoms would move more slowly than those closer in. Even though we can see these hydrogen atoms out to distances three times and more the distance from the center of the galaxy to the end of the spiral arms, no one has ever seen the predicted slowing down.

The only way to explain this phenomenon is to say that those hydrogen atoms are still in the middle of the gravitational influence of the galaxy. That means that luminous matter—the bright stars and spiral arms—is not the only thing that is exerting a gravitational force. Something else, something that makes up at least 90% of the mass of the galaxy and that extends far beyond the stars, exerts a gravitational force and affects the motion of the hydrogen we observe. Studies of many galaxies show the same effect, and scientists are now convinced that at least 90% of the universe is made of this mysterious dark matter. Scientists also find evidence that dark matter exists in between galaxies, in clusters, and in other places in the universe.

Dark matter is strange, indeed. It does not interact through the electromagnetic force. If it did, it would absorb or emit photons and it wouldn't be "dark" in the sense we're using the term here. Yet, because we know that it exerts a gravitational attraction, we can conclude that this unseen "stuff" must be a form of matter—matter that interacts with ordinary matter only through the gravitational force. Detecting dark matter, and finding out what it is, remains a very active research field today.

The existence of dark matter might help us understand a key event in the early history of the universe, because dark matter could have formed into clumps *before atoms formed*. In the first several hundred thousand years after the big bang, photons blew apart collections of luminous matter that were trying to form galaxies, but light would not have affected the clumping of dark matter. Therefore, when atoms formed and luminous matter could clump together, it found itself in a universe in which large clusterings of dark matter already existed. The luminous matter would simply have fallen into these clusters and would not have had to form under the influence of its own gravitational attraction. Thus, if dark matter exists, and if dark matter formed clumps early in the history of the universe, the problem of structure is solved.

In 1992 important new results from the COBE satellite supported the notion of dark matter. Examining the microwave background in great detail, astronomers found evidence that some regions in the sky emit microwave background at a slightly higher temperature than the surrounding regions. These so-called "ripples at the beginning of time" were not very large—they correspond to temperature increases of the tiniest fraction of a degree. Nevertheless, they could have been emitted only from regions that were more dense and slightly hotter than their neighbors. These ripples appear to mark the beginning of the collection of luminous matter immediately after the formation of atoms.

In 1993 astronomers announced the discovery of one kind of dark matter, called MACHOS (massive compact halo objects). These bodies form a swarm of large dark objects, each a fraction of the size of the Sun, that orbit the Milky Way. Thus these recent data support both the existence of dark matter in the early universe and its role in producing the structure we see in the sky.

This particular area of study will continue to be important in the coming years, and you're sure to see information about it in the media.

THINKING MORE ABOUT COSMOLOGY

The Future of the Universe

Once we understand the big bang, a simple question comes to mind. Will the expansion we see going on around us today continue in the future, or will all those outgoing galaxies someday slow down, reverse their motion, and fall together in an event that astronomers, half in jest, call the "Big Crunch"? This question, simple as it seems, is one of the most fundamental inquiries we can make, for it involves nothing less than the future of the universe itself.

We do not, as you might suppose, have to resort to esoteric theory or deep philosophy to answer the question. It can be answered (at least in principle) by observing the universe today. The only force we know that is capable of reducing the speed of a receding galaxy is gravity. If the universe holds enough matter, even those quasars at the edge of the observable universe will someday come falling back in. So

the question of the fate of the universe comes down to the question of whether the universe contains enough matter to exert that force. If it does, then we say the universe is *closed*, and the expansion will someday reverse. If it doesn't, then we say the universe is *open*, and it will go on expanding forever. The boundary between these two, in which the expansion slows and comes to a halt after infinite time has passed, is called a *flat* universe.

If you count only luminous matter (the stuff we can actually see), then we observe only about 1% of the matter needed to close the universe, and the expansion will go on forever. Dark matter, however, adds to this total. So far, astronomers have found perhaps 20% to 30% of the matter required, and the search for more continues.

One feature of the debate about the future of the universe that you might want to think about is this: The inflationary theories—the same theories that explain the behavior of the universe at 10^{-35} second—predict that exactly enough matter exists for the universe to be flat. Some theoretical astrophysicists have taken this to mean that the universe must be flat, and that observational astronomers should work harder to find the rest of the dark matter. Observers, on the other hand, say that if the universe is open, that's all there is to it.

How much faith do you think should be put in theoretical predictions of this sort? How hard should observers look for the "missing" matter?

Summary

Early in the twentieth century, Edwin Hubble made two extraordinary discoveries about the structure and behavior of the universe—the science we call *cosmology*. First, he demonstrated that our home, the collection of stars known as the *Milky Way*, is just one of countless *galaxies* in the universe, each containing billions of stars. By measuring the *redshift* of galaxies, he also discovered that these distant objects are moving away from each other. According to *Hubble's law*, the farther the galaxy, the faster it is moving away. This relative motion implies that the universe is expanding.

One theory that accounts for universal expansion is the *big bang*—the idea that the universe began at a specific moment in time and has been expanding ever since. Evidence from the *cosmic microwave background radiation* and the relative abundances of light elements, in addition to the expansion, support the big bang theory.

At the moment of creation, all forces and matter were unified in one unimaginably hot and dense volume. As the universe expanded, however, a series of six "freezings" led to the universe we see today. Freezings at 10^{-43} second, 10^{-35} second, and 10^{-10} second caused a single unified force to split into the four forces we observe today: the gravitational, strong, electromagnetic, and weak forces. At that early stage of the universe, when all matter and energy were contained in a volume no larger than a grapefruit, matter was in its most elementary form of quarks and leptons.

At 10^{-5} second the quarks bonded together to form heavy nuclear particles such as protons and neutrons. Subsequent freezings saw these particles first fuse to nuclei at 3 minutes, and ultimately to join with electrons to form atoms at 500,000 years. Stars, which formed from those atoms, then could begin the processes that provided all the other chemical elements.

The search for *dark matter*—mass that we cannot see with our telescopes—is a frontier in cosmological research that may help us determine whether or not the universe will continue expanding forever.

Key Terms About Cosmology

Milky Way	big bang theory
galaxy	cosmic microwave background radiation
cosmology	
redshift	dark matter
Hubble's law	

Review Questions

1. What is cosmology? How does cosmology differ from astronomy?

2. What is a galaxy? How does a galaxy differ from a star?

3. How are galaxies distributed in the universe?

4. How did Edwin Hubble discover that there are galaxies in the universe other than the Milky Way?

5. Describe Hubble's law. How did Hubble discover it?

6. What kinds of evidence support the big bang theory?

7. Why is interstellar matter composed mostly of hydrogen, helium, and lithium?

8. What event occurred when the universe was 500,000 years old?

9. What event occurred when the universe was 3 minutes old?

10. What event occurred when the universe was 10^{-5} second old?

11. What event occurred when the universe was 10^{-10} second old?

12. What event occurred when the universe was 10^{-35} second old?

13. What event occurred when the universe was 10^{-43} second old?

14. What is the significance of the discovery by Penzias and Wilson of cosmic microwave background radiation?

15. What are "ripples at the beginning of time," and why are they important?

16. What is dark matter and what evidence exists for it?

Discussion Questions

1. Why does the Earth seem to be at the center of the Hubble expansion?

2. If the universe is closed, describe the results that some future Hubble would get when he looked through a telescope during the period of contraction. Would he still see other galaxies? Would he still see a redshift?

3. Why was the steady-state theory of the universe abandoned? How does this episode fit into the discussion of the scientific method in Chapter 1?

4. How will we know whether the universe is open or closed? What measurements or observations could answer this question?

5. Louis Pasteur once said that "chance favors only the mind that is prepared." Apply this saying to the discoveries of Edwin Hubble, and of Penzias and Wilson.

6. Some advances in our knowledge have been made possible through better equipment, such as Hubble's discoveries at Mount Wilson. What other major discoveries in cosmology have relied on improvements in existing apparatus?

Problems

1. Assuming a Hubble constant of 75 km/sec/Mpc, what is the approximate velocity of a galaxy 100 Mpc away? 1000 million Mpc away?

2. If a galaxy is 500 Mpc away, how fast is it receding from us?

3. An observer on one of the raisins in our bread dough analogy measures distances and velocities of neighboring raisins. The data look like the following:

Distance (cm)	Velocity (cm/hour)
0.5	1.02
0.9	2.00
1.4	2.90
2.1	4.05
3.0	5.90
3.4	7.10

Plot these data on a graph and use the plot to estimate a "Hubble constant" for the raisins.

4. From the data in Problem 3, estimate the time that has elapsed since the dough started rising. Estimate the largest and smallest values of this number consistent with the data.

5. In the inflationary period, some theories say that the scale of the universe increased by a factor of 10^{50}. Suppose your height were to increase by a factor of 10^{50}. How tall would you be? Express your answer in light-years and compare it to the size of the observable universe.

6. Suppose a proton (diameter about 10^{-13} cm) were to inflate by a factor of 10^{50}. How big would it be? Convert the answer to light-years and compare it to the size of the observable universe.

7. How fast will a galaxy 5 billion light-years from Earth be moving away from us? What fraction of the speed of light is this?

Investigations

1. The Milky Way is a band of stars that, seen from Earth in the summer months, stretches all the way across the sky. Given what you know about galaxies, why do you suppose that our own galaxy appears this way to us? Who was the first natural philosopher to figure this out?

2. Will the constellation of Andromeda be above the horizon tonight? If so, go out and try to spot the Andromeda galaxy.

3. Look up the "Great Attractor." How does the existence of such an object fit in with the concept of the Hubble expansion? How would you modify the raisin-bread dough analogy to put in the Great Attractor?

4. Investigate the cosmologies of other societies. How do they think the universe began?

5. What agencies or organizations fund cosmological research? What was the role of the Carnegie Institution of Washington in Edwin Hubble's research?

Additional Reading

Ferris, Timothy. *The Red Limit*. New York: William Morrow, 1977.

Hawking, Stephen. *A Brief History of Time*. New York: Bantam, 1988.

Silk, Joseph. *The Big Bang*. New York: W.H. Freeman, 1989.

Trefil, James. *The Dark Side of the Universe*. New York: Scribners, 1988.

———. *Reading the Mind of God*. New York: Scribner's, 1989.

The Science of Life

Living things, the most complex systems studied by science, always obey the laws of physics and chemistry.

A Random Walk
The Life Around You

When you leave the room where you're reading this book and go outside, we'd like you to notice some things that you see every day but usually ignore. We'd like you to see how many living things you can find in your immediate neighborhood.

If you are at a traditional college campus, chances are the first things you'll notice are grass, bushes, and trees. How many different kinds of trees do you see? How many different kinds of grasses and weeds are in the square foot of lawn at your feet? How many kinds of birds do you see flying around? How many insects? Even if you find yourself in a concrete jungle downtown, living things are all around. Look at the fugitive blades of grass peeking up between cracks in the sidewalk, the decorative trees on main streets, the pigeons and sparrows eking out their existence.

There's not a place on the surface of the Earth where you won't find living things. The frigid Arctic seas teem with life, while the frozen wastes of Antarctica yield up hardy lichens and single-celled organisms.

19

Living things may be found in the deepest, darkest parts of the ocean and the most dry and forbidding deserts.

In fact, you could argue that what makes the Earth different from every other place in the solar system (and perhaps even any other place in the galaxy) is the existence of life.

The Organization of Living Things

Biology is the branch of science devoted to the study of living systems. Living things are by far the most complex systems that scientists study. Compared to a single living cell, the complexity of a supernova fades into insignificance, and a single human being has trillions of cells. Yet, despite this complexity, living things operate according to the same laws of nature as everything else we've studied. The great principles of nature we have explored up to this point—energy conservation, the laws of electricity and magnetism, chemical bonding of atoms, and others—operate in, and govern the behavior of, all living things. We have underscored this point by using examples from biology to illustrate all of the preceding chapters.

What Is Life?

We all have a sense of what it means to be alive. We recognize living things by their complexity, their use of energy to grow and repair themselves, their response to external stimuli, and their ability to reproduce. But the more we learn about the world around us, the harder it becomes to provide a precise definition of life. Some objects such as the mule are clearly alive but are unable to reproduce. Certain seeds or spores may lie dormant for centuries without any sign of life, but then suddenly awaken. Furthermore, as technology advances, machines take on more and more of the qualities we usually associate with life. As a result of these sorts of problems, most scientists prefer not to try to define what life is in the abstract, but to describe the properties of living systems at some level, as we do below.

Other medical, legal, and ethical questions relating to the definition of life concern when life begins and when it ends. How should our society define life? Is it possible that we could develop machines that are in some way alive? These are questions that must take scientific knowledge into account, but their ultimate answers lie outside the realm of scientific inquiry.

Despite the wide variety of living things we see around us, and despite the difficulty of defining life in the abstract, we can still list a few general characteristics that all living things have in common. We will return to each of these at length in the following chapters, but we introduce them here to give you a sense of these essential characteristics of life.

1. **Living things use energy.** Sometimes, as in plants, the Sun's radiant energy is tapped directly to produce energy-rich molecules. Sometimes, as with animals and fungi, energy-rich molecules are taken in from outside. In all cases, however, organisms need energy to continue living.

2. **Living things are made from cells, the chemical factories of life.** Cells, the highly organized building blocks of life, are complex structures in their own right. Many complete organisms, such as bacteria and blue-green algae, are single-celled, while in large organisms such as human beings, trillions of cells combine in a single organism.

3. **Living things reproduce.** Sometimes the process of reproduction can be as simple as the splitting of a single cell into two offspring, other times as complex as human sexual reproduction. In all cases, though, life consists of a chain of parents and offspring moving through time.

4. **Living things regulate their energy use and respond to their environments.** When you get too hot, for example, you sweat, and the heat needed to evaporate the water cools you off. When you get cold, you shiver and the extra heat generated by your muscles warms you up.

Science in the Making

Measuring Plant Growth

We can see the similarity between biology and the other sciences by looking at an example of how the scientific method was used to answer an old question. All animals derive their nourishment from the food they eat, but where do plants get theirs? Throughout most of recorded history, people thought that it must come from the soil—that plants "ate" soil in the same way that animals eat meat or fruit.

In 1648, Joannes Baptista Van Helmont (1579–1644), a Flemish physician and the discoverer of carbon dioxide, reported an experiment that changed this view. He planted a willow tree weighing 5 pounds in a pot containing 200 pounds of soil, and watered it for five years as it grew. At the end of that time he weighed the tree and found that its weight had increased to 169 pounds while the amount of soil in the pot had decreased by at most a few ounces. He concluded that all of the material used to construct the fabric of the willow tree came from the water.

He was wrong, of course. We shall see in Chapter 25 that much of the tissue in plants comes from carbon dioxide in the air. The point, however, is that in science a good quantitative experiment, even one with an incorrect conclusion, can clear the way for progress. In this case, Van Helmont's proof that plants did not take their fabric from the soil led eventually to our present understanding on the role of plants in removing carbon dioxide from the air and producing oxygen. ■

Ways of Thinking about Living Things

An ant provides a useful way to begin thinking about the study of living things. We can study the ant in many ways. We can, for example, examine the ant as an individual organism, in which case we ask questions like "How big is it?" "How much energy does it consume?" "Where does it get its energy?" We could take a more microscopic view and consider the individual ant as a collection of specialized organs. In this case we might ask about how the ant moves oxygen from the air to its cells, or how its hard outer covering serves to protect it and support its weight. Penetrating still deeper, we could look at the ant as a collection of cells, and ask how a single one of those cells operates. Questions like "How does this cell carry out its chemical functions?" and "What are the pieces from which this cell is made?" would then be appropriate. Finally, we could look inside the cell and think about its ultimate constituents—the atoms and molecules that combine and react chemically to make the cell what it is. Here we would ask "What are the molecules that operate in a cell?" and "How do they interact as chemicals?"

We could also choose to look at the big picture. Instead of probing ever-smaller parts of the ant, we could choose to view the ant as part of larger and larger structures. The single ant, for example, is part of the social organization of an ant colony, which in turn forms part of a small community of living and nonliving things—a lawn or a field or a patch of forest. This small community, in turn, represents part of the great global system that encompasses all living things on the planet.

Learning about a single living thing such as an ant, then, can take place on many levels. Different branches of biology deal with these levels, which we will discuss in the following chapters. The important point to remember, however, is that none of these ways of viewing the ant is right or wrong. Each complements, and is complemented by, the others.

Having said this, however, we should note that in the last half of the twentieth century a profound change has occurred in the life sciences. Traditional biology concerned itself largely with discovering, understanding, and cataloging organisms and their interactions. In terms of our analogy, the field concerned itself largely with the ant and, perhaps, the anthill. Starting in the nineteenth century, however, chemists began to study the interactions of molecules found in living systems. In the 1950s, the singular role of the molecule deoxyribonucleic acid (DNA) in carrying the genetic message of living things was uncovered (see Chapter 22) and an entire new world was opened in the life sciences. As a result, during the past few decades the study of living things has undergone a fundamental realignment. Today the great majority of biologists are studying living systems at the level of molecules, rather than at the level of the organism. Why this should be so will become clear in the following chapters.

Unity within Diversity

The diversity of living things is truly staggering. You have only to think of whales, palm trees, mosses, and mosquitoes to recognize this fact. Yet biologists have found that beneath this diversity there is a great deal that links all life together. In fact, the general principles that govern all living things will occupy us for most of the rest of the book. We list them here,

A sand scorpion poised to strike with its poisonous tail. Animals must obtain food energy in the form of other living things.

A colony of fire ants, with their white eggs.

however, to make the point that despite their seeming diversity, living things are really not that different from each other.

1. **All life is based on chemistry.** What makes one living thing different from another—what makes you different from a kumquat—are the chemical reactions that go on in cells. All life depends on these chemical reactions, and, as we shall see in Chapters 21 and 22, most living things share a basic set of molecular building blocks and chemical reactions.

2. **All living things share the same genetic code.** The chemical reactions in a cell are governed by a code written in the language of the molecule DNA (see Chapter 22). Just as all books in English are written with the same alphabet, so too the genetic messages for all living things are written in a single language.

3. **All living things descended from a common ancestor.** In Chapter 23 we will review the long chain of evidence that led Charles Darwin and others to recognize the evolution of life. We will see that the many similarities among living things arise from their common ancestry.

4. **All living things are part of larger systems of matter and energy.** Living things and their surroundings form complex systems called ecosystems (see Chapter 25). Matter continuously recycles in a given system, while energy flows through it.

Classifying Living Things

In Chapter 1 we saw that science begins by observation—people must start their exploration of a new area by looking to see what's there. Astronomers watched the skies for thousands of years and named many celestial objects before Newton could explain the motion of the solar system. Alchemists mixed chemicals and observed countless chemical reactions long before modern ideas of the atom or chemical bonding were developed. Similarly, humans spent thousands of years observing living things before they could begin to bring order to the proliferation of organisms they encountered.

Classification plays a central role in science, because our search for patterns in the universe depends on recognizing similarities and differences. Centuries ago, naturalists began the task of cataloging everything from rocks to chemical reactions, from subatomic particles to stars, to provide a framework for study and learning.

Classification provides us with a way of ordering our world, of course, but it does something else as well. It provides us with a way of fitting new phenomena into an existing framework. When a biologist finds a new organism, for example, he or she will group it with similar organisms and thereby is able to make a number of intelligent conjectures about that organism—about its diet, for example, or mating habits. The same holds true of classification schemes for nonliving things. Astronomers who find a star like the Sun, for example, know that it will have a lifetime of about ten billion years, will become a white dwarf (see Chapter 17), and so on.

Cataloging Life

Biologists, confronted by the amazing variety of living things, realized that the first thing they had to do was find some systematic way of cataloging the diversity; of imposing some sort of conceptual order on the diversity of life. The most successful systematic attempt to do this was begun by Swedish naturalist Carolus Linnaeus (1707–1778). His work was based on observation. When you look at living things, it becomes obvious that some characteristics are shared, while others are not. A human being, for example, is more like a squirrel than a blade of grass; a sparrow is more like a fish than algae in a pond. The purpose of the **Linnaean classification** was to group all living things according to their shared characteristics so that each organism is as close as possible to those other things that it resembles, and as far as possible from those it does not.

The basic unit of the classification scheme is the **species,** which we define today as an interbreeding population of individual organisms. Linnaeus, who worked at a time when European scientific institutions were being flooded with new types of plants from around the world, realized that simply describing species would not be enough. Instead, he introduced a system in which similar species are grouped together into a unit called a *genus* (plural: genera). When you go to a zoo, you will probably notice that the scientific names of plants and animals are given in terms of two Latin words. These two names are the genus and species. This so-called "binomial" nomenclature is one of the important legacies of Linnaeus's work.

Linnaeus's original goal was to organize the entire structure of nature, and he published classification schemes for the three "kingdoms" of plants, animals, and minerals. As you might expect in the first attempt to achieve such an ambitious goal, there were mistakes in his work. He thought, for example, that closely related plants had the same "marrow" (which he believed they inherited from the mother), but different "bark" (which he believed they inherited from the father). He also classed the rhinoceros as a rodent. But despite these sorts of flaws, Linnaeus's work showed that it is possible to find order among living things—an order that, as we shall see, arises from the ancient origins of biological systems.

The task of classifying living things is something like the problem of pinpointing one particular building in the entire world, a problem you face every time you mail a letter. You might start by specifying the continent on which the building stands, then the country, then the state or province, the town, the street, and finally the street number. Each step in this list of designations represents a separation, a splitting off. Houses on a different continent are more distant than those on the same continent, houses in a different town more distant than those in the same town, and so on. Eventually, you get to an exact designation—a street address in a given town—that specifies the building uniquely.

The modern biological classification scheme works the same way, except that instead of political and geographical distinctions, it uses biological ones. The categories that biologists use, going from the broadest to narrowest, are *kingdom, phylum, class, order, family, genus, and species.* (A useful mnemonic to help you remember this sequence is the sentence "*King Phillip Came Over For Good Spaghetti*"). You can think of this

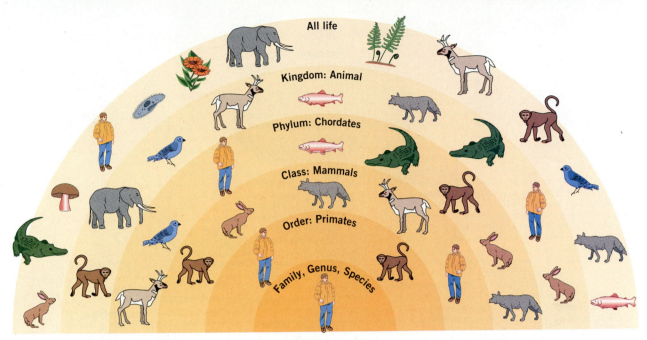

All life

Kingdom: Animal

Phylum: Chordates

Class: Mammals

Order: Primates

Family, Genus, Species

Figure 19–1

classification as being represented by a series of ever-narrowing circles, with the specifications of kingdom, phylum, class, order family, genus, and species serving to guide you to ever-smaller groupings until, at the end, you come to a single organism sitting in a circle all of its own. Each step in this narrowing process involves a judgment about what organisms are like each other and what organisms are not (Figure 19–1).

The **kingdom** is the broadest classification, corresponding to the coarsest division of living things. Until fairly recently, biologists recognized only two kingdoms—plants and animals. This fact, incidentally, explains the query in the game of "twenty questions" about whether an unknown object is animal, vegetable (i.e., plant), or mineral (nonliving). Today, most biologists classify living things into five separate kingdoms (see Figure 19–2):

1. **Monera.** Single-celled organisms without cell nuclei. Monera are the most primitive living things.

2. **Protista.** Mainly single-celled organisms with nuclei, but a few multicelled organisms that have a particularly simple structure.

3. **Fungi.** Multicelled organisms that get their energy and building blocks by absorbing materials from their environment.

4. **Plants.** Multicelled organisms that get their energy directly from the Sun through photosynthesis.

5. **Animals.** Multicelled organisms that get their energy and building blocks by eating other organisms.

One aspect of this classification scheme should be emphasized from the outset. Cataloging living things, describing them, and giving them names—the field of **taxonomy**—is not an exact science. Like the drawing of political boundaries, a certain amount of arbitrariness is built into

The Linnaean classification scheme recognizes the relationship of each living thing to every other. We can arrange species into groups that are more and more closely related. Although human beings are not treated differently from other organisms in this classification system, we are naturally more familiar with our own species than with any other.

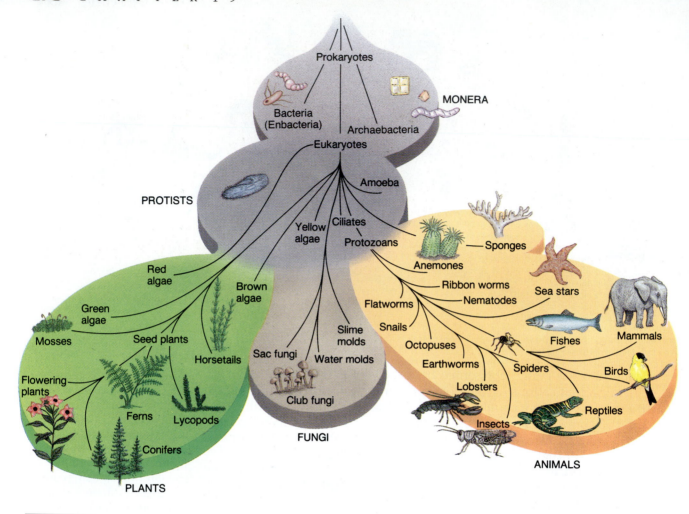

Figure 19–2

The five kingdoms of the biological classification system are illustrated by this diagram, which also suggests the progression from simple monera to more-complex life forms.

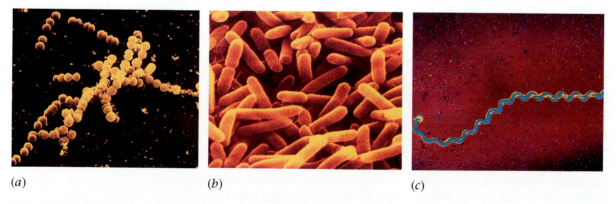

(a)   (b)   (c)

Monera—single-celled organisms without a nucleus—come in a variety of shapes. These microscope photographs show bacteria in the shape of (a) spheres, (b) rods, and (c) spirals.

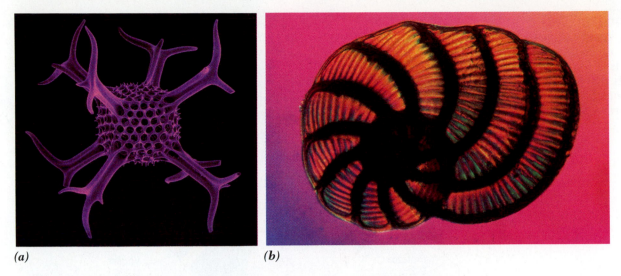

(a)   *(b)*

Some protista (single-celled organisms with a nucleus) develop beautiful microscopic shells. These shells of (*a*) radiolaria, and (*b*) formanifera house amoeba-like cells.

the Linnaean system. Whether we choose to identify two kingdoms or five, whether we have five subdivisions or ten, whether two organisms occupy the same or closely related families, are somewhat subjective decisions, voted on periodically by international panels of taxonomists. To outsiders, these decisions can seem quite political at times. It is not surprising, however, that a scientist who spends his or her life studying one particular kind of organism—a distinctive worm, for example— would much rather have that beast placed in a separate phylum than in a separate class.

Science by the Numbers

How Many Species Are There?

Given the fact that human beings have been observing the natural world and recording their thoughts about it for thousands of years, you might expect that we would by now have a pretty good idea of what's out there. In fact, we don't. Biologists of today engage in a lively debate on one of the most fundamental questions you can ask about the world: How many different species inhabit our planet? Estimates range from 3 to 30 million.

For large animals, the birds and mammals that command human attention, the situation is not so bad. All experts agree, for example, that the Earth is populated by about 9000 species of birds and 4000 of mammals, and only a few new ones are discovered each decade. For plants and insects, however, we are in a state of almost total ignorance. For one thing, there is no central depository where information on new species is collected, so we can't even make a firm statement

Figure 19–3

A pie graph of species distribution among different kinds of living things reveals that insects account for more than half of all known species.

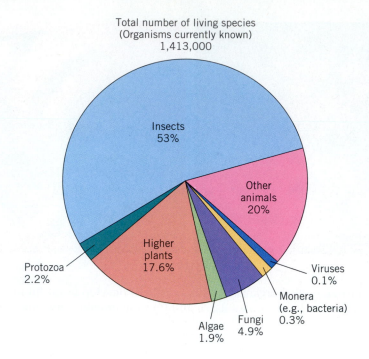

Total number of living species
(Organisms currently known)
1,413,000

Insects
53%

Other
animals
20%

Higher
plants
17.6%

Protozoa
2.2%

Viruses
0.1%

Monera
(e.g., bacteria)
0.3%

Algae
1.9%

Fungi
4.9%

about the number of species that have been *discovered*, much less about ones that haven't been. At the moment, the number of known species—species that have been examined and recorded by scientists somewhere—is estimated to be between 1.5 and 1.8 million. The distribution of these species is shown in Figure 19–3.

Scientists can use a number of methods to extrapolate from current data to a total expected number of species. If we make a graph of the number of known species as a function of time, for example, we should get a curve like the one in Figure 19–4. As time goes by and more species become known, the curve should flatten. If we assume that the discovery curve for insects, for example, will follow this same pattern, then by making a guess as to where we are on the curve, we can estimate what the final level will be. Using this notion, estimates of the total number of species of from 5 to 7 million can be obtained.

Another estimation technique is to do an exhaustive survey of organisms in a small area, determine the ratio of known to unknown species, and assume that this ratio applies worldwide. A study of a particular type of insect in the Indonesian rain forest, for example, came up with a total of 1690 species, of which 1065 were previously known. The ratio of unknown to known, then, is

$$\text{known species/unknown species} = 1690/1065$$
$$= 1.6.$$

If we assume that this number applies to all insects worldwide (and that's a very big if), and accept that there are about 900,000 recorded species of insects, the total number of insect species in the world would then be about

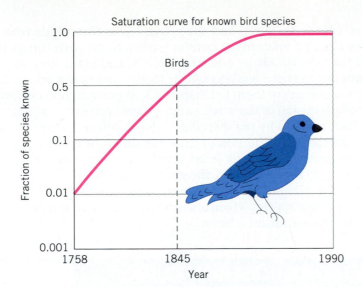

Saturation curve for known bird species

Figure 19–4
A curve of the number of known species of a given kind of living thing versus time should eventually reveal a flattening, called saturation. In the case of birds, we know when in the past a given fraction of species had been discovered.

total species = 1.6 × known species
= 1.6 × 900,000 species
= 1,440,000 species.

Other studies of this type, making longer strings of assumptions about how to apply local numbers to the global system, have produced estimates for the number of insect species as high as 30 million. Obviously, we will have to know a lot more about the kinds of species that actually exist before we can be comfortable with these sorts of estimates. ■

Classifying Human Beings

Human beings clearly are one member of the kingdom of animals. With that decided, we can go on to work out the rest of our address in the classification scheme. Among animals, there is one phylum called *chordates* whose members have a thickened set of nerves down their back. Most chordates (and most of the organisms that come to mind when we hear the word "animal") encase the nerve in bone. These animals form the subphylum of **vertebrates**—animals with backbones. Since human beings have backbones, we belong to this subphylum, as do fish, reptiles, and birds.

Among vertebrates, one group is made up of individuals that are warm-blooded (i.e., whose metabolism maintains the same body temperature regardless of the temperature of the environment), that have hair, and whose females nurse their young. Humans belong to this class of **mammals**, as do wolves, rabbits, and antelopes.

Among mammals we belong to the order of **primates**, which have grasping fingers and toes, eyes at the front of their heads, large brains, and fingernails instead of claws. Monkeys and apes are also members of this order.

At this point in the process of zeroing in on our own species, we encounter an anomaly. Human beings are members of the *hominid* family (primates who walk erect), and the genus *Homo* (hominids who satisfy still more detailed anatomical criteria having to do with things like the shape of the face, the shape and size of teeth, and brain size). In the past there have been other members of this family and genus on the Earth— we will discuss some of them in Chapter 23. At present, however, we have no near relatives—all of those who might have been our near relatives are now extinct, perhaps by our own doing. Most animals have several close cousins—members of the same genus but different species. The grizzly bear (*Ursus horribilis*) and the polar bear (*Ursus maritimus*), for example, are two separate species within the genus *Ursus* (bear).

In any case, all branches of the human race can interbreed, so we are all part of a single species, **Homo sapiens**, and we all inhabit one small circle in the classification scheme.

THINKING MORE ABOUT LIFE

Implications of Linnaean Classification

As we have presented it here, and as Linnaeus originally conceived it in the eighteenth century, biological classification concerns itself solely with segregating living things according to their degree of similarity to each other. No attempt is made to explain *why* living things can be grouped in this way.

Today, however, most scientists would argue that there are compelling reasons why a classification scheme like the one outlined above should exist. For one thing, we no longer have to be content with describing organisms purely in terms of their physical structure. In Chapter 22, for example, we will see that the code carried on molecules of DNA provides an alternate way of describing the structure and functioning of any organism. One way of talking about the differences between two organisms, therefore, is to discuss how closely their DNA molecules resemble each other. In essence, this scheme uses differences at the molecular level (rather than at the level of the organism's structure) to govern the classification system. The main idea is that the closeness of the relationship between two living things— the property that tells us where to draw the circles in Figure 19–1—can now be determined in a quantitative way by comparing the structures of specific molecules.

More importantly, however, the fact that living things can be grouped in successive lay-ers strongly suggests that what Linnaeus and his followers discovered was, in fact, a family tree. If, as we shall argue in Chapter 23, all living things descended from the same primordial ancestor, then the kind of groupings we see would follow naturally. The amount of similarity or difference between two organisms, then, would simply depend on the amount of time that has elapsed since the two shared a common ancestor, and each of the classification groups would result from real events in the past when species split off from each other.

The idea that descent and ancestry, rather than present structure, is what should be used in classifying living things is the basis of two other systems of classification. One, which goes under the general name of *phylogeny*, simply involves presenting the family trees, along with estimates of when specific splitting took place. Another, called *cladistic analysis*, records branchings in the family tree without reference to when they occurred. At the moment, these alternate methods of classification are used in specialized research situations, but there is no general move to replace the venerable Linnaean scheme with something else.

What then, ultimately, is the role of a classification scheme? Is it merely a convenient way to label things? Or should a classification scheme reflect the origins and evolution of the things being classified?

Summary

Biology, the study of living systems, began with efforts to describe the great variety of organisms on Earth. *Taxonomy*, the grouping of living things by their distinctive characteristics, has been aided by the *Linnaean classification* scheme, which groups organisms according to similarities in structure. Scientists now recognize five *kingdoms*: *monera* and *protista*, the two kingdoms of one-celled organisms; *fungi*, which eat dead organic matter; *plants*, which make their own food by photosynthesis; and *animals*, which eat other organisms. A given *species* is placed in groups comprising a series of increasingly specialized labels. The categories are, respectively, phylum, class, order, family, genus, and species.

The *species*, defined to be an interbreeding population of individuals, is the basic unit of classification. *Homo sapiens* (the human species) is in the phylum chordata, the subphylum of *vertebrates*, the class of *mammals*, the order of *primates*, the family of *hominids*, genus *Homo*, and species *Homo sapiens*.

The fact that living things can be grouped in this way can be interpreted as evidence that they all descended from a common ancestor.

Key Terms About Life

biology	taxonomy
Linnaean classification	vertebrates
species	mammals
kingdoms	primates
(monera, protista, fungi, plants, animals)	*Homo sapiens*

Review Questions

1. What are the categories in the modern system of biological classification?
2. Describe the complete Linnaean classification for humans.
3. What are the five kingdoms of living things?
4. How do plants get their energy?
5. How do fungi get their energy?
6. How do animals get their energy?
7. What are four characteristics of all living things?

Discussion Questions

1. How does the fact that living things can be classified in ever more specialized groups support the idea that they all descended from a common ancestor?
2. What is the range of estimates of the number of species on Earth? Why is the number so uncertain?
3. What are some of the reasons that two scientists might disagree on the Linnaean classification of a new kind of worm? Is the Linnaean system an "exact" science, with only one possible correct answer?
4. Why do you suppose that biologists are more confident of the total number of bird species than of insect species? Do you think we know a greater percentage of butterfly species or of worm species? Why?

Investigations

1. Think about courses at your university other than those in science. In what ways do classification play a role in those fields? Is there any field of study that does not use classification in one way or another? Why?
2. Examine recent issues of scientific journals and find an article that describes a new species of animal or plant. How was this new species classified?
3. In the eighteenth and nineteenth centuries, many well-educated amateur naturalists made significant contributions to the study of plants, animals, and fossils. Statesman Thomas Jefferson, artist John James Audubon, and explorers Lewis and Clark are among the famous Americans who described unusual plants and animals. Read a biography of an amateur naturalist and discuss how his or her work contributed to the science of biology.

Additional Reading

Attenborough, David. *Life on Earth.* Boston: Little, Brown, Co. 1979.

Thomas, Lewis. *The Medusa and the Snail.* New York: Viking, 1979.

Weiner, Jonathan. *Planet Earth.* New York: Bantam, 1986.

Wilson, E. O. *The Diversity of Life.* New York: Norton, 1992.

Molecules of Life

A cell's major parts are constructed from a few simple molecular building blocks.

A Random Walk
Seeing the Patterns

Next time you're outside, look closely at a tree. The trunk and limbs divide over and over again, a branching that is mirrored by the hidden root system. The tree has countless almost identical leaves on every limb, as well as myriad seeds in their season. The complex structure of the tree is made by repeating the same simple structures over and over again.

Structures in your city show the same pattern. Buildings feature stacks of identical bricks, row after row of identical windows, and numerous identical shingles, slates, or other roofing materials. The sidewalk is made of slab after slab of concrete, while street lamps and signs also repeat over and over again.

Indeed, almost any complex structure designed by nature or humans is modular, composed of a few simple pieces that combine to form larger objects. The chemicals of life are no different. A few very simple molecules combine to create the wonderful complexity of life around us.

20

Organic Molecules

Four Basic Characteristics

Wood. Leather. Hair. Cotton. Skin. All of these materials originated in living systems on our planet. And like all other materials found in living things, they share some basic chemical characteristics.

1. Most molecules in living systems are based on the chemistry of carbon. In Chapter 9 we saw that carbon atoms possess the ability to form long chains, branches, and rings. In fact, chemists usually refer to molecules containing carbon as **organic molecules,** whether or not they are part of a living system. The branch of science devoted to the study of such carbon-based molecules and their reactions is called *organic chemistry.*

2. Life's molecules form from very few different elements. In terms of the percentages of atoms, just four elements—hydrogen, oxygen, carbon, and nitrogen—comprise 99.4% of our bodies. Phosphorus and calcium each represent 0.2% of our atoms, while all the other elements make up the remaining 0.2% (see Table 20–1). These elements combine to form the molecules that control chemical reactions in all living things.

3. The molecules of life are modular, composed of simple building blocks. You might imagine that large and complicated molecules could be put together in two contrasting ways. One way would be to build each one from scratch, so that no piece of one molecule would be part of another. Another, very different, way would be to make the molecules modular—that is, to build them from a succession of simpler, widely available parts so that each large molecule differs from another only in the arrangement of those parts. Nature, for the most part, chooses the latter strategy in designing the molecules of living systems. We say that the molecules of life are built from simple molecules, or that they are modular.

This technique of modularity is how we construct most modern buildings. Both a humble cottage and a skyscraper can be built from the same basic parts—bricks, windows, doors, stairs, and so on. The skyscraper and

Table 20–1 • Atoms in the Human Body

Element	Percent
hydrogen	62.4
oxygen	24.0
carbon	11.9
nitrogen	1.1
phosphorus	0.2
calcium (mostly in bones)	0.2
all others	0.2

the cottage differ from each other both in the amount of material in them and in the arrangement of those materials, but they contain many of the same basic modules.

In a sense, the decision about which of these two extreme modes to use when making molecules boils down to one of economics. It takes a great deal of work, time, and money to custom-design every door and window and other component of your home. You might end up with a better-designed structure, but at a very high price. By building your home with widely available parts, you save money and still end up with a very satisfactory dwelling.

As it happens, almost all of the large molecules that are crucial in forming and operating living systems are built in a modular manner. Though life's molecules come in an extraordinary variety of shapes and functions, they are made from collections of just a few smaller molecules. This modularity does not mean that the final products are simple, just as there's nothing particularly simple about a skyscraper. It merely means that if we wish to understand how large molecules behave, we first have to talk about the simple pieces from which they are built.

4. Shape determines the behavior of organic molecules—in other words, molecular geometry controls the chemistry of life. The connection between geometry and the behavior of organic molecules can be understood if you remember one important thing about chemical bonds. All chemical bonds are formed by the shifting of electrons among specific pairs or groups of atoms. This bonding property is particularly true of atoms that tend to form ionic, covalent, and hydrogen bonds (see Chapter 9).

A very large and complex molecule may have millions of atoms arranged in a complicated shape. If this large molecule is to take part in chemical reactions—if it is to bind to another molecule, for example— then that binding must take place through the actions of the outer-shell electrons of atoms near the outsides of the two molecules. Specific atoms in each molecule must be able to get near enough to each other so that their electrons can form the bond. Consequently, the geometrical shape of a molecule plays a crucial role, because it determines whether atoms that can form bonds in each molecule will be able to get close enough together for the bonds actually to form.

In principle, an infinite number of molecules could be constructed according to these four rules. In fact, when we examine natural systems we find that only four general classes of molecules govern most of life's main chemical functions. We'll discuss each of these classes—proteins, carbohydrates, nucleic acids, and lipids—in this chapter. Throughout this discussion you should keep in mind that all of these molecules conform to the four rules: they are carbon-based, they form from just a few elements, they are modular in structure, and their behavior depends on their shapes.

Chemical Shorthand

It's not hard to see that if we kept showing all of the atoms and bonds, as we do for an amino acid in Figure 20–1, diagrams of molecules would get very cluttered as the molecules become more complex. Consequently, chemists have adopted a standard shorthand way of representing organic molecules. The rules are:

1. **No hydrogen atoms or bonds to hydrogen atoms are shown in the diagram.**
2. **Carbon atoms are not shown explicitly.**

As an example of how this notation works, look at the following diagram. Both of these drawings show molecules of benzene (a volatile fluid sometimes used as a fuel). On the left, all the atoms are shown, and you can see that each carbon atom forms four bonds to its neighboring atoms. On the right, the carbon atoms are not shown, but we know they are located at the points where the bonds come together. Similarly, the hydrogen atoms are not shown, but we see that each carbon atom has only three bonds shown. We infer the existence of the hydrogen by the "missing" bond.

All atoms and bonds are shown.

Only carbon–carbon bonds are shown. Carbon atoms, hydrogen atoms, and carbon–hydrogen bonds are implied.

 ## Science in the Making

The Synthesis of Urea

In the early 1800s, scientists were not convinced that molecules in living systems are formed according to the same chemical rules that govern those in nonliving systems. Such molecules, after all, had not been produced in the laboratory. In 1828, a German chemist by the name of Friedrich Wohler (1800–1882) performed a series of experiments that were crucial in establishing the ordinariness of organic molecules.

Like many scientists at that time, Wohler had a breadth of experience that is unusual today. Before he began his career as a chemistry teacher, for example, he became a medical doctor and qualified in the specialty of gynecology. He was also interested in the practical aspects of chemistry, and collected minerals from the time he was a child. He described his crucial experiments this way: "I found that whenever one tried to combine cyanic acid [a common laboratory chemical] and ammonia a white crystalline solid appeared that behaved like neither cyanic acid nor ammonia."

After extensive testing, Wohler found that the white crystals were identical to urea, a substance routinely found in the kidneys and (as the name implies) urine, where it plays the role of removing nitrogen wastes from the body. In other words, the appearance of that "white crystalline solid" showed that it is possible to take ordinary chemicals off the shelf and produce a substance found in living systems. He had demonstrated that organic molecules may be formed by the same chemical processes as other materials.

With humor some might find uncharacteristic of academicians, Wohler announced his findings in a letter as follows: "I can no longer, as it were, hold back my chemical urine: and I have to let out that I can make urea without needing a kidney, whether of man or dog." ■

Proteins: The Work Horses of Life

The molecules we call proteins play many key roles in living systems. Some proteins form building materials from which large structures are formed. Your hair, your fingernails, the tendons that hold your muscles in place, and much of the connective tissue that holds your body together, for example, are made primarily of protein molecules. Proteins also serve to regulate the movement of materials across cell walls, and thus control what goes into and out of each cell in your body. In addition, proteins serve as **enzymes** (see below), molecules that control the rate of complex chemical reactions in living things.

Because of these and many other functions, proteins are vital components of your diet, as you have probably heard. You must regularly take in proteins to supply your body with building materials to effect repairs and growth.

Amino Acids: The Building Blocks of Proteins

Proteins are modular just like all complex biological molecules. They are made up of strings of a few basic building blocks called **amino acids.** A typical amino acid molecule is sketched in Figure 20–1. All amino acids incorporate a characteristic backbone of atoms. One end terminates in a *carboxyl group* (COOH), a combination of carbon, oxygen, and hydrogen. On the other end is an *amino group* (NH_2), a nitrogen bonded to two hydrogen atoms (this group gives the molecules their name). Between these two ends a carbon atom completes the backbone.

Branching off the central carbon atom is another atom or cluster of atoms—the "side group" that makes amino acids unique and interesting. Hundreds of different amino acids can be made in the laboratory, each with its characteristic side group. A few common amino acids are sketched in Figure 20–2 to give you a sense of the kind of diversity that is possible within this basic structure.

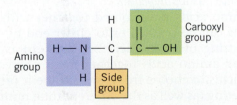

Figure 20–1

An amino acid, showing the amino and carboxyl groups and the side group. The side group varies from one type of amino acid to another, and gives that particular amino acid the chemical properties that distinguish it from any other.

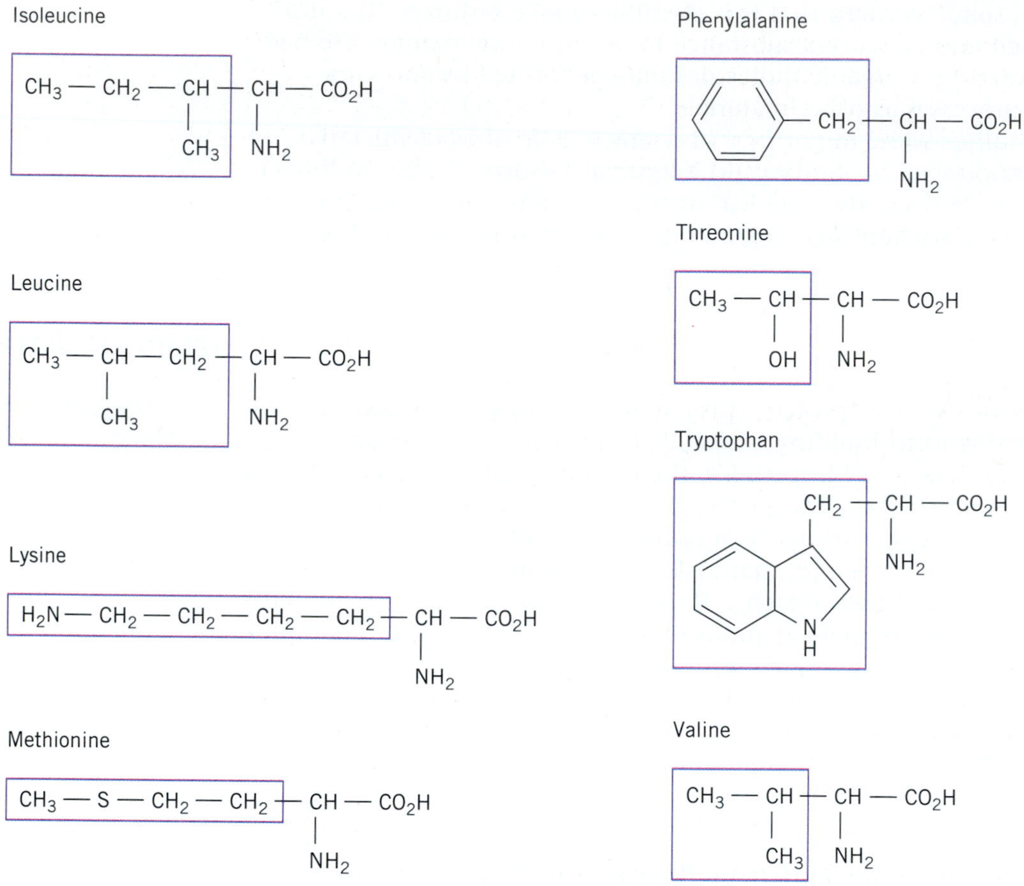

Isoleucine

$$CH_3 - CH_2 - CH | CH - CO_2H$$

Leucine

$$CH_3 - CH - CH_2 | CH - CO_2H$$

Lysine

$$H_2N - CH_2 - CH_2 - CH_2 - CH_2 | CH - CO_2H$$

Methionine

$$CH_3 - S - CH_2 - CH_2 | CH - CO_2H$$

Phenylalanine

Threonine

$$CH_3 - CH | CH - CO_2H$$

Tryptophan

Valine

$$CH_3 - CH | CH - CO_2H$$

Figure 20–2

Several amino acids. Each has a distinctive side group.

Two amino acids can bond together in a very simple way. The hydrogen (H) from one end of one amino acid will connect to the hydroxyl (OH) from the end of another amino acid to form a molecule of water (H_2O). This water molecule moves off (you can think of this process as "squeezing out" the water), leaving the two amino acids hooked together in what is called a *peptide bond*. This process is identical to the *condensation polymerization* reaction that is often used to manufacture plastics and other polymers (see Chapter 9). Indeed, chemists often refer to a bonded chain of amino acids as a *polypeptide*.

Once two amino acids have joined together with a peptide bond, more amino acids can be hooked onto either end by the same process to form a long string of amino acids. A **protein** is a large molecule formed by linking amino acids together in this way. There are many different amino acids to choose from, and different proteins correspond to a different ordering (as well as the different total number) of the amino acids in the string.

One of the great surprises that came out of the study of biochemistry in the early part of the twentieth century was that although chemists can synthesize hundreds of different amino acids, only a small number actually appear in the proteins of living systems on Earth. Only 20 different amino acids are produced in cells, although some of these are modified once the amino acid string is put together. The mystery of why only 20 amino acids are found in living things has a possible explanation in terms of the theory of evolution, which we will explore in Chapter 23. For the moment, however, we note that, even with only 20 basic building blocks, an almost infinite variety of different strings or proteins can be formed (see *Science by the Numbers* in this chapter).

The Structure of Proteins

Proteins are extremely complex molecules, sometimes consisting of many thousands of amino acids and millions of atoms. One of the great triumphs of modern science has been the determination of the exact atomic structures of many of these large molecules. To accomplish this feat, biochemists first have to isolate quantities of pure protein, and then form delicate single crystals in which the large molecules line up in a regular array. These crystals are examined by X-ray crystallography, which reveals the distribution of atoms in space (see Chapter 6).

A protein structure is usually described in four stages, each representing an increasing order of complexity (see Figure 20–3).

1. Primary Structure.   The exact order of amino acids that go into a given protein is called its *primary structure*. Every distinct protein has a different primary structure from others—that is, it has a different order of amino acids along its string.

2. Secondary Structure.   Once a string of amino acids has been formed, however, the story of the protein is not over. Depending on the arrangements of amino acids in the primary structure, hydrogen bonds can form that give the final protein a specific shape. Some proteins, for example, take the form of a long helix or long spring. Others fold back on themselves repeatedly to form rough spheres. Shapes taken by the string of amino acids that makes up the primary structure of a protein are called its *secondary structure*.

When you cook an egg, you can see the effect of secondary structure. The proteins in egg white are wrapped up into tiny spheres scattered throughout the fluid. This is why normal egg white is transparent. When you cook the egg white, you break the hydrogen bonds that keep the protein wrapped up and allow the molecules to unfold. The tough mat they form when they interlock gives the cooked egg white its characteristic texture.

3. Tertiary Structure.   As parts of the amino acid chain fold back on itself, atoms in the side groups can come into contact with each other. As a result, additional cross-linking chemical bonds form between side groups in amino acids in different parts of the chain. One common link occurs between sulfur atoms in different side groups. The distinctive shape of human insulin, for example, arises because of bonds that form between sulfur atoms in the amino acid cystine.

Figure 20–3

The structure of a protein can be described in four steps. Primary structure (*a*) is the sequence of amino acids. Secondary structure (*b*) is the way the sequence kinks or bends. Tertiary structure (*c*) is the shape of the completely folded protein. Quaternary structure (*d*) is the clustering of several proteins to form the active structure; in this case, hemoglobin which carries oxygen in your bloodstream.

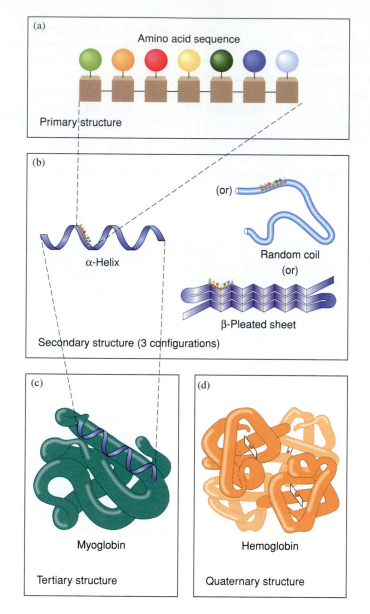

As a result of these links, a protein will twist around, kink up, and fold itself into a complex shape, much as a string will fold itself into a complex shape if it's dropped on a table. This complex folding is the *tertiary structure* of the protein.

4. Quaternary Structure. Finally, two or more long chains, each with its own secondary and tertiary structure, may come together to form a single larger unit. This joining of separate protein chains determines the *quaternary structure* of the protein.

The prediction of the exact shape that will be assumed by a given sequence of amino acids remains one of the great goals of modern biochemistry, a goal that we are still far from reaching. But whether or not we can predict the ultimate shape of a protein, the fact remains that each different sequence of amino acids will produce a large molecule with a different three-dimensional shape. This fact will become important when we consider the role of proteins in the cell's chemistry.

Science by the Numbers

How Many Proteins Can You Make?

At first, it may seem that with only 20 different amino acids available to build proteins in living systems, the number of different kinds of proteins you could make would be rather limited. Let's do some calculations to see why this isn't the case.

Suppose we start by asking a simple question: How many different proteins, each 10 amino acids long, can we make from amino acids found in living systems? (Once you've seen the answer to this question, you'll be able to figure out for yourself how to calculate the answer for a protein with any number of amino acids in it.)

One way of thinking about this question is to imagine that each protein consists of 10 amino acids on a string, and each amino acid can be selected from one of 20 different boxes. We could choose the first amino acid of the string from any one of 20 boxes. For each of these 20 choices for the amino acid in the first position, there are 20 choices we could make for the amino acid in the second position. Thus we could choose amino acids to fill the first two vacancies on the string in $20 \times 20 = 400$ different ways.

Following this logic, the number of ways to arrange 10 amino acids in a string is

$$20 \times 20 \times 20 \times 20 \times 20 \times 20 \times 20 \times 20 \times 20 \times 20 = 1.028 \times 10^{13}.$$

Thus we could make about 10 trillion different proteins that contain 10 amino acids.

This number is huge—100 million times larger than the number of proteins used in the human body. And, of course, this is just the number of different proteins you could make from only 10 amino acids. Typical proteins in living systems contain many more amino acids, sometimes up to hundreds of them. The number of different enzymes that could be made from such proteins is almost unimaginably large.

The bottom line of this calculation is that, although only 20 different amino acids appear in living systems, this number still allows for a tremendous diversity of proteins. ■

Proteins as Enzymes

One of the key roles that proteins play in living systems is to act as enzymes in chemical reactions in cells. An enzyme is a molecule that facilitates bonding between two other molecules, but which is not altered or taken up in that overall reaction. Because of the presence of the enzyme, the chemical reaction takes place at a much faster rate than it otherwise would.

Figure 20–4

An enzyme in action joins
two molecules (designated
A and B) and produces a
product (C). The product is
released and the enzyme is
free to repeat the process.

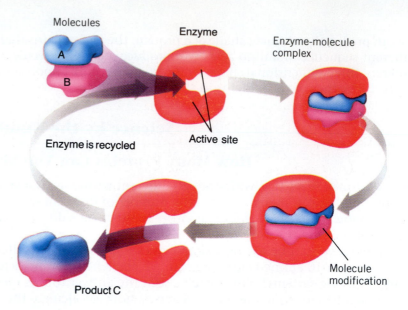

Enzymes play a role in chemical reactions similar to a broker or an
agent in a business deal. The role of the broker is to bring together a
buyer and seller; but brokers, themselves, do no buying or selling. The
buyer and seller eventually might find each other without the help of
the broker, but the deal goes through much quicker if the broker is there.
In the same way, a molecule that plays the role of an enzyme may bring
together two other molecules and facilitate their forming a bond, or it
may tear a molecule apart, without itself being included in the chemical
reaction. Because of the enzyme, the reaction takes place relatively
quickly.

Enzymes illustrate the primary importance of geometrical shape in
determining how chemical reactions take place among large molecules.
You can easily visualize the workings of an enzyme (see Figure 20–4).
Each large molecule has places on it—atoms or small groups of atoms—
where chemical bonding can take place. Think of these locations as sticky
spots somewhere on a large, convoluted molecular shape. In order for
two or more molecules to interact, their respective sticky spots have
to come into contact. More precisely, the atoms whose electrons will
eventually form the bonds must be brought close enough together for
the electrons to interact. This proximity is unlikely to happen at random.

Imagine each molecule as a large pile of string that is dumped in the
corner of a room, and think of the spots that could form chemical bonds
as patches of glue located at random places along the string. If you took
two such pieces of string and tossed them together into the corner, the
chances are very slight that any of the sticky spots would come near each
other. If, on the other hand, you picked up the string and arranged it so
that the sticky spots were next to each other, you could make sure that
they formed a bond. In this "chemical reaction," you are playing the
role of the enzyme. You cause bonds to form that probably wouldn't
form without you, but you do not become a part of the bond.

Enzymes perform an analogous function in organic reactions. Typically, an enzyme is a large molecule that has particular spots on its surface
into which reacting molecules will just fit. The enzyme attracts first one
of the molecules and then the other. In some enzymes, a pair of specific

molecules attach themselves in only one way, so it is guaranteed that their "sticky spots" will be near each other. The chemical bonds that hold them together will form. Once the bonds have formed, the overall shape of the resulting composite molecule suddenly becomes different from that of either of the two molecules that went into it. Consequently, the new large molecule no longer fits into the appropriate grooves and valleys of the enzyme, and it spontaneously breaks free and wanders off by itself. This separation leaves the enzyme free to mediate the same reaction again, each time with two new pieces. Another kind of enzyme, like those in your stomach, performs the opposite function—breaking apart large molecules into smaller units, over and over again.

If you think about the way an enzyme works, you will realize why molecular shape is so important. The hills and valleys on the surface of an enzyme serve as resting places for the molecules that interact on the enzyme's surface. Versatile protein molecules, which adopt many different shapes and, can therefore provide resting places for many different kinds of interacting molecules, are ideally suited to function in this way. For precisely this reason, proteins participate in most of the chemical reactions in living organisms.

The reactions of complex molecules in living systems are mediated by enzymes. Thus the molecules that act as enzymes in living systems play a crucial role in determining the properties of those systems.

The Human Body

Proteins and Diet

Because proteins play such an important role in living systems, the cells in your body need a constant supply of all 20 amino acids. In adults, 12 of the 20 amino acids are synthesized in the body. The rest, the so-called *essential amino acids*, have to be taken into the body in the proteins and other foods that we eat. Since amino acids, unlike fat, cannot be stored by the body, the essential amino acids have to be present in every meal in roughly the proportion they occur in the body's own proteins.

Foods vary widely in their total protein content, from about 1% in bananas and carrots to almost 30% in peanuts and some cheese. Foods that supply amino acids in roughly the same proportion as those in human proteins are called *high-quality proteins*, while those that supply too little of one or more amino acid are called *low-quality proteins*. In general, meat and dairy products supply high-quality proteins, while plant products supply low-quality proteins.

In Figure 20–5, we make this point by comparing the amount of each of the eight essential amino acids found in eggs (the food whose amino acids most closely match human protein proportions) and cornmeal. You should note, however, that it is not necessary for each food we eat, in and of itself, to supply all essential amino acids. It is possible to plan meals so that amino acids from one food make up for deficiencies of that amino acid in the other. Many traditional meals have this property. Milk, for example, provides the lysine that breakfast ce-

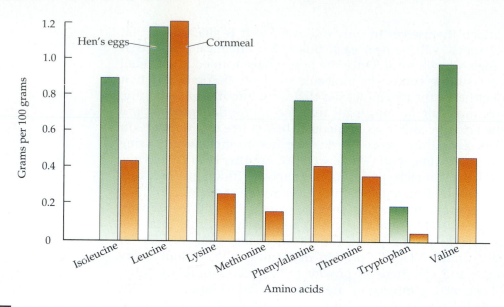

Figure 20–5

Proportion of essential amino acids in egg and cornmeal. The amino acid proportions for egg are close to those in the human body, while the cornmeal contains too little of the amino acids lysine, methionine, and tryptophan.

real lacks. The American staple, the peanut butter and jelly sandwich, provides the same kind of matching proteins. In fact, many traditional foods from around the world do pretty well in supplying complementary sources of amino acids. Some examples: corn tortillas and beans (Mexico), rice and tofu (Japan), and rice and ground nuts (West Africa). Because of the fact that plants provide low-quality protein, however, individuals on vegetarian diets must plan their food intake carefully to compensate for possible protein deficiencies. ■

 Technology

Designer Drugs

Many of the drugs we take produce their effect because of the shape of their molecules. Some drugs, for example, alter our body chemistry by blocking the action of enzymes. You can understand how such a drug might work by looking at Figure 20–4. The efficacy of the enzyme depends on the fact that the shape of its surface matches the shape of the molecules involved in the reaction. A drug molecule that attached itself to one of those crucial sites on the enzyme would block that site, preventing one of the molecules involved in the original reaction from occupying it. As a result, the enzyme would not be able to facilitate the reaction as it normally does, and the chemical balance of the cell would be changed. When you take an aspirin, for example, you are blocking the action of an enzyme that facilitates the production of molecules called prostaglandins. These molecules, among other things, affect the transmission of nerve signals.

Other drugs work in similar ways on other cellular processes. We shall see in Chapter 21, for example, that part of the process of moving materials in and out of a cell across the cell membrane involves the fit between the molecules being moved and specialized proteins, called receptors, in the membrane. A drug that attaches to the receptor or to the material being brought in or out will block this match and alter the traffic in and out of the cell. Similarly, in Chapter 5, we noted that nerve impulses are transmitted from one nerve cell to the next by special molecules called neurotransmitters. These molecules are shaped so that they fit into specific sites on the "downstream" nerve cell. Many drugs, including alcohol and Valium, gain their effect because they have the right shape to bind to the synapses and alter their operation.

Throughout most of the history of medicine, the search for new drugs has been carried out in a somewhat random way. If you think of the place where the drug is supposed to bind as a lock, and the drug molecule as a key that will fit the lock, then the traditional search for new medicines would correspond to trying one key after another until you find one that fits. It is not unusual, for example, for drug companies to test 10,000 different molecules to produce one effective drug.

As our understanding of the geometry of organic molecules has increased, however, scientists are increasingly able to consider producing a molecule with the right shape from scratch. In terms of our lock-and-key analogy, this process would correspond to studying the lock and then making a key to fit it. Products made in this way have been nicknamed "designer drugs." One such drug (called Captopril) has been in use since 1975. This drug blocks the action of an enzyme that produces molecules that contribute to hypertension, and so is used to control that condition. Designer drugs for treating psoriasis, glaucoma, AIDS, and some forms of cancer and arthritis are in advanced stages of testing and may be on the market soon. Given the rate of progress in understanding the molecular basis of life, it seems clear that many more designer drugs will be marketed in our future. ■

Carbohydrates ▬

Carbohydrates, the second important class of modular molecules found in all living things, are made up of carbon, hydrogen, and oxygen. They play a central role in the way that living things acquire and use energy, and they form many of the solid structures of living things. You use carbohydrates every day—in many of the foods you eat, the fuels you burn, the clothes you wear, and even the paper of this book. They are truly a diverse group of molecules.

The simplest carbohydrates are **sugars,** molecules that usually contain five, six, or seven carbon atoms arranged in a ringlike structure. *Glucose*, an important sugar in the energy cycle of living things, is sketched in Figure 20–6. Glucose figures prominently in the energy metabolism of every living cell. The "burning" of glucose in our cells supplies the energy that we use to move and grow.

The general chemical formula for sugar is $C_nH_{2n}O_n$ or $C_n(H_2O)_n$. Glucose, for example, has the formula $C_6H_{12}O_6$. As often happens with organic molecules, other forms of the molecule have the same chemical composition, but have the components arranged differently. In Figure 20–7, for example, we show the sugar fructose. As the name implies, this sugar is commonly found in fruit. It has the same number of carbon, hydrogen, and oxygen molecules as glucose, but the atoms are arranged slightly

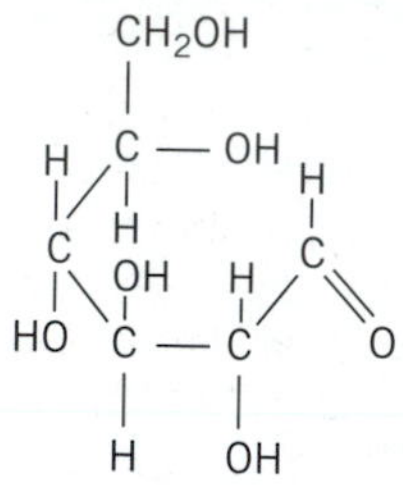

Figure 20–6
The structure of glucose.

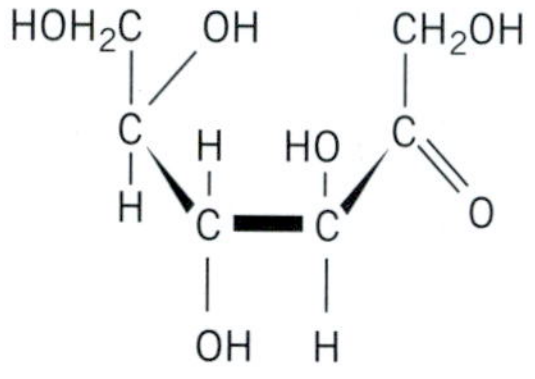

Figure 20–7
The structure of fructose. It has the same number and kinds of atoms as glucose, but is a different arrangement.

differently, and this different arrangement gives fructose a different chemical behavior.

Chemists call individual sugar molecules *monosaccharides*, meaning "one sugar." (The same root word is used when we describe an overly sentimental story as "saccharine.") The carbohydrates that we eat, however, are usually formed from two or more sugar molecules strung together in a covalently bonded chain to form a sugar polymer. Ordinary table sugar, for example, is made from two sugars, glucose and fructose, linked together by covalent bonds.

When many sugars are strung together in a chain, the resulting molecule is called a *polysaccharide* ("many sugars"). The two most familiar polysaccharides are starch and cellulose. Both of these kinds of molecules are made from long chains of glucose molecules. They differ from each other only in the details of the way the glucose molecules bind to each other.

Starches, a common component of the human diet, are a large family of molecules in which the glucose constituents link together at certain points along the ring. It is found in many plants, such as potatoes and corn. Humans break down starch molecules with an enzyme in the digestive system, thus releasing individual glucose molecules, which provide the fundamental energy fuel used by cells. Cellulose, a long, stringy polymer that provides the main structural element in plants, from stems and leaves to the trunks of trees, also forms from glucose molecules. Because the glucose molecules are linked in a different way, however, human beings cannot digest cellulose—we do not manufacture an enzyme that can separate individual glucose molecules from the cellulose polymer. Consequently, humans don't go out and graze on the lawn at lunchtime, though people on diets often eat celery and other "roughage" or "fiber." On the other hand, cellulose can be digested and broken down by many bacteria. Cows, for example, have bacteria in their stomachs that perform this function for them, as do wood-eating termites.

The wood fibers in the paper on this page are made from glucose molecules bonded into cellulose, basically the same chemical as in the stalk of a celery stick. The same glucose molecules, bonded in a different way, form the flour in the spaghetti you ate the last time you had an Italian dinner. An amazing diversity can be built into organic molecules through modular construction.

◼ Nucleic Acids

The discovery of the nature and function of nucleic acids has fundamentally transformed the study of biological systems in the past three decades. **Nucleic acids,** so called because they were originally found in the nucleus of cells, include **DNA** and **RNA,** the molecules that carry and interpret the genetic code. As we shall see in Chapter 23, these molecules govern both the inheritance of physical traits by offspring and the basic chemical operation of the cell. In that chapter we will explore how nucleic acids carry out these functions. Here we simply describe how the molecules are put together, with special attention to the way they conform to the twin principles of modularity and geometry followed by all other organic materials.

Ribose   Deoxyribose

Figure 20–8
Ribose and deoxyribose.
There is one "missing" oxygen in deoxyribose. As explained earlier, the hydrogen atoms are omitted from
this figure to help clarify
the difference between
these two molecules, but
the bonds to the missing
atoms are shown for com-
pleteness.

Nucleotides: The Building Blocks of Nucleic Acids

Proteins (chains of amino acids) and carbohydrates (clusters of sugar molecules) can form large structures from a single kind of building block. Nucleic acids, on the other hand, are assembled from subunits which are themselves made from three different kinds of smaller molecules. The assemblage of three molecules is called a *nucleotide*, and nucleic acids are made by putting nucleotides together in a long chain.

The first of the smaller molecules that go into an individual nucleotide is a sugar. In DNA it's *deoxyribose* (it is this sugar that gives DNA its complicated name, *deoxyribonucleic acid*), while in RNA (*ribonucleic acid*) the sugar is *ribose*. Figure 20–8 shows these two closely related sugars. Ribose is a standard sugar containing five carbons, while deoxyribose, as the name implies, is missing one oxygen atom (in the place shown by the dotted box in the figure).

The second small molecule of the nucleotide is the phosphate group—one phosphorus atom surrounded by a tetrahedron of four atoms of oxygen.

Finally, each nucleotide incorporates one of four different kinds of molecules that are called *bases*. The four different base molecules are often abbreviated by a single letter—A for adenine, G for guanine, C for cytosine, and T for thymine.

Each nucleotide combines the three basic building blocks—a sugar, a phosphate, and a base (see Figure 20–9). These three molecules bond together with the sugar molecule in the middle. You can think of a nucleotide as something like a prefabricated wall in a house. Both DNA and RNA are made by putting nucleotides together in a specific way.

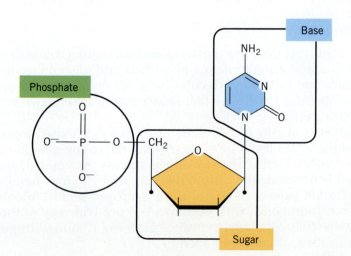

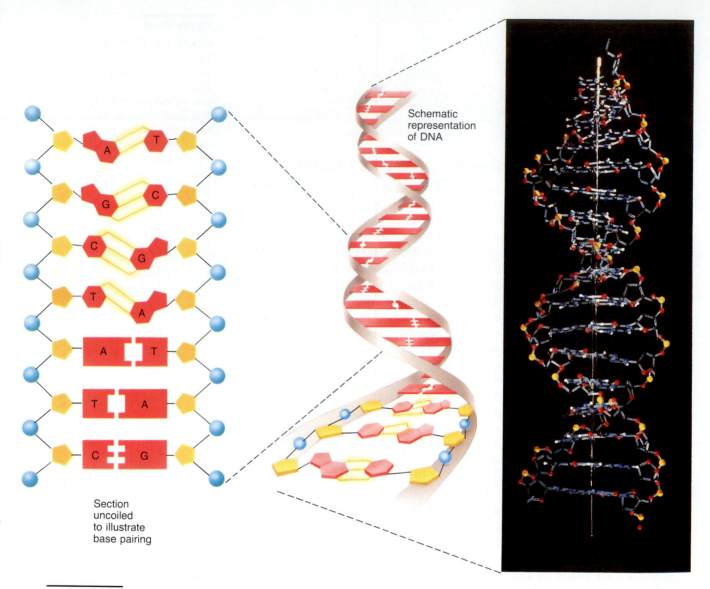

Schematic
representation
of DNA

Section
uncoiled
to illustrate
base pairing

Figure 20–10

The structure of DNA illustrated schematically uncoiled (*a*), as a double helix (*b*), and in an atomic model (*c*).

DNA Structure

We can start putting DNA together by assembling a long strand of nucleotides. In this strand, the alternating phosphate and sugar molecules form a long chain, and the base molecules hang off the side (see Figure 20–10). The whole thing looks like a half-ladder that has been sawn vertically through the rungs.

DNA consists of two such strands of nucleotides joined together to form a complete "ladder." The bases sticking out to the side provide the natural points for joining the two single strands. As you can see from Figure 20–11, however, the distinctive shapes of the four bases ensure that only certain pairs of bases can form hydrogen bonds. Adenine, for example, can form bonds with thymine, but not with any of the other bases or with itself. Similarly, cytosine can form a bond with guanine, but not with itself, thymine, or adenine.

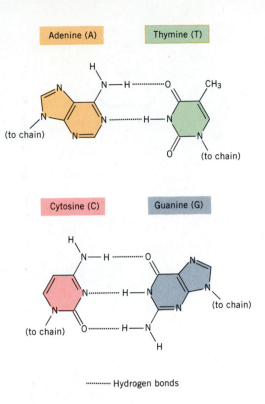

Figure 20–11
AT and CG base pair
linkages.

As a consequence, there are only four possible rungs that can exist in a DNA ladder. They are:

AT
TA
CG
GC

With the bonding of these *base pairs*, the complete DNA molecule is formed into a ladder-like double strand. It turns out that, because of the details of the shape of the bases, each rung is twisted slightly with respect to the one before it. The net result is that this ladder comes to resemble a spiral staircase, as shown in Figure 20–12—a helical shape that gives DNA its common nickname, the **double helix.** In Chapter 22 we will show how the sequence of bases along the DNA molecule governs both inheritance and the everyday chemistry of the cell.

RNA Structure

RNA is built in a manner similar to DNA with three important differences. First, RNA is only half the ladder—that is, it consists of only one string of nucleotides put together. Second, the sugar in the RNA nucleotide is ribose instead of deoxyribose. And third, the base thymine is replaced by a different base, uracil, abbreviated U. The shape of uracil is such that, like thymine, it will bond to the base adenine. We shall see in Chapter 23, however, that the ability of uracil to bond to adenine plays an important role in regulating chemical reactions in the cell. Several different

Figure 20–12

James Watson next to a model of the double helix structure, which he discovered with Francis Crick.

Figure 20–13
The structure of RNA.

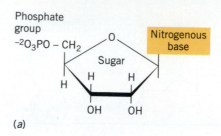

(a)

RNA

(b)

kinds of RNA operate in the cell at any given time. All of them, however, have the same basic structure, shown in Figure 20–13.

Lipids

The final important set of organic molecules that go into making up living things is something of a grab-bag category called lipids. **Lipids** include a variety of molecules that will not dissolve in water, and they are all around you. Lipid molecules form fats in food, waxes in candles, greases for lubrication, and a wide variety of oils. If you think of drops of oil or bits of fat floating around on top of a pot of soup, you have a pretty good picture of what large clumps of lipid molecules are like.

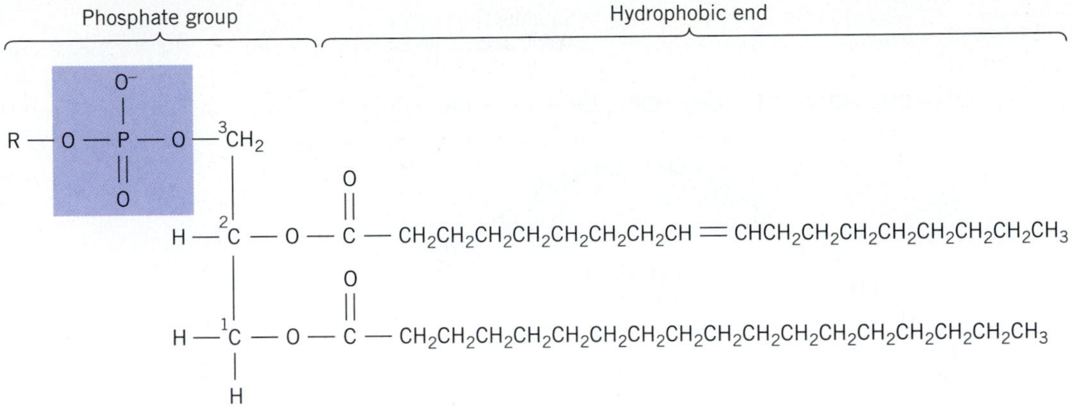

Phosphate group | Hydrophobic end

Figure 20–14
A phospholipid molecule, showing the negatively charged phosphate group at one end, and ordinary hydrocarbon chains at the other. The end with the phosphate group is attracted to water, and the hydrocarbon end is repelled by it. Different collections of molecules in the group labeled "R" correspond to different kinds of phospholipids.

At the molecular level, lipids play two important roles in living things. First, they form the cell membranes that separate living material from its environment, as well as separate one part of the cell from the other. They are also used to store energy. In fact, in the human body, excess weight is usually carried in the form of fat, which is a different kind of lipid from those in cell membranes. Lipids are extremely efficient storehouses for energy. A typical gram of fat, for example, contains twice as many calories as a gram of either protein or carbohydrate.

Like proteins, carbohydrates, and nucleic acids, numerous lipid molecules can come together to form large modular structures in every cell. An important class of these molecules, called *phospholipids*, are long and thin with a carbon backbone, as shown in Figure 20–14. In phospholipids, a phosphate group (one phosphorus and four oxygen atoms) is incorporated into one end of the molecule. The oxygen atoms in this group tend to be negatively charged, so that this end of the molecule is attracted to water (we say it is hydrophilic). The other end of the molecule, however, is repelled by water (we say it is hydrophobic). These particular types of lipids play an extremely important role in living systems because, as we shall see, they are the materials from which cell membranes are made.

Saturated and Unsaturated Fats

Every carbon atom in a lipid chain forms exactly four bonds to neighboring atoms (see Chapter 9). In a straight chain, each carbon bonds to two adjacent carbons along the chain and two hydrogens on the sides. Carbons of this type are *saturated*—fully bonded to four other atoms.

In some lipids, adjacent carbon atoms will have only three neighbors—two carbon and one hydrogen atoms. A kinked "double bond" will thus form between the two carbons (see Figure 20–15). A chain with one double bond is *unsaturated*, while two or more double bonds yield a *polyunsaturated* lipid.

Saturated fats in the diet provide the raw materials from which the body can synthesize *cholesterol*, an essential component of all cell membranes. Unfortunately, high levels of cholesterol in the blood can also lead to fatty deposits that clog your arteries. For these reasons, many food producers emphasize their use of cholesterol-free foods, rich in poly-

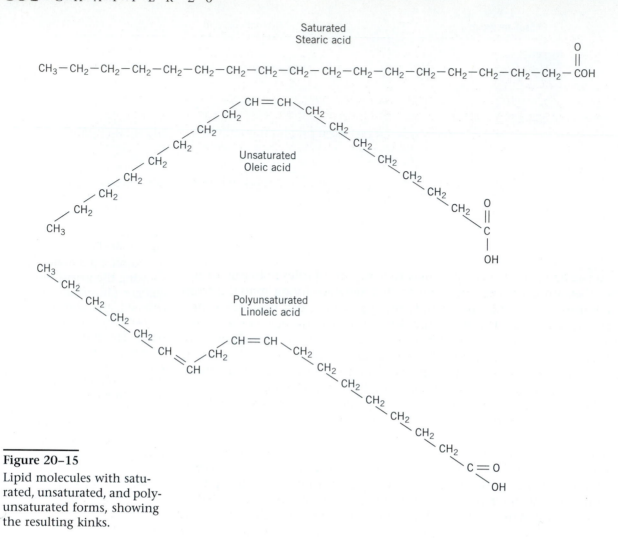

Saturated
Stearic acid

$CH_3-CH_2-CH_2-CH_2-CH_2-CH_2-CH_2-CH_2-CH_2-CH_2-CH_2-CH_2-CH_2-CH_2-CH_2-CH_2-CH_2-COH$

Unsaturated
Oleic acid

Polyunsaturated
Linoleic acid

Figure 20–15
Lipid molecules with saturated, unsaturated, and polyunsaturated forms, showing the resulting kinks.

Figure 20–16
The partial hydrogenation of liquid vegetable oil (*a*) converts it to a solid product (*b*).

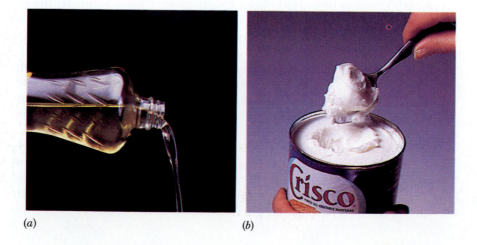

(*a*)   (*b*)

unsaturated fats. Food advertising often suggests that unsaturated foods are "good for you."

Be warned, however. Many of these polyunsaturated products require a further processing step, called *hydrogenation*, to give them a pleasing texture and consistency. Popular chocolate candies, for example, must be made from fats and oils that soften near body temperature. Hydrogenation—the addition of hydrogen atoms back into the carbon chains—eliminates carbon-carbon double bonds. The hydrogen atoms pull one of the carbon bonds apart and thus partially saturate the lipid chains. Many popular cooking oils that begin with highly unsaturated lipids are "partially hydrogenated for freshness and consistency" at the manufacturing plant, a process that undoes the good of the unsaturated bonds (Figure 20–16).

Cell Membranes

The most important single function of lipids in our bodies is the formation of cell membranes. Phospholipids, with their hydrophobic and hydrophilic ends, perform this function because, when placed in water, these molecules typically adopt a double-layered structure like the one shown in Figure 20-17. The hydrophobic ends of the molecule line up facing each other, while the hydrophilic ends face to the outside. In this way, water is kept away from the hydrophobic ends and is kept nearer the hydrophilic ends. A double-layered structure of molecules like this func-

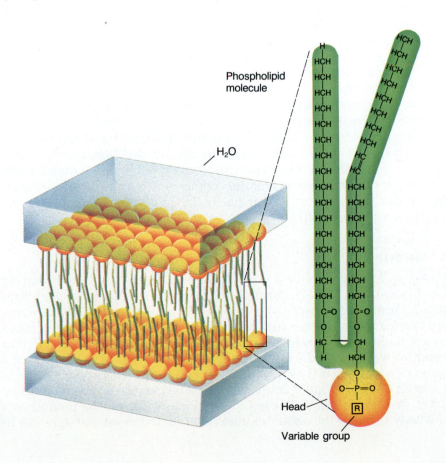

Phospholipid molecule

H_2O

Head

Variable group

Figure 20–17
The structure of a lipid bilayer. The hydrophobic ends of the molecules face each other, while the hydrophilic ends are in the surrounding water.

tions very well as a membrane of a cell. It is flexible and can change its shape, but it also provides a tough barrier. Special mechanisms have to be in place in order for anything to elbow its way through the bimolecular layer.

The structure of the lipid bilayers that make cell membranes is remarkably similar to liquid crystals (see Chapter 9). The lipid molecules are highly ordered in their orientation and spacing, but they are somewhat disordered in their exact positions side-to-side. This loose structure provides an important measure of flexibility to cell membranes.

Vitamins and Minerals

While most of our body is made and operates with proteins, carbohydrates, nucleic acids, and lipids, other chemicals are also vital to life. These essential chemicals include the familiar minerals and vitamins that must form part of our diet.

Minerals

Minerals, in a nutritional context, include all of the chemical elements in our food other than carbon, hydrogen, nitrogen, and oxygen. The most abundant mineral in our bodies is calcium, which is concentrated in bones and teeth and comprises almost 2% of our total weight. Even though bones appear to be solid, permanent structures, your calcium is constantly being replenished. Gradual loss of calcium, particularly in older women, is a major cause of bone disease and injury.

Cellular fluids require small amounts of the elements potassium, chlorine, sodium, and magnesium to maintain proper body acidity and control electrical charges in nerve processes. A grab bag of minor or *trace elements*, from iodine in the thyroid gland to iron in the blood, are also involved in the body's chemistry.

Every few years the National Academy of Sciences and the National Research Council publish a list of the Recommended Dietary Allowances (RDA) for minerals. The RDA represents the best judgment of the scientific community about how much of the various minerals need to be included in the diet to meet nutritional needs. In Table 20–2, we show some examples of mineral RDAs.

Vitamins

Biologists have discovered a host of complex organic molecules that, in small quantities, play an essential role in good health. These chemicals, though unrelated to each other in any chemical or physiological sense, are known collectively as **vitamins**, and they are generally designated by a letter, such as vitamin A. For historical reasons, a number of different vitamins were grouped together under vitamin B and have been given a series of numbers such as B_1 (thiamine) and B_2 (riboflavin).

With one exception, vitamins are not made in the body and must be taken in with our food. The exception, vitamin D, can be produced in the body through the action of ultraviolet radiation on the skin, but in

Table 20–2 ● Selected Recommended Dietary Allowances (RDAs) for Minerals[a]

Gender	Age	Weight kg	Weight lb	Height cm	Height ft'in"	Ca (mg)	P (mg)	Fe (mg)	Mg (mg)	Zn (mg)	I (μg)[b]	Se (μg)[b]
Both	0–0.5	6	13	60	2'	400	300	6	40	5	40	10
	0.5–1	9	20	71	2'4"	600	500	10	60	5	50	15
	1–3	13	29	90	2'11"	800	800	10	80	10	70	20
Men	15–18	66	145	176	5'9"	1200	1200	12	400	15	150	50
	19–24	72	160	177	5'10"	1200	1200	10	350	15	150	70
	25–50	79	174	176	5'10"	800	800	10	350	15	150	70
	51+	77	170	173	5'8"	800	800	10	350	15	150	70
Women	15–18	55	120	163	5'4"	1200	1200	15	300	12	150	50
	19–24	58	128	164	5'5"	1200	1200	15	280	12	150	55
	25–50	63	138	163	5'4"	800	800	15	280	12	150	55
	51+	65	143	160	5'3"	800	800	10	280	12	150	55
—Pregnant						1200	1200	30	320	15	175	65
—Lactating												
(First 6 months)						1200	1200	15	355	19	200	75
(Second 6 months)						1200	1200	15	340	16	200	75

[a] Condensed version of Recommendations by the Food and Nutrition Board of the National Academy of Science, National Research Council. Published in 1989.

[b] 1 μg (one microgram) = 10^{-6} g = 10^{-3} mg.

most parts of the world exposure to sunlight is normally too low to produce enough. Thus, as a practical matter, all vitamins must be taken in as part of the diet.

The vitamins in the B category, along with vitamin C, are *water-soluble vitamins*. As the name implies, these vitamins dissolve in water and hence are not retained by the body. The supply of water-soluble vitamins must be renewed daily. Vitamins A, D, E, and K, however, are *fat-soluble vitamins*. They can be stored in the body (in the liver, for example). In some cases, taking in too much of a fat-soluble vitamin can have unwanted or even harmful consequences. Too much vitamin D, for example, can lead to calcium deposits forming in the heart and kidneys, while too much vitamin A can be seriously toxic. (Too much beta-carotene, a precursor of vitamin A, can even turn your skin orange—but only temporarily.)

Vitamins serve a wide variety of functions in the body. Many of them assist enzymes in mediating the body's chemical reactions. In fact, most vitamins were discovered through the study of diseases that are caused by a chemical deficiency. The disease scurvy, for example, causes degeneration of tissues when the body fails to obtain enough vitamin C, a chemical abundant in citrus fruits. Scurvy was particularly common among sailors on long ocean voyages until the connection between diet and disease was made. Subsequently, sailors on British naval vessels were fed a regular diet of limes—hence the nickname "limeys." Similarly, the bone disease rickets results from a deficiency of vitamin D. The modern diet, with vitamin-enriched foods and vitamin supplements, can virtually eliminate these deficiency diseases. As with minerals, the National Academy of Sciences publishes RDAs for vitamins. The RDAs for some vitamins are shown in Table 20–3.

Table 20–3 • Selected Recommended Daily Allowances (RDAs) for Vitamins[a]

Gender	Age	Weight kg	Weight lbs	Height cm	Height ft'in"	Fat-Soluble Vitamins A (RE[b])	Fat-Soluble Vitamins D (μg^c)	Fat-Soluble Vitamins E (mg α-TE[d])	Water-Soluble Vitamins Folacin (μg^c)	Water-Soluble Vitamins Niacin (mg NE[e])	Water-Soluble Vitamins Riboflavin (mg)	Water-Soluble Vitamins Thiamine (mg)	Water-Soluble Vitamins B_6 (mg)	Water-Soluble Vitamins B_{12} (μg^c)	Water-Soluble Vitamins C (mg)
Both	0–0.5	6	13	60	2'	375	7.5	3	25	5	0.4	0.3	0.3	0.3	30
	0.5–1	9	20	71	2'4"	375	10	4	35	6	0.5	0.4	0.6	0.5	35
	1–3	13	29	90	2'11"	400	10	6	50	9	0.8	0.7	1.0	0.7	40
Men	15–18	66	145	176	5'9"	1000	10	10	200	20	1.8	1.5	2.0	2.0	60[f]
	19–24	72	160	177	5'10"	1000	10	10	200	19	1.7	1.5	2.0	2.0	60
	25–50	79	174	176	5'10"	1000	5	10	200	19	1.7	1.5	2.0	2.0	60
	51+	77	170	173	5'8"	1000	5	10	200	15	1.4	1.2	2.0	2.0	60
Women	15–18	55	120	163	5'4"	800	10	8	180	15	1.3	1.1	1.5	2.0	60
	19–24	58	128	164	5'5"	800	7.5	8	180	15	1.3	1.1	1.6	2.0	60
	25–50	63	138	163	5'4"	800	5	8	180	15	1.3	1.1	1.6	2.0	60
	51+	65	143	160	5'3"	800	5	8	180	13	1.2	1.0	1.6	2.0	60
—Pregnant						800	10	10	400	17	1.6	1.5	2.2	2.2	70
—Lactating															
(First 6 months)						1300	10	12	280	20	1.8	1.6	2.1	2.6	95
(Second 6 months)						1200	10	11	260	20	1.7	1.6	2.1	2.6	90

[a] Published in 1989.

[b] RE represents the number of retinol equivalents.

[c] 1 μg (one microgram) = 10^{-6} g = 10^{-3} mg (see Section 14.7).

[d] α-TE represents the number of α-tocopherol equivalents.

[e] NE represents the number of niacin equivalents.

[f] These represent recommended RDAs for nonsmokers. The RDAs of vitamin C for smokers are 67% greater than those for nonsmokers.

THINKING MORE ABOUT THE MOLECULES OF LIFE

Dietary Fads

The realization that the functioning of the body depends on the foods we eat is an old one, and is bolstered by the understanding that the cell's basic structures are built from molecules brought in through the digestive system. This understanding, coupled with the current preoccupation with health and fitness in the United States, leads occasionally to fads in which one food or another is touted as a new cure-all. It's hard to get enough information to analyze a fad while it's in full swing, but studies of fads after the fact can teach us a lot about them. The rise and fall of oat bran is a particularly enlightening case.

In the mid-1980s, people began to understand that high levels of cholesterol in the blood were correlated to the incidence of heart disease. Studies available at the time indicated that the inclusion of fiber in the diet, particularly oat bran, helped to lower blood cholesterol levels. Oat bran became a fad food, and for a time it was virtually impossible for stores to keep it in stock.

Then, in 1990, newspaper headlines blared that a study in the prestigious *New England Journal of Medicine* [vol. *322*, 147 (1990)] had shown that oat bran did not, in fact, lower cholesterol levels. The oat bran industry, running at $54 million a year, collapsed. Processing plants closed and people lost their jobs.

Was this a reasonable response to the *New England Journal* paper? Let's look at the study that was reported and try to find out. The study took 20 people, all healthy hospital employees of ages 23 to 49 with low cholesterol levels, and tested them on diets with high-fiber oat bran and low-fiber diets for six-week periods. The result? The mean cholesterol levels of the subjects was 172 ± 28 milligrams per deciliter on the low-fiber diet, and 172 ± 25 on oat bran. (Physicians usually start to worry when your cholesterol level gets to the neighborhood of 200.) This inconclusive result, based on 20 healthy people, provided the basis for the headlines.

Does this study tell you anything about what would happen if someone with high cholesterol went on an oat bran diet? How representative of the entire population are 20 hospital employees in Boston? Given the spread of cholesterol levels in the group, could the actual levels have gone down (or up) without the researchers being able to detect it?

Summary

Organic molecules share the following characteristics: they are based on carbon, they usually form from only three or four different kinds of atoms, they are generally modular structures (that is, no matter how large or complex they are, they are formed from a few simple building blocks), and their chemical function is largely determined by their geometrical shape. The four main types of biological molecules are proteins, carbohydrates, nucleic acids, and lipids.

Proteins form from chains of *amino acids* to make many of the body's physical structures such as hair and muscle. Proteins in cells also function as *enzymes*—molecules that increase the rate of reaction between other molecules, but are themselves unaffected by the reaction. Proteins thus mediate many of life's chemical reactions.

Carbohydrates are modular molecules built from *sugars*, which are relatively simple molecules built from carbon, oxygen, and hydrogen. Carbohydrates provide an essential source of energy for all animals, and they provide much of the solid structure in the cellulose of plants.

The *nucleic acids*, *DNA* (deoxyribonucleic acid) and *RNA* (ribonucleic acid), are polymers consisting of a chain of small molecules called nucleotides. Each nucleotide, in turn, incorporates three smaller molecules—a sugar (deoxyribose or ribose), a phosphate group (PO_4), and a base (one of four different molecules, each with a distinctive shape). RNA is a single strand of nucleotides, while DNA forms a twisted double strand—a *double helix* structure.

Lipids, including fats and oils, are molecules that will not dissolve in water. If the carbon atoms form single bonds, the lipid is said to be saturated, while molecules on which adjacent carbon atoms form double bonds are said to be unsaturated. All cell membranes are constructed from bilayers of lipids, which are terminated by one end that attracts water and the other end that repels water.

In addition to the major nutrients—proteins, carbohydrates, and lipids—humans also require small amounts of other chemicals—*vitamins* and *minerals*—that perform specialized chemical functions in the body. The National Academy of Sciences publishes RDAs for both minerals and vitamins.

Key Terms About Life's Molecules

organic molecules   DNA

enzymes   RNA

amino acids   double helix

protein   lipids

carbohydrates   minerals

sugars   vitamins

nucleic acids

Review Questions

1. What is an organic molecule?

2. What is an enzyme? How does it work?

3. What is a protein? An amino acid? What groups of atoms are common to all amino acids?

4. How is a protein constructed from amino acids?

5. What is the primary structure of a protein?

6. What is the secondary structure of a protein?

7. What is the tertiary structure of a protein?

8. What is a carbohydrate?

9. What is the difference between saturated and unsaturated fats? Give examples of each.

10. What is the difference between cellulose and starch at the molecular level? Give examples of each substance.

11. What is a nucleotide?

12. Name the four bases that occur in DNA.

13. Describe the construction of the double helix structure of DNA.

14. How does RNA differ from DNA?

15. What roles do lipids play in the human body?

16. In terms of diet, what are minerals? Give examples of dietary minerals. What are RDAs for minerals?

17. What are vitamins? Which are water soluble, which are fat soluble? What are RDAs for vitamins?

Discussion Questions

1. Marathon runners are often advised to "carbo load" (i.e., to eat a lot of carbohydrates) before a race. Why do you suppose they are given this advice?

2. There are several different forms of vegetarianism. Some people simply avoid red meat, but will eat fish and chicken. Others avoid all meat, but will consume milk and eggs. Still others avoid all animal products. What care would each group have to take to see that it had the proper supply of amino acids?

3. Given what you know about the structure of protein molecules, why might a single mistake in the sequence of amino acids make a significant difference in the shape of a protein?

4. Look at a typical menu at your campus food service. Does it provide a balanced source of nutrition? If not, how would you change it to do so?

5. Regulations require that drugs meet certain advertising standards—if they claim to stop hair loss or promote weight gain, for example, there has to be evidence that they actually do so. Should vitamin and mineral supplements have to meet the same standard? Why or why not?

6. Is there a difference between the vitamin C in freshly squeezed orange juice and in a vitamin C tablet?

7. Almost any table salt you buy in the store has iodine added to it. Why do you suppose this is done?

8. In the novel *Jurassic Park*, the scientists who create modern-day dinosaurs had, as a last-ditch safety measure, what they called the "lysine defense." The idea was that the metabolism of the dinosaurs had been manipulated so that they could not manufacture the amino acid lysine. How would this defense work? Can you think of some way the dinosaurs might get around it?

Investigations

1. Read *The Double Helix* by James Watson, co-discoverer with Francis Crick of the DNA structure. What data did they use to unravel the structure? What were the key steps in solving the double helix structure?

2. Make a detailed record of one week's intake of food, vitamins, and other supplements. Consult nutrition charts and determine the percentage by weight of protein, carbohydrate, fats, and other substances that you consumed. What percentage of the fat consumed was saturated? What changes in your diet could reduce the total percentage of fat consumed, and lower the ratio of saturated to unsaturated fat?

3. Read about a vitamin-deficiency disease. How was it discovered and how was it cured?

4. Visit a health food store in your neighborhood. How do the claims made on labels square with what you know about the chemical basis of nutrition?

5. Read the nutritional information on a box of standard cereal and on one that claims to be "natural." Which actually supplies more of the RDA of minerals and vitamins?

Additional Reading

Fruton, Joseph. *A Skeptical Biochemist.* Cambridge, MA: Harvard University Press, 1992.

Stent, Gunter. *Paradoxes of Progress.* San Francisco: W.H. Freeman, 1978.

The Living Cell

Life is based on chemistry, and chemistry takes place in cells.

A Random Walk
The Chemical Plant

Have you ever driven by an oil refinery or other kind of chemical plant? It's a place of bewildering complexity, laced with a maze of pipes and towers, bustling with diverse activity as energy and materials constantly flow in and out.

As complex and diverse as chemical plants might seem, they all share a few basic features. Every chemical plant must have walls, a control room, and facilities to provide and distribute energy. There must be loading docks to bring in supplies and remove trash, conveyer belts or pipes to move materials from one place to another, and storage rooms to keep a stock of critical items at the ready. And at the heart of every plant must be pieces of machinery that control chemical reactions and manufacture the desired chemical products.

A chemical plant is, in fact, a lot like a living cell.

21

513

The Nature and Variety of Cells

What are things made of? What are the most basic units of matter that form the world around us? For centuries scientists have tried to answer these questions. In a sense, the answer depends on your point of view. For the nuclear physicist, the isotope is of central concern, while particle physicists focus on quarks and leptons. Chemists deal with atoms and molecules, while astronomers see the universe in terms of stars or clusters of stars. Like so many things in life, no one simple answer exists to the question "What are things made of?"

Dealing with living things is particularly complex. Although all living things are made out of atoms and molecules, as we saw in the previous chapter, living things are more than just collections of atoms. An entity must do many things for us to say that it is alive. It must reproduce, take in and process energy, get rid of wastes, respond to its environment, and maintain some sort of structural integrity. These functions cannot be performed by a random collection of atoms; rather they result from the collective behavior of large numbers of atoms organized into some kind of system. In fact, in our search for the basic unit of life, we need not study quarks and leptons, nor atoms and isotopes, nor even molecules. For our purposes, the **cell** can be considered to be the smallest identifiable unit capable of carrying on the basic chemistry that we associate with living things.

An enormous number of different kinds of cells can be found in nature. Cells come in all sizes and shapes, and perform all sorts of functions. While typical animal cells are about one-hundredth of a millimeter (a thousandth of an inch) in diameter, they can range in size from bacteria only a few hundred-thousandths of a centimeter across (much too small to see in most light microscopes, and smaller than some large molecules), to an ostrich egg that would require two hands to hold, and which is much larger than most species of animals. The great majority of cells are too small to be seen with the naked eye, but are easily studied in a microscope.

Cells also come in a wide variety of shapes. Plant cells are often rectangular or polygonal, while most egg cells are usually spherical. Bacteria may be rod-shaped or spiral in form, muscle cells are extremely elongated, nerve cells sport a complex array of filaments, and sperm cells have tadpole-like tails. The recognition of common characteristics in this extraordinary zoo represents one of the great advances in biology.

These differences in shape reflect the differences in the functions that the cells perform. Muscle cells are elongated so that they can exert forces when they contract. Nerve cells (see Chapter 5) have their unique shapes so that they can transmit impulses over a distance, and so on. To fulfill their functions, cells constantly require raw materials and energy to live and reproduce. Two very different strategies help living things maintain an energy supply. Some cells (such as bacteria and protozoa) operate as separate entities, ensuring their survival by reproducing in vast numbers. Multicellular organisms such as plants and animals, on the other hand, employ cells collectively. In these more complex life forms, different groups of cells serve very different functions, with each group depending on others in a complex web of interdependence.

Science in the Making

The Discovery of Cells

An essential and distinctive feature of all cells is the membrane that isolates and protects the insides from the outer environment. In 1663 Robert Hooke (1635–1703), a brilliant and contentious contemporary of Isaac Newton, used one of the first microscopes to observe honeycomb structures in a thin slice of cork. It was Hooke who first called the tiny boxlike units "cells."

Within the next few years, Dutch merchant and self-taught scientist, Anton van Leeuwenhoek (1632–1723) employed superb microscopes of his own design and construction to discover a rich variety of cells, including those in blood, saliva, semen, and the intestines. On October 9, 1676, van Leeuwenhoek sent a letter to the president of the Royal Society in London. He wrote, "In the year 1675 I discovered living creatures in Rain water, which had stood for a few days in a new earthen pot..." and he went on to describe a series of "animacules" that were visible through his microscope.

In spite of Leeuwenhoek's colorful descriptions, it was not until the nineteenth century that scientists finally accepted the idea that animals and plants are essentially aggregates of cells. By the 1860s, biologists recognized that only cells can produce other cells, and that these tiny objects represent the indivisible units of life—a discovery as fundamental as the discovery of atoms in chemistry and quanta in physics. ■

How a Cell Works

Early microscopes were rather primitive affairs (see the following *Technology* section). Consequently, scientists were unable to see a lot of the details of cell structure. Indeed, until the middle part of the twentieth century, science textbooks often spoke of something called *protoplasm*. This substance was supposed to be a kind of uniform, molasses-like fluid that filled cells. Today, with much better instruments, we know that cells have a very complex structure indeed. Advanced cells are full of specialized structures, and even an amoeba is quite complicated; complicated in its own way as larger life forms such as human beings. In fact, we shall see that the molecules of life introduced in Chapter 20 play a crucial role in the complex workings of the cell, with each performing a separate (but vital) function. Today, we refer to the fluid that takes up the spaces between all this complexity as *cytoplasm*.

Technology

The Microscope

Early microscopes and their modern high-tech descendants all operate on the same basic principle shown in Figure 21–1*a*. Ordinary visible light focuses on a sample between two transparent layers of glass or

Figure 21–1

Light (*a*) and electron microscopes (*b*). In the light microscope, waves of ordinary visible light travel through a series of lenses to form an image. The electron microscope uses the wavelike nature of electrons (see Chapter 8) to accomplish the same purpose.

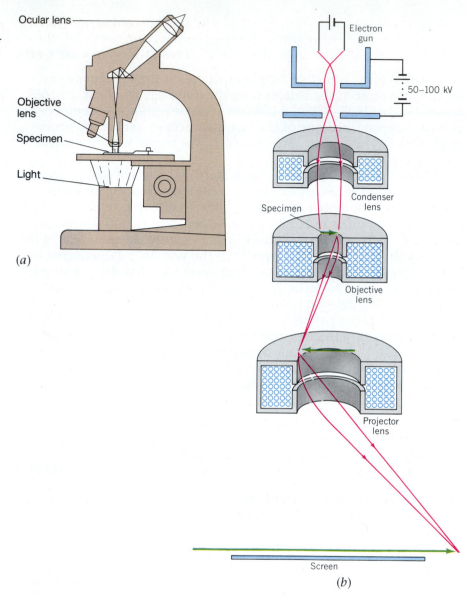

(*a*)

(*b*)

plastic. The light is brought up through a series of lenses so that a magnified image is presented in the eyepiece. This kind of apparatus is called an *optical microscope*, and today these instruments can magnify more than 1000 times and resolve details less than a ten-thousandth of a centimeter across—enough to make an ordinary cell look as big as a quarter. In many modern optical microscopes, a miniature video camera in the eyepiece allows the image to be displayed on a television screen for recording or group viewing.

Special dyes that are taken up by only one part of a sample are often used to increase contrast in the image, and hence to make internal structures clearer. For example, we shall see that parts of cells called chromosomes play an important role in reproduction. These features were first seen as colored structures in cells that had been stained, a fact that is reflected in their name.

The ability of a microscope to differentiate objects that are close to each other is called its resolving power. The resolving power of all microscopes is limited by the wavelength of the light used: objects about the size of one wavelength appear blurred. Light microscopes, for example, typically employ electromagnetic waves with wavelengths on the order of a ten-thousandth of a centimeter; thus objects about that size or smaller will appear as undifferentiated blurs in even the most perfectly designed instrument. This means that many of the smaller structures in the cell cannot be seen with this sort of microscope.

In the 1930s, German scientist Ernst Ruska, working at a university in Berlin, introduced the *electron microscope*, a major new advance in microscopes that uses electrons instead of light to illuminate objects. In Chapter 8, we discussed the notion that quantum objects such as the electron can be thought of as tiny particles, but also as energetic waves. The electron's wavelength depends on its energy—the greater the energy, the shorter the wavelength—and typical electron wavelengths in modern instruments are comparable to the size of a single atom. In a situation such as that shown in Figure 21–1*b*, where an electron beam is shot at a target, we are in effect examining the target with very short wavelengths, and thus achieving resolving powers up to 100,000 times that of optical microscopes. Modern electron microscopes are often used to resolve atomic-scale features.

The electron microscope works this way: An electric current heats a tungsten filament in a strong electric field—typically about 100,000 volts—to produce an electron beam. Electrons boil off the negatively charged tungsten wire and accelerate toward the positive end of a tube. This beam of electron "waves" is focused by specially designed ring-shaped electromagnets, the analogs of glass lenses in a conventional microscope. The focused beam strikes the sample, and the electrons then hit a detector that converts the beam into an image, as shown in Figure 21-2.

Modern electron microscopes are very expensive—it is not unusual for one to cost more than $500,000. Nevertheless, they are an invaluable tool in all areas of science and industry that require examination of objects at extremely high magnification. ■

A scientist operates an electron microscope.

Cell Membranes

Every cell must be isolated from its environment—that is, there must be an "inside" that is distinct from an "outside." A boundary must separate the living from the nonliving. In addition, we shall see that complex cells also require different parts of the cell to be separated from each other by inner partitions.

Both of these requirements point to the need for **cell membranes**. A cell membrane is a structure that separates the inside of a cell from the outside, or separates one part of a cell from another. The simplest cells will have only one external membrane. More-complex cells have many internal membranes in addition to the outer envelope. The basic molecular structure of cell membranes is the lipid bilayer, of the type we discussed in Chapter 20. The chainlike lipid molecules are arranged in a double layer with the hydrophobic ends of the molecules pressed against each

Figure 21–2
Electron microscope photograph of a fruit fly with a far-wing mutation.

other, and the hydrophilic ends on the outside. Because cells almost invariably exist in a watery fluid, this arrangement is the lowest possible energy state in which the molecules can arrange themselves. Water is squeezed out from the space between the layers, thereby ensuring that the hydrophobic parts of the membrane are not in contact with it. In the same way, the hydrophilic parts find themselves in a watery environment, as they should.

One way to visualize the cell membrane is to think of a technique that is often used in moderate climates to protect swimming pools from freezing in the winter. Instead of draining water from the pool—an expensive and time-consuming operation—owners simply throw a large number of Styrofoam balls into the pool. These balls float next to each other. They cover the water, constantly touching, but also constantly jostling and moving around. They lift up and down when waves move across the water, but the covering retains its integrity. In the same way, cell membranes are made up of molecules stacked or arranged next to each other. These molecules can change shape and move around according to the dictates of their environment, but they retain their integrity and do not rupture. Thus they perform the function of separating a cell from its environment.

Transporting Material across a Cell Membrane

If a cell membrane was totally impermeable, life would not be possible. In order to be alive, a cell must take in raw materials from its surroundings, and pass wastes and other chemicals into the environment, just like the loading docks and pipelines that run to and from a chemical plant.

Materials are transported across cell membranes in many ways. The simplest are the movement of individual molecules across the cell membrane by diffusion and osmosis. Diffusion refers to the transfer of molecules by ordinary random thermal motion. When perfume moves from one side of a room to another, for example, it does so by the process of diffusion. Similarly, helium leaks out of a party balloon as individual helium atoms diffuse through the balloon. Osmosis, a special case of this sort of molecular movement, refers to a process that transfers materials such as water across a membrane while at the same time blocking the passage of molecules dissolved in the water. One example of osmosis with which you are familiar can be seen in your supermarket. Produce managers frequently spray fresh vegetables with water. They do not do this for decoration—they put water on the carrots, lettuce, and other vegetables so that it will move through the membranes into the cells, making up for water that has diffused out and evaporated. In this way, vegetables maintain a fresher appearance for longer periods of time.

Cell membranes also incorporate various kinds of channels and molecular-sized openings, which allow specific materials to go back and forth. When nerve signals move through the human body, as we saw in Chapter 5, sodium and potassium ions move back and forth across nerve cell membranes to create the signal. The various kinds of channels that exist in different kinds of cells allow different kinds of atoms and molecules to come through.

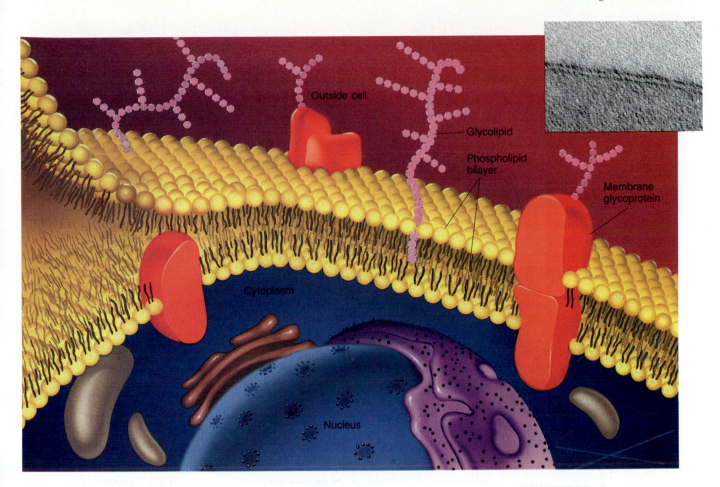

Outside cell

Glycolipid

Phospholipid bilayer

Membrane glycoprotein

Cytoplasm

Nucleus

One important means of bringing material through a cell membrane, however, depends on the notion that "geometry is destiny," as we discussed in Chapter 20. Interspersed here and there in the cell membrane are large structures made of folded proteins, called **receptors** (see Figure 21–3). Receptor molecules each have a specific geometrical shape, and each molecule will make a good fit and bind to a specific type of molecule in the environment. When that molecule is present, it, and only it, can bind to the receptor. Thus the receptor molecules can be thought of as the cell's "door guard" looking over prospective entrants and picking out only those whose shape is exactly right. When a receptor "recognizes" a particular molecule on a particle, the molecule and receptor form a chemical bond because they fit together. A sequence of events such as that sketched in Figure 21–4 thus takes place. The receptor binds to the particle in question and holds it while the cell membrane deforms. Once the particle is inside, as shown, the membrane nips off, enclosing the particle in its own special wrapping, and the cell membrane reforms behind it. The tiny container, called a *vesicle*, then becomes the vehicle by which the particle moves around inside the cell. A similar process that works in reverse is used when molecules from inside the cell are moved out.

Figure 21–3

All cells possess a lipid bilayer membrane, which is studded with protein receptors (also called membrane glycoproteins). The membrane separates the inside of the cell, with its nucleus, cytoplasm, and various organelles, from the outside. In an electron microscope photograph (*inset*), the membrane appears as two dark lines (the outer surfaces) separated by a less-dense region.

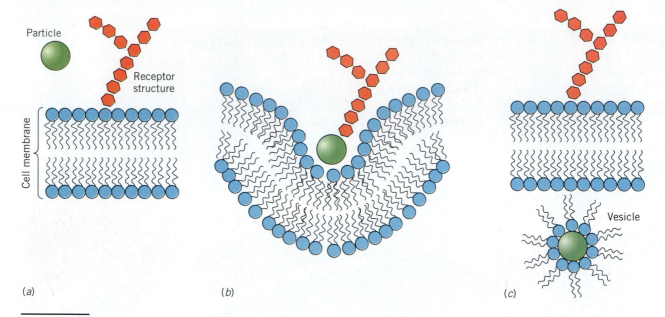

Particle

Receptor
structure

Cell membrane

(a)

(b)

Vesicle

(c)

Figure 21–4

The action of receptors in a
cell.

The outer covering of plant cells contains an additional structure in
addition to the kind of flexible membranes described above. Plant cells
may also be connected to each other by a **cell wall,** a solid framework
made of cellulose molecules and other strong polymers. Cell walls, which
often account for about a third of a living plant's mass, give tree trunks
and leaves the strength to grow upright against the force of gravity. In
fact, it was cell walls (as opposed to cells per se) that Robert Hooke saw
when he looked at his piece of cork in 1663 (see Figure 21–5).

The Nucleus

In most cells, the most prominent and important interior structure is the
nucleus. The nucleus forms a relatively large enclosed structure that
contains the cell's genetic material—its DNA. This DNA contains the
instructions for the day-to-day chemical operation of the cell, as well as
the mechanism by which the cell reproduces itself. If we think of the
cell as being analogous to a large refinery, the nucleus can be thought
of as the front office, the place where blueprints are stored and from
which instructions for the operation of the entire system go out.

Not all cells have nuclei. In some cells, the DNA is coiled together
but not separated from the rest of the material in the interior. Presumably,
these sorts of primitive cells, called *prokaryotes* ("before the nucleus"),
evolved first. The kingdom of monera (see Chapter 19), made up of
bacteria and their relatives, includes all cells that do not have a nucleus.

The more advanced single-celled organisms, as well as all multicellular
organisms (including human beings), are made from cells that do contain
nuclei—the *eukaryotes* ("true nucleus"). The kingdom of protista includes
single-celled eukaryotic organisms. Virtually all of the organisms with
which we are familiar are made up of eukaryotic cells (see Figure 21–5).

One interesting feature of the nucleus—a feature that may contain a
good deal of information about the evolution of higher life forms—can

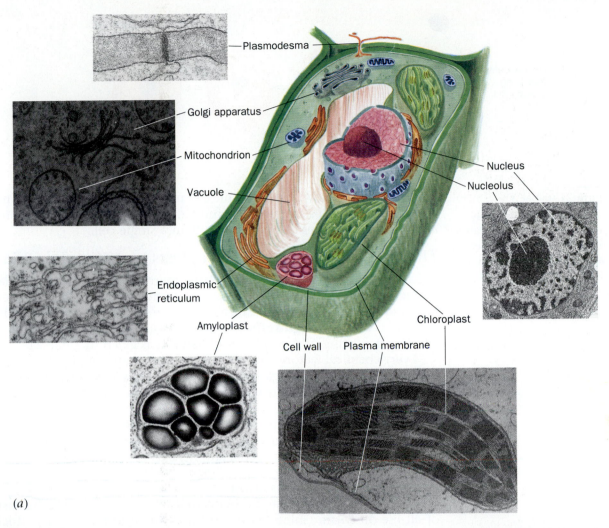

Plasmodesma

Golgi apparatus

Mitochondrion

Vacuole

Nucleus

Nucleolus

Endoplasmic
reticulum

Amyloplast

Cell wall

Plasma membrane

Chloroplast

(a)

(b)

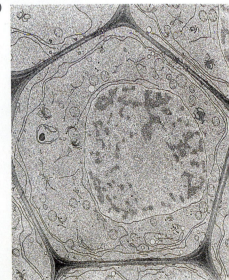

Figure 21–5
A typical plant cell. (a) A generalized drawing with electron micrographs of
some key organelles. (b) Photograph of a cell of maize rust. The dark boundary
is the cell wall, made from cellulose. The nucleus takes up almost half the area
of the central part of the cell.

be found in its confining membranes. The nucleus has not one, but two
membranes, as shown in Figure 21–6. The standard explanation of the
double membrane in the nucleus is that it is a vestige of an earlier stage
of development. The idea is that at some point in the past, a large cell
engulfed a small one, much as modern cell membranes use receptors to
engulf molecules. Over time, a symbiotic relationship developed between
those first two cells. Each cell was able to do better in a partnership than
it could do alone. The double-membrane nucleus is interpreted as follows:
The inner membrane is the descendent of the original membrane of the
swallowed cell, while the outer membrane is the descendent of the vesicle
that formed when the first cell was enveloped.

Other structures in the cell also have a double membrane, and we'll

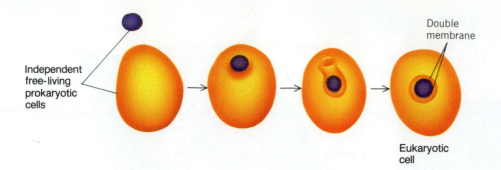

Independent free-living prokaryotic cells

Double membrane

Eukaryotic cell

Figure 21–6

The nucleus has a double membrane, perhaps because it evolved from an earlier stage in which a large cell engulfed a small one, as shown.

point them out below. These double cell membranes suggest that, while cells are indeed the basic unit of life, individual cells in complex organisms are more like colonies of smaller cells than a single cell.

The Energy Organelles: Chloroplasts and Mitochondria

Eukaryotic cells (those with nuclei) have many inner structures similar to the nucleus. Each of these structures carries out a special chemical function in the cell. Any specialized structure in the cell, including the nucleus, is called an **organelle**. Important organelles are shown in Figure 21–7, and a listing of their primary functions is given in Table 21–1.

Every chemical factory requires energy, and living cells have adopted two very different energy-gathering strategies. Plants rely on the Sun for their energy, while animals ultimately depend on the chemical energy stored in plants.

Chloroplasts are the main energy transformation organelles in plant cells and, as the name suggests, they are the places where molecules of *chlorophyll* are found. Chlorophyll absorbs the energy from sunlight and uses that energy to transform atmospheric carbon dioxide and water into energy-rich sugar molecules such as glucose, and, as a by-product, oxygen. The formation of glucose and other carbohydrates by plants is the basis of all life on Earth. Chloroplasts have a double cell membrane, which suggests that they once may have been independent cells.

Mitochondria, sausage-shaped organelles, are places where molecules derived from glucose react with oxygen to produce the cell's energy. Mitochondria are, in effect, the furnaces where fuels are oxidized. It makes sense to separate this energy production from the rest of the cell because the process is complex and must take place in a controlled environment. Furthermore, oxygen is a very reactive molecule, and oxidation reactions, like burning, are potentially dangerous operations. Just as we segregate the burning of fuel in our homes to the furnace, the cell segregates the burning of glucose to the mitochondria. A typical eukaryotic cell will have anywhere from a few hundred to a few thousand mitochondria.

Like the nucleus, mitochondria have a double cell membrane and even their own complement of DNA. Most scientists have concluded, therefore, that mitochondria were originally independent cells in the early history of life on Earth.

Cytoskeleton

Although it is not, strictly speaking, an organelle, the *cytoskeleton* gives the cell its shape, keeps things anchored in place, and, in some cases,

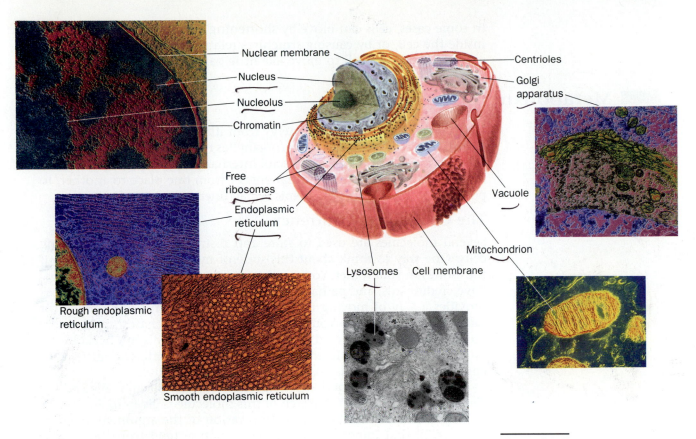

Rough endoplasmic reticulum

Smooth endoplasmic reticulum

Figure 21–7

A typical animal cell showing the nucleus, mitochondria, and various other organelles (see Table 21–1).

allows the cell to move. The cytoskeleton is a series of protein filaments that extend throughout the cell, more or less like a complex of spider webs.

Inside the cell, the cytoskeleton serves as the transport system along which the vesicles that carry material from one place to another move.

Table 21–1 ● Organelles and Their Functions

Organelle	Primary Function(s)
nucleus	stores DNA, controls cell's chemistry
endoplasmic reticulum	contributes to protein and lipid synthesis
mitochondria	production of energy for metabolism
ribosomes	site of protein synthesis
chloroplasts	site of photosynthesis (plants only)
Golgi apparatus	part of synthesis of molecules
lysosomes	digestion, breakdown of wastes
vacuoles	waste storage (plants only)
cytoskeleton	structure, internal transportation

In some cases, cells can move by shortening and lengthening filaments inside the cell, or by causing structures related to the cytoskeleton that extend outside the cell to move like little oars.

Metabolism: Energy and Life

The cell's process of deriving energy from its surroundings is called **metabolism.** In addition to specific organelles that convert external energy sources (food or sunlight) into useful internal energy supplies, cells must also have a means to transfer that energy from one place to another, to power the varied pieces of chemical machinery necessary for life.

The Cell's Energy Currency

Several molecules are used to store and distribute energy in all living cells. One way to think about this suite of molecules is to think about the money in your wallet. You may have some cash, but you also probably have credit cards, and perhaps even a check or two. Each of these items provides a way of moving money around, and each is appropriate for different kinds of uses. You pay cash for small purchases, make larger purchases with a credit card, and pay some of your bills by check. In the same way, a cell has different molecules that store different amounts of energy, each of which is appropriate to a particular use.

The most common of these energy carriers is a molecule called *adenosine triphosphate*, or *ATP*. A sketch of this molecule is given in Figure 21–8. The structure of ATP is a good illustration of the argument given in Chapter 20 that important molecules in nature tend to be built from the same simple building blocks. ATP contains three phosphate groups (collections of phosphorous and oxygen atoms) at the end of the molecule (these three phosphates are what give ATP the "tri" part of its name). The other parts of the molecule consist of the sugar ribose (part of the structure of RNA) and the base adenine (part of both DNA and RNA).

It requires energy to put phosphorous groups onto the ends of the tail of an ATP molecule. In the language of Chapter 9, putting the last phosphate group on the ATP tail is an endothermic process. If the phosphate group is removed in another chemical reaction, that energy is available to drive other chemical reactions. Thus the molecule can carry energy in the form of phosphorus-oxygen bonds from one point in the cell to another.

In one place in a cell—the mitochondria, for example—chemical reac-

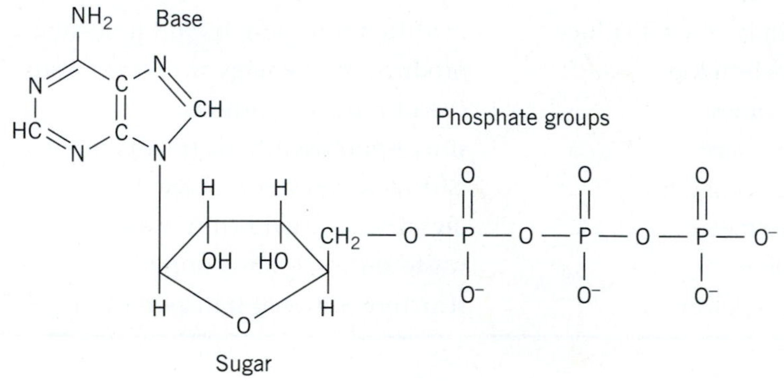

Figure 21–8

Sketch of ATP, the energy "money" of cells. Note that it is built from a sugar (ribose), a base (adenine), and three phosphate groups.

tions produce energy. This energy is used to produce a molecule of ATP. The molecule then moves out of the place where it is created to a place where energy is needed. At this point, the ATP acts something like a battery: the ATP molecule attaches to part of the cell's chemical machinery, a phosphate group is removed, and the released chemical potential energy becomes available to drive the desired chemical reactions. The *tri*phosphate (three phosphates) then becomes a *di*phosphate (two phosphates), ADP:

$$ATP \rightarrow ADP + PO_4 + energy.$$

ATP is the molecule that serves as the cell's "cash" in the cell's energy system. A typical cell will have several million ATP molecules doing their job at any given time. Other more-complex molecules (represented by the letters FADH and NADH) correspond to the "credit cards" in the cell's energy system. They store up energy and can be cashed in on short notice when extra energy is required. One way of thinking about the role of ATP is to say that when a chemical reaction adds the last phosphate group to ATP, the molecule picks up the equivalent of a pocketful of money. That money can be spent later for almost any purpose.

Photosynthesis

Photosynthesis, the mechanism by which plants convert the energy of sunlight into energy stored in carbohydrates, provides the chemical energy for virtually all life on Earth. In plants, this complex process operates as follows. Sunlight is absorbed by a large molecule, usually chlorophyll (see Figure 21–9). After a series of chemical reactions, this energy is ultimately stored in a set of molecules that includes ATP. Once the energy has been captured in this way, it is used in another complex series of reactions to produce carbohydrates. The most important carbohydrate molecule is glucose, an energy-rich sugar that also provides the module from which all cellulose and starch polymers are made (see Chapter 20). The end result is the conversion of the energy in the electromag-

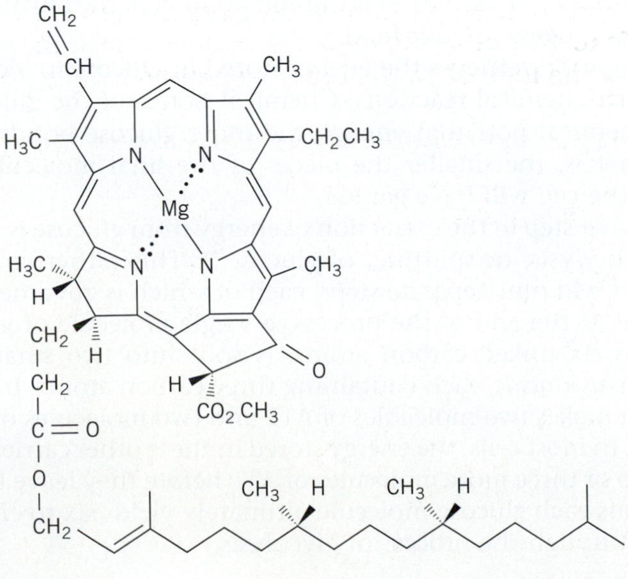

Figure 21–9

The structure of the chlorophyll molecule, which converts solar energy into chemical energy in plants. Many of the molecule's carbon and hydrogen atoms are not labeled.

netic radiation from the sun into chemical energy stored in the bonds holding the carbohydrate molecules together.

The shorthand way of thinking about photosynthesis is to say:

▶ **In words:**

(energy + carbon dioxide + water) react to produce (carbohydrate + oxygen).

▶ **In equation form:**

energy + CO_2 + H_2O → carbohydrate + O_2.

In most cases, the rate at which plants can produce oxygen and carbohydrates is limited by the amount of carbon dioxide in the air.

The most familiar molecule involved in photosynthesis is chlorophyll. The electrons in various forms of chlorophyll have energies that allow them to absorb red and blue light. The energy that falls as sunlight on a leaf is white light, roughly equal mixtures of all of the visible spectrum (see Chapter 6). Red and blue components of light from the Sun are absorbed by the leaf, while green light is reflected. This is why leaves appear green. In addition to chlorophyll, a number of secondary molecules are involved in photosynthesis. These molecules tend to absorb blue light, so they appear to be red and orange. A normal leaf contains much more chlorophyll than secondary molecules, so the color of the secondary molecules is masked. In the fall, however, when the leaf dies and chlorophyll is no longer produced, its underlying color can be seen. We say that the leaves "change color" although, in fact, the brilliant fall colors were there all the time.

Glycolysis: The First Step in Energy Generation in the Cell

The primary source of energy for living things comes from the oxidation, or "burning" of carbohydrates such as glucose. These sorts of reactions, called **respiration,** are taking place in all of your cells at this very moment. You breathe in oxygen produced by plants, and the carbon dioxide that you breathe out is the end product of the burning of carbohydrates you ingest in your food.

Respiration retrieves the energy stored in glucose in a complex series of cellular chemical reactions. Chemical bonds of the glucose molecule store chemical potential energy; the more glucose bonds that are broken—that is, the smaller the pieces of the final molecules—the more energy the cell will have gained.

The first step in the extraction of energy from glucose is called **glycolysis** ("the lysis, or splitting, of glucose"). This rather complex process takes place in nine separate steps, each of which is governed by a specific enzyme. At the end of the process, a single molecule of glucose, which contains six linked carbon atoms, is split into two smaller molecules called *pyruvic acids*, each containing three carbon atoms. In addition, the reaction makes two molecules of ATP and two molecules of other energy carriers. In most cells, the energy stored in these other carriers is converted into two or three more molecules of ATP before they leave the mitochondria. Thus each glucose molecule ultimately yields six to eight molecules of ATP through the process of glycolysis.

After the glucose has been split by glycolysis, energy can be generated in two separate and distinct ways, called respiration and fermentation. As we shall see, respiration requires the presence of oxygen, and is therefore said to be **aerobic**. Fermentation, on the other hand, can occur in the absence of oxygen, and is said to be **anaerobic**.

Fermentation: A Way to Keep Glycolysis Going

Pyruvic acid molecules still hold a great deal of chemical potential energy. In the absence of oxygen, however, that energy cannot be liberated to run cellular metabolism. In this situation, however, a process called **fermentation** can be used to provide energy to keep glycolysis going. As long as there is a supply of glucose, the cell can go on generating energy, albeit somewhat inefficiently. In some cases, as in single-celled yeasts, the end product of fermentation is ethanol—ordinary beverage alcohol. This sort of fermentation provides the basis for the production of wine and other alcoholic drinks.

When cells are forced to use glucose in the absence of oxygen, only a fraction of the available energy is used, and a great deal of energy is left stored in other molecules. Take yeast as an example. You know that alcohol, which is produced by yeast during the process of fermentation, contains a great deal of energy—it can, after all, be burned as a fuel. The chemical energy in alcohol is left behind by the yeast cells that made it; energy that the cells could not use because they were not able to metabolize it.

A somewhat different kind of fermentation takes place in cells of animals such as human beings. As we shall see in a moment, the energy in pyruvic acid is normally tapped by the process of respiration. In the absence of oxygen, however, our muscles can use fermentation to keep glycolysis going (often taking the extended supply of glucose from materials stored in the muscles themselves). The end product of this sort of fermentation is lactic acid, a three-carbon molecule that then accumulates in the muscle. When we undergo strenuous exercise, if our bodies are not prepared to deliver all the oxygen that is needed, the cells will eventually fall back on this simpler process. The stiffness that you feel in your muscles after unusually strenuous exercise is caused by the buildup of lactic acid that must be removed once you are through exercising.

The fact that some of our cells can operate both with and without oxygen—the fact that they have this reserve process to fall back on—is taken by some scientists to indicate that cells evolved fermentation reactions first, and only later developed the ability to burn oxygen. You can think of the body's use of fermentation as analogous to writing a term paper with a pencil when a power outage makes a computer unavailable. It's not the most efficient way to work, but it gets the job done. In a sense, then, the stiffness that you feel in your muscles is a reminder of the origin of our species in simple cells.

The Final Stages of Respiration

In human cells, the "burning" of the products of glycolysis takes place in the mitochondria, and all the energy that was stored in the glucose is retrieved. The end effect of these chemical reactions is that oxygen and the pyruvic acids take part in a complex series of reactions to produce carbon dioxide, water, and a large amount of energy stored in ATP molecules.

In cells where oxygen is available, pyruvic acids (three-carbon molecules) are first broken down into two-carbon groups, and then enter a complex series of chemical reactions called the *Krebs cycle*. In the course of this cycle, the carbon in the original glucose is broken down completely into carbon dioxide, some energy is released to ATP molecules, and the rest is stored in some of the other energy-carrying molecules. In the final stage of respiration, called *terminal electron transport*, the energy in these other carriers is used to produce even more ATP molecules.

The exact number of ATP molecules produced from a single glucose molecule depends on details of the structure of the mitochondrial membranes, and varies slightly from one cell to another. As a general rule, however, the metabolism of a single glucose molecule ultimately produces 36 to 38 molecules of ATP, which can then be used by the cell to run all the rest of its chemical machinery. Compare this production to the 6 to 8 ATP molecules produced by glycolysis.

Thus aerobic reactions yield significantly more energy per molecule of glucose than anaerobic ones yield. It can be argued that the large amounts of energy needed to maintain a multicelled organism would not have been available to organisms that had not developed respiration. The outcome of this line of thought is that before the atmosphere developed a significant amount of oxygen, complex life forms could not have developed. We will discuss the development of complex organisms in Chapter 23.

Cell Division

Individual cells do not last forever. As you read this, cells are dying and being replaced throughout your body. In order for this sort of replacement to take place, cells must be able to reproduce. There are two separate processes by which cells divide and reproduce.

Mitosis

In the great majority of cell divisions in living organisms, a single cell splits, so that two cells appear where there once was only one. This process is called **mitosis**. It is through mitosis that organisms grow and maintain themselves. Mitosis involves the reproduction of individual cells, but is not involved in sexual reproduction in higher plants and animals.

In Chapter 22, we will discover that the chemical working of any cell is governed by its DNA. In eukaryotes, DNA is contained within the cell nucleus in structures called *chromosomes*. When chromosomes were first discovered in the nineteenth century, there was an intense debate about their function. Today, we understand that each chromosome is a long strand of the DNA double helix, with the strand wrapped around a series of protein cores like tape around a spool.

Chromosomes come in pairs. Humans have 46 chromosomes (23 pairs), but the number of pairs varies from one species to another. Mosquitoes, for example, have 6 while dogs have 78. In other words, there does not appear to be any particular connection between an organism's complexity and the number of its chromosomes.

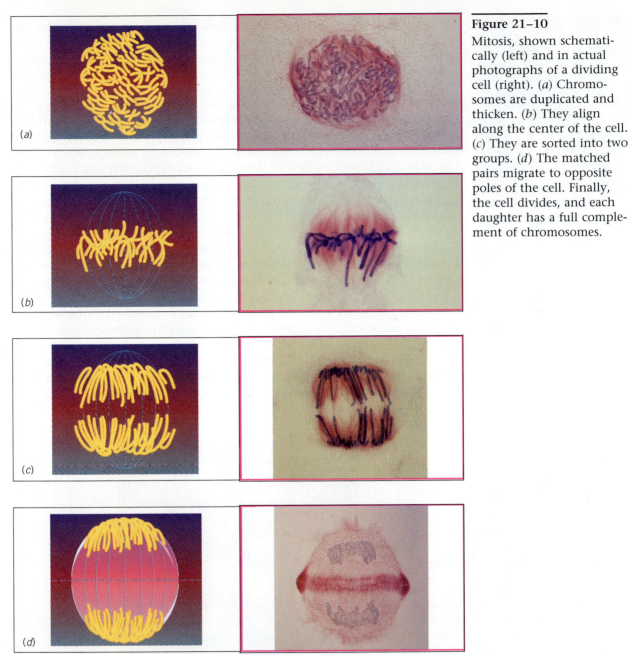

Figure 21–10

Mitosis, shown schematically (left) and in actual photographs of a dividing cell (right). (*a*) Chromosomes are duplicated and thicken. (*b*) They align along the center of the cell. (*c*) They are sorted into two groups. (*d*) The matched pairs migrate to opposite poles of the cell. Finally, the cell divides, and each daughter has a full complement of chromosomes.

The first step in mitosis is the copying of the chromosomes (a process we will describe in more detail in the next chapter). As shown in Figure 21–10, a cell that is about to divide has twice the normal number of chromosomes, neatly paired off like socks that have been sorted after going through the laundry.

After duplication of the chromosomes, the nuclear membrane dissolves and a series of fibers called spindles develops. The matched chromosome pairs are pulled apart and migrate to opposite ends of the cell. After this separation, the nuclear membranes reform and the cell splits down the middle. The result is two cells, each of which carries a set of chromosomes that are identical to the original.

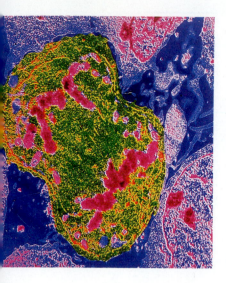

A human cell dividing. Genetic material shows as dark pink in this false-color photomicrograph.

Some cells in your body divide often—the cells in the lining of the small intestine, for example, are replaced every few days. Other cells, such as those of the nervous system, don't divide at all after maturity.

Meiosis

A more complex kind of splitting takes place in a few cells in the reproductive system of organisms that reproduce sexually. In this process, called **meiosis,** a single cell with a full complement of DNA splits to form four daughters, each with half the normal complement. The purpose of meiosis is to generate the sperm and eggs (or ova) that will combine to produce a new member of the species. Meiosis involves the duplication of chromosomes followed by two cell divisions (rather than one division as in mitosis).

As shown in Figure 21–11, the first steps of meiosis are the same as for mitosis—the DNA is copied. The chromosomes then pair up (a process similar to tying matching socks together when you put them in a drawer) and are pulled to opposite ends of the cell. The cell then divides once to form two daughters, each with a complete set of matched chromosomes. In the second stage of meiosis, the chromosome pairs in each of the daughters separate and are moved to opposite sides of each daughter cell. The daughter cells then split, producing a total of four cells, each of which has half the normal complement of chromosomes. These chromosomes eventually are incorporated into sperm or ova. We will discuss the reproductive system of which these cells are a part in Chapter 24.

Figure 21–11

Meiosis, shown schematically. As in mitosis, the chromosomes are duplicated and paired. The pairs separate and the cell divides for the first time. The chromosome pairs then separate and each daughter divides again, producing a total of four cells, each with half the normal complement of DNA.

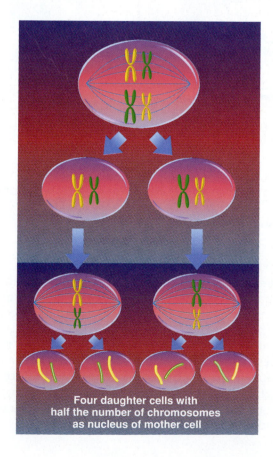

Four daughter cells with half the number of chromosomes as nucleus of mother cell

The Human Body

The Immune System

One of the most important but least-understood systems in the human body is the *immune system*, which allows us to defend ourselves against microorganisms that could do us harm. We are under constant attack by foreign organisms and objects—bacteria, viruses, and nonliving matter from cuts and scrapes, for example. The function of the immune system is to recognize these foreign invaders in the human body and destroy them, without at the same time harming body cells themselves. The body has some defenses that are quite general. Cells called macrophages, for example, engulf and devour entire foreign cells, regardless of their type. The immune system, on the other hand, depends on the geometrical nature of molecular interactions to attack specific kinds of invaders.

The main working cells of the immune system are called *lymphocytes*. Two important classes of lymphocytes, designated B and T, each operate in a different way against invaders.

B lymphocytes are produced in bone marrow and then migrate out into the body, where they are constantly on patrol (about two trillion exist in your body right now). Many different types of B cells occur, each of which has a different set of molecules called *antibodies* on its surface. Like receptors, these antibodies will bind to a specific foreign molecule. Any substance that elicits an immune response is called an *antigen*. In effect, each kind of B cell can "recognize" one specific foreign molecule, which may be free-floating or may be attached to the surface of a foreign cell. When the antigen binds to the antibody, the B cell starts to divide. Some of its offspring churn out huge amounts of the antibody molecule over a short period of time, while others remain in the body for long periods, perhaps even for a lifetime.

Antibodies quickly attack the invaders by coating their surfaces so that they stick together and are consumed by cells like macrophages, by blocking some vital function so that they die, or, in company with other components of the blood, destroying them outright. The longer-lasting descendants of the B cell provide a quick response against future invasions, and are the basis for acquired immunity to diseases such as mumps and measles. The body, then, does not produce tailor-made antibodies to counter a specific invasion. Instead, it selects from among a variety of "off-the-rack" antibodies available in the B cells, and retains those that work against the possibility of future invasion.

T lymphocytes also begin in bone marrow, but play a different role in the immune response. Their role is to destroy cells in the body that are already infected or abnormal. From the bone marrow, T cells migrate to the thymus, a small gland behind the upper breastbone (the T stands for thymus). A process goes on in the thymus that selects only those T cells whose receptors will recognize fragments of foreign antigens that are brought to the surface of the body's own cells. If a T cell has a receptor that binds to that protein, it differentiates into cells that can attack and kill the infected cell. The action of T cells enables the body to fight off cancers, and causes the body to reject tissue in organ transplants. ■

THINKING MORE ABOUT CELLS

Biochemical Evidence for Evolution

The most striking thing about the energy metabolism of cells is that every single living thing on the Earth, from the lowest bit of pond scum to the cells in our own bodies, uses part or all of the same chemical reaction cycles to obtain energy. They share, in other words, a common biochemical background.

Most cells get energy from glucose by the process of glycolysis. Cells in advanced organisms—those in your own body, for example—can get more energy by adding more oxidation steps to the process of glycolysis. Some cells cannot do this, and therefore obtain less energy from each glucose molecule. This difference turns out to be a universal feature of living things—that is, when you examine life's chemical reactions, you find that more specialized cells tend to use more chemical processes, but those specialized reactions are built up from chemical reactions present in more primitive cells.

You can think of the situation in cells as something like the way a complex chemical factory might develop. Long ago the factory may have made just one kind of chemical—perhaps white powdered lime for your lawn. As the factory grew and expanded, many new chemicals were produced—fertilizers and insecticides, for example. But the lime-making operation was still intact, ready to be used anytime. We could, in fact, deduce the history of the plant by taking it apart and seeing how various chemical operations have been added on.

Scientists suspect that cells, the chemical factories of life, behave the same way. Biochemical evidence demonstrates that older, simpler chemical reactions lie at the heart of the more complex operations of today's cells. Some biologists argue that this fact implies that all life descended from a common ancestor. How could you make such an argument based on what you know of respiration and fermentation? What do you think the more primitive ancestor must have been like? Do you think the ancestor must have been in an environment that was rich in oxygen? Why or why not?

Summary

Cells, complex chemical systems with the ability to duplicate themselves, are the fundamental units of life. All cells are bounded by a *cell membrane* consisting of a lipid bilayer. Most plants also have a *cell wall* made of cellulose and other strong polymers. Nutrients move into and wastes pass out of the cell membrane at protein *receptors*, which bind to specific molecules because of their distinctive shapes.

Cells possess a complex internal structure with many different kinds of chemical machinery. All but the most primitive cells have a *nucleus*, a structure surrounded by a double cell membrane that contains DNA. Other discrete structures, or *organelles*, in the cell perform various specialized functions.

Every cell must have a chemical mechanism for obtaining and distributing energy—the process of *metabolism*. Plants absorb light from the Sun and convert this radiant energy into chemical energy in *chloroplasts* by the process of *photosynthesis*. Animals must eat energetic chemicals, primarily carbohydrates such as glucose derived from plants. The first step in getting energy from glucose is *glycolysis*, a series of chemical reactions that takes place in *mitochondria*, by which the glucose molecule is split into pyruvic acids. In the *anaerobic* process of *fermentation*, pyruvic acids are broken down into molecules such as ethanol and lactic acid and the energy is used to keep fermentation going. In the *aerobic* process of *respiration*, pyruvic acids are broken down to carbon dioxide and water, liberating much more energy in the process.

Most cells divide by the process of *mitosis*, in which chromosomes are first duplicated, then separated. The cell then divides, producing two daughters, each of which has the same complement of DNA as the original cell. In *meiosis*, which produces sperm and ova for sexual reproduction, chromosome duplication is followed by two divisions that result in a set of cells that have half the normal complement of DNA.

Key Terms About the Cell

cell
cell membrane
receptor
cell wall
nucleus
organelle
chloroplasts
mitochondria
metabolism

photosynthesis
respiration
glycolysis
aerobic
anaerobic
fermentation
mitosis
meiosis

Review Questions

1. What is a cell?
2. What is the difference between a light microscope and an electron microscope? What are the smallest things that can be seen with each?
3. Describe the structure of a cell membrane.
4. How do materials move across cell membranes?
5. Explain how receptors work.
6. How do cell walls and cell membranes differ?
7. What is a prokaryote?
8. What is a eukaryote?
9. What is an organelle? Give some examples.
10. What does the double membrane of the nucleus tell us about the evolution of eukaryotes?
11. What is ATP? What role does it play in the energy balance of a cell?
12. What is a chromosome? Describe its structure.
13. Why do leaves appear green? What happens when they change color?
14. What are the products of photosynthesis? What molecules are involved in making photosynthesis happen in plants?
15. What is glycolysis? How does it provide energy for living cells?
16. What is fermentation? How does it provide energy for living cells?
17. What is respiration? How does it provide energy for living cells?
18. Does fermentation, glycolysis, or respiration yield the most energy per molecule of glucose?
19. Describe the steps in (a) mitosis and (b) meiosis.
20. How does an antibody work? How does its function depend on geometry?
21. What is the difference between B and T cell lymphocytes? What role does each play in the immune system?

Discussion Questions

1. Could wine ferment if it were left exposed to the air? (*Hint*: Think about the word "anaerobic.")
2. What does it mean to say that all life on Earth depends on photosynthesis?
3. In what ways do the cells of plants and animals differ? In what ways are they the same?
3. List all the ways you can in which cells are analogous to chemical factories.
4. The text says that all life on Earth depends on photosynthesis. Trace the foods you had for dinner last night back to specific plants.
5. List as many ways as you can in which the working of a cell illustrates the importance of geometry in molecular interactions.
6. Why is it that if you cut your hand it will heal, but if someone's spinal cord is severed in an accident that person is permanently paralyzed? (*Hint*: Can nerve cells undergo mitosis?)

Investigations

1. Locate an electron microscope on your campus or at a nearby laboratory. Arrange a visit to watch the microscope in action. All analytical equipment has three major components: hardware to produce and control a source of energy (in this case the electron beam); hardware to mount and manipulate the specimen; and hardware to detect the interaction of the sample with the energy. Sketch the microscope and control panels and indicate which parts are associated with which of these three components.
2. Read the novel *The Andromeda Strain* by Michael Crichton. Discuss the novel in the light of what you now know about the body's immune system.
3. We often hear of "aerobic" exercises. Is there any connection between these exercises and aerobic processes in cells?
4. Look at tap water under a microscope. If you were van Leeuwenhoek, seeing this for the first time, how would you describe it?

Additional Reading

Thomas, Lewis. *Lives of a Cell*. New York: Penguin Books, 1978.

Woods Schindler, Lydia. *Understanding the Immune System*. Bethesda, MD: National Institutes of Health Publication 92–529, 1992.

Classical and Modern Genetics

All living things use the same genetic code.

A Random Walk
The Family Album

Think about the last time you were leafing through your family photo album. Did you notice how much the people resembled each other? Do you look like any member of your family? Your mother, perhaps, or an uncle? We share an expectation, based on experience, that members of the same family are likely to bear physical resemblances toward each other. Why should this be so? Why should one generation bear any resemblance at all to its parents?

At a deeper level, we are all aware of the fact that offspring are invariably of the same species as their parents. Dogs do not give birth to kittens, nor do fish give birth to birds. In fact, we take this everyday fact so much for granted that when an author, in a flight of fancy, violates it, we see the ridiculousness of the situation at once. For example, in his novel *The Dictionary of the Khazars*, Serbian author Milorad Pavic offers the following scenario:

> In the beginning, to the horror of others, every living thing could create every other living thing; it was not until the Khazar god of salt that beings could give birth only to their own image.

On a more sobering note, the transfer of traits from parents to children can have tragic consequences. Medical science now recognizes thousands of diseases, some invariably fatal, that are passed from generation to generation. Almost everyone has a friend or relative who is afflicted by hereditary conditions such as cystic fibrosis, sickle-cell anemia, Tourette's syndrome, or varieties of diabetes, cancer, and heart disease. Scientists must understand the mechanisms of heredity if they are to combat these devastating genetic diseases.

Classical Genetics

Genetics, the study of ways in which biological information is passed from one generation to the next, was pioneered by an Austrian monk named Gregor Mendel (1822–1884). Perhaps more than any other prominent scientist, Mendel closely matches the popular image of the lonely genius conducting exacting research in isolation. Working at the monastery in Brno, in what is now the Czech Republic, Mendel began to ask the kinds of questions we have been posing: Why do offspring resemble parents? And why do offspring differ from parents?

Mendel attempted to answer these questions as any good scientist should—that is, by observing nature, doing experiments, and seeing what there was to see. In a series of classic studies with pea plants in his monastery garden, he delineated the basic laws that govern the inheritance of physical characteristics.

The technique that Mendel used is simple to describe, although it was somewhat more difficult and tedious to carry out. He cross-pollinated different varieties of peas. For example, he would fertilize the flowers of **purebred** tall pea plants—plants that always produced tall offspring—with the pollen from short ones, and then observe the characteristics of the "children" and "grandchildren," as shown in Figure 22–1. The offspring of two different strains, such as tall and short pea plants, are called **hybrids**.

When Mendel carried out these observations, he found that there were remarkable regularities in the characteristics of the offspring. All

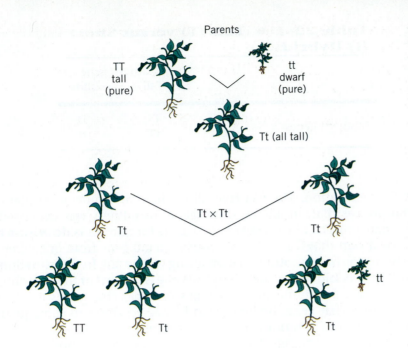

Parents

TT
tall
(pure)

tt
dwarf
(pure)

Tt (all tall)

Tt × Tt

Tt

Tt

TT

Tt

Tt

tt

Figure 22–1
The parents, first, and second generations of tall versus short pea plants. On average, the second generation shows a 3 : 1 ratio of dominant traits. Three-fourths of the plants, for example, will be tall, and one-fourth short. We have also shown the fourth generation.

offspring from the first generation of a tall-short cross were tall. If these offspring were bred with each other, however, the results were quite different. On the average, three-fourths of the offspring were found to be tall, while one-fourth reverted to being short. Thus, in hybridization, shortness disappears for one generation, only to reappear in the next. Mendel observed the same kind of behavior in half-a-dozen other pea plant traits: seed pod shape, flower color, and so on.

Mendel invoked something he called the "unit of inheritance"—what we now call the **gene**—to explain his findings. He had no idea what a gene might be, or even whether it had a real physical existence. Today, as we shall see shortly, the gene can be identified as part of a long molecule of DNA. For Mendel, however, the existence of DNA was unknown, and he deduced the presence of genes purely from the study of physical characteristics of his plants.

In the simplest version of Mendelian or **classical genetics**, we assume that every offspring receives two genes for every characteristic—one from the father and one from the mother. In the experiment with pea plants we just described, for example, every offspring in every generation received a gene for height (either tall or short) from its mother and another gene from its father. Mendel concluded that every characteristic he observed was determined by pairs of genes, and, furthermore, that the offspring may receive a different gene from the father than from the mother. So, if two genes are present, which gene "wins"—which characteristic is actually seen in the offspring? In the language of modern geneticists, we phrase the question by asking: Which gene is "expressed"?

Returning to our example of the pea plants, we recall that Mendel produced plants with what we might call a "tall gene" and a "short gene"

Table 22–1 • Tall (T) versus Short (t) Hybrids.

		Father's genes	
		Tall	short
Mother's genes	Tall	TT	Tt
	short	tT	tt

in the first-generation. The fact that all of these first-generation plants were tall means that, in all cases, the gene for tallness was expressed. Mendel stated this fact by saying that the gene for tallness is **dominant**. By this he meant that if an offspring receives a tall gene from one parent and a short gene from another, that offspring will be tall. In this situation, the short gene is said to be **recessive**. The gene is present in the offspring, but it does not determine the offspring's physical characteristics; it is not expressed. That gene, however, can be passed along to subsequent generations, even if it is not expressed in the present generation.

In this scheme of things, Mendel's experiment can be understood in very simple terms. In the first generation, every hybrid receives a tall gene and a short gene. Since the tall gene is dominant, all of the first generation of hybrid plants will be tall, even though each is carrying the recessive short gene.

In the next generation, there are four possible gene combinations, as shown in Table 22–1. Each plant in the second generation can receive either a tall or a short gene from each of its parents. On average, the distribution of genes will be random, so that we can argue as follows: In roughly one-fourth of the cases, the offspring will receive a tall gene from its mother and a tall gene from its father. In another one-fourth of the cases, the offspring will receive a short gene from its mother and a short gene from its father. In the remaining half of these cases, the offspring will receive a tall gene from its father and a short gene from its mother, or vice versa.

Consequently, in the second generation approximately three out of every four offspring will have at least one gene for tallness, and only one in four will have two genes for shortness. Given the fact that tallness is a dominant characteristic, this distribution means that three of four offspring in the second generation will be tall, while only one will be short. This situation is precisely what Mendel observed.

EXAMPLE 22–1: Breeding Peas

You are given two purebred pea plants. One plant has smooth pea pods and purple flowers (dominant traits). The other plant has wrinkled pea pods and white flowers (recessive traits). These characteristics are expressed independently of each other. What distribution of characteristics would you expect in the first generation of plants bred from these two parent plants? What distribution of traits would you see in the second generation?

Table 22–2 • SsPp Matrix with Resulting 9 : 3 : 3 : 1 Distribution

	SP	Sp	sP	sp
SP	SSPP	SSPp	SsPP	SsPp
Sp	SSpP	SSpp	SsPp	Sspp
sP	sSPP	sSPp	ssPP	sspP
sp	sSpP	sSpp	sspP	sspp

1 SSPP	2 SSPp	2 SsPP
4 SsPp	1 SSpp	2 Sspp
1 ssPP	2 ssPp	1 sspp

Smooth + Purple = 9
Smooth + white = 3
wrinkled + Purple = 3
wrinkled + white = 1

▶ **Reasoning:** Every plant in the first generation of offspring receives dominant genes for a smooth pea pod (S) and purple flower (P) from one parent, and recessive genes for a wrinkled pod (s) and white flower (p) from the other. Every plant in the first generation, therefore, has exactly the same gene combination, abbreviated SsPp. All of these plants will appear with smooth pods and purple flowers because S and P are dominant.

The second generation, however, will display a mixture of traits. The easiest way to predict the distribution of these traits is to set up a matrix, similar to the one shown in Table 22–1. In this case, however, we must deal with four different genes in each parent (SsPp), so the matrix must be 4 × 4, as shown in Table 22–2.

This table shows that there are 16 different possible combinations of the four genes. On average, 9 out of every 16 plants will appear with smooth pods and purple flowers—both dominant genes will be expressed. In addition, 3 of 16 on average will display wrinkled pods but purple flowers, and 3 of 16 will have white flowers but smooth pods. Finally, only 1 in 16 of the second generation will display both recessive traits: white flowers and wrinkled pea pods.

Mendel's observation of this characteristic 9:3:3:1 distribution of second-generation traits for two different genes was instrumental in his development of the genetic theory. ▲

The Rules of Classical Genetics

Mendel's research can be summarized by three rules that frame classical genetics:

Rule 1. Physical characteristics or traits are passed from parents to offspring by some unknown mechanism—we call it a gene.

Rule 2. Each offspring has two genes for each trait—one gene from each parent.

Rule 3. Some genes are dominant and some are recessive; the trait of a dominant gene will be expressed in preference to the trait of a recessive gene.

The rules of classical genetics were worked out in great detail during the early twentieth century. Careful records were kept on many kinds of organisms, from humans to cattle to agricultural plants, and large lists of dominant and recessive genes were compiled. In human beings, for example, dark hair and eye color are dominant over light, the ability to roll your tongue is dominant over inability, and hairy toe knuckles are dominant over hairless.

These sorts of lists can be used to analyze your own family. For example, it often happens that two dark-haired parents have a blonde child, or that two brown-eyed people have a blue-eyed child. In these situations, each of the parents, while carrying the dominant dark-hair or dark-eyed gene, also carries a recessive gene for the light hair or light eyes. The child in question has obviously received a recessive gene from each parent, and therefore may look quite different from the parents. Usually, if you search the family tree long enough, you will find other examples of the recessive coloring.

Qualitative versus Quantitative Genetics

Did Gregor Mendel really accomplish anything new? Were his results useful? In one sense, the qualitative aspects of Mendelian genetics have been understood for many centuries. Early human societies knew, for example, that if you saved the largest potatoes and planted them in the spring, the resulting crop would be better than if you just planted potatoes at random. They also knew that if you had a large bull that put on weight rapidly and produced a lot of meat, you should breed that bull to as many cows as possible so some of the offspring would share the characteristics of the father.

But Mendel's careful statistical analysis of pea plant traits carried genetics beyond the qualitative level. By discovering the distinctive $3:1$ and $9:3:3:1$ ratios of traits in second-generation plants, Mendel was able to propose a predictive model of genetics—a model that recognized the equal importance of both parents, and the distinction between dominant and recessive traits. When Mendel's rather obscure publications were "discovered" at about the turn of the century, they provided a model that allowed breeders to approach their work in a far more controlled and directed manner.

The traits of prize bulls and race horses, for example, are carefully documented, as are the pedigrees of their offspring. The success of plant and animal breeders in using and controlling the flow of genes from one generation to the next is attested to by the appearance of cattle such as Black Angus (which are little more than a rectangular block of beef on very short legs) and the many "perfect" varieties of vegetables and fruits that stock our supermarket shelves.

Black Angus cattle have been bred for centuries to produce beef.

Of equal importance, the laws of Mendelian genetics can now be used to trace cases of hereditary or **genetic disease**, such as the many famous cases of families with hemophilia—a rare disease in which the blood does not clot. The gene for hemophilia is recessive, but the royal families of Europe in the nineteenth century engaged in a lot of inbreeding. Because royalty tended to marry close relatives such as first cousins, the original gene (which was believed to have been introduced by Queen Victoria) spread to many members of the ruling families of Europe. The last heir to the Russian throne, for example, suffered from this disease.

Although we have chosen examples in which one physical characteristic is correlated to one gene, we should note that most cases of inheritance are not this simple. Human height and skin color, for example, are affected by the action of several genes and height can be influenced, by nutrition as well as genetics. Thus, while the general principles of classical genetics have widespread validity, the way that they work out in practice may be quite complex.

So long as genetics stayed at the level to which Mendel brought it, we could understand a great deal about inheritance, but we could not understand exactly what made inheritance operate the way it does. In particular, we had no idea of what a gene actually was. In fact, if you look in old textbooks on genetics, you see that the authors took great pains to distinguish between the gene (which they considered to be a concept that existed only in the human mind) and physical structures in the cell, such as chromosomes (see Chapter 21), that are involved in inheritance.

DNA and the Birth of Molecular Genetics

In Chapter 21 we learned that the functioning of the cell—the basic mechanism of all life—depends on chemical interactions between molecules. **Molecular genetics** is the study of how the mechanism that passes genetic information from parents to offspring functions on the basis of molecular chemistry. Chromosomes, the distinctive elongated structures that appear to divide just prior to cell division, became an obvious focus for genetic study. Could these cellular structures carry information and pass it from one generation to the next?

By the mid-twentieth century, biologists came to realize that chromosomes are made primarily of DNA, deoxyribonucleic acid (see Chapter 20). The discovery of the distinctive double-helix structure of this molecule provided the key to unlocking the genetic code. We now know that DNA carries our inheritance and governs all of the chemical functions of our cells.

The Replication of DNA

In Chapter 21, we described the processes by which cells divide. In both mitosis and meiosis, the first step is the copying of the DNA in the chromosomes. Thus DNA replication is one of the first steps in the passing of genetic information from one generation to the next.

The mechanism by which DNA reproduces itself depends crucially on the notion (see Chapter 20) that the geometry of the base pairs in DNA allows only certain kinds of bindings—that is, that adenine (A) binds only to thymine (T), and cytosine (C) binds only to guanine (G). No other pairings are allowed. When a cell is about to divide, special enzymes move along the DNA double helix, breaking the hydrogen bonds that link the bases—in effect, breaking the "rungs of the ladder"—as shown in Figure 22–2. The result of this splitting is that the two arms of the DNA ladder have exposed bases on them.

Consider, just for the sake of argument, an adenine (A) base that is no longer locked into its partner on the other side of the double helix. In the fluid around the DNA are many nucleotides, and some of these nucleotides contain an unattached thymine (T). This thymine will bind to the exposed adenine in the original strand of DNA. The binding process is aided by a special group of enzymes called DNA polymerases.

In the same way, an exposed cytosine (C) will bind to a molecule of guanine (G) in the fluid around the nucleus. No other type of nucleotide can bind to that particular site.

The net result of these preferential bindings along a single strand of exposed DNA is that the missing strand is reconstructed, base by base. The same thing happens in mirror image to the other half of the exposed DNA strand. Thus, once the DNA is unraveled, each strand replicates its missing partner. The end product is two double-stranded DNA molecules, each of which is identical to the original molecule.

When a cell divides through mitosis, the genetic information contained in the DNA of one cell is passed on to its daughters. Thus each daughter cell will have chromosomes identical to those of the parent. In meiosis, on the other hand, each daughter cell has only one chromosome from each pair of chromosomes in the original. When a sperm and an egg come together during fertilization, the resulting cell once again has a full set of chromosomes, but now one chromosome in each pair comes from the father, the other from the mother.

The simple chemistry of the base pairs gives us a mechanism for reproducing DNA. This feature of DNA molecular structure accounts for one of the striking facts we noticed earlier—that offspring do share many

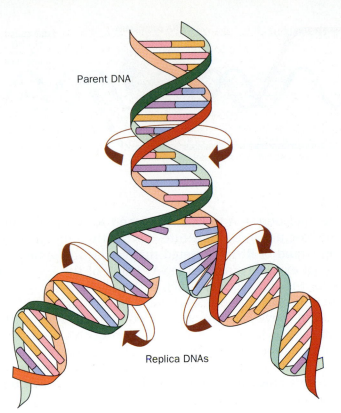

Parent DNA

Replica DNAs

Figure 22–2
A DNA double helix may be split, thus exposing bases on both strands. Two new identical double helices may then be formed from the original.

traits of their parents. Ultimately, the chemical binding of base pairs makes possible the inheritance of parental traits.

Transcription of DNA

In addition to replicating itself so that cell division can take place, DNA also supplies the information that runs the chemistry within each individual cell. This process depends on the fact that all chemical reactions in cells are governed by protein enzymes that run every cell's chemical reactions (see Chapter 20). Thus the question of how cell chemistry is regulated boils down to how the information in DNA can be used to produce proteins. If we understand this step, then we will understand how DNA governs the chemical functioning of every cell in our body.

DNA is a very large molecule. In eukaryotic cells it is found outside the nucleus only in mitochondria and chloroplasts. The first question we have to ask, then, is how information in the DNA gets out into the cell at large. The answer to this question involves a process called *transcription*, which uses the other nucleic acid we discussed in Chapter 20—RNA.

When it is time to fabricate a new protein to act as an enzyme in a cell, other enzymes "unzip" a section of DNA as shown in Figure 22–3. Nucleotides of RNA that are always floating in the nuclear material are then hooked, with the aid of enzymes, onto the appropriate bases by a process exactly analogous to that which occurs in the replication of DNA. Each of the exposed bases on the "unzipped" strand of DNA binds to its

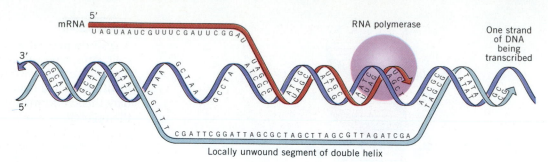

Figure 22–3

Transcription of DNA occurs when a segment of DNA is split and a single-stranded mRNA segment forms. The messenger RNA carries the same information that was on the original DNA segment.

appropriate nucleotide—A to U, C to G, and so forth. (Remember that in RNA, the base uracil, U, substitutes for the thymine in DNA). In this way a short strand of RNA is created that carries the same information as the original exposed strand of DNA. Think of the RNA as being the "negative" of the true picture, which is the DNA.

Because it is relatively short and not connected to anything else, the RNA strand can move out through tiny pores in the wall of the nucleus and into the cell at large. Thus the function of this kind of RNA is to carry the information that was contained in the central DNA molecule out into the region of the cell where chemical reactions are going on. Because it carries a message, this kind of RNA is called **messenger RNA,** or mRNA for short.

The Synthesis of Proteins

The exact sequence of base pairs on messenger RNA carries a coded message that contains chemical instructions. Once the mRNA arrives at the place where proteins are to be synthesized, it encounters a second type of RNA—a molecule called **transfer RNA,** or tRNA for short. The job of tRNA is to read that coded message. Transfer RNA, whose shape is shown in Figure 22–4, has a configuration at one end that attracts one of the 20 common kinds of amino acids found in living things (see Chapter 20). At the other end is a small loop of molecule with three exposed bases on it. One of four different bases can be found in each of the places on the top loop, so there are 64 ($4 \times 4 \times 4$) different kinds of tRNA molecules. Of these, 61 tRNA molecules attach to a specific amino acid at its other end, while the remaining 3 act as "stop" signs for protein building.

The sequence of bases along the mRNA is, as we have seen, a transcription of the information contained in the sequence of bases along the original DNA. Messenger RNA in effect carries a coded message, spelled out in four letters: A, U, C, and G. Each group of three exposed bases on the mRNA chain is like a word—a sequence of three letters that will bind to one, and only one, of the sets of bases on one of the 64 tRNA molecules. If a string along the mRNA reads A-A-U, for example, then the tRNA molecule that has as its unpaired bases U-U-A will bond to that particular spot. Thus, after time, the mRNA and tRNA will form a sequence as shown in Figure 22–4.

The set of three bases on the mRNA, called a *codon*, determines which of the possible tRNA molecules will attach at that point. Each codon on

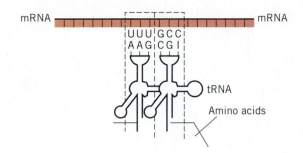

mRNA — | UUU | GCC | — mRNA
       | AAG | CGI |
       tRNA
       Amino acids

Figure 22–4
The interaction of mRNA and tRNA. One end of a tRNA molecule is attached to bases in the mRNA, the other end to a specific amino acid. Enzymes hook the amino acids together to form a protein molecule.

the mRNA determines a single amino acid, and the string of codons determines the sequence of amino acids—what we have called the primary structure of the protein that is being assembled. This connection between the codons and the amino acid they select is called the **genetic code,** as detailed in Table 22–3. This code is shared by all living things.

As the tRNA molecules attach themselves along the mRNA, a string of amino acids in a specific order—a protein—is assembled as shown in Figure 22–5. Once its amino acid has been incorporated into the protein, a tRNA molecule moves away to be replenished with another amino acid and used again.

Table 22–3 • The Genetic Code

THE GENETIC CODE

1st letter	U	C	A	G	2nd letter
U	Phenylalanine	Serine	Tyrosine	Cysteine	U
U	Phenylalanine	Serine	Tyrosine	Cysteine	C
U	Leucine	Serine	stop	stop	A
U	Leucine	Serine	stop	Tryptophan	G
C	Leucine	Proline	Histidine	Arginine	U
C	Leucine	Proline	Histidine	Arginine	C
C	Leucine	Proline	Glutamine	Arginine	A
C	Leucine	Proline	Glutamine	Arginine	G
A	Isoleucine	Threonine	Asparagine	Serine	U
A	Isoleucine	Threonine	Asparagine	Serine	C
A	Isoleucine	Threonine	Lysine	Arginine	A
A	(start) Methionine	Threonine	Lysine	Arginine	G
G	Valine	Alanine	Aspartic acid	Glycine	U
G	Valine	Alanine	Aspartic acid	Glycine	C
G	Valine	Alanine	Glutamic acid	Glycine	A
G	Valine	Alanine	Glutamic acid	Glycine	G

3rd letter

Examples of tRNAs

cys
ACG
Codon: UGC

his
GUG
Codon: CAC

gly
CCU
Codon: GGA

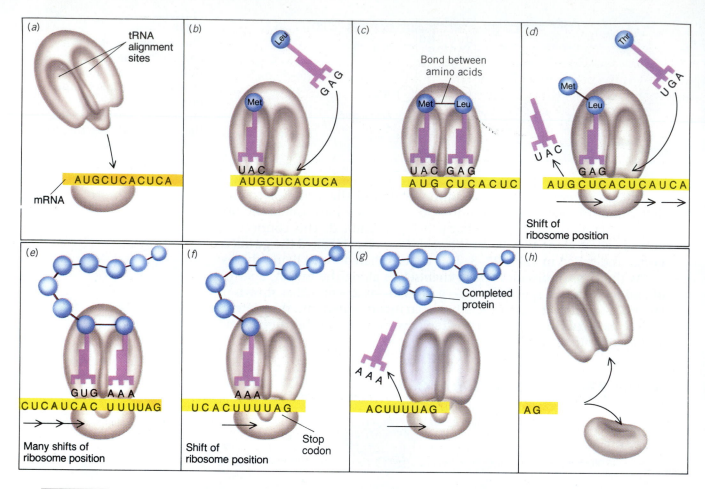

Figure 22–5

The formation of a protein requires three kinds of RNA. (*a*) A strand of messenger RNA fits into a groove in a ribosome (an organelle formed from proteins and ribosomal RNA). (*b*) The ribosome attracts the appropriate transfer RNA, which carries with it an amino acid. (*c*) A second tRNA attaches to the ribosome and the two adjacent amino acids are linked. (*d–g*) The ribosome begins to shift along the mRNA, attracting new tRNA molecules and adding amino acids to the chain. Once the amino acids and tRNA are disconnected, the tRNA floats off to find another amino acid. (*g–h*) The completed protein is assembled and released by the ribosome, and all the components are available to start the process over again.

The protein synthesis actually takes place on ribosomes, as large irregularly shaped organelles made of proteins and yet another kind of RNA—*ribosomal RNA*, or *rRNA* (see Chapter 21). As shown in Figure 22–5, the process of synthesis is somewhat more complex than the simple discussion we have given here. Ribosomes align the messenger RNA and transfer RNA during protein assembly. Thus three different kinds of RNA are involved in the synthesis of a single protein.

As a net effect of this rather complex molecular manufacturing process, the information encoded in the DNA molecule has been transferred to a sequence of amino acids that determines the identity of the appropriate protein enzyme. Thus a specific stretch of DNA located on one chromosome produces the enzyme that runs a particular chemical reaction in the cell. This stretch of DNA is what we have called a gene. That chemical might control skin color, hair texture, or any of the myriad other traits that we recognize.

One of the central rules of modern molecular biology is:

> ## One gene equals one protein.

That is, one stretch of DNA will code for one mRNA molecule, which will code for the sequence of amino acids in one protein, which will drive one chemical reaction in the cell. While a number of more complex processes produce a few specialized proteins, this rule holds for the vast majority of protein molecules. More than a century ago Mendel postulated the existence of a gene, without knowing what is was. Now molecular biologists can tell you exactly where many specific genes lie along a stretch of DNA, as well as the sequence of base pairs along them.

All living systems employ the genetic mechanism we have just described. The transfer of genetic information by DNA and the production of proteins by RNA is a process shared by every cell on Earth. Each species, and each individual within a species, has a slightly different message written on its DNA. The identity of every cell, as well as the organism of which the cell is a part, is determined by the chemical reactions that take place there, the chemical reactions are determined by the enzymes, and the enzymes are coded for in the DNA. Thus DNA is truly the molecule that contains the code of life.

What is perhaps most remarkable about this process is that all living things use essentially the same code to translate between the messages carried in the genes of DNA, the messages carried in RNA, and the string of amino acids in proteins. This relationship explains why biologists speak of "the genetic code" when they refer specifically to the relationship between a triplet of base pairs on the mRNA, and the corresponding amino acid in the protein. The basic "word" of the molecular world, then, is the triplet of bases along DNA—the codon. Each codon eventually codes to one amino acid in a string of proteins.

The fact that all living organisms, from single-celled yeast to human beings, use precisely the same biochemical apparatus and precisely the same technique for making proteins and running their chemistry is one of the great unifying ideas in the science of biology. Indeed, one of the great principles of science is:

> ## All living things use the same genetic code.

This great truth in no way limits the tremendous variety and diversity one can find in living things. Just as many different books can be written using the 26 letters of the English alphabet, so too can many different life forms be constructed using the four "letters" in the genetic code.

Mutations and DNA Repair

If DNA were copied faithfully from one generation to the next, no living thing could be much different from its ancestors. But mistakes do happen, and many agents in nature can damage the DNA molecule. Numerous chemicals (particularly those that cause oxidation reactions in cells), nuclear radiation, X-rays, and ultraviolet light (which also produces oxidizing chemicals) are all examples of such agents.

If the DNA of a parent is altered, then the alteration will be faithfully copied by the process we have just described. The offspring will inherit the change, just as they inherit all other genetic information from the parents. Such a change in the DNA of the parent is called a **mutation**. As we shall see in the next chapter, mutations have played a very important role in the development of life on Earth.

Recently, scientists have begun to realize that DNA is damaged at a far higher rate than had previously been thought. Careful chemical analyses indicate that damage to DNA in humans goes on at the rate of about 100,000 "hits" per cell per day. Fortunately, the body has developed repair mechanisms that take care of almost all of this damage as soon as it happens. The study of DNA repair, and the hope that it may help us deal with diseases such as cancer, represents a major frontier in science today.

Why Are Genes Expressed?

Every cell in your body contains an identical set of chromosomes—the exact same set of DNA molecules—yet your cells are not all alike. In fact, chemical reactions that are critical to one set of cells—those that produce insulin in your pancreas, for example—play no role whatsoever elsewhere. The genetic coding for making insulin is contained in every cell in your body, but only turned on in a few. How do the cells in the pancreas "know" that they are supposed to activate the particular gene for insulin, while the cells in the brain know they are not supposed to?

The mystery of DNA's operation runs even deeper than this. It now appears that only about 5% of all DNA in human beings is actually taken up by the genes. The other 95% used to be called "junk DNA," because nobody understood why it was there. Scientists are increasingly coming to believe, however, that at least some of the rest of the DNA contains instructions for turning genes on and off. The study of gene control is a frontier field, and we understand very little about how it works. We do know, however, that genes are activated at certain times in the growth of plants or animals, and the triggers for this activation appear to be enzymes or other chemical agents.

Many scientists also believe that the failure of these instructions leads to diseases such as cancer. If a cell is dividing and the mechanism that tells it when it's time to stop is faulty, the cell may continue to multiply and produce a malignancy or tumor. Damage to the control mechanisms in a cell, then, may be much more serious than damage to the genes themselves.

Viruses

If you ever have had flu or a common cold, you know something about viruses. They aren't alive in the sense that bacteria and other single-celled organisms are. Unlike the life forms we discussed in Chapter 19, viruses do not metabolize and are not capable of reproduction on their own.

A **virus** consists of a short length of RNA or DNA wrapped in a protein coating. The protein is shaped so that it fits cell receptors and is taken into a cell. Once inside the cell, a variety of events may occur, depending on the exact nature of the virus. The viral DNA may replicate itself, producing its own mRNA, or viral RNA may serve directly as messenger RNA. Thus the virus takes over the cell's machinery, using the cell's enzymes and tRNA to produce more viruses like itself, eventually killing the cell.

Alternatively, as in the HIV virus that causes AIDS, the virus contains an RNA sequence that can be transcribed back into DNA, along with some enzymes that insert the DNA into the cell's own DNA. Once that stretch of DNA is inserted, it acts just like any other gene and co-opts the cell into making more viruses. No matter what the mechanisms, however, the result is the same: the cell eventually dies.

The surface of the HIV virus contains protein structures that fit perfectly into one of the molecules in the outer membrane of one type of human T cell, a critical part of the human immune system (see Chapter 21). T cells are thus "fooled" into pulling the virus inside. The virus then attacks and ultimately destroys cells that are essential to the operation of the human immune system.

Note that a "computer virus" operates in the same way. This kind of virus is a set of instructions taken into a computer that co-opts all the computer's machinery to its own ends.

Science in the Making

Viral Epidemics

There is an old joke about someone who goes to a doctor with a cold and is told to take a shower and stay outside in the cold with wet hair and without a coat.

"But if I do that, I'll get pneumonia," the patent protests.

"Of course," says the doctor, "but I can cure pneumonia."

The medical profession has enjoyed a great deal of success in dealing with diseases such as pneumonia that are caused by invading bacteria. Antibiotics often work by blocking particular enzymes in the bacteria. Since these enzymes don't operate in human cells, antibiotics can destroy the bacteria without harming the human whose body they are invading.

Viruses, on the other hand, co-opt most of the cell machinery of the host. With a few exceptions, they will not be affected by antibiotics. This is why viral diseases such as the common cold cannot be treated with drugs. The most effective countermeasure for viral diseases has been to stimulate the human immune system to produce antibodies to combat the virus. Poliomyelitis, smallpox, and yellow fever have all been dealt with in this way.

Viruses not only hide inside cells, they change very rapidly, producing new forms as quickly as we find vaccines against them. One reason for this is that, while the copying of DNA in cell division is subject to the cell's "proofreading" mechanisms so that daughter cells are the same as the parents, viruses have no such proofreading. Consequently, viruses mutate at a rate up to a million times faster than normal eukaryotic cells. In addition, if two viruses invade the same host, they may swap sections of their nucleic acids, producing a new strain in the process. This rapid rate of mutation in viruses is the main reason that Americans are urged to get new flu shots each year, to counteract whatever virus has developed since last year's shot.

As the AIDS epidemic should remind us, viral diseases remain a very real threat to the human race. In particular, several features of modern life make human beings particularly susceptible to viral attack. For one thing, we now tend to live together in cities, providing a large host population for new viruses. We also travel a great deal, so that a virus that develops in one part of the world will quickly spread. Finally, humans are coming into more contact with isolated wilderness areas, and therefore into contact with whatever viruses are already living on hosts in those areas. One example is the virus responsible for AIDS, which is believed to have arisen from a virus affecting monkeys in remote African forests. A hunter cutting his finger while skinning an infected monkey, for instance, could have introduced the virus to the human population.

How much attention do you think governments and scientists should pay to the dangers of new viral diseases? Do you think an international medical center should be established to monitor the appearance of new diseases? What good would such an early warning system do? ■

The Human Body

The Human Genome Project

The sum of all information contained in the DNA for any living thing—the sequence of all the bases in all the chromosomes—is known as that organism's **genome**. Determining a complete genome is a vast project. With the exception of the simplest bacterium and a few kinds of virus, we aren't even close to doing this for any organism, but that situation may change in the next decade or so. One large-scale scientific effort that you are likely to read about in the coming years is called the **Human Genome Project**. This massive undertaking, expected to last at least a decade and cost many billions of dollars, will result in a complete knowledge of the entire human genome—all 46 chromosomes, all 3 billion base pairs. Researchers must accomplish two key tasks, mapping and sequencing, in the exploration of the human genome.

Scientists attempting to describe our genetic makeup must first find the location of every gene on every chromosome, a process called **DNA mapping**. Creating a complete genetic map is no mean feat. Human DNA, for example, contains about 100,000 genes divided among the 23 pairs of chromosomes. Each chromosome is a long chain of DNA. Mapping all the genes—discovering where they lie along each chromosome—is time-consuming, but it plays a vital role in dealing with many kinds of hereditary diseases. Medical researchers investigating genetic diseases, for example, must find the exact chromosome and location of the defective gene in order to identify genetic markers for the disease and, perhaps, find a cure. In recent years the locations of the defective genes for cystic fibrosis, sickle-cell anemia, and genetic forms of arthritis have been discovered on specific chromosomes, and many similar advances can be expected in the next few years. Mapping techniques are highly developed, and the exploration of human DNA is well underway. In a few years, scientists expect to have a reasonably complete map of human DNA.

The genetic map, like a good road atlas, tells us the general location of the most interesting places in our tour of the human genome. Genes are like villages and towns in our atlas. But these maps alone tell us little about the details of those places. We also need directories to each village and town if we are to really understand how the genome works. **DNA sequencing** is the process of determining, base pair by base pair, the exact order of bases along a DNA molecule. The net result of a sequencing operation is a string of letters (ATTGCG-CATT..., and so on); a sequence that tells us how the DNA is put together in that particular stretch.

One result of this sort of knowledge is that a gene sequence can be used to deduce the sequence of amino acids in a protein, and this information, in turn, may give some insight into the function of that protein in the organism.

Mapping the human genome is already well underway, and high-resolution maps are available for several chromosomes. Sequencing, however, is at present too time-consuming for a concentrated effort. A five-year objective for the Human Genome Project, therefore, is to develop automated procedures that will reduce the expense of sequencing to less than a dollar per base pair. Like any expensive government effort, the Human Genome Project is not without controversy. Many scientists fear it will divert funds into what is basically a repetitive mechanical operation, while taking resources away from other fundamental research. Others counter that knowledge of the human genome will provide the most fundamental basis for understanding human development and health. ■

Genetic Engineering

Every week headlines such as "Genetically Altered Tomato Now on Shelves" or "New Mouse Strain Patented" appear in your newspaper. All of these headlines refer to a technology known as **genetic engineering,** a technology in which foreign genes are inserted into an organism, or existing genes altered, to modify the function of living things. The basic technique of genetic engineering is very simple. Certain chemicals, called restriction enzymes, have the ability to cut a DNA molecule (as shown in Figure 22–6) so that the DNA has several unattached bases at the cut end. Think of these exposed bases as being something like pieces of Velcro at the ends of the strands of DNA.

If another strand of DNA is cut in the same way, and if the exposed base pairs on that second strand are complementary to the base pairs on

Figure 22–6

Restriction enzymes act something like a pair of scissors that break a DNA chain at specific sequences. Biologists can use this procedure to insert or remove segments of DNA.

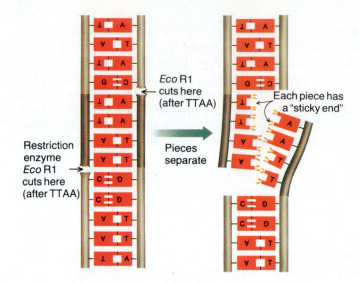

the original strand, then when the two strands are put together the bases will bind and the strands will stick together. This procedure allows researchers to take a stretch of DNA in a cell nucleus, cut it, and splice in another stretch of DNA that has the appropriate base pairs on its ends. The new DNA will contain an extra length in it.

If the new stretch of DNA is a gene, then the same mechanism that drives the chemistry in every cell on Earth will start expressing that gene in its new environment. The gene that makes human insulin, for example, has been spliced successfully into the single-celled *E. coli*—a common laboratory bacterium. Thus modified, a vat full of genetically engineered *E. coli* will begin producing human insulin, in most respects indistinguishable from the stuff produced in your own pancreas. In fact, most of the commercially available insulin used to treat diabetes is now made in this way. This genetically based process is a tremendous improvement over the old method, which involved laborious extraction of the insulin from the pancreas of slaughtered pigs.

The story of genetically engineered insulin is only a small part of the changes that are possible with genetic engineering. In 1993, for example, genetically engineered tomatoes appeared on supermarket shelves. These tomatoes have been altered by the incorporation of a single gene. Normally, when a tomato is picked from its plant, a gene is expressed that produces a chemical that speeds the rotting process. From the point of view of the tomato this accelerated decay is a good thing, because it means that the tomato seeds will be dispersed and put into the ground as quickly as possible. From the point of view of the tomato grower and consumer, however, it is not a good thing, because it means that the tomatoes have to be picked green if they are to last until they can be eaten. In the genetically engineered tomato, a gene is inserted that blocks the production of the rot-inducing chemical. Thus the tomato has a very long shelf life and can be allowed to ripen on the vine before it is picked.

Other applications of genetic engineering include strawberries that are highly resistant to frost, plants that are highly resistant to specific

Genetic engineering has produced a variety of modified organisms, including (*a*) frost-resistant strawberry plants; (*b*) tomatoes with long shelf lives; and (*c*) genetically identical mice for medical research.

(*a*)

(*b*)

(*c*)

diseases, and plants that manufacture natural insect repellents. Genetic engineering of animals has led to varieties that grow faster or have other agricultural advantages, while researchers have recently patented varieties of mice especially "designed" for medical research. There is even a patent for genetically engineering bacteria that eat petroleum. These sorts of bacteria might some day be used to clean up oil slicks from tanker accidents.

Underlying all of this technology is the central notion that all life is based on the same genetic code—the translation of DNA sequences in a gene to amino acid sequences in a protein. The exact same geometrical shapes of the four base molecules occur in every living thing. The fact that scientists routinely switch genes from plants to animals and back again is simply another indication of the underlying chemical unity of all life.

We have not even begun to imagine the changes that can take place through the use of this technology. The social issues involved in genetic engineering are profound. In 1974, scientists working in the field of genetic engineering voluntarily undertook a moratorium on further research until thought could be given to the question of how to keep potentially dangerous organisms from escaping into the environment. In most cases, such as the *E. coli* used in insulin production, the host bacteria are so specialized and so dependent on their laboratory culture that they could not survive in the wild. Nevertheless, the "Andromeda strain" scenario, in which a genetically altered, destructive life form escapes into the environment, is something we must be concerned about. Is it possible that a genetically engineered disease—a common human bacterium with an inserted gene for a lethal toxin, for example—could be turned loose on the human race?

These sorts of concerns are justifiable and need to be addressed whenever a new strain of plant or animal is developed. We should point out, however, that the introduction of new strains of plants and animals has been going on for centuries through carefully controlled selective breeding. What is new about genetic engineering is that, unlike the breeders of the past, the modern biologist can usually tell exactly what gene is being added to or deleted from the organism being developed.

Technology

DNA Fingerprinting

The analysis of DNA in human tissue is starting to become important in the judicial system in the United States. Except for identical twins, no two human beings in the world have the same DNA. Thus analysis of blood, skin, or semen samples from the scene of a crime can be used to identify criminals in much the same way that fingerprints do.

Analysis of DNA samples depends on the fact that all human DNA contains many regions called "variable number tandem repeat" (VNTR) sequences. In these regions, a particular nonsense phrase is repeated over and over again—up to 256 times. The number of repeats in a given sequence varies from one individual to the next.

To begin the analysis, an enzyme splits DNA at the start of these distinctive sequences, and then the numerous small segments of DNA are placed in a gelatin-like material. When subjected to an electric field, the DNA segments start moving through the gel, with the shorter and lighter strands, in general, moving faster. The net result is that after a certain time, the different strands of DNA will have moved different distances. A strand containing a VNTR with 10 repeats, for example, will have moved farther than one containing 200.

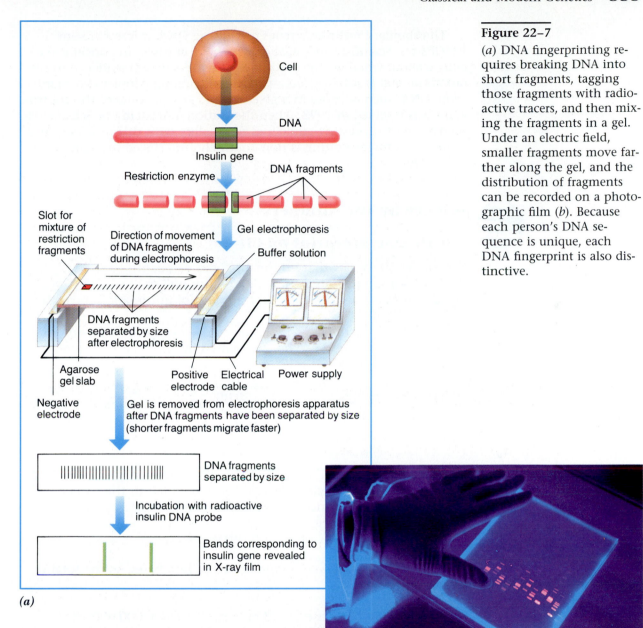

(a)

(b)

Figure 22–7
(*a*) DNA fingerprinting requires breaking DNA into short fragments, tagging those fragments with radioactive tracers, and then mixing the fragments in a gel. Under an electric field, smaller fragments move farther along the gel, and the distribution of fragments can be recorded on a photographic film (*b*). Because each person's DNA sequence is unique, each DNA fingerprint is also distinctive.

Technicians then splice a stretch of standardized DNA prepared with radioactive isotopes onto the pieces of DNA in the sample and a piece of photographic film is laid over the gel. The radioactive emissions are recorded on standard photographic emulsion.

The end result of this process is that the information in each person's DNA is reduced to something like a bar code (see Figure 22–7). The "bar code" obtained from a tissue sample at the scene of a crime can be compared to that obtained from a suspect to make an identification.

The standard technique relies on cutting DNA at five different VNTR sites. Scientists and legal experts have engaged in considerable debate about whether this procedure provides enough evidence to rule out accidental identifications—particularly among people who share some DNA, such as those in a given ethnic group. Some of the controversy was resolved in 1992, when the National Academy of Sciences issued a report suggesting guidelines for dealing with this problem. As a result, DNA fingerprinting is now routinely introduced in many courts. ■

Science by the Numbers

The Human Blueprint on CD

In Chapter 10 we saw that information can be quantified in units of the "bit"—a simple statement about "yes or no" or "on or off." We can use this notion to calculate the amount of information in the human genome.

Each site along the DNA molecule can be occupied by one of four bases. This information can be represented by two bits. We could, for example, set up a code as follows:

$$A = on\ on$$
$$T = on\ off$$
$$C = off\ on$$
$$G = off\ off.$$

Using this code, we could go down the molecule specifying two bits of information at each nucleotide, and this would tell us the sequence. For example, the sequence AGT would be rendered

on on; off off; on off.

The human genome contains about 3 billion bases, so the total information content is

(3,000,000,000 bases) × (2 bits/base) = 6,000,000,000 bits.

Let's compare this information content to that of a familiar object—a CD. A compact disc reproduces sounds by sampling sound 44,100 times each second, then representing the intensity of the sound by 16 bits. Thus each second of sound on a CD corresponds to

(44,100 samples/second) × (16 bits/sample) = 705,600 bits/second.

How long do you have to listen to a CD to get as much information as is contained in the human genome? The answer is approximately

$$\frac{6,000,000,000\ bits}{705,600\ bits/second} = 8,500\ seconds,$$

or about 140 minutes. Thus the entire blueprint for a human being could be contained on about two and a half hours of a compact disc. ■

THINKING MORE ABOUT GENETICS

The Ethics of Genes

Advances in genetic research are dramatically altering our understanding of human health and behavior. Scientists now can predict many characteristics of an individual, including the presence of life-threatening diseases, before birth. Every year we learn more about genetic characteristics, and thus are better able to foresee a child's future. But with this knowledge comes an ethical challenge that will face every American in the coming decades. What should we do with genetic information?

Eventually we may be able to test every fetus for a variety of incurable genetic diseases. Should those tests be mandated? Should parents be informed of their future child's fate? Should the prospect of an incurable disease provide grounds for abortion? It has also been suggested (though not proven) that alcoholism and other behavioral disorders may be related, at least in part, to genetic factors. Suppose a person were found to carry a particular gene or combination of genes that were thought to predispose individuals toward alcoholism? To whom should that information be conveyed? To the individual? To his or her doctor? His or her employer? His or her insurance company?

Taking these issues a step further, it may soon be possible to alter an individual's DNA *in utero*, perhaps even in the first weeks of pregnancy. Many people would probably agree to genetic manipulation if it could cure their child of a fatal disease, but where do we draw the line? Would you allow such a procedure to improve genetically defective eyesight, or perhaps prevent crippling arthritis in later years? Would you be willing to enhance your child's IQ, or make her more athletic? What about changing his height or hair color? As with many other aspects of science and technology, we must come to grips with the question of whether it is ethical to do something simply because we are able to do it.

Summary

Genetics, the study of the way in which biological information is carried from one generation to the next, is a field as old as the selective breeding of animals and the selection of seeds for crops. Gregor Mendel attempted to quantify aspects of this process by cross-pollinating *purebred* varieties of pea plants to produce *hybrids*. He discovered that all first-generation hybrids appeared the same, with the traits of just one parent plant, but the second generation displayed characteristics of both parents. Typically, three-fourths of the members of the second generation display one trait, one-fourth the other. Mendel explained his observations by developing laws of *classical genetics*: (1) traits are passed from parent to offspring by "atoms of inheritance" (we call them *genes*); (2) each parent contributes one gene for each trait; and (3) some genes are *dominant* and will be expressed, while others are *recessive* and will only appear if no dominant gene is present.

Modern *molecular genetics* seeks to understand the molecular basis for Mendel's observation. The key to understanding genetics lies in the unique structure of the DNA double helix, with its ladderlike sequence of base pairs. The four different bases, A, T, C, and G, which always come in the pairs AT or CG, act like letters of a coded message—the message of life. Because of its structure, DNA can replicate itself and store the information needed to make proteins.

Every cell has a set of chromosomes with the complete DNA blueprint in its nucleus. The process of copying DNA before cell division is called replication, and involves the splitting of the two sides of the DNA double helix, exposing apart the complementary base pairs.

Each exposed base binds to its complement, and so two complete DNA strands form where before there was only one.

The coded DNA message is read by RNA—a process called transcription. *Messenger RNA*, a single-stranded molecule, copies the sequence for one gene and carries it out of the nucleus to the part of the cell where proteins are made. *Transfer RNA* matches sequences of three base pairs to corresponding amino acids; thus a RNA gene sequence translates into a string of amino acids—a protein. The correspondence between base-pair sequences and amino acids is called the *genetic code*, which is the same for every living organism.

While the DNA message is resilient to most damage, errors in the coded sequence can occur and cause *mutations*. *Genetic diseases* are hereditary mutations that cause sickness or death. *Viruses*, on the other hand, cause sickness by usurping a cell's chemical factories with foreign RNA genetic instructions.

The complete description of an organism's genetic code is called its *genome*. Scientists determine a genome by first *mapping* the positions of every gene on every chromosome, and then *sequencing* the exact order of base pairs on every gene. The *Human Genome Project*, a decade-long, multi-billion-dollar effort, is devoted to determining the complete 100,000-gene, 3 billion-base pair sequence of the human genome. Our enhanced understanding of the human genome may lead to new advances in *genetic engineering*, which involves the insertion of foreign genes into an organism, or the alteration of existing genes, to create modified life forms.

Key Terms About Genetics

genetics

purebred

hybrid

gene

classical genetics

dominant

recessive

genetic disease

molecular genetics

messenger RNA

transfer RNA

genetic code

mutation

virus

genome

Human Genome Project

DNA mapping

DNA sequencing

genetic engineering

Review Questions

1. How did Mendel define the gene? How do we define it today?

2. What is a dominant gene?

3. What is a recessive gene?

4. How does DNA replicate itself?

5. How is the information of DNA copied onto mRNA?

6. What is the function of mRNA?

7. How does tRNA determine the primary structure of proteins?

8. What is a codon? What is the genetic code?

9. What is a mutation? What agents cause mutations?

10. What is a virus?

11. What is the difference between mapping and sequencing DNA?

12. What is genetic engineering?

13. What is the Human Genome Project?

14. How does DNA fingerprinting work?

Discussion Questions

1. Could a recessive trait skip more than one generation? How could this happen?

2. How might DNA evidence be used in a courtroom? Has it ever been used in your local courts?

3. If you were given two pea plants, one with waxy leaves and one with fuzzy leaves, how would you determine which leaf trait is dominant?

4. Why is DNA sequencing more time-consuming than DNA mapping?

5. What does the statement "one gene, one protein" mean?

6. State some of the arguments for and against the use of genetic engineering.

7. Inbreeding, or the mating of closely related individuals, tends to perpetuate both good and harmful traits. Why should this be so?

8. If every cell in your body has exactly the same DNA, how can the cells perform such different functions?

9. Would you say that viruses are alive? Why or why not?

10. People frequently worry that exposure to environmental chemicals or radiation will cause damage to children, even though those children have not even been conceived at the time of the exposure. Give a molecular explanation of how such harm could occur. What role should information on DNA repair play in such discussions?

Problems

1. In Chapter 20 we showed that the information content of a letter is six bits. The *Encyclopedia Britannica* has about 1500 words per page, 1000 pages per volume, and 30 volumes. How many sets of the *Encyclopedia Britannica* would it take to transmit the same amount of information as is contained in human DNA?

2. Suppose a particular breed of mice can be either white or gray, and have either clear or pink eyes. Suppose further that if true-breeding white mice are crossed with true-breeding gray mice, all the offspring are gray; and that if true-breeding clear-eyed mice and true-breeding pink-eyed mice are crossed, all the offspring have pink eyes. What will be the distribution of hair and eye color in the second generation if we start by crossing true-breeding gray, pink-eyed mice with true-breeding white, clear-eyed mice?

3. Scientists frequently use fruit flies, which breed every 10 days, to do genetics experiments. How long would it take to repeat one of Mendel's experiments on peas using fruit flies?

Investigations

1. Prepare a report on a genetic disease. What progress has been made in mapping the defective gene? Sequencing the gene? What kinds of medical treatments are now available?

2. Read Michael Crichton's recent novel *Jurassic Park*, which describes experiments in genetic engineering. Do you think the scenario is realistic? What precautions should the scientists have taken?

3. Some psychological disorders are now believed to be caused by defects in the body's chemistry. Read about one such disease, and summarize the argument between those who believe that psychological problems all have a molecular basis and those who believe that they are all due to the environment in which the individual has lived.

4. Make a family tree for your own family, recording characteristics such as eye and hair color, height, causes of death, and so on. Can you apply Mendel's rules to this tree?

5. A number of obscure viral diseases have arisen over the past few decades. Look up Korean hemorrhagic fever, dengue fever, Lassa fever, or the Ebola and Marburg viruses. What symptoms do these diseases or viruses produce? What danger do they pose to the larger human population?

Additional Reading

Crick, Francis. *What Mad Pursuit*. New York: Basic Books, 1988.

Gonick, Larry and Mark Wheelis. *The Cartoon Guide to Genetics*. New York: Barnes and Noble, 1983.

Freeland Judson, Horace. *The Eighth Day of Creation*. New York: Simon and Schuster, 1979.

Watson, James D. *The Double Helix*. New York: Athaneum, 1985.

Evolution

All life on Earth evolved from single-celled organisms by the process of natural selection.

A Random Walk
To the Zoo

Think about the last time you visited a zoo. You undoubtedly enjoyed seeing the elephants and watching the monkeys. You may have walked through the reptile house and looked at the alligators and tortoises. Perhaps you saw some endangered species such as the panda. Your zoo may even have had a "bug house," where various kinds of invertebrates could be seen. The zoo was probably located in a parklike setting, with all sorts of trees, bushes, and grasses around. And yet for all the diversity of life you saw, the zoo contains representatives of only a tiny fraction of the living things on our planet.

How could such an incredible array of living things have developed on a once-lifeless planet? How could living things on our planet include such extraordinary variety, from a slime mold to a sperm whale? The answer to these questions requires us to think not only about what life is, but how it got to be the way it is.

The Two Steps of Evolution

In Chapter 19 we surveyed the enormous diversity of life on our planet. The five kingdoms fill the world around us—you probably have seen hundreds of different species in the last 24 hours, whether you took note of them or not. Biologists estimate that there are tens of millions of species on our planet right now. When we examine the record of the past, we find that life has existed on our planet for billions of years. This record is written mainly in fossils—replicas in rock of dead organisms. It tells us that the diversity of life we observe around us today has characterized the Earth for hundreds of millions of years. In fact, for every species of life that lives on the planet today, perhaps 1000 have lived in the past but are now **extinct**—they have disappeared forever from the Earth. How could this tremendous diversity have arisen and been maintained?

The Earth started out as a hot, lifeless ball of molten rock (see Chapter 14). The first rocks were formed when the planet cooled, but even then the Earth looked nothing like it does today. Water filled the ocean basins, but no fish swam in it and no algae floated on it. All of the billions of different life forms that would someday develop were absent in this early stage.

The transition from a lifeless planet to one that teems with living things came in two stages. The first stage involved the appearance of the first living cell from the lifeless chemical compounds that existed in the early Earth, and was governed by the laws of chemistry and physics that we have learned.

This stage in the development of life saw the gradual transformation of a single living cell into a wide diversity of complex living things. This stage of evolution involved a new dynamic process—natural selection. This mechanism for evolution, not proposed until the middle of the nineteenth century, recognized that all living things must compete for a limited supply of resources.

Chemical Evolution

The central question of **chemical evolution** is how one can start with the simple chemical compounds that were most likely present in the Earth's early atmosphere and wind up with an organized, reproducing cell. This relatively new area of research is one in which we still have large gaps in our knowledge. Perhaps the most important experiment relating to chemical evolution was performed in 1953 by Stanley Miller (b. 1930) and Harold Urey (1893–1981) at the University of Chicago. The novel apparatus of the **Miller-Urey experiment** is sketched in Figure 23–1.

Based on analysis of gases that are released by volcanoes today, scientists have argued that the Earth's early atmosphere contained water vapor, hydrogen (H_2), methane (CH_4), and ammonia (NH_3). Miller and Urey mixed these materials together in a large flask. Then, realizing that powerful lightning would have laced the turbulent atmosphere of the early Earth, they caused electric sparks to jump between electrodes in the flask.

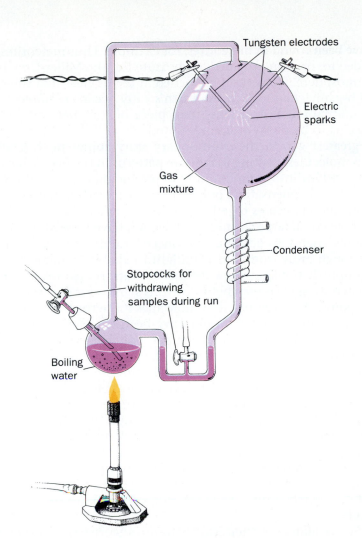

Tungsten electrodes

Electric sparks

Gas mixture

Condenser

Stopcocks for withdrawing samples during run

Boiling water

Figure 23–1
The Miller-Urey experiment. Several of the chemical compounds thought to have been present on the early Earth were mixed and subjected to electrical discharges. Within a few weeks, amino acids had formed.

After a period of a couple of weeks, they noticed that the liquid in the flask became cloudy and started to turn a dark brown color. Analysis revealed that this brownish liquid contained a large number of amino acids, one of the basic building blocks of life (see Chapter 20).

Thus, as early as the 1950s, scientists had found that the modules of at least one of the important molecules of life—proteins—could be generated quite easily by natural processes in the atmosphere of the Earth. Since that time, it has been found that energy sources such as ultraviolet radiation (from the Sun) and heat (for example, that supplied by volcanoes) will also produce amino acids. In subsequent experiments at the University of Chicago and other laboratories, scientists have used modified Miller-Urey devices to make other organic molecules including lipids and bases, as well as complex substances such as long protein chains.

Some scientists suggest that this scenario has important implications for the early Earth. For perhaps several hundred million years, they say, the amino acids and other molecules created by the Miller-Urey process were concentrated in the ocean, producing a rich broth, sometimes called the *primordial soup*. Additional amino acids may have been added to this chemical mixture in the early oceans by other sources as well. Amino acids have been found in meteorites, for example. Thus meteorites could have added to, or even provided a substitute for, the Miller-Urey process in the atmosphere. However it happened, it seems clear that the enrichment of organic chemicals in the Earth's early oceans required nothing beyond normal chemistry. This small piece of the chemical evolution puzzle seems to be pretty well understood.

Our greatest gap in the evolutionary story comes next. Could the countless molecules, floating in random patterns in the ocean, have organized themselves into a functioning, reproducing cell? While mechanisms such as condensation polymerization (see Chapter 9) can join simple organic molecules together, sunlight tends to break these bonds apart. Where could large clusters of molecules have formed near enough to the Sun's energy to accumulate concentrations of organic material, but far enough away from direct sunlight to avoid destruction?

We simply do not know how the first cell formed. However, a number of very interesting ideas may help close this significant gap in our knowledge. We know, for example, that molecules more than ten meters or so deep in the primeval ocean would have been shielded from the Sun's ultraviolet radiation. In this environment, polymers and other complex molecules could have grown and diversified. If the concentration of organic molecules was high enough, the sunlight breaking up these molecules would not have been able to overcome their formation rate, and the concentrations would have grown even more.

Alternatively, some scientists have proposed that the first cell may have evolved in a tidal pool. If it turns out that something like a protected tidal pool is necessary before the development of life, then life may be a relatively rare thing in the universe, since our present understanding is that the formation of a large terrestrial planet together with a large moon (and, consequently, the generation of significant tides) is an unlikely event.

A major weakness of chemical evolution theories is that they tell us little about how organic molecules could isolate themselves from the environment. How could a cell membrane form, for example? Early optimism that research on the primordial soup would lead quickly to an understanding of life's origins has faded, and scientists now understand that major questions remain. On the other hand, this work has shown clearly that the basic building blocks of life were available very early in the history of the Earth, perhaps within a few million years after the first oceans formed. Here are a couple of ideas being put forward now:

The Primordial Oil Slick Scientists who study Miller-Urey-type experiments have deduced that key organic chemicals in addition to amino acids—lipid molecules, for example—probably were created in those early chemical reactions. In this way of thinking, the early ocean not only would have contained the primordial soup, but would have been covered, at least in part, by a layer of lipids—a "primordial oil slick."

In this situation, you would expect some of the lipids—those with one hydrophilic end and one hydrophobic end—to organize themselves into layers, and the layers to curl up into bubbles, through the actions of much the same forces that cause water droplets to "bead up" on a newly waxed car. Recalling our discussion of lipids (see Chapter 20), you would expect the oil slick to form in a double layer with the hydrophilic tails of the molecules turned out, and the hydrophobic ends head-to-head on the inside.

In this scheme, lipids would form themselves into little globules (think of them as the prototypes of the cell membrane), enclosing within them some pieces of the organic soup. This notion has a clear advantage: whatever chemistry might be necessary to change the primordial soup into a living system will take place in an environment that is already separated from the outside.

You would expect that literally billions of globules would form, and you would not be far wrong to imagine each of them as a tiny chemistry lab. Each one would enclose different mixtures of different kinds of molecules that react chemically in different ways. You can think of the early oceans as a place where one arrangement after another is tried until one that is able to reproduce itself—the very first cell—is found.

RNA Enzymes   Today's cells run most of their chemistry by using protein enzymes, which, in turn, are coded for in DNA. On the other hand, in order to turn the DNA into a "working" protein, other enzymes are necessary. This cycle gets us into a kind of chicken-and-egg controversy. You need DNA to make the proteins, but you need the proteins to make the DNA. How could the first living cell have solved this dilemma?

Scientists have attempted to resolve this problem with a number of intriguing solutions, all of which share one assumption: that the very earliest life forms had a rather different (and much simpler) chemistry than the ones we see around us today. This early chemistry, being rather inefficient, was completely replaced when the DNA-RNA protein system evolved later. Thus we do not see it in living systems today, but, perhaps we can deduce its properties from studies of biochemistry.

One particularly interesting observation is that some kinds of RNA molecules have been found to act as enzymes for chemical reactions, in addition to their usual role as nucleic acids. This behavior suggests one way that the present system of cell chemistry could have evolved. RNA molecules catalyzed reactions that created proteins, and, over time, the proteins necessary for the development of the more-complex (and presumably more-efficient) DNA coding system were developed. Alternatively, other scientists have suggested that some kind of clay or other inorganic mineral may have provided sites for chemical reactions, as well as catalytic properties to help those reactions along.

None of these very speculative ideas is necessarily exclusive. It may very well be that the chemistry that went on in the primordial oil slick involved RNA as an enzyme, and that the best place for the enzyme to do its work was in a tidal pool. Scientists are hard at work in all these areas, but no clear answers have yet been found.

The Window of Opportunity

Whatever the chemical processes were that led to the first organism capable of reproducing itself, we know that those processes had to take place rapidly. We know that the Earth, like all the other planets, went through the period of the great bombardment (see Chapter 14). During this period, large chunks of debris fell onto the planets from space, bringing enormous amounts of energy with them. An impact involving an asteroid several hundred miles across would heat the Earth enough to boil the oceans—the planet would literally be sterilized by such an impact. Any life that might have developed would be wiped out. Therefore, the process that led to the ancestors of all present life on Earth could not have begun until after the last big impact. The best estimate for this date is about 4 billion years ago.

On the other hand, recent discoveries have made it clear that by about 3.5 billion years ago, life was not only present on Earth, but flourishing. In 1993, William Schopf of UCLA discovered evidence for colonies of primitive bacteria in rocks from that period. Schopf's evidence is pretty clear. When he takes thin slices of these old rocks and looks at them under a microscope (see Chapter 20), he sees the imprints left behind by these prokaryotic organisms (see Figure 23–2).

The 3.5-billion-year-old organisms were similar to modern green pond scum. Schopf identified no fewer than 11 different species in his initial survey. The rocks that contain these records were laid down as sediment in a shallow bay in what is now Australia. The bay supported a complex ecosystem, similar in many ways to systems on the Earth today. It must have taken quite a while for life to develop from the first cell to this complex ecosystem, which means that the first cell must have appeared soon after the last big impact.

There is, then, a well-defined window of opportunity during which the first cell could have developed. One side of the window is fixed by the time of the last big impact, the other by the appearance of fossils. The window extended roughly from 4.0 to 3.5 billion years ago, and the origin of life may well have occurred in the earlier part of that interval.

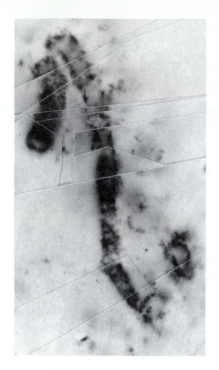

Figure 23–2
The earliest evidence of life, showing clusters of bacteria that lived about 3.5 billion years ago. These bacteria were part of a complex ecosystem, so the earliest life must have appeared considerably before this time.

The First Cell

Think about the unique status of the very first cell on Earth. It need not have been particularly efficient in using the chemicals found in its environment—after all, it had no competition. There were no predators, and no other life forms to compete for the abundant stock of organic molecules that enriched the early ocean. Once the first cell formed it would have been able to multiply rapidly.

Biologists have suggested that the special characteristics of the first cell may explain one of the great mysteries of modern biochemistry—the fact that living things today contain only 20 different kinds of amino acids (see Chapter 20). The notion is that, perhaps by chance, or perhaps because this particular combination gave them a competitive edge, some early cells contained only these amino acids, and it was the descendants of these cells that managed to dominate the Earth. In this way of thinking, the combination of these 20 amino acids is something of a "frozen accident," one of perhaps many chance events in the formation of life on Earth. Like so many other questions about the origin of life, this one will remain unanswered until we know more about the subject than we do now.

Science by the Numbers

Cell Division

The first cell was a microscopic organism, but it may not have taken long for that first bit of life to spread great distances around the globe. To get a feel for this process, imagine how long it would take to fill up the Mediterranean Sea starting with a single cell that divides once a day, assuming all cells survive and continue to divide.

To get an answer we must estimate the volume of an ordinary bacterium and compare it with the volume of the Mediterranean Sea. In Chapter 21 we learned that a typical bacterium is about a thousandth of a centimeter across, so its volume is approximately

$$(1/1,000 \text{ cm})^3 = 10^{-9} \text{ cubic cm.}$$

A recent world atlas gives the surface area of the Mediterranean Sea as about 2.5 million square kilometers, with an average depth of 1.4 kilometers, for a total volume of

$$2.5 \text{ million square kilometers} \times 1.4 \text{ kilometers} = 3.5 \text{ million km}^3$$

The question thus boils down to how many times would you have to double a 10^{-9} cubic centimeter bacterium to make 3.5×10^6 cubic kilometers. To make things easier, we convert cubic kilometers to cubic centimeters:

$$1 \text{ kilometer} = 10^5 \text{ centimeters}$$
$$\text{so,}$$
$$1 \text{ km}^3 = 10^{15} \text{ cm}^3$$

The total volume of the Mediterranean in cubic centimeters is

$$\frac{3.5 \times 10^6 \text{ km}^3 \times 10^{15} \text{ cm}^3}{\text{km}^3}$$
$$= 3.5 \times 10^{21} \text{ cm}^3$$

How many bacteria would it take to fill this volume? We divide the immense volume of the Mediterranean Sea by the tiny volume of a single bacterium:

$$\frac{3.5 \times 10^{21} \text{ cm}^3}{10^{-9} \text{ cm}^3} = 3.5 \times 10^{30} \text{ bacteria}$$

Starting with a single bacterium on the first day, there would be two on the second day, four on the third, eight on the fourth, and so on. After about three weeks there would be more than a million bacteria, taking up only about a thousandth of a cubic centimeter. Day-by-day, however, the number would increase geometrically. After two months there would be more than 10^{18} bacteria; after three months 10^{27} individuals, occupying more than 10,000 cubic kilometers. And in just 10 days more—only 100 days after the first cell began to divide—the Mediterranean Sea would be completely filled with bacteria.

Naturally, no body of water could be "completely filled" with bacteria. Early life probably did not spread this fast, nor was the process so regular and predictable. But the implication is clear. While it may have taken hundreds of millions of years for the first cell to evolve, a large number of descendants of that first cell could have spread throughout the world's oceans relatively quickly. ■

Natural Selection and the Development of Complex Life

Once the first cell has formed, our understanding of how life developed becomes much more detailed and precise. This understanding is largely due to the work of one man—Charles Darwin (1809-1882). His book *Origin of Species*, published in 1859, is arguably one of the most influential books ever written, and certainly is one of the world's most influential books about science. In it he sets forward a view of the development of living things that, since his time, has been expanded and developed to the point where it is fair to say that it is the single theory that unifies all of biology. A biochemist working on the transport of particular molecules across the cell membrane and a zoologist working on the ecology of a tropical lake will both share the central ideas of Darwin's theory, and hence will have a common vocabulary and a common way of attacking problems. Darwin's theory was influential and controversial because it identified a simple mechanism for evolving complex multicellular life forms from single-celled life. His entire theory is built around one central concept—the concept of natural selection.

Natural Selection

The easiest way to understand what Darwin meant by natural selection is to think first about the process that he called *artificial selection*. Farmers have known for millennia that the way to get bigger fruit, healthier plants, or animals with more meat on them is to carry out a conscious process of breeding. If you want large potatoes, you should plant only the eyes from the largest potatoes in any given crop. Over long periods of time, this practice will give you a variety of potatoes that is significantly different from the one you started with. Since human choice, not nature, drives this process, it is given the name of artificial selection. It explains how you can get animals as diverse and different from each other as longhorn and Angus cattle, or Chihuahuas and Great Danes, from the same basic ancestral stock.

If human beings can introduce such wide-ranging changes in living things, Darwin reasoned, then nature should be able to do the same. The mechanism he proposed, which he called **natural selection**, depends on two basic facts for its operation:

1. Every population contains genetic diversity—that is, the individual members of the population possess a range of characteristics. Some

(a)

(b)

Contrasting breeds of dogs, including (a) St. Bernard, (b) Dalmation, and (c) Shih Tzu, illustrate the changes possible with artificial selection

(c)

are able to run a little faster than others, some have a slightly different color than others, and so forth.

2. Many more individuals are born than can possibly survive. Therefore, those characteristics that make it more probable that a given member of the population will live long enough to reproduce will tend to be passed on to a greater percentage of the population's subsequent generations.

Let's look at a hypothetical example to see how this works. Suppose an island supports a certain number of birds, and suppose that the environment of this island is such that having a color that blends in with the local vegetation makes it easier for those birds to avoid their predators. Just by chance, some members of the bird population will have colors that match the colors of local leaves and trees better than others. As we saw in Chapter 22, this property would be determined by the DNA in the cells of those particular birds.

Better-camouflaged birds will be less likely to be eaten by predators, and, therefore, will be more likely to survive to adulthood and mate. You would expect, then, that the particular genes that give this advantage will be more likely to be passed to the next generation. In effect, the genes that are propagated in the next generation are influenced by natural forces. In this case, the selection is based on the color of the feathers.

If this process goes on for a long period of time, the entire population would eventually begin to share those advantageous genes for feather color. Natural selection works this way to modify a gene pool, just as populations of farm animals now share genes for rapid growth and meat production. Nature "selects" those characteristics that will be propagated in any given species. A structure, process, or behavior that helps an organism survive and pass on its genes is called an **adaptation**.

Natural selection, it should be remembered, is neither as controlled nor as rapid as artificial selection. It is always possible that birds who do not carry the selected gene will, in some generations, be more successful at mating than those who do. Over the long haul, however, the selective advantage granted by color will win. Thus Darwin envisioned natural selection as a process that operates over long periods of time to produce gradual change in populations—not a process that can explain short-term variations in a few individual traits.

About the time that Darwin's book came out, a rather extraordinary example of the power of natural selection appeared in his native England.

Figure 23–3
Light and dark versions of the peppered moth. The population shifted toward dark colors when coal burning became widespread in central England, then started to shift back when environmental controls were instituted.

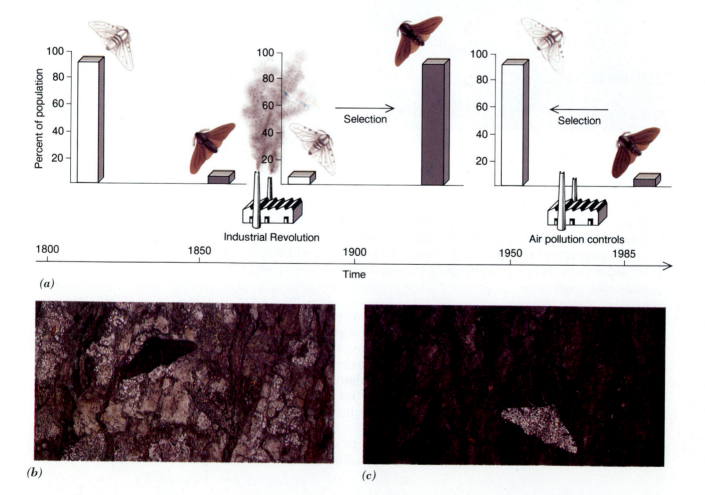

The Midlands area of England in the nineteenth century had become a vast industrial belt. Factories poured smoke into the air (this was before the time of environmental awareness) and covered much of the countryside with dark soot. The peppered moth, a small moth living in this region at the time, featured black-and-white wing patterns that blended in well with the light-colored lichen that had grown on trees in this area for centuries. At the time, lighter individuals predominated, though an occasional dark one was seen. While this coloration gave the moth protection from predators in an environment that did not include soot, the common lighter moths stood out clearly on trees darkened by industrial activity. Over a period of less than a century, the selective pressure on the moth population shifted in favor of darker individuals (see Figure 23–3). By the turn of the century, most of the peppered moths to be found were almost black. Scientists point to this rapid shift in moth coloration as a modern example of natural selection in operation.

We should add that, since environmental controls have been adopted in England, soot emissions from Midlands factories have been cut significantly. As Darwin would have predicted, the peppered-moth population has started to lighten again, as the original mossy tree trunks become more common.

Science in the Making

The Reception of Darwin's Theory

Charles Darwin formulated the basic outline of his theory of natural selection in 1838, but he waited more than 20 years to publish his findings. This delay was not simple procrastination. He realized that the central precepts of his theory would cause a furor. Eventually, learning that another British naturalist, Alfred Russell Wallace, had developed similar ideas, Darwin hastened to get his *Origin of Species* into print.

Written in accessible prose and published in a widely available edition, his theory evoked intense reactions. Naturally, some theologians denounced the book for its denial of a miraculous creation and the relatively short chronology demanded by a literal reading of the Bible. More disturbing to Darwin was the reaction of the majority of readers who embraced the "theory of evolution" as scientific evidence for God's hand in the progress of nature, and thus proof of man's moral and spiritual superiority. They seized Darwin's discovery as a shining example of God's wisdom and beneficence. Some intellectuals of the late nineteenth century even went so far as to cite Darwin in their defense of an economically and socially stratified society—the most "fit" individuals were those who rose to the top, they claimed.

Ironically, Darwin never intended his theory to suggest the idea of inevitable "progress" in nature, only inevitable change. Indeed, Darwin never even used the word "evolution"—a word that connotes improvement—in his book, nor did he address the question of human origins in *Origin of Species*. Far from being guided by a divine hand, he saw natural processes as violent and amoral—a constant struggle for survival in which the ability to reproduce fertile offspring was the only

The "progression" from chimpanzee to human is often used as an icon of Darwinian evolution. Darwin, however, did not see natural selection as leading inevitably from lower forms to humans.

measure of success. He observed successful natural strategies that his contemporaries would have viewed as repulsive in any moral sense—species whose females devour their mates, species whose offspring eat each other until just a few survive, and parasites and predators that kill without thought in the frantic quest for energy to survive. To Darwin, human ascendancy seemed an evolutionary accident rather than a divine plan, and he saw no sign of God in the brutal process of natural selection. Nevertheless, in his own concluding words:

> There is grandeur in this view of life, with its several powers, having been originally breathed by the Creator into a few forms or into one; and that, whilst this planet has gone cycling on according to the fixed law of gravity, from so simple a beginning endless forms most beautiful and most wonderful have been, and are being, evolved. ▪

The Story of Life

As soon as the first cell split into two competing individuals, natural selection began to operate. In that early environment, where the first cells were surrounded by energy-rich molecules and very few neighbors, competition would not have been very intense. Before long, however, mutations would have started to occur and some cells would have been different from others. Some of those differences involved the efficiency with which cells were able to utilize the molecules that they found in their environment. Certain cells, for example, might have been able to get energy more quickly from those molecules (and therefore reproduce faster) than others. Over time, the beneficial mutations would come to be shared throughout the entire population by the process of natural selection.

At this early stage, just as in today's life forms, the vast majority of mutations and the resultant differences were not beneficial. Random changes in DNA, after all, are not likely to produce organisms that can interact with their environment more efficiently than their fellows. Non-beneficial mutations died out quickly, and only beneficial mutations remained. This process is a little like our view of movies from the 1930s and 1940s. A great many poor films were made in those days, but we don't see them anymore. What we remember and preserve are the most successful films, such as *Citizen Kane* and *Casablanca*. In the same way, only the "greatest hits" of all the mutations survived into the future.

Over time, you would expect the descendants of that first cell to spread around the Earth and occupy most of the oceans. Some scholars have suggested this spread may have taken as little as a few years, given the lack of competition for the environmental resources. That this might be a reasonable result is suggested by the results of the *Science by the Numbers* section of this chapter. In this process of spreading, some cells would wind up in different environments than others. Some, for example, would be in tropical waters, while others would be in the Arctic. Some would be in deep oceans, while others would be next to the shore. Each of these environments would exert slightly different pressures on the cells. An adaptation that might be very advantageous in the tropics, for example, might not be advantageous near the pole, and vice versa. The driving force of natural selection, coupled with the fact that many different environments exist on our planet, would quickly have produced a number of very different living things. Thus we would expect the creation of diversity—the process of speciation—to have begun quite early in the history of life.

Our knowledge of this early period of life is limited by the fact that we have very little in the way of hard physical evidence that pertains to it. The fossil record does not contain a lot of single-celled organisms. This is not to say, however, that we have no record at all. As Figure 23–2 shows, we have found examples of fossil bacteria, and even a few cases of fossil bacteria caught in the act of dividing.

The best guess as to what went on until about a billion years ago is that the new living things spread around the world and differentiated, driven all the while by natural selection and changes in the Earth's climate. We suspect that, in the early part of this evolutionary process, all cells were prokaryotic, and scientists often speak of this as the period dominated by cyanobacteria—single-celled life forms that produce oxygen as a byproduct of photosynthesis. To an outside observer, the Earth would have looked remarkably sterile. There was no life at all on land, but the margins of the oceans were covered with collections of green scum that was going about the business of taking in carbon dioxide and returning oxygen to the atmosphere.

About a billion years ago, symbiotic relationships were set up between cells that eventually led to the development of eukaryotes. At some point, smaller cells found that they did better living inside their larger neighbors than they could do on their own, and cells whose genetic materials were carried inside a nucleus were born. These cells, like their neighbors and ancestors the prokaryotes, remained as single-celled organisms.

Sometime during this period as well, cells began to come together to form large colonies. At first, these objects were probably nothing more than clumps of single-celled organisms living next to each other. Later, however, they developed into larger bodies. Indeed, by about 600 million years ago, the seas were probably full of large multicelled animals and plants. You can think of some of them as resembling modern jellyfish. The stage was set for one of the most important developments in the history of life—the invention of the skeleton.

A wide variety of fossils are found in rocks from every continent. (*a*) A soft-bodied, multicellular animal from 650 million years ago; (*b*) the fossil bird *Archaeopteryx* from 150 million years ago; (*c*) a fossil relative of the scorpion preserved in amber (hardened tree sap).

About 570 million years ago, a crucial development took place in living systems. By a process that we don't fully understand, but which probably involved a new enzyme that converted calcium in the ocean water into shell material, some animals began to grow hard shells. This new chemical trick turned out to be such an advantageous development that the seafloor was soon teeming with many different kinds of hard-shelled animals. As always happens when a new evolutionary technique is developed, there was a great deal of competition and experimentation among living things as they tried to find the best type of outer shell, body design, and metabolism for each environment.

From the scientist's point of view, one of the most important aspects of this development was that, for the first time, living things left large numbers of fossils. In fact, for most of the nineteenth and twentieth centuries, before discovery of the fossils that indicated the presence of primitive forms of life, it looked to scientists as if life suddenly exploded at the beginning of this period. This sudden change in life on Earth, therefore, is often referred to as the *Cambrian explosion* (geologists refer to the time during which skeletons developed as the *Cambrian period*, after Cambria, the old Roman name for Wales, where rocks from this period were first studied).

Following the momentous development of shells, the last half-billion years or so have been a period of enormous growth in both the complexity and diversity of living things. A short summary of major developments is given in Table 23–1, see Appendix B for more details.

Geological Time

Before the development of radiometric dating in this century (see Chapter 11), scientists knew about the existence of fossils, and could see that some fossils were older than others, but had no way of attaching numbers to any of the changes they observed in the fossil record. As the above discussion implies, there are several landmarks in the process of evolution; each was used as a boundary in the delineation of past times.

In the nineteenth century, scientists were not aware of fossils of bacteria, or even fossils of soft-bodied organisms. To them, the evidence seemed to indicate that life suddenly appeared at the beginning of the Cambrian (when, as we know now, fossils of hard-bodied organisms first appeared). The era from the beginning to 570 million years ago was therefore called the *Proterozoic* ("before life"). Next was the *Paleozoic* ("old life") era from about 570 to 245 million years ago. This era saw a marvelous diversification of life, including the development of fish, amphibians, land plants and animals, and rudimentary forms of reptiles. The third great era (245 to 65 million years ago) was the *Mesozoic* ("middle life"), also known as the age of dinosaurs, when the major vertebrate life forms on Earth were large reptiles. Finally, the *Cenozoic* ("new life") era began with the extinction of the dinosaurs some 65 million years ago and continues to the present day. This is the time when mammals proliferated and began to dominate the Earth. The human species arose at the very end of the Cenozoic.

Throughout this long and intricate process of change, the principle of natural selection was always at work, shaping and molding life forms (see Figure 23–4).

Table 23–1 • **Major Steps in the Evolution of Life**

Time (millions of years ago)	Event
4000–3500	First cell
	Development of cyanobacteria
1000	Eukaryotes
700	Multicellularity
450	Vertebrates
400	Amphibians
350	Reptiles
100	Placental mammals
65	Primates
3–now	Hominids

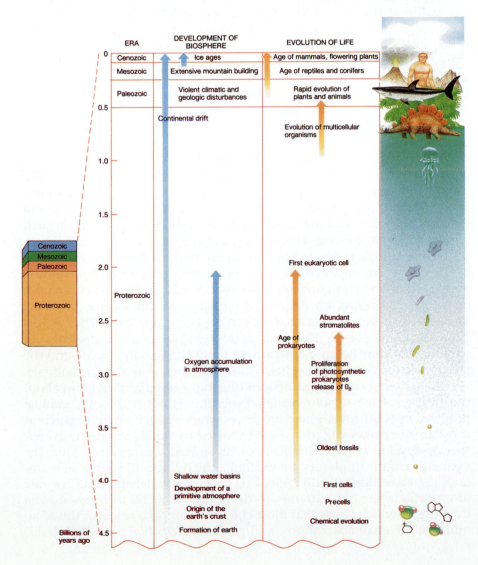

Figure 23–4

The geological time scale with representative living things illustrated.

Mass Extinctions and the Rate of Evolution

Under normal circumstances, the rate of extinction seems to be such that roughly 10 to 20% of the species at any given time will be extinct in a matter of 5 or 6 million years. The fossil record shows, however, that not all extinctions are "normal." Rare catastrophic events in the past have caused large numbers of species to become extinct suddenly. These events are called **mass extinctions.**

By "large numbers of species," we mean anywhere from 30 to 90% of the species alive at the time. By "suddenly," we mean a time too short to be resolved by standard geological techniques. The extinction may have taken place over a period of a few tens of thousands of years, or over a couple of days.

The best known of these mass extinctions is the one in which the dinosaurs perished some 65 million years ago, at the end of the Cretaceous period, which was at the end of the Mesozoic era. In that extinction, about two-thirds of all living species disappeared. In some cases, as with ocean plankton, this number may have climbed as high as 98%. But the extinction at the end of the Mesozoic was neither the largest nor the most recent mass extinction. About 250 million years ago, near the end of the Paleozoic era, about 80% of existing species disappeared in a single extinction event. A somewhat milder extinction, which wiped out 30% of existing species, appears to have taken place about 11 million years ago. In fact, geologists who study the past history of life in detail distinguish as many as 11 of these mass extinctions.

One of the most interesting explanations for how these mass extinctions could occur was put forward in 1980 by the father-and-son team of Walter Alvarez (a geologist), and Luis Alvarez (a Nobel laureate in physics). Based on evidence they accumulated, they suggested that the impact of a large asteroid killed off the dinosaurs and other life forms. Such an impact would have raised a dust cloud that would have blocked out sunlight for several months. This catastrophe would have been such a shock to the world ecosystem that it is a wonder anything survived at all.

Most scientists today accept that an asteroid hit the Earth at the end of the Cretaceous, and agree that it was at least partly responsible for the mass extinction. This conclusion was bolstered in 1992, when a crater over 100 miles across was discovered buried under the sea floor near the Yucatan Peninsula in Mexico. Less certain is the role that other factors played in these events. The world ecosystem was under a great deal of stress at that time because of relatively rapid changes in climate and the recent creation of mountain chains, both of which were altering habitats.

The existence of mass extinctions illustrates an important point about the history of life on our planet. Evolution is not a smooth, gradual progress through time. There are times when sudden changes (such as those in the mass extinctions) are followed by rapid evolution, as new species develop to take the place of those that disappeared. After the extinction of the dinosaurs, for example, the number of species of mammals increased dramatically. Scientists continue to debate about the rate of evolution. The two extremes in the debate have been the *gradualism* hypothesis, which holds that most change occurs as a result of the accumulation of small adaptations; and *punctuated equilibrium*, which holds that changes usually occur in short bursts, separated by long periods of stability. It now appears that both of these extremes, and probably any rate of evolution in between, have occurred at some time in the Earth's past.

Evidence for Evolution

Today scientists accept evolution as a given, something as well established as the fact that the Earth goes around the Sun. Like every scientific theory, however, the theory of evolution must be supported by experimental and observational data. A tremendous body of scientific literature is devoted to this subject, and we will discuss here only three of the most important pieces of evidence for the theory of evolution: the fossil record, the evidence of biochemical similarity, and the occurrence of vestigial organs.

The Fossil Record

When a plant or animal dies, the remains are usually lost. A tree will rot, the carcass of an animal will be torn apart by scavengers and dispersed, a crab shell will be broken up by the action of the surf. Occasionally, however, an organism is removed quickly from the environment, typically by being buried in sediments and sealed off. The hard parts of such an organism may remain underground for long periods of time.

As time goes by, two things may happen. First of all, the material around the organism may go through the rock cycle as described in Chapter 15, and be turned into rock. Second, minerals in the water flowing through the surrounding area may gradually replace the calcium and other atoms in the buried hard parts, thus creating a replica in stone of the original organism—a **fossil**.

When we think of fossils, we usually think of large dinosaur skeletons in a museum. Such fossils are the rock replicas we have just described. The term is also used to refer to other kinds of records of past life, such as the imprint of a leaf on mud that changes into rock, or an insect preserved in amber.

The term *fossil record* refers to all of the fossils that have been found, catalogued, and analyzed since human beings first began to study them in a systematic way in the early part of the nineteenth century. This fossil record, which is most copious since the Cambrian explosion, gives us the best indication of how different organisms came to be what they are. The fossil record of horses, for example, includes a series of precursor animals beginning with one about the size of a cat some 50 million years ago, and changing through many intermediate forms up to modern times. Throughout this sequence of fossil mammals, you can see gradual transitions from a small quick animal to a large grazing one.

The fossil record also contains some examples of actual changes in species. In order to do this, the fossil record has to be very complete, with many thousands of years of continuous sediments. Such continuity is rare, but in some instances the transitions from one species to another can be documented.

Even so, the major problem with the fossil record is that it is very incomplete. It is estimated that only one species (*not* one individual) out of every 10,000 early life forms is represented in the fossil record. Thus, in interpreting the past, we must always be aware that we are dealing with a very small and select sample of what was actually there. This sample is strongly biased toward organisms that were more likely to have been buried soon after death. We have a much better record of mollusks and clams that lived on the continental shelf, therefore, than we do of insects that flew around primeval forests. Nevertheless, the fossil record was the first (and for a long time the only) evidence that backed up the notion that life is constantly changing and evolving.

Three key ideas quickly emerged from studies of fossils. First, the older the rocks, the more their animal and plant fossils differ from modern forms. Mammals in 5-million-year-old rocks are not terribly different from today's fauna, but few species that existed 50 million years ago would be recognizable today, and dinosaurs rather than mammals were dominant 150 million years ago. Similar patterns occur in shells, plants, fish, and all other forms. Often the earlier forms appear to combine characteristics of later organisms. Ancient insects preserved in amber, for example, show some forms that may be intermediate between ants and wasps. Early mammals, similarly, have general mammalian characteristics, but few of the specialized structures that have evolved in bats, whales, tigers, and rabbits.

Fossils also display general trends in overall complexity of form. All known fossils from before the Cambrian explosion are either single-celled organisms or simple invertebrates like jellyfish. Marine invertebrates with hard parts—mollusks, coral, and crustacea, for example—dominate the record for the next 200 million years or so. Very simple land animals and plants appear next, followed by flowering plants and a much greater variety of large land animals. This long-term trend toward increasing complexity of organization is consistent with all theories of evolution.

Finally, the fossil record proves beyond a doubt that most species that have lived on Earth have died out and are now extinct. Scientists estimate that for every species on the planet today, as many as 999 species have become extinct at some time in the past. In fact, the average lifetime of a species in the fossil record seems to be a few million years. Species, like individuals, are born, live out their life, and die. This fact alone is ample evidence that some natural mechanism must exist to produce new species as the old disappear.

The Biochemical Evidence

We all carry within us the record of our descent from the first primeval cell. The DNA of each living organism represents the sum of all the changes in the DNA that connect that organism back to the DNA in the first forms of life. It should follow, then, that if we knew how to "read" the DNA in living organisms, then we should be able to deduce some things about the way those organisms developed.

According to the standard picture of evolution, for example, human beings and the great apes had a common ancestor about 8 million years ago. Much farther back in time, about 250 million years ago, human beings and reptiles shared a common ancestor. DNA changes slowly under the influence of mutation and natural selection. The fact that we've had only 8 million years for human DNA and ape DNA to differentiate, compared to 250 hundred million years during which human DNA differentiated from the reptiles, suggests that there ought to be more similarities between human and primate DNA than there are between DNA of humans and reptiles.

DNA strands from two different organisms cannot always be compared directly, but from Chapter 22 we know that the proteins in cells are related to DNA in much the same way as a negative is to a photograph. You can get the same kind of information by comparing sequences of amino acids along proteins as you can by comparing base pairs along the DNA molecule. In a protein called cytochrome C that every living cell uses in its energy metabolism, for example, the difference between humans and chimpanzees (our closest relative) is nonexistent. The cytochrome-C molecules from the chimpanzee and human are exactly the same. As you move farther and farther away from human beings in the main classification scheme (see Chapter 19), the differences become greater. In a rattlesnake, for example, there is an 86% overlap in the molecules, while in common brewer's yeast only 58% of the molecules are the same. The fact that our DNA is very similar to those organisms with whom we shared the most recent ancestors has been cited as one of the most important pieces of evidence for evolution. If, for example, each plant and animal were created separately and specially, there would be no reason at all to see this kind of progression.

Evidence from Anatomy: Vestigial Organs

Each of us carries within our own body compelling evidence for evolution. We have a number of internal features that serve no useful function whatsoever—**vestigial organs** that are reminders of our mammalian ancestors. Organs that are well adapted to their environment, surprisingly, *do not* provide unambiguous proof of evolution. The perfection of the human eye, for example, was often claimed in the nineteenth century as proof of God's special creation of human beings. The evidence for evolution comes instead from considering organs that have no use or are even harmful to the organism in which they are found.

Consider the human appendix, a thin closed tube connected to the upper part of the large intestine. This three-inch-long organ is actually a threat to every one of us. Before the invention of surgery, inflammations of the appendix were often fatal. The pressures of natural selection would never lead to the development of an appendix in a human being, yet it sits there at the end of each of our intestines. How could this organ come to be?

Vestigial organs such as the appendix can be explained by recognizing that they once had a function which they no longer have, and are in the process of disappearing. In a sense, modern human beings are a snapshot in a continuous process in which the appendix perhaps once served as an important part of the digestive system, but is no longer needed.

Numerous examples of vestigial characteristics have been identified, and they provide important evidence against the argument that every organ in every creature is part of a grand design. Penguins have vestigial wings now used to swim, some whales have tiny internal vestigial legs, and humans have vestigial tail bones and muscles to wag them. A striking example of vestigial organs is found in species of cave-dwelling worms that have no need of eyes, but have eye sockets *under their skin*.

The Evolution of Human Beings

The evolution of our own species is no different, at least in principle, than that of any other species. Nevertheless, the subject holds an intrinsic interest for us that the evolution of, say, the mayfly does not. The general picture of human evolution is that our branch of the family tree broke off from that of other primates about 8 million years ago. A possible human family tree is sketched in Figure 23–5.

The oldest hominid, *Australopithecus ramidus* ("southern ape, root of humans"), is known from fossils about 4.5 million years old. It appears to be midway in form between later hominids and modern great apes. The best known of the early human fossils are bones of **Australopithecus** afarensis ("southern ape from the Afar triangle region of Ethiopia"). Better known as **Lucy**, after the name given by paleontologists to a near-complete skeleton of the species. This fossil was discovered relatively recently, in 1974. The name arose because the paleontologists celebrated their discovery around a campfire while playing tapes of the Beatles' song "Lucy in the Sky with Diamonds." Radiometric dating established Lucy's age as about 3.5 million years. *Australopithecus afarensis* is a different species and different genus from our own *Homo sapiens*, but it is closer to us than to any other primate.

Lucy and her family walked erect, but had brains about the size of a modern chimpanzee's. They were also rather small (the adults probably weighed no more than 60 to 80 pounds), and may well have been covered with hair. If you saw an example of *A. afarensis*, the only thing that might make you think it is related to modern human beings would be its upright gait.

Scientists used to think that the development of large brains made human beings special, and that the brain's development led to upright walking. In fact, the evolutionary story seems to be the other way around. We walked upright first, which freed the hands for use, and then the large brain developed. Some scientists have suggested that the evolutionary advantage bestowed by hand-eye coordination provided the competitive edge for *Australopithecus*, and that led to the large brain.

The gap between the last common great-ape ancestor at 8 million years ago and the first member of the hominid family at 4.5 million years ago is unlikely to be filled in any time soon. The period from 3 to 8 million years ago happens to be a time interval that is not well represented by geological formations at the Earth's surface, and so very few fossils are found from this period.

Following Lucy, the line of *Australopithecines* developed larger and larger brains, and at various times there were several different species within the genus. From our point of view, however, the most important event happened about 2 million years ago, when the first member of the genus *Homo* appeared. Fossils of **Homo habilus** ("man the toolmaker") were discovered in East Africa in the mid-twentieth century. *Homo habilus*

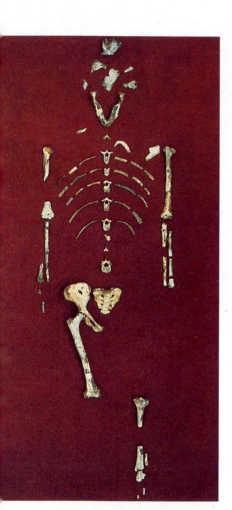

"Lucy," a 40-percent complete skeleton of *Australopithecus afarensis*, lived 3.5 million years ago in what is now northern Africa. This skeleton proved that hominids of this period walked erect.

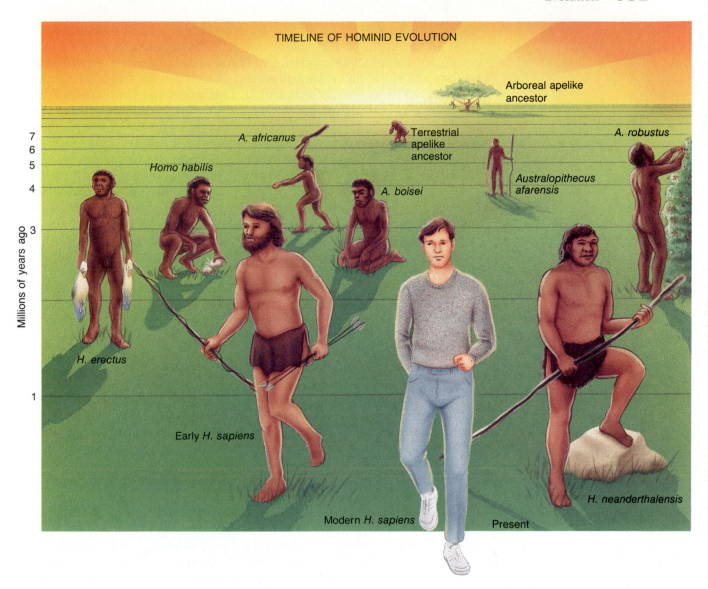

TIMELINE OF HOMINID EVOLUTION

Arboreal apelike ancestor

A. africanus

Terrestrial apelike ancestor

A. robustus

Homo habilis

A. boisei

Australopithecus afarensis

Millions of years ago

H. erectus

Early H. sapiens

Modern H. sapiens

Present

H. neanderthalensis

Figure 23–5

A chart of human evolution. The earliest hominid appeared about 3 million years ago. For most of the intervening period, there was more than one kind of "human" on the earth. The disappearance of the Neanderthals 35,000 years ago left *Homo sapiens* as the sole hominid on Earth.

was larger than Lucy, and had a larger brain. More importantly, *habilus* fossils are found with crude stone tools, so the association of human beings with toolmaking starts with this species. Shortly thereafter, another member of our genus, **Homo erectus** ("man the erect") appeared. *Homo erectus* fossils are found not only in East Africa but in Asia and the Middle East as well. Many of the famous fossil humans you may have heard of—Java Man and Peking Man, for example—were members of this species. *Homo erectus* lived at the same time as some of the later *australopithecines,* and survived until about 500,000 years ago. *Homo erectus* was the first in the line of human ancestors to use fire.

Fossils that we recognize as anatomically modern humans begin to appear in rocks about 200,000 years old. A major debate has emerged among paleontologists as to whether anatomically modern humans began in Africa and spread to the rest of the world, or whether they emerged at several different points around the world. This question is still the subject of serious discussion and, unless major new finds of fossil hominids are unearthed, the debate is not likely to be resolved soon.

About the same time yet another type of human being appeared on the scene—the so-called **Neanderthal man**. We sometimes use "Neanderthal" to denote something stupid. This use of the word comes from the fact that early studies of Neanderthal fossils concluded that this species walked stooped over, knuckles swinging, and had the thick brow ridge we associate with gorillas. These early suggestions were based on the study of a single skeleton of an old man who had a severe case of arthritis. Modern studies on other fossils reveal that Neanderthals, although far from being identical to modern human beings, were not all that different. They tended to be short, with thick, powerful arms and legs, and a skull that is much more elongated and pulled forward than that of modern *Homo sapiens*. On the other hand, Neanderthal had a large brain, on the average 10% larger than those of modern humans. They had a complex social structure, cared for elderly and infirm members of their tribe, and performed burials with ritual—a fact that suggests the presence of both a religion and a language. Thus Neanderthal was not too different from his contemporaries among the anatomically modern humans.

Several major mysteries and controversies surround Neanderthal. The first puzzle is how closely Neanderthals were related to modern human beings. Were they, as some scientists claim, merely a subspecies of *Homo sapiens*? Scientists who adopt this view classify Neanderthal as *Homo sapiens neanderthalensis*. On the other hand, the traditional view (and the view of many modern scientists) is that Neanderthal, although our nearest relative, was a separate species. People who hold this view classify Neanderthal as *Homo neanderthalensis*, to indicate that it is the same genus, but a different species, than modern humans. This debate will continue as more fossil evidence becomes available.

The second great mystery about the Neanderthals—some would say *the* mystery about Neanderthals—is the question of what happened to them. In Europe, where the fossil record is most complete, it appears that Neanderthals flourished until 35,000 years ago, and then disappeared rather suddenly. Their disappearance coincided with the entry into Europe of modern *Homo sapiens*. Several very different theories have been put forward to explain Neanderthal's disappearance. Some suggest Neanderthals were wiped out by the invading members of our own species in what might be described as a prehistoric instance of genocide. Other scientists have proposed that Neanderthals intermarried with the invaders so that a certain percentage of the genes of modern human beings are Neanderthal in origin. In order for this to be true, Neanderthal, by definition, would have been a subspecies of (as opposed to a separate species from) *Homo sapiens*. Finally, some have suggested that Neanderthal couldn't compete with the more technologically advanced newcomers and simply died out. In this case, Neanderthal was not wiped out by acts of war, but was simply moved away from the desirable settlement locations and eventually disappeared. Such a situation would be an example of the displacement of one species by another, a common phenomenon in the history of life.

From this description of the descent of modern human beings, one important point emerges. In the past, many different beings could be classed as "human." Many members of the hominid family have walked the face of the Earth. For whatever reason, none of them survived to this day except ourselves. The branch of the family tree leading to human beings, in other words, has been extensively pruned by the processes of natural selection. This fact, we believe, made it easy for people in the nineteenth century to discount or misinterpret Darwin, and to believe that the human race was special and not related to the rest of the web of life that exists on our planet.

THINKING MORE ABOUT EVOLUTION

Creationism

Opposition to Darwin did not end in the nineteenth century. As recently as March of 1981, the legislature of the State of Arkansas passed a law requiring that the Biblical story of the creation of the Earth be taught alongside the theory of evolution in public schools. The law was contested in the courts, and the federal courts eventually ruled that this law was an attempt to impose religious beliefs in the public schools—something expressly forbidden by the United States Constitution.

What is creationism, and why does it have such a strong hold in modern America? Three central beliefs of creationism, all based on a literal reading of the Bible, are:

The 1960 movie "Inherit the Wind" dramatized the debate between creationists and Darwin's theory of evolution by natural selection. Spencer Tracy (*left*) and Frederick March (*right*) acted the roles of the protagonists.

1. The Earth and the universe were created a short time ago, perhaps 6000 to 10,000 years in the past
2. All life forms were created by God in a miraculous act, in essentially their modern forms.
3. The present disrupted surface of the Earth and the distribution of fossils is primarily the consequence of a great catastrophic flood.

From the creationist's point of view, all of the diversity we see among modern life forms has existed from the start, and no subsequent evolution (or very little of it) has happened since then.

Evidence in favor of evolution, which requires a very old Earth and a means for transforming one species into another, has been presented throughout this book. Darwin's idea of

natural selection, particularly as applied to the origin of human beings, is not comforting. In the scientific picture, human beings do not warrant a special history. We are simply one more species that happened to evolve and is at this moment having its day in the Sun.

To what extent do you think that parents should have the right to decide what scientific theories and ideas are presented in schools? To what extent do you think parents ought to have the right to demand that opposing religious views be taught as well? Should the views of creationism, which are primarily based on one particular branch of Christianity, be given special consideration?

Summary

Life on Earth evolved in two stages. The first period of *chemical evolution* was characterized by the gradual buildup of organic chemicals in the primitive oceans. The *Miller-Urey experiment* showed that simple compounds, including water, methane, ammonia, and hydrogen, subjected to electrical sparks or some other energy source, combine to make the building blocks of life: amino acids, lipids, and other molecules. Through a sequence of events not yet well understood, a primitive but complex self-replicating chemical system developed. All subsequent life evolved from that first cell.

The first cell, free from competitors, quickly multiplied in the nutrient-filled oceans. As the oceans became crowded and competition for resources increased, a new stage of evolution—*natural selection*—began. The theory of natural selection, introduced by Charles Darwin in his 1859 monograph *Origin of Species*, recognized that every species exhibits variations in traits, and that some traits enhance an individual's ability to survive and produce offspring. Just as breeders develop new varieties of animals by selecting desirable traits artificially, Darwin argued, nature selects traits through the struggle for survival. In this way, over immense spans of time, new species arise. Geological time is divided into Proterozoic, Paleozoic, Mesozoic, and Cenozoic eras, depending on the kinds of *fossils* found from the period when the rocks formed.

Many types of evidence support the concept of evolution. A rich record of fossils demonstrates that life began simply and increased in complexity over time. The older a rock, the more its fossils are likely to differ from modern forms, because the vast majority of life forms have become *extinct*. While extinction is a continuous process, there have been a number of catastrophic episodes of *mass extinction*, when most species disappeared in a brief time interval. Asteroid impacts may account for some of these events.

Our biochemistry also supports the idea of evolution. Not only do all life forms employ the same biochemical mechanism for translating DNA into proteins, but many of those proteins are similar in very different species. Comparison of structural details reveals that closely related species, like humans and chimpanzees, have nearly identical proteins, while those of more distantly related animals show more protein differences. *Vestigial organs*, such as the appendix, provide yet another piece of evidence in the evolution story.

Human evolution can be traced back approximately 4.5 million years to *Australopithicus*, a hominid that walked erect but had a brain about the size of a chimpanzee's. Our knowledge of *Australopithicus* is based to a great degree on the partial skeleton of a female, known as *Lucy*. *Homo habilis*, the first member of our genus that appeared about 2 million years ago, was distinguished by a larger brain and the first appearance of stone tools. *Homo erectus*, who learned to use fire, evolved at about the same time but disappeared about 1.5 million years ago. Modern humans of the species *Homo sapiens* are recognized in fossils as old as 200,000 years. The status of the so-called *Neanderthal man* is still under debate: some say Neanderthal is a separate species, now extinct, while others argue that it is merely a subspecies of *Homo sapiens*.

Key Terms About Evolution

extinct	vestigial organ
chemical evolution	Australopithecus
Miller-Urey experiment	Lucy
natural selection	Homo sapiens
adaptation	Homo habilis
mass extinction	Homo erectus
fossil	Neanderthal man

Review Questions

1. What is chemical evolution? How does it differ from natural selection?

2. Describe the Miller-Urey experiment. Why are its results important?

3. What is the primordial soup? What role might it have played in the origin of life?

4. Where is the greatest gap in our knowledge of the evolution of life?

5. What is natural selection? Give examples of natural selection at work.

6. How does natural selection differ from artificial selection?

7. State the two basic facts that govern the operation of natural selection.

8. How old are the earliest known fossils?

9. What is the Cambrian explosion?

10. What is the fossil record? How does it support the theory of evolution?

11. What is a mass extinction? Give an example. What are some possible reasons for mass extinctions?

12. How does the overlap of human DNA with that of other living things support the theory of evolution?

13. When did the first members of the hominid family appear on the Earth?

14. Who was Lucy?

15. Which came first in human evolution, large brains or upright posture?

16. Give an example of a theory used to explain the disappearance of Neanderthal.

17. What is creationism?

Discussion Questions

1. Discuss the possible connection between plate tectonics and the constant process of natural selection. How might a study of fossils and plate motions be used to test Darwin's theory?

2. All scientific theories must be able to make testable predictions. What are some predictions that follow from Darwin's theory of natural selection?

3. Why do most scientists think all life evolved from a single cell? What evidence do we have to support this hypothesis? What alternate hypotheses can you propose?

4. Is the development of intelligent life an inevitable consequence of natural selection? Why or why not?

5. Fossils are usually found in sedimentary rocks. Why aren't they likely to be found in igneous rocks? What biases might this fact introduce into the fossil record?

6. Some diseases, such as adult-onset diabetes and Parkinson's, begin to affect individuals only in middle age. Why hasn't natural selection eliminated individuals with these diseases?

7. How fast are species disappearing from the Earth right now? What is the main reason for these extinctions?

8. Some people have argued that the progress of modern medicine has stopped the workings of natural selection for human beings. What basis might there be for such an argument? Should this argument be taken into account in formulating public policy? Why or why not?

Investigations

1. Read *The Panda's Thumb* by Stephen Jay Gould. Discuss the book in the context of evidence for evolution.

2. Consult your geology department and find out locations of the nearest fossil-bearing rocks. Visit a fossil location and collect a variety of samples. What kinds of life forms did you find, and what kind of environment did they live in? How old are these fossils? What living organisms most resemble these fossils? What kinds of rocks were they found in?

3. Read accounts of one of the recent trials on evolution versus creationism. What legal arguments were advanced on both sides? How did the courts rule?

4. Investigate the concept of social Darwinism. What was this doctrine? What government policies did it encourage? When was it in fashion?

5. Read an account of Charles Darwin's visit to the Galapagos Islands while be was on the *Beagle*. What did he see there that led him to natural selection and evolution?

6. Read an account of the development of the hypothesis that an asteroid was responsible for the extinction of the dinosaurs, such as *Nemesis*, by David Raup. What role did the chemistry of the element iridium play in this hypothesis? Does the history of this idea support our argument in Chapter 1 that scientists must believe the data that results from their observations?

7. Read an account of the Scopes "Monkey Trial," or see a movie based on it—for example, *Inherit the Wind*. How was the conflict between science and religion portrayed in these writings and movies? Is such a conflict inevitable?

Additional Reading

Darwin, Charles. *On the Origin of Species: A Facsimile of the First Edition.* Cambridge, MA: Harvard University Press, 1975.

Johanson, Donald and Maitland Edey. *Lucy.* New York: Simon and Schuster, 1981.

Mushrat Frye, Roland (*ed.*) *Is God a Creationist?* New York: Scribner, 1981.

Gould, Stephen Jay. *The Panda's Thumb.* New York: W. W. Norton, 1982.

Rudick, Martin. *The Meaning of Fossils.* New York: Science History Publications, 1976.

Wilson, E. O. *Sociobiology: The New Synthesis.* Cambridge, MA: Harvard University Press, 1975.

Strategies of Life

Living things use many different strategies to deal with the problems of acquiring matter and energy, adapting to the local environment, eliminating wastes, and reproducing.

A Random Walk
A New Look at the Life around You

We began our discussion of living systems by asking you to go outside and look at things that you see every day but usually ignore. We asked you to see how many living things you could find in your immediate neighborhood. We'd like to ask you to do the same thing now, but to look at those things in a different light.

Instead of just looking at the trees, bushes, and grass on your campus, for example, think of them as enormously complex organisms, each made of trillions of cells, with the DNA in each cell driving the production of enzymes that run thousands of chemical reactions. We'd also like you to see each plant as the product of a long line of evolution that has suited it for the particular role it plays. Think about how powerful the force of natural selection must be to have produced both an oak tree and a blade of grass from those first plants that moved onto land millions of years ago.

And what about the animals? The ants and beetles beneath your feet, the bird flying through the air, and your fellow students are also enormously complex things, each with its own genetic message written in its DNA, each the product of millions of years of natural selection. So, despite the incredible diversity of what you see, common features can help us understand the organization of the living world.

Problems Living Things Must Solve

Relatively speaking, life for a single-celled organism is pretty simple. Materials necessary for maintaining the chemistry of the cell can be taken in through the cell membranes, either by osmosis, through channels, or through the mediation of receptor molecules. Inside the cell, genes on the cell's DNA code for enzymes that drive chemical reactions that produce the cell's energy, move materials around inside the cell, prepare the cell for mitosis, and perform the other chores needed to keep the cell alive. The waste products produced by these reactions can leave the cell in the same way that materials come in, directly through the cell membrane. For most of the history of the Earth, the operation of single cells (be they prokaryotes or eukaryotes) was all there was to life.

With the advent of multicelled life, however, strategies became more complicated. Molecules that provided the cell's chemical energy could no longer simply diffuse into individual cells from the environment (most cells in a multicelled organism, after all, are shielded from the environment by other cells). Instead, those molecules must be carried from distant parts of the organism to the cells where they are to be used. Similarly, the waste products of processes such as respiration must be carried away from individual cells to the external environment.

Some very primitive multicelled organisms (sponges, for example) are merely collections of cells that could just as well survive on their own. In virtually all higher organisms, however, jobs such as moving energy-rich materials to cells and removing wastes are taken care of by specialized cells grouped together into organs and organ systems. In humans, for example, the heart and blood vessels form the circulatory system (which we will discuss later), whose task it is to circulate the blood.

One way to look at living things, then, is to identify the tasks that every organism must perform, and then ask what strategy a particular organism has evolved to perform that task. The important tasks that every multicellular organism has to perform are

1. obtain and distribute molecules needed to supply energy and build the fabric of the organism;
2. regulate internal processes in response to changes in the environment;
3. eliminate wastes;
4. support itself against the force of gravity; and
5. reproduce itself.

In some cases, more than one organ or organ system may be involved in carrying out these tasks. In human beings, for example, the carbon dioxide produced in the burning of carbohydrates is carried in the blood (the circulatory system) to the lungs (the respiratory system), where it is exhaled. Other kinds of wastes are carried by the blood to the kidneys, where they are removed and eliminated in urine. This particular way of dealing with the problem of wastes is common in higher mammals, but is far from universal even among animals, as we shall see.

Let's look again at the marvelous diversity of living things that natural selection has produced, with an eye toward seeing what strategies they have evolved to deal with the five tasks listed above.

Strategies of Fungi

Fungi, including such diverse organisms as molds, mushrooms, and yeast, were once classed as plants, but they are so different from true plants that they are now given their own kingdom. Some fungi (yeasts, for example) are single-celled, but most of the ones with which we are familiar, such as mushrooms, are multicellular. They grow from spores that can travel long distances through the air and water, but, once growth starts, fungi are as immobile as plants. Multicelled fungi grow by sending out filaments and absorbing food directly through the cell walls, thereby playing an important role in nature by breaking down dead organic material. You often see fungi, for example, growing like shelves out of the sides of fallen trees in a forest.

In fact, the structure of multicelled fungi is fairly simple, consisting of little more than a mass of filaments. Because of this structure, complex systems to move materials around aren't necessary. No cell is far from the material on which the fungus lives, and hence that material can be absorbed directly.

Fungi can reproduce by having their filaments break off and grow, but more often they produce spores. These spores can come from single organisms, which simply grow them by having certain special cells divide; or they can be produced by "mating," the fusion of two cells in the filaments from different plants. The spores are often held in small containers that grow on top of stalks—in effect, they get a "running start" by being above the ground when the container breaks. The fuzzy appearance of the molds on the old food in your refrigerator often comes from the spore containers on top of the stalks.

Lichens, which play an important role in the weathering of rock, are actually a combination of a fungus with a single-celled organism that can use the Sun's energy in photosynthesis. Lichens typically absorb most of the minerals they need from the air and rainfall, and hence can grow in inhospitable places such as mountaintops and deserts (Figure 24–1). In addition to breaking up dead organisms, then, fungi also play a role in creating soil from rocks, an essential first step in the creation of new habitats.

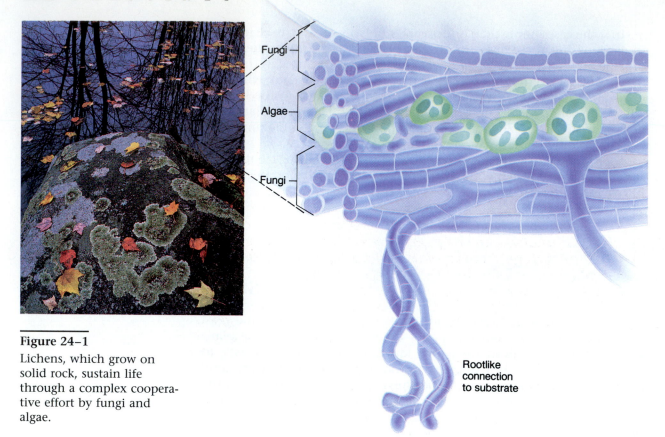

Fungi

Algae

Fungi

Rootlike
connection
to substrate

Figure 24–1
Lichens, which grow on
solid rock, sustain life
through a complex coopera-
tive effort by fungi and
algae.

In summary, fungi carry out the five tasks of life by

1. absorbing nutrients directly into cells;
2. passively accepting whatever changes in environment occur;
3. secreting wastes through the cell walls;
4. possessing a structure that is a mass of filaments; and
5. usually reproducing by spores, but sometimes sexually as well.

Science in the Making

The Discovery of Penicillin

When you get an infection, or even just a bad cut, your doctor will
very likely prescribe an antibiotic—a medicine that is capable of destroying foreign bacteria that otherwise would flourish in your body.
It's hard for us, living in an age where antibiotics are common, to realize that a half-century ago bacterial diseases like pneumonia and minor cuts that became infected were major killers.

The discovery of the first of the modern antibiotics in 1928 was the
result of a botched experiment in the London laboratory of bacteriolo-
gist Alexander Fleming (later Sir Alexander). He was growing cultures
of *Staphylococcus* (a common infectious bacterium) in dishes when he

noticed that one of his experimental dishes had been contaminated by a common mold called *Penicillium*, and that the bacteria didn't grow in a ring around places where the mold was located.

Other scientists had probably seen the same thing, but had just thrown the contaminated plates away and continued with their experiments. Fleming, however, realized that the mold must have been secreting a substance that killed the bacteria. When he finally isolated that substance, he named it penicillin, after the mold.

For a period after Fleming's discovery, penicillin could not be made in large enough quantities or in pure enough form to have any medical effect. In 1938, however, Howard Florey and Ernst Chain at Oxford succeeded in producing relatively pure forms of the substance and, under the pressure of World War II, a major development program had the drug in mass production by 1943. It saved countless lives on the battlefields of that war, and is now one of an array of substances used to maintain human health around the world. Fleming, Florey, and Chain shared the Nobel prize for medicine in 1945.

The working of penicillin is a perfect example of the "geometry is destiny" rule we discussed in Chapter 20. The cell walls of bacteria are constructed from molecules that have short tails made from a string of amino acids, and there is a special enzyme that links amino acids in adjacent molecules to give the cell wall structural integrity. Penicillin binds to this enzyme, preventing it from binding to the amino acids, and hence prevents the growth of the bacterial cell wall. Since human cells do not use this enzyme, the drug does not have this effect on us (although it can trigger allergic reactions in some people). ■

Strategies of Plants ▬

Plants and algae are the energetic basis for all life on Earth. Through the process of *photosynthesis* (see Chapter 20), they take energy from the Sun and lock it up in the form of chemical energy in their tissues and cells. In the process, they remove carbon dioxide and water from the air and produce oxygen, including the oxygen you are breathing right now. The combined weight of all the plants in the world is probably at least 10 times that of all other living things combined.

Biologists have not reached a consensus about how plants should be classified, or even on how to draw the boundaries of the plant kingdom. *Algae*, for example, are single-celled organisms (or simple multicelled ones) that carry out between 50 and 90% of the Earth's photosynthesis. Blue-green algae, single-celled photosynthetic organisms, are normally classified as monera; but other kinds of algae, including primitive multicelled organisms such as seaweeds and kelp (including green, red, and brown algae) are called plants in some schemes and protista in others. For our purposes, we will define plants to be multicelled organisms that perform photosynthesis, along with a smattering of closely related non-photosynthetic organisms. Plants are primarily found on land, and the main divisions between them have to do with the way they reproduce and the way they acquire and circulate water.

The Simplest Plants

The most primitive terrestrial plants are in the phylum of *bryophytes*, whose most familiar members are the mosses. These plants do not have roots, as more advanced plants do, but absorb water directly through above-ground structures. This is why they are found in moist environments. They are anchored to the ground by filaments called rhizoids, as are fungi. Unlike fungi, however, they use the Sun's energy to produce their own food by photosynthesis.

Like all plants, mosses reproduce by a process known as the "alternation of generations." In this process, half of the life cycle, which occurs after fertilization, involves cells that contain half the normal complement of DNA, and the other half of the cycle involves cells with a full complement of DNA. Mosses usually reproduce by manufacturing spores, which contain half the normal complement of DNA and can lie dormant for long periods of time. When the spores germinate, they grow into the familiar velvety green plant. Some branches in this plant produce sperm (see Chapter 21), while others produce flowerlike organs that contain eggs. When it is wet enough, the sperm are released and manage to swim to the eggs, where fertilization occurs. The cells that result from fertilization, of course, have a full complement of DNA. They divide to form a tall tendril in which the spores are produced by meiosis. When

Figure 24–2

Life cycle of a moss. *n* refers to stages with half the full complement of chromosomes; 2*n* refers to the full complement.

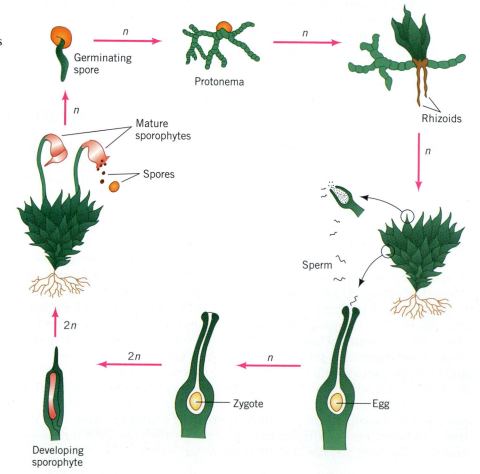

the spores are mature, they are released and the entire cycle starts again (Figure 24–2).

As we saw in Chapter 23, life on the Earth began in the oceans and migrated to land only about 400 million years ago. This history is evident in today's mosses, both in the fact that they are not equipped to get moisture directly out of the soil, and that they cannot reproduce in a dry environment.

Vascular Plants

By far the greatest number of plants that play a role in our everyday life are in the phylum of **vascular plants**. These plants have internal "plumbing" capable of carrying fluids from one part of the plant to another. They also have ways of controlling water loss and protecting the sperm and egg so that they can survive outside of water. The development of internal plumbing not only adapted plants to life in the dry terrestrial environment, it began the process of providing internal structure to overcome the downward pull of gravity. Both of these abilities are obviously crucial for plants living on land.

The most primitive vascular plants, a group whose most familiar members are ferns, still reproduce by producing sperm that must swim through water to fertilize eggs and generate spores. You can see the small yellow spore containers on the underside of the leaves of ferns. In this way they differ from the more advanced plants which, as we shall see below, reproduce by means of seeds.

Although seedless vascular plants such as ferns play a relatively minor role on the Earth today, they were the main form of plant life 300 million years ago. Huge forests of ferns and related trees blanketed the land, storing the Sun's energy in plant tissues. When the plants died, they were buried and, over millions of years, turned into *coal*. Thus you could say that the industrial revolution depended on seedless vascular plants, as does much of modern technological civilization.

In some modern classification schemes, vascular plants constitute a separate phylum, with separate classes for each different type. Ferns, for example, are accorded their own class. The most common vascular plants today are in the class of **gymnosperms** (plants that produce seeds without flowers; e.g., fir trees) and **angiosperms** (flowering plants). The distinguishing feature of these plants, which have dominated the plant kingdom for the last 250 million years, is that they reproduce by means of seeds. All seeds contain a fertilized egg and some nutrient, wrapped in a protective coating. Like spores, seeds are capable of lying dormant for long periods, and hence can wait through times of cold or drought before they sprout. Unlike spores, seeds contain the full complement of DNA. One way that plants have become fully adapted to life on land is that, unlike ferns, the sperm of seed plants, which are familiar to us as pollen grains, typically move through the air or are carried by insects. Thus, while seed plants may need water to grow, they do not need to be near standing water in order to reproduce.

The name gymnosperm means "naked seed" and refers to the fact that the seeds grow unprotected from the elements. The most familiar gymnosperms are evergreen trees, or conifers, such as the pine. On these

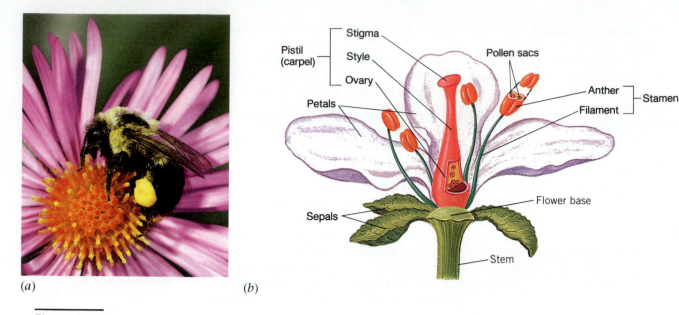

(a)   (b)

Figure 24–3

Flowers produce sperm and ova for sexual reproduction. (*a*) pollinators such as bees transport pollen from one flower to the next. (*b*) The structures of a flower are represented schematically.

trees, some cones produce pollen, which is dispersed by the wind. When pollen grains land on cones that contain unfertilized eggs, fertilization takes place. The seed develops on the cone, and, when conditions are right, it is released and carried by the wind to a new location. Typically, hundreds or even thousands of seeds have to be released to get one new seedling.

Flowering plants, which have been around for at least 100 million years and have dominated the Earth's plant life for the last 65 million years, comprise the great bulk of known species of modern plants. These plants reproduce through the complex structure of their flowers. In a flower, the pollen grains containing the sperm grow on stalks known as stamens (see Figure 24–3). The grains are carried by wind or insects either within the same flower or to other flowers. They grow tubes and enter the ovary, where the eggs are found. After fertilization, a seed forms within the ovary, and the ovary itself develops into fruit. When you cut into an apple or a peach, you can see the seeds and ovary quite clearly. When the fruit is ripe, it detaches itself from the plant. Many adaptations allow seeds to be carried long distances, from winged seeds (like those of some trees) to small seeds on berries that are eaten and passed through the digestive tracts of animals.

Although flowering plants can reproduce through the fertilization process, they can also reproduce by sending out runners or shoots. Grass on a lawn, for example, reproduces this way if it is prevented from producing seeds by constant mowing. In the first case we say that the reproduction is sexual (because it requires that a sperm and an egg come together), and in the second case we say that it is asexual.

The Structure of a Flowering Plant

As with all living systems, flowering plants feature a wide assortment of structures and specialized organs. Nevertheless, we can point out a few generalized structures shared by most of them (see Figure 24–4).

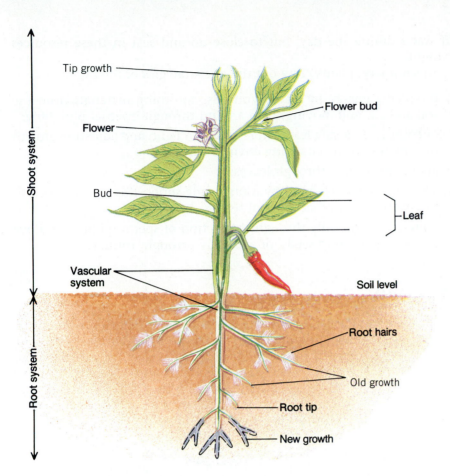

Tip growth

Flower bud

Flower

Bud

Leaf

Vascular system

Soil level

Shoot system

Root system

Root hairs

Old growth

Root tip

New growth

Figure 24–4
The basic design of all plants includes a root system below ground and a shoot system above ground.

Flowering plants possess roots, which are specialized multicellular structures that extend down into the ground. Water and dissolved nutrients and minerals enter the plant through the roots and are carried upward. Roots also serve to anchor the plant to the ground, and can grow at the surface in fibrous form (as in matted grass), extend along the surface for some distance (as in many weeds), or penetrate deep into the soil as a taproot (as in many tall trees). Roots play an important role in modern forests and meadows, as they hold down soil and prevent erosion. The growth of roots into rocks is also one way that rocks are broken down into soil.

The stem of the plant holds it upright, so that the leaves can be exposed to the Sun. Tissue at the center of the stem acts as a plumbing system to carry water and minerals from the roots to the rest of the plant. You can think of the center of vascular plants as being something like a handful of straws capable of carrying fluids. The other part of the plumbing system, made from living tissue, distributes the nutrients produced in the leaves to the rest of the plant. In a tree, the wood at the center of the trunk is the inert part, while the cambium, the slippery layer just under the bark, is the living component. It is very likely that you spend part of your time in buildings made from wood, so, in a sense, you have been sheltered by plant plumbing all your life.

Photosynthesis occurs in leaves—complex structures designed to give off water during the day, but to close up and seal in these resources at night.

In summary, plants carry out the five strategies of life by

1. producing energy by photosynthesis, absorbing materials from the soil and air, and moving them around through internal plumbing;
2. opening and closing leaves in response to light, and adjusting growth in response to the changing environment;
3. giving off oxygen through leaves;
4. anchoring in the ground by roots, while supporting upward growth by cellulose and other internal tissues; and
5. reproducing sexually through production of sperm (pollen) and ova (in flowers) to form seeds, or asexually through runners.

Strategies of Animals

Animals are multicelled organisms that must get their nourishment by capturing and consuming molecules produced by other life forms. The variety of organisms in this kingdom is truly staggering, ranging from sponges to tiny worms to beetles to elephants and, of course, human beings. Although the mass of all the plants on the Earth far exceeds that of all the animals, the animal kingdom, with over 1.3 million documented species, takes the prize for diversity. Depending on the details of how lines are drawn, over 30 phyla can be included in the kingdom. This diversity is summarized in Figure 24–5.

It is not easy to make generalizations about animals, a group that includes sponges, earthworms, eagles, and professors. As we pointed out earlier, however, a number of tasks must be carried out by every animal. Each phylum solves these problems in different ways.

Invertebrates

When we think about animals, we usually visualize large vertebrates such as birds or elephants or seals. Most animal species, however, are **invertebrates**—organisms without backbones. The simplest invertebrates, like sponges, exhibit characteristics somewhere between an aggregate of individual cells and true multicellularity. If a sponge is passed through a sieve, the individual cells not only survive, but eventually reorganize themselves into a sponge. Corals are colonies of countless separate microscopic organisms—break up the collection and each individual can function on its own.

Most of the phyla of the animal kingdom consist of things such as worms, mollusks, and various microscopic organisms. Some of these represent animals that have become incredibly specialized—one phylum of worms, for example, contains over 70 species that are found as parasites in the noses, sinuses, and lungs of vertebrates. Sea cucumbers, jellyfish, earthworms, mollusks, snails, and tapeworms are all examples of the diverse invertebrate forms to be found in the animal kingdom.

Arthropods are by far the most successful phylum in the animal kingdom, both in terms of number of species and total mass. Arthropods

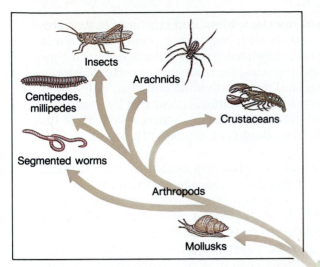
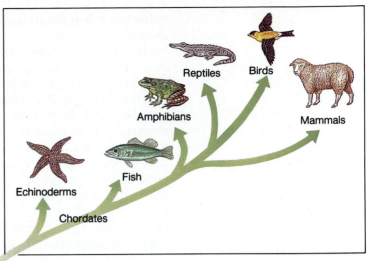
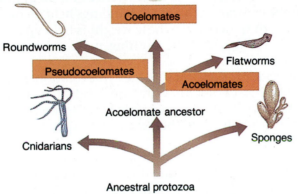

Insects

Arachnids

Centipedes, millipedes

Crustaceans

Segmented worms

Arthropods

Mollusks

Reptiles

Birds

Amphibians

Mammals

Echinoderms

Fish

Chordates

Coelomates

Roundworms

Flatworms

Pseudocoelomates

Acoelomates

Acoelomate ancestor

Cnidarians

Sponges

Ancestral protozoa

Figure 24–5
The family tree of animals, showing some of the major phyla.

(a)

(b)

(c)

Most known animal species are arthropods, with segmented bodies and jointed limbs. They include (a) trilobites, known only as fossils; (b) spiders, with eight legs and often several pairs of eyes; and (c) crustacea, with specialized feeding appendages and two pairs of antennae.

include familiar forms such as insects, spiders, and crustacea (e.g., crabs, shrimp, and lobsters)—all animals with segmented bodies and jointed limbs. The more than 900,000 recognized species of insects account for at least 70% of all known animal species. There are more species of beetles than any other type of animal, and more individual ants than individuals of any other type. These facts prompted a famous comment by evolutionary theorist J. B. S. Haldane (1860–1936) who, when asked what his studies of biology had taught him about God, replied, "He has an inordinate fondness for beetles."

One of the problems that all land-dwelling organisms have to deal with is finding a way to support their structures against the pull of gravity without support from the water in which life originally evolved. Arthropods solve this problem with a hard external covering known as an exoskeleton. This strategy is different from that of the vertebrates, whose weight is supported by an internal skeleton and whose outer coatings are soft. You can see this same difference in strategy illustrated architecturally in the contrast between old buildings held up by massive stone walls, and modern skyscrapers, whose weight is held up by a steel skeleton and whose outer skin may be nothing more substantial than thin sheets of glass.

The hard exoskeleton provides an evolutionary advantage to arthropods whether or not they live on land: it is a "coat of armor" that protects the animal from predators. Because the exoskeleton cannot grow, arthropods usually shed their exoskeletons periodically, a process known as molting.

Insects are the most numerous arthropods. A typical insect (see Figure 24–6) has a body divided into three segments. Three pairs of legs originate from the central segment, or thorax, while a pair of antennae adorns the head. Many insects also have one or two pairs of wings. Insects have a heart, but no lungs—they bring oxygen into their bodies through a set of tubes, as shown in the figure.

Vertebrates

When most of us hear the term "animal," we think of familiar creatures such as rabbits, birds, frogs, or fish. All of these animals have spinal chords encased in a backbone. The easiest way to understand the connection between the earliest branches of the vertebrate family tree is to think of each new branching up to reptiles as another development on the way from water-dwelling animals to those fully adapted to life on land. As we saw in Chapter 23, this line of thought follows the actual evolution of modern vertebrates, whose ancestors developed in the ocean and later made the move to a terrestrial environment.

The earliest fish, which became common about 400 million years ago, fed by moving water through their mouths and filtering materials out—they had no jaws. Lampreys (which look like eels without jaws) are modern descendants of these primitive fish. Over the next 100 million years, jaws evolved, and the evolutionary advantage of speed over heavy protection led to a loss of bone. Modern fish such as sharks and sting rays, for example, have no bone at all, but skeletons made from cartilage. Modern fish absorb oxygen from the water they bring in through their gills, and

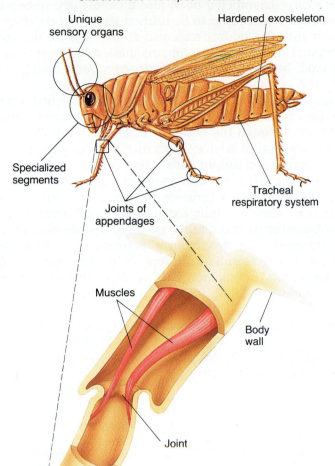

Characteristic Arthropod Features

Unique sensory organs

Hardened exoskeleton

Specialized segments

Joints of appendages

Tracheal respiratory system

Muscles

Body wall

Joint

Figure 24–6
The generic insect.

the oxygen enters the bloodstream by diffusing directly into the blood vessels in the gills as water flows through them.

Bony fish is a category that includes salmon, perch, trout, and most other organisms we think of when we hear the word "fish." Their skeletons are, as the name implies, made from bone. Oddly enough, the ancestors of these fish developed in fresh water and had both lungs and gills. In modern fish, these lungs have evolved into the swim bladder, an internal sac the fish can inflate to control its buoyancy, while the gills function to supply oxygen. A few species of lungfish survive today. These fish live in stagnant water and gulp air occasionally to supplement what they can extract from the oxygen-poor water in which they live. Lungfish can also live out of water for short periods, allowing them to move from one pool to another during droughts. This ability almost certainly led to the movement of vertebrates to land.

The first vertebrates that were clearly adapted to live at least part of their lives on land were *amphibians*, whose modern descendants include frogs, toads, and salamanders. Most amphibians spend part of their life cycle in water, part on land. Frogs, for example, hatch as tadpoles, complete with gills and fins, and develop into adults with lungs and legs.

They have a three-chambered heart and a circulatory system in which blood is pumped to the lungs to be infused with oxygen, then returned and sent out into the body. Frogs also absorb a good deal of oxygen directly through their skin. Amphibians mark a halfway stage between water and land, both in terms of their anatomy and their place in evolution.

Reptiles, such as lizards, turtles, and snakes, are the first animals fully adapted to life on land. They are covered with hard scales, cutting down the loss of water through the skin. The egg is fertilized within the body of the female (instead of relying on chance unions of egg and sperm in water, as with fish and amphibians). The young develop in eggs surrounded by a shell that can retain water and thus survive on land. Like amphibians, reptiles have a three-chambered heart, but with divisions in the chambers that allow fully oxygenated blood to be sent out more efficiently. Amphibians and reptiles are *cold-blooded*—that is, they must absorb heat from their environment to maintain body temperature.

Birds are modern descendants of reptiles, possibly direct descendents from the dinosaurs. Their anatomy differs from that of reptiles because of their adaptation to flight. In birds, the scales of the reptiles have

Figure 24–7
The mammalian family tree.

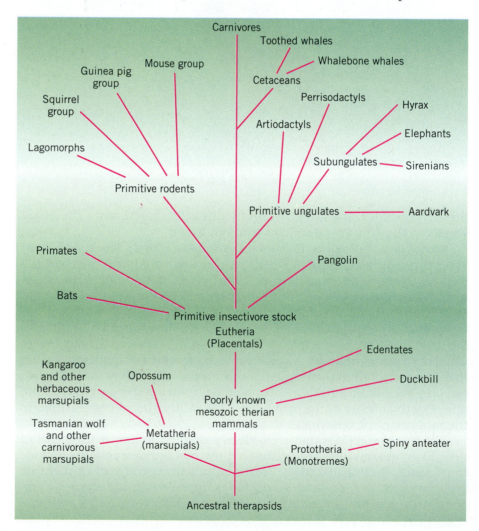

evolved into feathers. Birds require high levels of energy to sustain flight, so their respiratory and circulatory systems are more complex than in reptiles and amphibians. They have a four-chambered heart, as do humans (see below). One side of the heart pumps blood to the lungs to be oxygenated, and the other side pumps oxygenated blood around the body. This feature allows the bird to use the energy in its food with maximal efficiency, so that it can maintain a constant body temperature in any environment. Thus birds are *warm-blooded*.

Taxonomists recognize 18 orders of living *mammals* in the world today (see Figure 24–7). Like birds, mammals maintain a constant body temperature by burning food. In almost all cases, the development of the fertilized egg takes place entirely inside the body of the mother, so that the young are born live. They are nourished by milk from special glands in the female, and have a number of other adaptations to the terrestrial environment. These include hair to aid in temperature regulation, specialized teeth for breaking up food, legs located under the body, and, perhaps most importantly, an enlarged brain that makes possible the development of behaviors that allow mammals to cope with their environment in a flexible and intelligent way.

In summary, mammals carry out the five strategies of life by

1. ingesting food for energy, digesting it chemically, and sending materials around the body in the circulatory system, as in all other vertebrates;
2. maintaining a constant body temperature;
3. secreting liquid and solid wastes, while breathing out carbon dioxide from respiration processes in cells;
4. supporting tissues by an internal skeleton and system of muscles, tendons, and ligaments; and
5. reproducing sexually, bearing live young, and nursing infants.

Human Anatomy

Think about your body as you woke up this morning. You stretched your muscles, and noticed the first sensations from nerves in your eyes, ears, and skin. As you lay in bed, summoning the energy to get up, you may have noticed your heart beating. You took a few deep breaths, swung your legs to the ground, and headed to the shower. After you dressed, enjoyed a hearty breakfast, and made a final trip to the bathroom, you were ready to begin the day. In that short hour before leaving home you engaged in all of the essential activities of life (with the probable exception of reproduction).

From a strictly biological standpoint, there is no reason to single out *Homo sapiens* for special attention in a discussion of life strategies for either animals in general or mammals in particular. Nevertheless, as members of this species, we have an understandable and justifiable interest in how our own bodies work. All of the remarks made earlier about the various tasks that must be done to support multicellular life apply to the human body. Like other complex organisms, our bodies are made of cells, cells make up organs, and organs are incorporated into organ systems to do the necessary work.

In what follows, we give a brief description of the organ systems in our body, as an example of how one particular multicellular life form works. You may wish to refer to Figure 24–8 as we discuss many of the human organ systems.

Skeleto-Muscular System

As in all vertebrates, the weight of the human body is supported by an internal **skeleto-muscular system** (Figure 24–8). The rigid structure consists of *bone*, while movement is produced by the action of *muscles* that are attached to the bone by *tendons*, as shown. Muscles are made of long cells that can contract when they are stimulated. When you move your arm or leg, the simultaneous contraction of many of these cells produces the movement.

Bones are held together at joints by tough tissues called *ligaments*. Unlike muscles, ligaments are fairly inelastic and, once torn or injured, can take a long time to heal. Padding between bones is provided by cartilage, which you may know as the "gristle" in the joints of your Thanksgiving turkey.

There are two kinds of muscle in the human body, with names derived from the way they look under a microscope. The muscles that are under your control (such as the ones in your arms and legs) are called striated muscles; muscles that perform their function automatically (such as those that open and close the iris in your eye) are called smooth muscles. The muscles in the heart are similar to the striated muscles that move the skeleton, but have a different metabolism that allows them to pump continuously and cannot be controlled voluntarily.

Respiratory and Circulatory Systems

Oxygen, essential for the operation of the chemical reactions that produce energy in our bodies, is distributed by the **respiratory and circulatory systems**. When you breathe in, air is pulled into your *lungs* and, ultimately, into tiny thin-walled sacs called *alveoli* (Figure 24–8). *Blood vessels* are in close contact with the walls of the alveoli, so that oxygen from the air diffuses across the thin membranes into the blood, while carbon dioxide from the blood can diffuse back. The carbon dioxide is the result of the burning of energy-rich molecules in the cells (see Chapter 21). When you breathe out, you exhale the carbon-dioxide-rich air. On the next breath, the whole process starts again.

The blood with its load of oxygen goes back to the *heart* (see Figure 24–9). Like the heart of birds, the mammalian heart has four chambers, each of which carries out a separate function. Oxygenated blood from the lungs enters the left upper chamber, called the left atrium, from which it is pumped into the left lower chamber (left ventricle) and then out into the body through a series of smaller and smaller arteries. Eventually, it enters a network of thin-walled tubes called capillaries, where the oxygen diffuses out into cells and carbon dioxide diffuses back in. The deoxygenated blood is then returned to the right upper chamber of the heart through a series of larger and larger veins. It is then pumped into

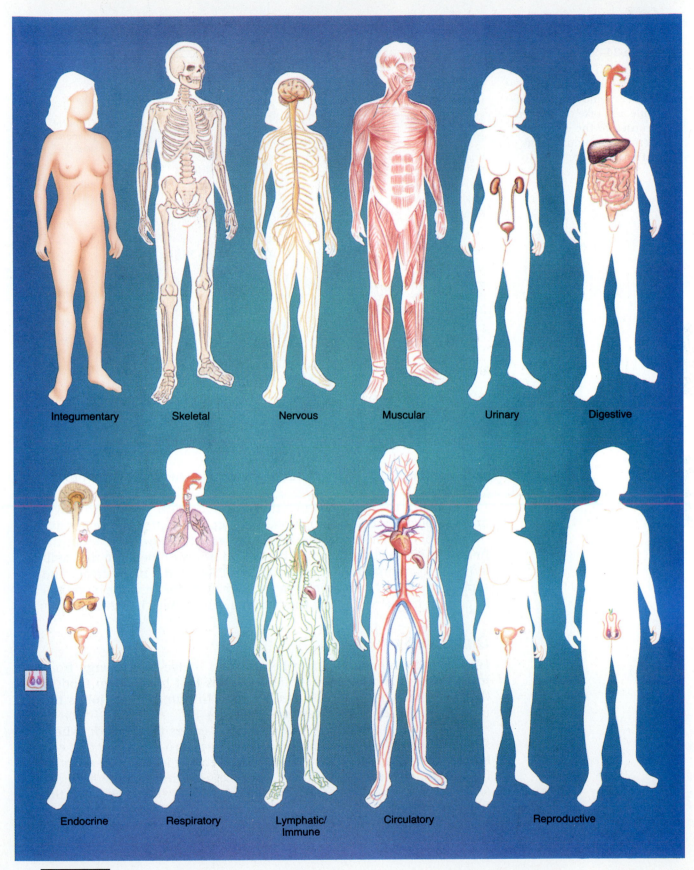

Integumentary Skeletal Nervous Muscular Urinary Digestive

Endocrine Respiratory Lymphatic/ Circulatory Reproductive
Immune

Figure 24–8
The organ systems of the human body.

Figure 24-9
The human heart. Blood is sent by the heart to the lungs to remove carbon dioxide and take on oxygen, returned to the heart, and then pumped out to the rest of the body.

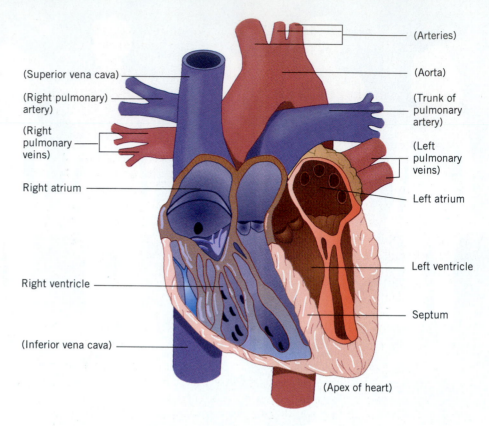

(Superior vena cava)

(Right pulmonary artery)

(Right pulmonary veins)

Right atrium

Right ventricle

(Inferior vena cava)

(Arteries)

(Aorta)

(Trunk of pulmonary artery)

(Left pulmonary veins)

Left atrium

Left ventricle

Septum

(Apex of heart)

the right lower chamber and out to the lungs, where the entire process starts again.

Paralleling the blood system is the *lymphatic system*, which also moves fluids through the body. The lymphatic system, consisting of an extensive network of capillaries and veins linked to about 500 *lymph nodes* in the human body, removes material that fails to be reabsorbed into the capillaries. It also transports fat from the intestines to the bloodstream and supplies the blood with *lymphocytes*, which are a kind of white blood cell important in the working of the immune system (see chapter 21).

Digestive System

Like all other animals, human beings must derive their energy from ingested food, and hence must have systems that break down food so that its stored energy can be used by cells. In humans, this job is the function of the **digestive system** (Figure 24-8).

After being broken up by chewing, food enters the *stomach*, where strong acids and other chemicals break it down into molecules that can be used by individual cells—a process that actually is begun by saliva in the mouth. In the *small intestine*, this process continues as the *liver, gall bladder,* and *pancreas* secrete specialized substances that break down the starches, carbohydrates, proteins, and fats in the food to small molecules. These molecules then pass through the walls of the intestine and enter the bloodstream. They are carried to the cells, where their energy is released in chemical reactions analogous to burning (see Chapter 20).

Muscles along the intestinal wall propel the undigested food into the *large intestine*, where water is removed and feces are formed and, ultimately, voided.

Waste products from the metabolism of the cells is carried by the blood to the *kidneys* (Figure 24–8). In a complex series of chemical reactions, the blood is filtered and then materials are selectively reabsorbed. Whatever isn't reabsorbed becomes part of the urine, which, after being collected in the urinary bladder, is voided.

The kidneys also maintain the balance of salt and water in the human body. Their structure is such that they cannot produce urine with a salt concentration of more than 2%. If the waste has a higher concentration of salt, then the kidneys have to take water from elsewhere in the body to dilute it. This limitation explains why drinking sea water (which is roughly 3% salt) always increases your thirst. As Samuel Coleridge wrote in *The Rime of the Ancient Mariner*:

> Water, water everywhere and all the boards did shrink,
> Water, water everywhere, nor any drop to drink.

Sensing and Control Systems

Human beings become aware of their environment through the action of five senses—sight, hearing, touch, taste, and smell. The *eye*, arguably our most important sense organ, was discussed in Chapter 6. The *ear* (see Figure 24–10) contains a membrane that senses the arrival of sound waves and vibrates. This vibration, in turn, is transmitted through three small bones to the inner ear, where it is converted into signals on the auditory nerve.

Smell and taste both involve specialized chemical receptors. Molecules of the material being sensed actually come into contact with the receptors, which then change their shape and generate a nerve signal. Touch is a sense generated by specialized cells in the skin that respond to pressure, and the skin also has specialized sensors for temperature and pain.

Our response to the environment is mediated through the two control systems in our body. The more familiar of these is the **nervous system,** whose main component, the nerve cell, was discussed in Chapter 5. Chains of nerve cells run throughout the human body. Some of these comprise the *autonomic nervous system*, which controls actions such as the beating of the heart and contractions of the gut. We aren't aware of the effects of the autonomic nervous system—we don't have to think about every breath or heartbeat, for example. The contraction of muscles that are part of ordinary volitional motion are controlled by nerves in the *somatic nervous system*.

The central organ of the human nervous system is the **brain**, which receives the signals from sense organs as well as signals that keep it apprised of the status of internal organs. The brain constantly sends out signals to the body to keep it functioning, and serves as the seat for all higher functions such as speech and thought. Scientific study of the brain is still in its infancy; we know relatively little about the brain and its functions. Nevertheless, we can make a few general remarks about this remarkable organ.

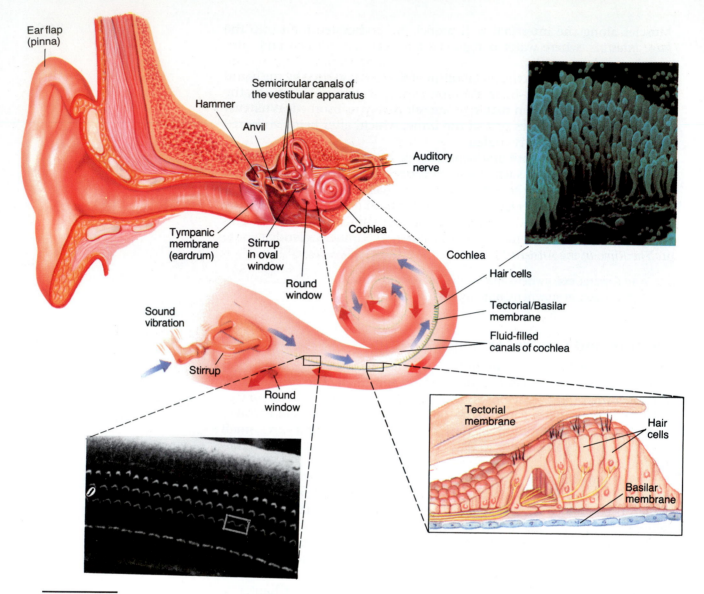

Figure 24–10

The human ear.

The brain is composed of interconnected nerve cells, and cells that serve to support and nourish them. It can be split roughly into three parts: hindbrain, midbrain, and forebrain or cerebrum, as shown in Figure 24–11. The hindbrain, which is located at the top of the spinal cord, controls basic body functions such as breathing, balance, and blood pressure. The midbrain controls eye movements (in lower animals it is also the place where visual information is processed) and processes auditory signals. The midbrain and hindbrain, respectively, are often referred to as the brain stem.

The lowest part of the forebrain controls basic body metabolism and, through the *pituitary gland*, the body's system of hormones. The part of the brain that you usually see in illustrations—the wrinkled "gray matter"—is the outer layer of the *cerebral cortex*. All of the activities we normally associate with the higher human faculties—speech, rational thought, and memory, for example—are carried out here. The cerebral cortex also acts as the seat for processing of sensory information and conscious movement.

In addition to the nervous system, the human body has a second control mechanism in the *endocrine system*. The body's several glands secrete specific molecules, called *hormones*, that travel through the bloodstream. When they encounter cells that have a specialized receptor that fits their particular shape, they are taken into that cell and produce specific chemical effects. The adrenals located on top of the kidneys, for example, secrete substances that raise blood pressure and heart rate and send blood to the skeletal muscles. This surge of adrenalin prepares the body for "fight or flight." The pituitary gland secretes growth hormones that are crucial in human development—if they are not present in sufficient amounts, the individual will not reach full height. Human growth hormone is one substance that can now be produced by genetic engineering. Other hormones control basic metabolism, the maintenance of secondary

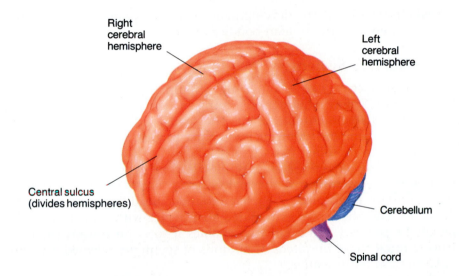

Figure 24–11
The human brain.

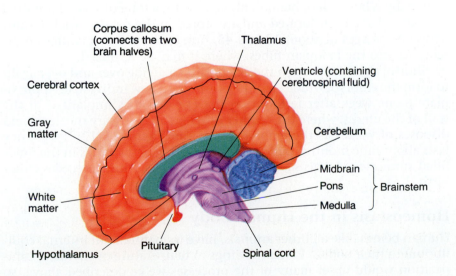

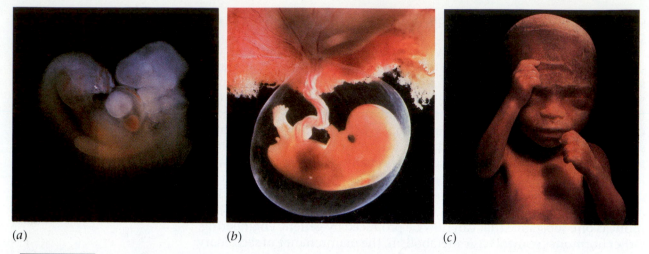

(a)   *(b)*   *(c)*

Figure 24–12

Stages in the development of the human embryo and fetus. (*a*) At 4 weeks (length 0.7 centimeters) structures that will become the heart, eyes, and legs are evident. (*b*) At 2.5 months (3 centimeters) the fetus, floating in the amniotic cavity, is connected to the mother by an umbilical cord. All the major organs are defined. (*c*) At 5.5 months (30 centimeters) the fetus is identifiably human, with all organs in their permanent positions.

sexual characteristics, the development of sexual organs in fetuses, and a wide variety of other human bodily functions.

Reproductive System

The human species, like all other higher animals, reproduces sexually. Sexual reproduction requires the joining of two sex cells, or gametes, each with 23 single chromosomes. The female's ova, or egg, and the male's sperm combine following intercourse to form a fertilized egg with 23 pairs of chromosomes (see Chapter 23).

The male reproductive system is illustrated in Figure 24–8. Sperm are formed in vast numbers in the male's testes. Each sperm, less than a thousandth of a centimeter long, has a rounded head that contains chromosomes and the enzymes required to interact with, and ultimately penetrate, the egg, and a tail that enables it to swim toward the egg. During intercourse, hundreds of millions of sperm are released, though only one can fertilize an egg.

Eggs are produced in the female's ovaries (Figure 24–8). By the time a girl baby is born, all of her potential egg cells, a total of several hundred thousand, have been formed and are stored in the ovaries. Each month between the ages of about 12 and 45, one of these cells matures and is released into the fallopian tubes, where it may be fertilized.

Shortly after fertilization the egg begins to divide over and over again to form the embryo, which is the beginning of a new individual. Approximately one week after fertilization the egg becomes implanted in the wall of the uterus, where the mother provides all necessary nutrients and disposes of wastes as the embryo grows (see Figure 24–12). Pregnancy lasts about nine months, during which time the embryo lies in the fluid-filled amniotic sac. This sac ruptures during labor, which precedes childbirth.

Homeostasis in the Human Body

Human beings, like all other animals, have to maintain an internal equilibrium in their bodies. Extreme swings in temperature or chemical composition would upset many of the processes we've described above, so the body must have systems that can detect departures from equilibrium and move to correct them. The process of maintaining equilibrium in this way is known as *homeostasis*. We will look at two examples of homeostasis in the human body: the maintenance of water balance and temperature.

The bodies of animals are about 70% water, and approximately one-third of that water is found outside the cells. From the point of view of a cell in your body, that intercellular water plays roughly the same role that the Earth's oceans did for the earliest cells. Long ago, animals evolved their own internal "ocean" of intercellular fluid, along with mechanisms for maintaining its chemical composition, as well as the chemical composition of other fluids. To accomplish this end, the system must remove wastes, regulate the amount of various ions and molecules, and control the amount of water.

As we pointed out above, the main organ concerned with internal chemical regulation and waste removal in humans is the kidney. Here materials such as nitrogen compounds and excess water are removed from the blood and excreted in urine, while useful materials such as glucose, amino acids, and various ions are returned to the blood. The kidney function is controlled by hormones that attach to membranes in the kidney and alter their ability to allow water to pass through. When the body is dehydrated, more of this hormone is released and the body retains whatever water it has. When there is too much water (if you have drunk a lot, for example), hormone production drops and more water is excreted. In this way, the concentration of various materials in the body's water is maintained at a roughly constant level.

Human beings produce heat to maintain their body temperature internally. As such, they are said to be warm-blooded, or endothermic. By contrast, animals such as reptiles must absorb their heat from the environment, and are said to be cold-blooded, or ectothermic. Humans, unlike reptiles, do not have to spend large amounts of time sunning themselves to generate internal warmth.

In addition to generating its heat internally, the human body includes a number of mechanisms to maintain its internal temperature at a constant value despite the temperature of the environment. The hypothalamus in the brain, which acts as the body's thermostat, receives signals from temperature receptors in the skin as well as from its own receptors that monitor blood temperature directly. When the external temperature goes up, signals from the hypothalamus trigger sweat glands in the skin to secrete water, whose evaporation cools the body. Sweating also affects the body's water balance, which brings the kidneys into the act. When the temperature falls, blood vessels near the surface of the skin are contracted so that the body's heat loss drops and internally generated heat is conserved.

Science in the Making

Maintaining Body Temperature

One of the pioneers in the study of the human body's ability to maintain a constant temperature was a man named Charles Blagden (1748–1820). Trained as a physician in Scotland, he served as an army doctor before eventually becoming Secretary of the Royal Society in London. There he conducted a series of rather striking experiments that demonstrated that, despite enormous ranges of external temperatures, the human body temperature never varies by more than a few degrees F from its normal value.

In the most dramatic demonstration of this fact, he put himself, Sir Joseph Banks (then President of the Royal Society), a small dog, and an uncooked steak into a room that was heated to 260°F (126°C) for 45 minutes. At the end of the experiment, Blagden, Banks, and the dog were fine and had experienced only small changes in body temperature. The steak, however, was completely cooked!

Blagden's feat became the model for popular circus side-show acts by so-called fire-resistors in the nineteenth century. Audiences marveled at performers who climbed into fire-engulfed ovens and proceeded to calmly cook and eat dinner while sitting in the oven! ■

THINKING MORE ABOUT LIFE'S STRATEGIES

Eating through the Phyla

Human beings consume an amazing variety of foods. One way to make the diversity of life more real to yourself is to think about the variety of kingdoms and phyla your meals come from. Consider, for example, a slice of mushroom pizza. The crust derives from wheat, a vascular plant. The mushrooms, of course, are fungi. The cheese comes from milk produced by a cow, a vertebrate animal, and the milk is converted into cheese through the action of a single-celled member of the kingdom monera. The remains of these organisms are still in the cheese when you eat it, so with one bite of pizza you are consuming representatives of four kingdoms. If you could imagine adding a dash of seaweed or kelp, which are protista (at least in some classification schemes), you could get all five kingdoms in one bite.

Biologist Harold Morowitz, in his book *En-*

tropy and the Magic Flute (Oxford University Press, 1993), points out that traditional Japanese cuisine incorporates more phyla than any other, since it includes such delicacies as seaweeds, and animals such as sea cucumbers (in the same phylum as starfish), as well as the usual crustaceans (arthropods), shellfish, and various bony fish. But even a Western meal such as seafood pasta accompanied by a salad with oil and vinegar dressing contains a wide variety of phyla. It has vascular plants (lettuce, tomato, and olive oil), a phylum of monera (which ferments the vinegar), clams and squid (phylum mollusca), and bony fish (vertebrates), not to mention flour (from angiosperms), and perhaps eggs (from vertebrates) in the pasta.

How many phyla combine to make your favorite meal?

Summary

Every multicelled organism must develop ways to obtain and distribute molecules for energy and structure, regulate internal processes, eliminate wastes, and reproduce itself. Different kinds of life forms have evolved different ways of dealing with these problems.

Fungi absorb materials through filaments and reproduce through the production of spores. The most abun-

dant plants are *vascular plants*, which use thin tubes to distribute water to their leaves. The two largest groups of vascular plants are *gymnosperms*, such as evergreens with exposed seeds, and *angiosperms*, which have flowers. The most abundant animals are *invertebrates*, which are dominated by *arthropods* such as insects and spiders.

The human body incorporates several important

systems of organs. The *skeleto-muscular system* provides a sturdy framework of bones for support and muscles for motion. The *respiratory and circulatory systems* distribute oxygen to cells for energy production, and remove the carbon dioxide and other wastes that are produced in these chemical reactions. The *digestive system* processes food by breaking it down into the small molecular building blocks required by cells. The *nervous system*, including the *brain*, controls the complex interworkings of all the parts of the body. The body also has a chemical control system that operates through the use of hormones. It regulates its internal environment by a process known as homeostasis.

Key Terms About Life

vascular plants

gymnosperms

angiosperms

invertebrates

arthropods

skeleto-muscular system

respiratory system

circulatory system

digestive system

nervous system

brain

Review Questions

1. How do plants get their energy? How do fungi obtain energy? How is the energy-gathering strategy of animals different?

2. What are bryophytes? Give some examples.

3. What characteristics distinguish vascular plants from others? Give examples of different kinds of vascular plants.

4. What are the differences between gymnosperms and angiosperms?

5. What are arthropods? How do they support their weight?

6. What are vertebrates? Give several examples.

7. What are the main differences between amphibians and reptiles?

8. What classes of animals are warm-blooded?

9. Describe the human digestive system.

10. Describe the human circulatory system.

11. Describe the human respiratory system.

12. Describe the human nervous system.

13. Describe the functioning of the kidneys.

14. What are the main parts of the human brain? What does each control?

15. What is the endocrine system? What is a hormone?

Discussion Questions

1. What environmental conditions might favor the development of ferns instead of cherry trees?

2. Speculate on why vascular plants are more widely distributed than bryophytes.

3. Why are there so many species of insects compared to species of birds or mammals?

4. What characteristics might distinguish two species of trees?

5. In what ways might changes in the nature of land plants have affected the evolution of land animals?

6. What are the major organs of the human body, and to which system do they contribute?

7. Where are the main coal deposits of Earth? What does this tell you about the location of forests hundreds of millions of years ago?

Investigations

1. Take a walk around your campus and see how many of the four major plant groupings—bryophytes, seedless vascular plants, gymnosperms, and flowering plants—you can identify. On most campuses you should be able to find all four, and even in the middle of a city you should be able to find two or three.

2. Go to your favorite supermarket and try to count how many different phyla of plants and animals are for sale. What is the average number of phyla per aisle? Does this number vary much from one store to another?

3. What are some of the most common diseases associated with the major systems of the body? Study one of these diseases in detail. What disruption does the disease cause? How can it be treated?

4. Many medicines were derived originally from plants. List some of them. Are any of the plants that produced those medicines now endangered?

Additional Reading

Restak, Richard. *The Brain*. New York: Bantam, 1984.

Zinsser, Hans. *Rats, Lice, and History*. Boston: Little Brown, 1963.

In addition, please refer to the reading list at the end of Chapter 19.

Ecosystems

Ecosystems, interdependent collections of living things, recycle matter while energy flows through them.

A Random Walk
On a Cold Winter's Day

Walking outside on a cold winter's day, you can see your breath—water vapor from your lungs condensing into tiny droplets when it meets the cold air. But you can't see a lot of the gases in the air you breathe out. Your breath, for example, always contains carbon dioxide molecules, one of the waste products from the burning of carbohydrates in the cells of your body. That carbon dioxide will move off into the atmosphere and eventually be incorporated into the cells and cell-wall fibers of plants. Some day, you may again encounter some of those atoms of carbon when you eat an ear of corn or a tomato.

Indeed, from the point of view of materials, the Earth is an enormous closed system in which all the atoms of carbon, oxygen, and other elements play a constant game of musical chairs, cycling through plants, animals, the atmosphere, and other parts of our dynamic planet for billions of years. In a very real sense, an understanding of our living planet can be attained by seeing it not as a series of isolated systems, but as one vast interconnected machine.

Ecology and Ecosystems

The word **ecology**, derived from the Greek word for household or housekeeping, is the branch of science that focuses on natural living systems in the broadest sense. An **ecosystem** includes the plants and animals that live in a given area together with their physical surroundings. An ecosystem can be as small as a single community of organisms on and near a bush in a tropical forest. It can be a lake, including the fish, the insects, and all the plants in it. Or an ecosystem can be a mountain meadow, a salt marsh, a continent, even an entire planet. No matter what size ecosystem we talk about, however, the emphasis of ecology is to look at the system—its matter and its energy—as a whole, rather than as a group of independent parts.

Parts of the Whole

The **ecological niche**, a key concept in ecology, refers to a particular mode of survival—a particular way of obtaining energy—within an ecosystem. In a forest ecosystem, for example, there may be a niche that can be filled by warm-blooded, insect-eating, nocturnal vertebrates—bats. Another niche may be filled by fungi growing in shaded wooded areas—mushrooms. Each plant or animal in an ecosystem fills an ecological niche, and, in the context of natural selection, different organisms compete for dominance in their preferred ecological niche.

Much can be learned from looking at living things as part of integrated natural systems, but this realization is relatively recent. Throughout the nineteenth century, for example, biologists were concerned with cataloging living things, and paid little attention to how living things were affected by (and, in turn, affected) their environment. Only within the last 20 or 30 years have many of the insights discussed in this chapter come to be recognized as different aspects of the study of ecology.

The complex interweaving of living things in their environment leads to what is perhaps the central insight in the science of ecology: It is virtually impossible to change one thing in an ecosystem without affecting other parts of the system, often in as-yet unpredictable ways. The entire system is interdependent, so the whole responds to every stimulus.

Complex coral reef ecosystems are extremely sensitive to changes in local conditions, such as salinity and water temperature. Ocean warming in equatorial regions has killed large areas of coral near the coasts of North and South America.

This phenomenon makes it extraordinarily difficult to study ecosystems, since you cannot do a classic physics-type experiment in which everything is held constant except for one or two carefully monitored variables.

One way of illustrating this property of ecosystems is to speak of the **law of unintended consequences**. This "law" tells us that whenever we do one thing to an ecosystem, something will happen that we didn't anticipate.

The Lake Victoria Disaster

Millions of people who live on the shores of Lake Victoria, the largest freshwater lake in Africa, know firsthand about the law of unintended consequences. What was once a rich fishing ground and source of most of their protein has been devastated by the introduction of a single new species—the Nile perch.

About 30 years ago, sport fishermen, seeking a greater challenge for the growing tourist market, introduced this large, aggressive predator into the lake. The perch thrived and rapidly decimated populations of smaller fish that not only provided an essential part of the local diet but also controlled populations of algae and parasite-bearing snails. Unchecked, live algae spread over the lake's surface, while dead algae sank, decayed, and destroyed oxygen in deepwater fish habitats. Snails have also multiplied and become a serious health hazard because they carry parasites that can affect humans.

Native fishermen now rely on Nile perch, which weigh up to several hundred pounds each, rather than smaller fish, but this change carries its own ecological consequences. Unlike the small fish that were sun dried, Nile perch must be roasted over fires. Lake Victoria's shoreline, each year stripped of more trees for this purpose, is suffering extensive soil erosion and further damage to the lake's ecosystem.

A single new species has thus drastically altered a vast ecosystem.

Energy in Ecosystems

As we saw in Chapter 3, energy constantly flows through systems of all kinds. Virtually all the energy for the Earth's ecosystem comes from the Sun (the only exception is a small amount of heat from radioactive decay that leaks out from the interior of the Earth). This energy stays on the Earth for longer or shorter periods of time, but is eventually sent back into space in the form of infrared radiation. A closer look at ecosystems tells us a great deal about how that energy is used while it is here.

Trophic Levels

The concept of trophic levels is particularly useful when tracking the movement of energy around an ecosystem. A **trophic level** consists of organisms that get their energy from the same source. In this concept, all plants that produce energy from photosynthesis are in the *first trophic level*. Plants in the first trophic level absorb energy from sunlight and use it to drive chemical reactions that make the tissues of the plants themselves, as well as complex carbohydrates that are subsequently used as energy sources by organisms in higher trophic levels.

Although you might expect it to be otherwise, the efficiency with which solar energy is used in our ecosystem is very low, despite the pressures of natural selection and the competition among living things to use energy efficiently. When sunlight falls on a field of corn in the middle of Iowa in August—arguably one of the best situations in the world for plant growth—only a few percent of the energy in that sunlight actually winds up in the plants. All the rest of the radiant energy is either reflected, heats up the soil, evaporates water, or performs some other function. This situation is a general rule—no plants anywhere use as much as 10% of the available solar energy.

The *second trophic level* includes all herbivores—animals that get their energy by eating plants. (The word herbivore comes from Latin words that mean "plant eaters.") Cows, rabbits, and many different kinds of insects occupy this level. As is the case with the plants themselves, animals in the second trophic level use energy very inefficiently. Typically, only 10% of a plant's chemical potential energy winds up as tissues in the animal that eats it. That is, only about 1% (10% of 10%) of the original energy in sunlight makes it into the second trophic level.

Incidentally, you can do a rough verification of this statement in your supermarket. Whole grains (those that have not been processed heavily) typically cost about one-tenth as much as fresh meat. Examined from an energy point of view, this cost differential is not too surprising. It takes 10 times as much energy to make a pound of beef as it does to make a pound of wheat or rice, and this fact is reflected in the price.

The *third trophic level*, as you might expect, consists of animals that get their energy by eating organisms in the second trophic level. This level includes a wide variety of animals—the familiar carnivores ("meat eaters") such as wolves, eagles, and lions, as well as insect-eating birds, blood-sucking ticks and mosquitoes, and many other organisms. Once again, only about 10% of the energy available on the second trophic level is used by animals on the third trophic level.

A few more groups of organisms fill out the scheme of trophic levels on the Earth. Some carnivores, such as killer whales that eat other carnivores, occupy a fourth trophic level. Termites, vultures, and a host of bacteria and fungi get their energy from feeding on detritus—dead organisms—and they generally are placed in a separate trophic level. (The usual convention is that the trophic level associated with detritus feeding is not given a separate number, since detritus can come from any of the other trophic levels.) Finally, a number of animals and plants span the trophic levels. Human beings, raccoons, and bears, for example, are omnivores that gain energy from plants and higher trophic levels as well.

One of the most interesting examples of the notion of energy flow through trophic levels can be seen in the fossils of dinosaurs. In many museum exhibits, the most dramatic and memorable specimen is a giant carnivore—a *Tyrannosaurus* or *Allosaurus* with six-inch dagger teeth and powerful claws. So often are these impressive skeletons illustrated that you might get the impression that these finds are rather common. In fact, fossil carnivores are extremely rare and represent only a small fraction of known dinosaur specimens.

Our knowledge of the fearsome *Tyrannosaurus*, for example, is based on only about a dozen skeletons, and most of those are quite fragmentary. By contrast, paleontologists have found hundreds of skeletons of plant-eating dinosaurs. This distribution is hardly chance. Carnivorous dinosaurs, like modern lions and tigers, were relatively scarce compared to their herbivorous victims. In fact, statistical studies of all dinosaur skeletons reveal a roughly 10:1 herbivore-to-carnivore ratio, a value approaching that of warm-blooded mammals and much higher than the ratio observed in modern cold-blooded reptiles. This pattern is cited by many paleontologists as evidence that dinosaurs were warm-blooded.

Science in the Making

Island Biogeography

Because ecosystems are so complex, it is often difficult to draw unambiguous conclusions from field studies. In 1963, however, American ecologists Robert MacArthur and Edward O. Wilson looked at the populations on islands, small ecosystems separated from the rest of the terrestrial world by water. From their study, they were able to frame the following hypothesis: Whenever a new species migrates to an island that already has a thriving and stable ecosystem, it will flourish only if another species becomes extinct. According to this so-called equilibrium hypothesis, only a fixed number of species exist in any ecosystem, and if a new species invades, one of the old species will be driven to extinction.

The hypothesis was supported some time later by Wilson and Daniel Simberloff in a classic ecological experiment. They chose a series of small mangrove islands off the coast of southern Florida and did a complete survey on each one, counting the species of arthropods. They then removed all the living animals on the islands by draping large plastic sheets over them and fumigating. Over a period of years, they watched the islands undergo the process of repopulation as new animals migrated from the mainland or from other islands. As expected, the total number of species on each island at the end was about the same as it had been at the beginning. Perhaps less expected, however, was the fact that the kinds of animals on the repopulated islands could be significantly different from those that had been there before. If a particular species happened to be carried to an island on the tides, it may have been able to establish itself in a particular ecological niche, which was then unavailable to competitors that arrived later. ■

Materials in Ecosystems

Energy flows in from the Sun, passes through the Earth's ecosystem, and eventually radiates out into space. Most matter, by contrast, continuously cycles from one part of the Earth to another. The breath of air you just took, for example, contains molecules that were in many completely different parts of the world a few years ago. The oxygen, nitrogen, water vapor, and other molecules in that breath have been cycling through the atmosphere, oceans, rocks, and living things for billions of years.

An ecosystem in which materials are free to move in and out is called an *open ecosystem*. If we consider the Earth as a whole, materials almost never leave. An individual atom may, over the course of many years, be part of many different structures, but it always remains in the ecosystem. (For our purposes, we can ignore the very small leakage of gas into outer space from the top of the Earth's atmosphere.) Thus, while energy flows through the Earth's ecosystem on its way from the Sun to outer space, materials do not. The Earth is an example of a *closed ecosystem*.

Tracing the Chemical Cycles

Perhaps the easiest way to understand the cycling of atoms through the Earth's biosphere is to follow a single atom of carbon that leaves your lungs the next time you breathe out a molecule of carbon dioxide. This carbon atom enters the atmosphere, where many different things can happen to it (see Figure 25–1). It can, for example, be taken up by a plant during photosynthesis and then be incorporated into the tissues of a tree or a blade of grass. The plant can then be eaten so that the carbon atom becomes part of the tissue of a herbivore. Alternatively, the carbon can simply return to the atmosphere if the plant dies and rots without being eaten.

If the carbon atom is taken into the tissue of a herbivore, it may show up on your dinner plate one day, and be taken into your body as part of some food you eat. It might even be incorporated into your own body to stay there until you die, or to move through the chemical cycles as described in Chapter 22. In either case, the carbon, in time, will enter the atmosphere again. Another possible track for carbon is shown in Figure 25–1. It can enter the ocean by being added to a mollusk shell or the skeleton of a microscopic organism. Upon the death of the organism, these hard parts sink to the ocean bottom, where, in the form of calcium carbonate, they are turned into limestone. In this case, the carbon atoms can remain locked up for hundreds of millions of years until the limestone is weathered and the carbon is released into the atmosphere.

A single atom of carbon, in other words, may have gone through many different chemical reactions during the 4.5-billion-year life of the planet, and will continue to do so as long as the Earth has living things on it. The one thing it will not do, however, is leave the planet. This is what we mean when we say that:

In closed ecosystems, materials cycle.

A similar multibranched story can be told for every chemical element in the Earth. The Earth's atmosphere forms an enormous reservoir of nitrogen, for example. Nitrogen in the air is converted into nitrate (NO_3^-) by bacteria in the soils and is taken in that form into the structure of

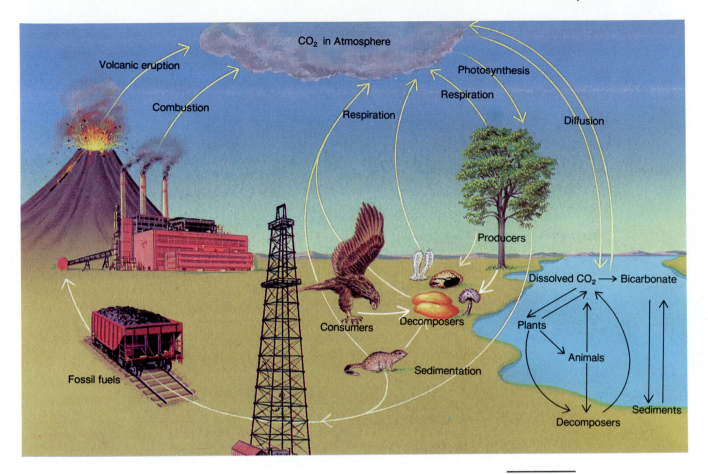

Figure 25–1

The carbon cycle. Carbon cycles through the different parts of the Earth's ecosystem. It is found in carbon dioxide in the air, then taken into plants to become part of the plant's structure. If the plant is eaten, the carbon may be returned to the air through respiration or become part of an animal's tissue. When the animal dies, the carbon dioxide returns to the air.

plants. Animals get their nitrogen by eating plants or other animals, and return the nitrogen to the soil in their wastes. From there, it can be incorporated into plants again or be returned to the atmosphere by the action of other bacteria.

The point of both these examples, as with any other of life's essential elements, is that no matter in an ecosystem is ever lost. In colloquial language, the general rule above can be stated:

> **You can never throw anything away forever.**

The Problem of Urban Landfills

The fact that nothing is ever really thrown away has become very much a concern in urban America. The problem is that garbage (so-called solid waste) is generated at an enormous rate in American cities today. New York City alone adds 17,000 tons of solid waste to its landfill on Staten Island every day. Environmental engineers estimate that, at current rates, every individual American will generate solid waste equivalent in volume to the Statue of Liberty in only about five years.

To make matters worse, the nature of modern landfills is such that the normal process of breakdown and decay in the carbon and nitrogen cycles is slowed enormously. In a landfill, solid waste is dumped on the ground and compacted, then covered with a layer of dirt, then another layer of compacted waste, then another layer of dirt, and so on. Material in such a landfill is cut off from air and water, and the bacteria that normally operate to decompose the waste cannot thrive. Archaeologists digging into landfills have discovered, for example, that newspapers from the 1950s are still readable after having been buried for 40 years! This means that, unlike an ordinary garden compost pile in which materials are quickly broken down by the action of bacteria, the landfill is really more like a burial site than a location for recycling.

One reason that recycling has become so important in America in the 1990s is the simple fact that all the obvious places to dump solid waste near major cities are being used up, and that no replacement sites seem feasible. Something as simple as recycling newspapers may not be a paying business proposition, for example, but most municipalities are realizing that it is a lot cheaper to pay whatever little is needed to recycle newspapers than to find a new waste disposal site in which to dump them.

Science by the Numbers

Trash

How much solid waste is produced in the United States every year? Engineers estimate that the average American each year is responsible for about 40 tons (80,000 pounds) of trash, including everything from disposable containers, newspapers, and mail-order catalogs to old automobiles and appliances, as well as the industrial wastes necessary to manufacture the things we buy. What is the total volume of this waste? Well-compacted trash weighs perhaps 80 pounds per cubic foot—somewhat denser than water, but less dense than rock (of course, it takes up much more volume before it's compacted). The volume of 40 tons, therefore is equivalent to a volume of

$$\frac{80{,}000 \text{ pounds}}{80 \text{ pounds per cubic foot}} = 1000 \text{ cubic feet.}$$

That's enough compacted trash to fill two large dump trucks for every man, woman, and child in the United States every year. Thus 250 million Americans produce a total annual volume of trash of

$$250{,}000{,}000 \text{ people} \times 1000 \text{ feet}^3/\text{person} = 2.5 \times 10^{11} \text{ feet}^3.$$

That's almost two cubic miles of trash every year—enough to build a solid 500-foot-wide wall across the Grand Canyon at its widest and deepest point. ■

The Science in Recycling

Because land that can be used for dumps near big cities is growing ever more scarce, and because the environmental cost of using materials once and then throwing them away is growing steadily, governments have recently begun to pay more attention to recycling. Not only does every recycled plastic milk jug or sheet of paper mean less material in landfills, it also means less petroleum taken from the ground or fewer trees cut down.

But recycling is not as simple as it sounds. A great deal of science and engineering has to be done before even the simplest materials can be reused. In addition, the processes that have to take place to recover one kind of material are, in general, different from those needed to recover another. The recycling of different kinds of plastics, for example, requires different kinds of chemical reactions, and processes that work for plastic soft-drink bottles will not necessarily work for ketchup containers. As a result, each kind of material that is to be recycled poses its own unique problems to the engineer.

Take white paper, for example. The average office worker generates about 250 pounds of high-grade paper waste per year, and many offices around the country have paper recycling programs. The first step in this process is simple: the paper is ground up into a pulp and added to water to make a slurry. Ink particles from typewriters and pens rise to the surface of the slurry and can be skimmed off, leaving a material that can be added to fresh pulp to make new paper. As we pointed out in Chapter 5, however, copying machines and laser printers work by melting bits of carbon mixed with resins onto the paper. That sort of ink makes heavier particles when the paper is ground up, and those particles sink to the bottom along with the paper fibers. Until quite recently, such paper could only be recycled into things such as cardboard or tissue paper, for which color quality is not important.

The new technology for dealing with this problem involves the addition of substances called surfactants to the pulp. The molecules in these substances are shaped so that they bind to the heavier ink particles on one end, and to bubbles of gas on the other. Once the molecules are attached to the ink, various gases are bubbled through the slurry. The surfactants and their load of ink rise to the surface with the bubbles and can be skimmed off, leaving clean paper fibers for reuse.

The national recycling effort, then, will involve hundreds of different processes such as this, each geared to a specific material, but each doing its part to make a coherent whole. ■

The Global Ecosystem and the Environment

The global ecosystem is closed, and nothing is ever really thrown away. These two inescapable facts, coupled with the growing pressure of human populations for energy and material goods, have led to a number of large-scale environmental problems. We are going to look at three of those problems—the degradation of the ozone layer, acid rain, and the greenhouse effect. All of these problems are serious, but their solutions entail different levels of national and international commitment. Taken together, the three give a very good sense of the sorts of difficulties that will have to be solved in order for our modern industrial society to keep functioning.

The Ozone Problem

In Chapter 17 we pointed out that, although the Sun gives off most of its radiation in visible light, a certain amount of that radiation comes in the form of ultraviolet light from the higher-energy part of the spectrum. Ultraviolet radiation can be very damaging to living organisms; indeed, it is routinely used to sterilize equipment in hospitals. If the surface of the Earth were not shielded in some way from the Sun's ultraviolet rays, life on land would be very different, if not impossible.

Ozone, a molecule made up of three oxygen atoms instead of the usual two, absorbs ultraviolet radiation. If enough ozone molecules exist in the atmosphere, they will absorb the ultraviolet radiation from the Sun and keep it from reaching the ground in any appreciable quantities. In fact, a protective shield of ozone formed high in the Earth's atmosphere several hundred million years ago, and it was only after this shield had formed that life could move onto land.

The Ozone Layer   Scientists detect ozone in the atmosphere by using several techniques. One is simply to fly specialized aircraft into the region where ozone is common and collect samples. For the past decade or so, this kind of sampling has been done routinely by organizations such as

Figure 25–2

The ozone layer. Although ozone is found everywhere in the atmosphere, even at ground level, it is concentrated in a layer 20 to 30 miles above the Earth's surface. The labels on the right are the standard terms scientists use to describe different levels of the atmosphere.

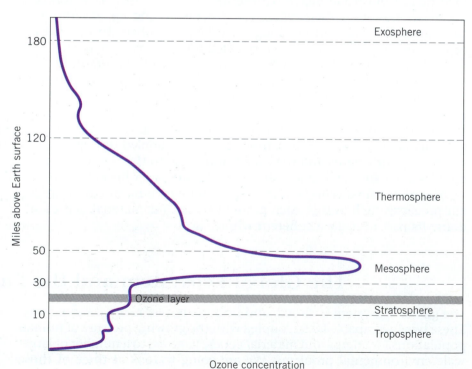

the National Oceanic and Atmospheric Administration (usually called "Noah" after its acronym NOAA) and its counterparts in other countries. Another way to detect ozone is to measure characteristic spectral lines given off by the ozone molecule (see Chapter 6). These measurements can be made from satellites, from aircraft, or by ground-based observers. In general, all these techniques are now used to give us a picture of the health of the ozone layer.

In Figure 25–2, the concentration of ozone as a function of height reveals several important characteristics. First, ozone is a *trace gas*—it constitutes less than one molecule in a million in the Earth's atmosphere. Second, while ozone is found at every altitude (you are breathing a small amount even as you read this), most of the ozone is found some 20 to 30 miles up. In this region, concentrations of ozone are significantly higher than they are in other parts of the atmosphere, although even here the amounts are very small. This region of enhanced ozone concentration is called the **ozone layer**. Most of the absorption of ultraviolet radiation goes on in this layer, but it should not be thought of as anything analogous to a cloud bank in the sky.

The Ozone Hole   In 1985, British scientists working in Antarctica noticed that during the Antarctic spring (roughly the months of September through November) the amount of ozone in the ozone layer dropped significantly. Later studies from satellites and ground-based experiments confirmed these results. During this period of the year, the concentration of ozone above Antarctica falls by different amounts in different years. This phenomenon was dubbed the **ozone hole** (see Figure 25–3).

Figure 25–3
The ozone hole over the South Pole. The dark region right over the pole is one where the ozone concentrations are significantly lower than normal; the bright band represents normal concentrations. This pattern persists only for a few months during the Southern Hemisphere spring.

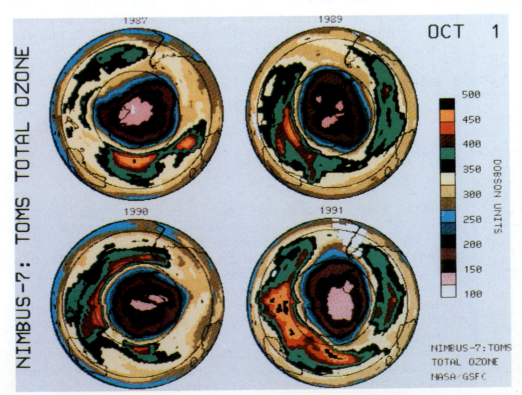

The ozone hole is not a place where the atmosphere has deteriorated. It is simply a volume of the atmosphere in which the concentration of the trace gas ozone has declined significantly. The reason that scientists became concerned about the appearance of the ozone hole is that the ozone layer worldwide is vital to the existence of life on our planet.

Once the annual appearance of the ozone hole was firmly established, scientists began to ask what caused it and whether it could be prevented from forming. Due to the lack of previous measurements, they were uncertain whether the appearance of the ozone hole was a normal natural phenomenon, or one that was caused by human activities. This scientific inquiry quickly narrowed the causes down to a widely used class of chemicals—the so-called *chlorofluorocarbons*, or *CFCs*.

Up until the 1950s, the fluids that were used in appliances such as refrigerators and air conditioners were rather nasty chemicals like ammonia. If they were released into the atmosphere, they would cause a noticeable decrease in the quality of the air. Chlorofluorocarbons, on the other hand, are very stable and generally nonreactive materials—they last a long time and do not break down readily when they are released into the atmosphere. The introduction of CFCs not only made possible the great boom in air-conditioning that has made so much of the southern part of the United States comfortably habitable during the summer months, but these chemicals seemed to be an environmental boon as well. Unlike what they replaced, they did not present an immediate danger in the case of accidental leaks.

In a classic example of the law of unintended consequences, however, CFCs turned out to present a very real danger to the Earth's ozone layer. Over periods of time that range into the decades, molecules of CFCs work their way into the upper regions of the atmosphere, where they can be broken apart by sunlight. The chlorine atoms that are freed in this way act as a catalyst in a reaction that can be written as follows:

▶ **In words:**

(ozone + chlorine + sunlight) become (ordinary oxygen + chlorine).

▶ **In symbols:**

$$2O_3 + Cl + \text{sunlight} \rightarrow 3O_2 + Cl.$$

While this reaction proceeds very slowly, each chlorine atom liberated from a CFC can, over time, destroy millions of ozone molecules before it is safely locked into another chemical species in the atmosphere.

Over most of the Earth, the effect of the chlorine is not striking because new ozone molecules are being created all the time. In the Antarctic, however, a number of unusual circumstances come together to create the ozone hole. For one thing, during the months immediately preceding the hole, no sunlight falls in the Antarctic region of the Earth. This period of darkness leads to the appearance of high clouds made entirely of ice crystals, the so-called polar stratospheric clouds. Crystals of ice in these clouds provide sites on which chlorine atoms undergo a series of chemical reactions. These chemical reactions proceed up to the final step before ozone molecules are actually broken down. As soon as energy in the form of sunlight returns in the Antarctic spring, the destruction of ozone proceeds very quickly. The ozone is destroyed in a matter of days or weeks, and the ozone hole results.

You might think that the disappearance of the ozone shielding in the Antarctic spring would not be a major environmental problem. After all, almost no life exists on the Antarctic continent itself, and very few people go sunbathing there in any case. The real danger of the ozone hole, however, is that it points to chemical reactions that could have long-term effects on the entire ozone layer. Not only has the ozone hole grown larger over the past decade, but some measurements suggest that the ozone layer has been depleted by a few percent worldwide.

Dealing with the Threat to the Ozone Layer In 1986, an international congress meeting in Montreal produced a treaty by which all the industrial nations of the world agreed first to limit, then to eliminate, their production of CFCs. This decision triggered a lot of activity in major chemical companies, where people are looking to find replacement substances. In 1992, the reduction of CFCs was proceeding so quickly that the target date for elimination of most CFCs was set at 1996. Current calculations suggest that the ozone layer will return to normal early in the 21st century.

We view the environmental problems posed by the ozone hole as an example of a serious environmental concern, but one that has a relatively straightforward solution. Scientists have clearly established the cause of the problem. The effects of the ozone hole, while serious, are not totally devastating. And the cost of solving the problem is relatively low. Even so, the problem of ozone depletion appears to be well on its way to being solved.

Acid Rain and Urban Air Pollution

Burning—the chemical reaction of oxidation—inevitably introduces chemical compounds into the atmosphere. Carbon dioxide and water vapor, the oxidation products of hydrocarbons (see Chapter 10), for example, are always released. But burning produces three other significant sources of pollution: nitrogen oxides, sulfur compounds, and hydrocarbons.

Trees near Lake Tahoe, California, have been damaged by a combination of ozone pollution, acid rain, and road salt.

1. **Nitrogen oxides.** Whenever the temperature of the air is raised above about 500°C, nitrogen in the air combines with oxygen to form what are called NO_x compounds: nitrogen oxide, nitrogen dioxide, and so on. The "x" subscript refers to the fact that these compounds have different numbers of oxygen atoms in them.

2. **Sulfur compounds.** Petroleum- and coal-based fossil fuels usually contain small amounts of sulfur, either as contaminants or as an integral part of their structure. The result is that chemical combinations of sulfur and oxygen, particularly sulfur dioxide (SO_2), are released into the atmosphere as well.

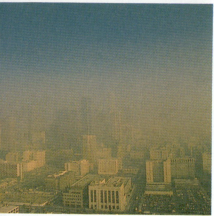

A comparison of Los Angeles during a clear day (top) and on a smoggy day (bottom).

3. **Hydrocarbons.** The long-chain molecules that make up hydrocarbons are seldom burned perfectly in any real-world situation. As a result, a third class of pollutants—bits and pieces of unreacted hydrocarbon molecular chains—enters the atmosphere.

The Effects of Air Pollution and Acid Rain The emission of NO_X compounds, sulfur dioxide, and hydrocarbons gives rise to a number of serious environmental problems. One of these, which has immediate consequences for urban residents, is **air pollution.** Sunlight hitting nitrogen compounds and hydrocarbons in the air triggers a set of chemical reactions that, in the end, produce ozone. And, while ozone in the stratosphere is essential to life on Earth, ozone at ground level is a caustic, stinging gas that can cause extensive damage to the human respiratory system. This "bad ozone" is a major product of modern urban air pollution associated with photochemical *smog*—the brownish stuff that you often see over major cities during the summer.

Urban air pollution is a serious problem, but it is also an immediate and transitory one. If the air quality in a city declines, people know about it immediately. On the other hand, the arrival of a thunderstorm or stiff winds can solve the problem for a particular city quickly (although only at the cost of moving the problem somewhere else).

However, long-term problems are associated with the presence of nitrogen and sulfur compounds in the air—problems that may not have an immediate effect on the place where the emissions occur. When these compounds are in the air, they interact with water to form tiny droplets of nitric and sulfuric acid. (The latter is the type of acid normally used in automobile batteries.) When it rains, these droplets of acid wash out. They become, in effect, a rain of dilute acid rather than water. This phenomenon is known as **acid rain**.

We should note that rain is normally slightly acid because carbon dioxide dissolves in raindrops to make a weak solution of carbonic acid. When we use the term acid rain, then, we are referring to the considerable extra acidity produced by human activities.

You can see one effect of this sort of acid rain in cities. Many of the great historical monuments in European cities, for example, are made from limestone, which is particularly susceptible to the effect of acid. Over the years, the acid rain simply dissolves the fabric of the building (see Figure 25–4).

In the mid-twentieth century, the local effects of acid rain and other kinds of pollution in the United States were dealt with by the construction of tall smokestacks, particularly in the industrial parts of the Midwest. The effect was to put the pollutants high enough in the atmosphere to be taken away by the prevailing winds. But, in keeping with our dictum, "you can't throw anything away," that approach didn't really solve the problem. It merely displaced it. The nitrogen and sulfur compounds emitted in the smokestacks fell as acid rain on the forests of New England.

We must point out that the evidence that acid rain actually destroys forests and lakes is somewhat controversial. Other agencies, such as local climate changes, may be responsible in some cases. In only a few cases

Figure 25–4
The effects of acid rain. Over a period of 60 years, this sandstone statue on a castle in Germany has been completely destroyed.

has acid rain been identified definitively as the main agent of the destruction of a forested area or lake ecosystem. Nevertheless, in a society that is becoming increasingly concerned with preserving nature, the added stress on forests and lakes due to acid rain receives a great deal of attention.

Dealing with Acid Rain   The response of governments to urban air pollution has centered on reducing the levels of emissions associated with the burning of fossil fuels. In California, for example, by 1998 two percent of the cars sold in the state will have to be emission-free. In effect, they will have to be electrical cars. Because of the size of the California automobile market, it is expected that this will lead to the rapid development of electrical cars. At the same time, large facilities such as power plants, which emit huge amounts of pollutants, are required to

Smokestacks equipped with an electrostatic precipitator create an electric field that attracts ash and soot and collects them before they can pollute the atmosphere (*a*). When the precipitator is turned off (*b*) thick clouds rise from the stacks.

(*a*)   (*b*)

install complex engineering devices known as scrubbers, whose job it is to remove the sulfur compounds from the smokestack before they become part of the atmosphere.

In our view, acid rain and air pollution are examples of moderate environmental problems. We understand in a general way what the problems are, what the consequences of pollution are, and what has to be done to prevent the pollution. The costs of dealing with these problems, however, are considerably higher than the costs of reversing the depletion of the ozone layer. Political and economic questions become very important. How much are we willing to pay for clean air? How much is the preservation of a mountaintop in New England worth compared to the jobs that might be lost by closing down an outmoded factory in the Midwest? These are not easy questions, nor are they questions answerable by science alone.

The Greenhouse Effect

The temperature at the Earth's surface is determined in large measure by the extent to which gases in the atmosphere absorb outgoing infrared radiation (see Chapter 6). The atmosphere is largely transparent to the Sun's incoming visible and ultraviolet radiation, which warms the surface; but it is somewhat opaque to the infrared (heat) energy that radiates out into space (see Figure 25–5). If it were not for the trapping of heat by the atmosphere, the average temperature of the Earth (i.e. the average day–night, winter–summer temperature) would be about −20°C. Thus, like a greenhouse, the atmosphere raises the temperature of the Earth from about −20°C to its present temperatures. This temperature increase is the so-called **greenhouse effect**.

Many atmospheric gases contribute to the greenhouse effect. Most people emphasize the role of carbon dioxide, certainly an important greenhouse gas, but CO_2 accounts for only about 10% of the total infrared absorption. Water vapor, especially in clouds, is the dominant greenhouse gas, while the trace gases methane and CFCs, which make up less than a few millionths of the atmosphere, are molecule-for-molecule the most efficient infrared absorbers.

Figure 25–5

The greenhouse effect. Just as the Sun's energy passes through the glass of a greenhouse and becomes trapped inside as heat, the atmosphere acts as a greenhouse to warm up the Earth.

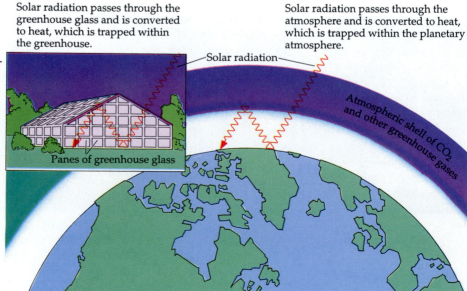

Solar radiation passes through the greenhouse glass and is converted to heat, which is trapped within the greenhouse.

Solar radiation passes through the atmosphere and is converted to heat, which is trapped within the planetary atmosphere.

Solar radiation

Panes of greenhouse glass

Atmospheric shell of CO_2 and other greenhouse gases

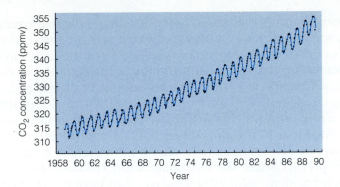

Figure 25–6
Measurements of atmospheric CO_2 concentrations reveal an annual cycle, as well as a gradual increase.

When people talk about the greenhouse effect these days, they usually are concerned about possible increases in average global temperatures, which could result from the increasing concentration of human-generated carbon dioxide and other industrial gases in the atmosphere. While it is straightforward to scrub smokestacks to reduce the sulfur and nitrogen compounds that cause acid rain, it is not possible to remove carbon dioxide created when fossil fuels are combined with oxygen. The best possible outcome for the burning of the fossil fuel, after all, is the liberation of chemical potential energy, accompanied by the production of water and carbon dioxide.

In Figure 25–6, we show a graph of the amount of carbon dioxide in the Earth's atmosphere as measured at a high mountaintop observatory in Hawaii during the past four decades. This graph shows that since the detailed measurements were begun, the amount of carbon dioxide has increased steadily year by year. The small wiggles in the graph correspond to annual cycles by which carbon dioxide is taken into leaves in the spring, and then returned to the atmosphere in the fall. In a sense, the wiggles in this graph show the Earth breathing.

Debates about Global Warming   Without question, the burning of fossil fuels by human beings has increased the amount of carbon dioxide in the Earth's atmosphere. All scientists agree on this fact, and all scientists agree that carbon dioxide absorbs infrared radiation. Once we get past these two basic statements, however, the agreement among scientists becomes much less unanimous. From the point of view of policy, only one pair of questions needs to be answered about the greenhouse effect: Will the increased concentration of carbon dioxide lead to an increase in the Earth's temperature, and, if so, what will be the social and economic consequences? Unfortunately, it is extraordinarily difficult to answer either of these questions with the scientific knowledge we now have.

It has become standard practice in the greenhouse debate to make all calculations and predictions about the Earth's temperature for the case in which atmospheric carbon dioxide doubles. In 1990, the Intergovernmental Panel on Climate Change conducted a major study that represented the best calculations based on the combined judgment of the world's meteorological community on this question. Their prediction: a doubling of CO_2 would lead to a global warming of 1 to 5°C.

Scientists are engaged in many ongoing arguments on the nature and consequences of the greenhouse effect. Many scientists are skeptical of conventional warming wisdom. Some fear more catastrophic levels of warming will occur, while others doubt that warming will take place at all. The reasons for these uncertainties stem from the fact that our only way of describing the Earth's atmosphere is through global circulation models (see Chapter 15). These models are, at best, imperfect ways of predicting changes in climate. For one thing, they break the atmosphere into unrealistic uniform chunks several hundred miles on a side. Such a coarse-grained look at the atmosphere cannot possibly hope to deal with effects of clouds that typically are only a few miles on a side.

Clouds play a critical role in climate. More clouds in the atmosphere reflect more incident sunlight into space. You have direct experience of this fact because you know that when a cloud comes between you and the Sun, you often feel cooler than you did when the cloud wasn't there. If the effect of increased carbon dioxide is to increase the temperature of the atmosphere slightly, one consequence might be increased evaporation of water from the ocean, and, hence, increased formation of clouds. Thus, some scientists argue, the clouds produce an automatic feedback that counteracts the effects of the carbon dioxide. Indeed, when a group at the British meteorological office tested their predictions using global circulation models and the ordinary description of clouds, versus a prediction using a somewhat better description of how clouds interact with solar radiation, they found that the predicted warming increment dropped from 5 to 2.5°C.

Another important effect that is difficult to incorporate in global circulation models is that of the world's oceans. A constant interplay takes place between water and atmosphere at an ocean's surface, and carbon dioxide moves into and out of the oceans all the time. In fact, the amount of carbon dioxide locked in the oceans and their sediments is much greater than that stored in the atmosphere. Even small changes in the way that oceans interact with atmospheric carbon dioxide can thus have huge effects on the world's climate. In addition, as we saw in Chapter 15, ocean currents are instrumental in spreading heat around the surface of the Earth. Small changes in those currents could have enormous effects on the Earth's climate.

Many unknown effects might, in the ways just described, mitigate the effects of added carbon dioxide in the atmosphere and lower the warming due to the greenhouse effect. It is also possible, however, that unknown effects might work in just the opposite way—they might increase the warming. At the moment, we simply do not understand the workings of the atmosphere well enough to know.

Why do so many conflicting arguments arise about the effects of additional greenhouse gases on the climate? The main reason is that the Earth already has a large greenhouse effect. What we are trying to predict with our models, then, is not whether a greenhouse effect will occur, but rather how the existing greenhouse effect will be modified by a small change in atmospheric composition due to the addition of human-made pollutants to the atmosphere. Such a minor change in the global climate is difficult to predict, and it requires a much more detailed understanding of the way the Earth works than we have at the present time.

The range of possible consequences of greenhouse warming is also the subject of much debate. As a general rule, it appears that for every half degree Celsius of greenhouse warming, a global line of a given temperature will move about 100 miles northward. Thus, for 2°C warming, temperatures in Washington, D.C., will be comparable to those in Atlanta, and temperatures in Minneapolis will be comparable to those in St. Louis. Depending on how large you expect the greenhouse warming to be and how quickly you expect it to happen, the effects on the Earth's biosphere and ecosystems can be large or small. The total warming in the Northern Hemisphere after the last ice age, for example, was about 5°C, and took place over a period of several thousand years. We know from studies of pollen deposited in the bottom of lakes that this warming, though large, was sufficiently gradual that plant populations were able to adapt and migrate north with the retreat of the glaciers. More recently, studies of the northern Atlantic have indicated that there have been periods in which the temperature in that region has changed by 5°C over a much shorter period—perhaps as little as a few decades. No known ecological disasters appear to be associated with these events. The predictions of consequences of greenhouse warming, should it occur, thus are also surrounded with a great deal of uncertainty.

Dealing with the Greenhouse Effect   In the opinion of the authors, the greenhouse effect is both the most difficult and the most potentially alarming of the many environmental problems that face the global ecosystem. On the one hand, it is the most difficult to model because the effect of adding carbon dioxide to the atmosphere is uncertain, and the cost of doing something about it is very high. The idea that you can take the world economy, which runs almost entirely on fossil fuels, and change it over to other sources of energy in a very short period of time is unrealistic. In the past, it has taken many decades to make similar changes in a society's energy use and consumption. Figure 25–7, for example, shows

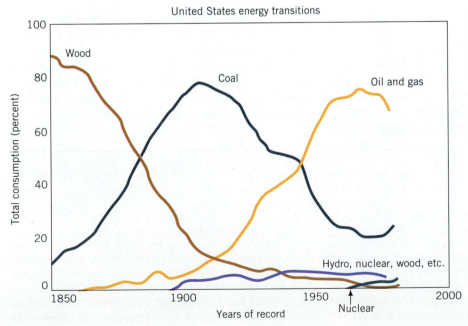

Figure 25–7

U.S. energy transitions. In the past, it has taken 30 to 50 years to make the transition from one type of fuel to another. We could expect a transition to solar energy (as an example) to take about this long.

the transitions from wood to coal and from coal to oil and gas in the U.S. economy. As you can see, it takes 30 to 50 years for a new fuel to work its way into the economy. If the more disastrous predictions of greenhouse warming are true, in about 50 years the warming will already have occurred, and it will be too late to do anything about it.

In addition, the best scientific estimates now indicate that if warming occurs, it will not become evident for several decades, far beyond the planning horizon of corporations, governments, and other major institutions in any society. The question comes down to something like this: Are you willing to give up driving your car three days a week because a possibility exists that global warming will happen sometime during the lifetime of your grandchildren? Human beings find it difficult to suffer real hardship in the present to prevent an uncertain event from happening in the future.

THINKING MORE ABOUT ECOSYSTEMS

Disposable Diapers

One of the great controversies that erupted in the environmental movement in the early 1990s concerned the advisability—some even said the morality—of using plastic disposable diapers. The critics argued: In order to make plastic diapers, we must spend energy to extract petroleum from the ground, and more energy to make the plastic. Once the plastic is thrown into a landfill, it does not decompose, and remains there as a monument to the profligacy of twentieth-century humanity. For a time, a very vocal group asserted that plastic diapers should not be used and that parents should go back to using washable cloth diapers. A very strong call was made for biodegradable plastic diapers that would not fill waste disposal sites.

This argument was countered by plastic-diaper defenders who pointed out that it takes a significant amount of energy to heat water to wash and dry cloth diapers, and that laundry detergents add to water pollution; thus the environmental cost of a cloth diaper is actually much higher than that of a plastic diaper. At the time the whole issue faded from public view, things stood at this impasse.

Given what you now know about the cycling of materials in ecosystems, can you add something to the debate on disposable diapers? If a plastic diaper was made that was indeed biodegradable, where would the carbon atoms in that plastic diaper eventually wind up? The carbon atoms that went into making the plastic diaper came from petroleum under the ground. Is it better to have those carbon atoms return to the ground in the form of a permanent landfill, or to have them dumped into the atmosphere in another form? What effects do you think that disposable plastic diapers might have on the greenhouse problem?

Summary

Ecology is the branch of science that studies interdependent groups of living things, called *ecosystems*. In every ecosystem many different organisms, each competing for matter and energy, occupy their own *ecological niches*. Energy from the Sun is used by photosynthetic plants in the first *trophic level*; these plants provide the energy for animals in higher trophic levels. Roughly speaking, only about 10% of the energy available at one trophic level finds its way to the next. In this way energy flows through an ecosystem. In a closed ecosystem, matter is constantly recycled as atoms are used over and over again—you never really throw anything away. While these principles seem simple, the actual behavior of ecosystems is extremely complex and unpredictable. It is virtually impossible to change one aspect of such a complex interdependent network without affecting something else, often inadvertently—a phenomenon often called the *law of unintended consequences*.

Human actions are causing changes in the Earth's global environment. The use of chlorofluorocarbons, or CFCs, during the past several decades, for example, is having a pronounced effect on *ozone*, a molecule of three oxygens found as a trace gas in the *ozone layer* of the upper atmosphere. Ozone provides important protection on Earth from the Sun's harmful ultraviolet radiation, but chlorine atoms from CFCs hasten the breakdown of ozone molecules, thus creating a growing *ozone hole*.

Burning untreated coal and other fossil fuels releases sulfur and nitrogen compounds into the atmosphere—chemicals that contribute to *air pollution* and *acid rain*. Dealing with acid rain will require cleaning up emissions from automobiles and generating plants.

Carbon dioxide, a necessary product of all combustion of carbon-based fuels, adds to the atmosphere's store of infrared-absorbing gases, and thus contributes to the *greenhouse effect*. At present, scientists are not able to predict the consequences of these global changes with much certainty.

Key Terms About Ecosystems

ecology	ozone
ecosystem	ozone layer
ecological niche	ozone hole
law of unintended consequences	air pollution
	acid rain
trophic level	greenhouse effect

Review Questions

1. What is ecology?

2. What is an ecological niche? Give an example.

3. State the law of unintended consequences. Give an example of the law in operation.

4. What is the difference between an open and closed ecosystem? Give examples of each.

5. How does carbon cycle through the Earth's ecosystem? How does nitrogen cycle?

6. What is a trophic level? Give some examples.

7. Why doesn't trash decompose in a modern landfill?

8. What is ozone? Why is it important to life on Earth?

9. What is the ozone layer? Where is it located?

10. What is the ozone hole? How have governments responded to the discovery of the ozone hole?

11. How does ozone become part of the urban air pollution problem?

12. What is acid rain? What steps would have to be taken to solve the acid rain problem?

13. What is the greenhouse effect?

14. Why are predictions of global warming so uncertain?

15. What steps would have to be taken to reduce the severity of the greenhouse effect?

Discussion Questions

1. Would you expect the number of species present on an island to depend on the size of the island? Why or why not?

2. Cite five examples of environmental changes that have been made in your area. Can you identify any unintended consequences of these changes?

3. If plants use only a few percent of the Sun's radiant energy, where does the rest of the energy go?

4. What environmental changes might result from a global warming of 2°C? What countries might be most affected?

5. Do you think Wilson and Simberloff's experiment was ethical? Why or why not? Was there any other way to get this information?

6. One of the problems of understanding the workings of ecosystems is that it is not possible to hold everything constant and change only one variable. Why is this true? What effects might this have on the interpretation of observations or experiments?

7. How did you affect your environment today? How did it affect you?

8. What would the political, social, and economic consequences be in your community if serious steps were taken to reduce acid rain or air pollution?

9. Suppose we decided today to make a massive conversion to solar energy. How long do you think it would take to make a complete transition? What are the factors that would make such a transition possible?

10. Of what ecosystem is the squirrel on your lawn or the bird flying across your campus a part? Are there any threats to the well-being of that ecosystem?

11. Where do electrical cars obtain their energy? Is it true that they do not generate pollution because they have batteries?

Problems

1. Estimate what the price of lion meat might be, and give reasons for your prediction.

2. Assuming that there is a global warming of 1°C, what kind of weather will your town have after the warming has taken place? (*Hint*: Find a town to the south of you that has that sort of weather now.)

Investigations

1. Do commercial jet airliners fly through the ozone layer? What effects might this have?

2. How are solid wastes such as plastics handled in your community? Does that method represent a long-term or short-term solution to the problem? What other options are available to your community?

3. Write a short story chronicling the passage of a carbon atom through ten different stages.

4. What is an environmental impact statement? Are there any projects in your area that have required environmental impact statements? Where should they be required?

5. What happens to your garbage? Is there a recycling program in your town? What materials can be recycled?

6. Read about the "pea soup fogs" that used to plague London. What caused them? How many people could be killed in one?

Additional Reading

Benedick, Richard. *Ozone Diplomacy: New Directions in Safeguarding the Planet*. Cambridge MA: Harvard University Press, 1991.

McPhee, John. *Control of Nature* New York: Farrar, Strauss, and Giroux, 1989.

Michaels, Patrick. *The Sound and the Fury: The Science and Politics of Global Warming*. Washington, D.C.; Cato Institute, 1992.

Lee Ray, Dixie. *Trashing the Planet*, Washington DC: Regnery Gateway, 1990.

Schneider, Stephen. *Global Warming: Are we Entering the Greenhouse Century?*, San Francisco: Sierra Club Books, 1989.

Appendix A
Units and Numbers

A Random Walk
The Hardware Store

Walk into any hardware store in the United States and immediately you will notice that the things for sale are measured in many different ways. You buy paint by the gallon, grass seed by the pound, and insulation in terms of how many BTUs will leak through it. In some cases, the units are strange indeed—nails, for example, are ranked by "penny" (abbreviated "d"). A 16d nail is a fairly substantial object, perfect for holding the framework of a house together, while a 6d nail might find use tacking up a wall shelf.

But no matter what the material, there is a unit to measure how much is being sold. In the same way, all areas of science have systems of units to measure how much of a given quantity there is. We've encountered some of these units in the text—the newton as a measure of force, for example, and the degree as a measure of temperature. Every quantity used in the sciences has an appropriate unit associated with it.

Systems of Units and Standards

We customarily use certain kinds of units together, in what is called a system of units. In a given system, units are assigned to fundamental quantities such as mass, length, time, and temperature. Someone measuring with that system will use only those units and ignore the units associated with other systems. Once units of mass, length, and time have been specified, a whole series of derived units (for force, for example, or energy) follow from them.

Two systems of units are in common use in the United States. The one encountered most often in everyday life is the English system. This traditional system of units has roots that go back into the Middle Ages. The basic unit of length is the foot (which was defined in terms of the average length of the shoes of men outside a certain church on a certain day), and the basic unit of weight is the pound.

Throughout this book, and throughout most of the world outside the United States, the preferred system of units is the metric system; or, more correctly, the International System (or Systéme Internationale, SI). In this system the unit of length is the meter (originally defined as a certain fraction of the distance around the Earth at the longitude of Paris), and the unit of mass is the kilogram. In both the SI and English systems, the basic unit of time is the second.

Systems of units are one case where governments become intimately involved with science, since the maintenance of standards has traditionally been the task of governments. In the Magna Carta, a document signed in England in 1215 and generally considered to be one of the founding instruments of modern democracy, King John agreed that "There shall be a standard measure of wine, corn, and ale throughout the kingdom," and to establish measures of length (for cloth merchants) and weights. Since that time, governments have maintained standards for use in industry and commerce. When you buy a pound of meat in a supermarket, for example, you know that you are getting full weight for your money because the scale is certified by a state agency, which relies, ultimately, on international standards of weight maintained by a treaty between all nations.

Originally, the standards were kept in sealed vaults at the International Bureau of Weights and Measures near Paris, with secondary copies at places such as the National Institute of Standards and Technology (formerly National Bureau of Standards) in the United States. The meter, for example, was defined as the distance between two marks on a particular bar of metal; the kilogram as the mass of a particular block of iridium-platinum alloy; the second as a certain fraction of the length of the day.

Today, however, only the kilogram is still defined in this way. Since 1967, the second has been defined as the time it takes for 9,192,631,770 crests of the light emitted by a certain quantum jump in a cesium atom to pass by a given point. Since 1960, the meter has been defined as the length of 1,650,763.73 wavelengths of the radiation from a particular quantum jump in the krypton atom. In both these cases, the old standards have been replaced by numbers relating to atoms—standards that any reasonably equipped laboratory can maintain for itself. Atomic standards have the additional advantage of being truly universal—every cesium or krypton atom in the universe is equivalent to any other. Only mass is still defined in the old way, in relation to a specific block of material kept in a vault, and scientists are working hard to replace that standard by one based on the mass of individual atoms.

The International System of Units

Within the SI system, units are based on multiples of 10. Thus, the centimeter is one-hundredth the length of a meter, the millimeter one-thousandth, and so on. In the same way, a kilometer is 1000 meters, a kilogram is 1000 grams, and so on. This organization differs from that of the English system, in which a foot equals 12 inches, and 3 feet make a yard. A list of metric prefixes follows.

Metric Prefixes

If the prefix is:	Multiply the basic unit by:
giga	billion (thousand million)
mega-	million
kilo-	thousand
hecto-	hundred
deka-	ten

If the prefix is:	Divide the basic unit by:
deci-	ten
centi-	hundred
milli-	thousand
micro-	million
nano-	billion

Units of Length, Mass, and Temperature

Next we give the conversion factors between SI and English units of length and mass.

Length and Mass Conversion from SI to English Units

To get:	Multiply:	By:
inches	meters	39.4
feet	meters	3.281
miles	kilometers	1.609
pounds	newtons	0.2248*

* Recall that the weight of a 1-kilogram mass is 9.806 newtons.

Thus, for example, a distance of 5 miles can be converted to kilometers by multiplying by the factor 1.609:

$$5 \text{ miles} \times 1.609 = 8.05 \text{ kilometers}$$

Length and Mass Conversion from English to SI Units

To get:	Multiply:	By:
meters	inches	0.0254
meters	feet	0.3048
kilometers	miles	0.6214
newtons	pounds	4.448

To convert from Celsius to Fahrenheit degrees, use the following formula:

$$°F = 1.8(°C) + 32,$$

where °F and °C stand for degrees Fahrenheit and Celsius, respectively. To find temperatures in the Kelvin scale, simply add 273.15 to the temperature on the Celsius scale.

Units of Force, Energy, and Power

Once the basic units of mass, length, time, and temperature have been defined, the units of other quantities such as force and energy follow. Recall that the energy units we have defined in the text are:

joule: a force of 1 newton acting through 1 meter

foot-pound: a force of 1 pound acting through 1 foot

calorie: energy required to raise the temperature of 1 kilogram of water by 1 degree Celsius

British Thermal Unit, or BTU: energy required to raise the temperature of 1 pound of water by 1 degree Fahrenheit

kilowatt-hour: 1000 joules per second for 1 hour

Power units are:

watt: 1 joule per second

horsepower: 550 foot-pounds per second

Conversion factors between SI and English units for energy and power follow.

Energy and Power Conversion from SI to English Units

To get:	Multiply:	By:
BTUs	joules	0.00095
calories	joules	0.2390
kilowatt-hours	joules	2.78×10^{-7}
foot-pounds	joules	0.7375
horsepower	watts	0.00134

Energy and Power Conversion from English to SI Units

To get:	Multiply:	By:
joules	BTUs	1055
joules	calories	4.184
joules	kilowatt-hours	3.6 million
joules	foot-pounds	1.356
watts	horsepower	745.7

Powers of Ten

Very large or very small numbers may be written conveniently in a compact way that doesn't involve writing down a lot of zeroes. The so-called "powers of ten" notation accomplishes this goal. The basic rules for the notation are:

1. Every number is written as a number between 1 and 10 followed by 10 raised to a power, or an exponent.
2. If the power of 10 is positive, it means "move the decimal point this many places to the right."
3. If the power of ten is negative, it means "move the decimal point this many places to the left." Thus,

$$3.56 \times 10^3 = 3560, \text{ and } 7.87 \times 10^{-4} = 0.000787.$$

Multiplying or dividing numbers with powers of ten requires special care. If you are multiplying two numbers, such 2.5×10^3 and 4.3×10^5, you multiply 2.5 and 4.3, but you add the two exponents:

$$(2.5 \times 10^3) \times (4.3 \times 10^5) = (2.5 \times 4.3) \times 10^{3+5}$$
$$= 10.75 \times 10^8$$
$$= 1.075 \times 10^9$$

When dividing two numbers, such as 4.3×10^5 divided by 2.5×10^3, you divide 4.3 by 2.5, but you subtract the denominator exponent from the numerator exponent:

$$\frac{4.3 \times 10^5}{2.5 \times 10^3} = \frac{4.3}{2.5} \times 10^{5-3}$$
$$= 1.72 \times 10^2$$
$$= 172$$

THINKING MORE ABOUT UNITS

Conversion to Metric Units

The reasons that the United States still uses English units long after most of the rest of the world has converted to SI have to do with non-scientific factors such as the geographical isolation of the country, the size of our economy (the world's largest), and, perhaps most importantly, the expense of making the conversion. Think, for example, of what it would cost to change all of the road signs on the Interstate Highway System so that the distances read in kilometers instead of miles.

To understand the debate over conversion, you have to realize one important point about units. There is no such thing as a "right" or "scientific" system of units. Units can only be convenient or inconvenient. Thus U.S. manufacturers who sell significant quantities of goods in foreign markets long ago converted to metric standards to make those sales easier. Builders, on the other hand, whose market is largely restricted to the United States, have not.

By the same token, very few scientists actually use the SI in their work. United States engineers use English units almost exclusively—indeed, when the federal government was considering a tax on energy use in 1993, it was referred to as a "BTU tax." Hospital and medical professionals routinely use the so-called "cgs" system, in which the unit of length is the centimeter and the unit of mass is the gram. Next time you have blood drawn, take a look at the needle. It will be calibrated in "cc"—cubic centimeters. Even scientists doing basic research choose non-SI units for convenience. Astronomers, for example, talk about light-years or parsecs instead of meters. Nuclear physicists measure distances in "fermis"—roughly the distance across a proton.

Given this wide range of units actually in use, how much emphasis should the U.S. government give to metric conversion? How much should the government be willing to spend on the conversion process: how many new signs as opposed to how many repaired potholes on the roads?

Appendix B
The Geologic Time Scale

A Random Walk
In Time

As you drive through the countryside, you may come across an abandoned farmhouse, its windows boarded up, its roof open to the elements, its only inhabitants, animals and insects. The surrounding fields, once cultivated, may have been overtaken by the prairie grasses that were there long before the farmhouse was built. Within a very short time span, perhaps as short as 25 years, that farmhouse may be razed and a subdivision built on the same site. Over a longer time span, the same site may have been at the bottom of a prehistoric ocean, then lifted to the top of a mountain, and finally eroded to the level at which the farmhouse was built. All things change over time.

All things also affect others as they change. In the following pages, we show you the *geologic time scale,* a chronological arrangement of geologic time units as currently understood. In addition, we show you the major steps in evolution that were made possible, in part, by the conditions that existed at every step of that time scale.

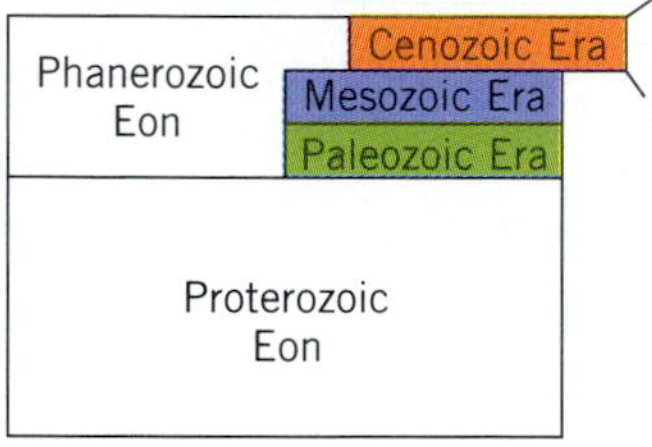

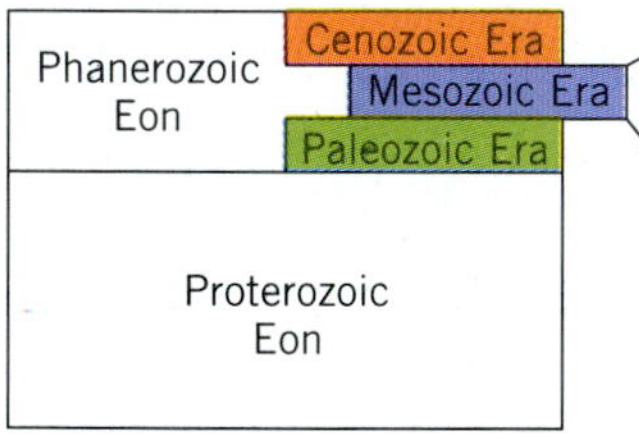

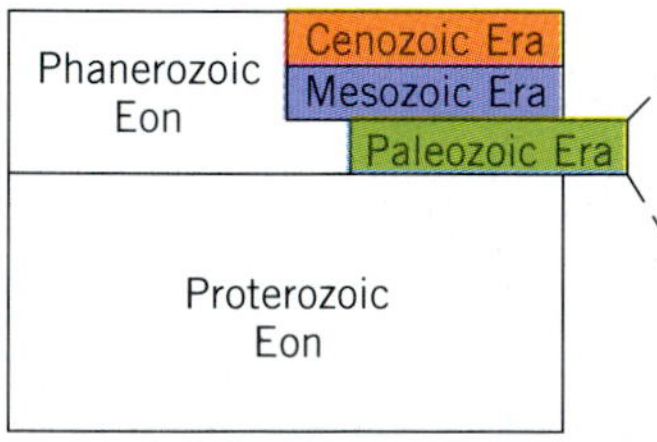

PERIOD	EPOCH	PLANT EVOLUTION
Quaternary	Holocene	Repeated glaciation
	Pleistocene	
Tertiary	Pliocene	Decline of forests, spread of grasslands
	Miocene	
	Oligocene	
	Eocene	Explosive radiation of flowering plants
	Paleocene	
Cretaceous		First flowering plants
Jurassic		Forests of gymnosperms and ferns over most of the Earth
Triassic		Gymnosperms dominant
Permian		Widespread extinction Decline of nonseed plants
Carboniferous Pennsylvanian		Gymnosperms appear
Mississippian		Widespread forests of giant club moss trees, horsetails and tree fern — create vast coal deposits
Devonian		First seed plants Development of vascular plants: club mosses and ferns
Silurian		First vascular plants First land plants
Ordovician		
Cambrian		Algae dominant

Phanerozoic Eon
Cenozoic Era
Mesozoic Era
Paleozoic Era
Proterozoic Eon

Phanerozoic Eon
Cenozoic Era
Mesozoic Era
Paleozoic Era
Proterozoic Eon

Phanerozoic Eon
Cenozoic Era
Mesozoic Era
Paleozoic Era
Proterozoic Eon

2.5
65
135
195
240
285
375
420
450
520
570
Millions of years ago

ANIMAL EVOLUTION	MAJOR GEOLOGICAL EVENTS
Appearance of *Homo sapiens* First use of fire Appearance of *Homo erectus*	 Worldwide glaciations Linking of North America and South America
Appearance of hominids Appearance of first apes All modern genera of mammals present In seas, bony fish abound Rise of mammals First placental mammals	Opening of Red Sea Formation of Himalayan Mountains Collision of India with Asia Separation of Australia and Antarctica Opening of Norwegian Sea and Baffin Bay
Dinosaurs extinct Modern birds	Formation of Alps Formation of Rocky Mountains
First birds Age of dinosaurs	
Explosive radiation of dinosaurs First dinosaurs First mammals Complex arthropods dominant in seas First beetles	 Opening of Atlantic Ocean
Widespread extinction Appearance of mammal-like reptiles Increase of reptiles and insects Decline of amphibians	 Final assembly of Pangaea
Early reptiles First winged insects Increase of amphibians	 Formation of coal deposits
Amphibians diversify into many forms First land vertebrates — amphibians	
Golden Age of fishes First land invertebrates — land scorpions	
First vertebrates — fishes Increase of marine invertebrates	
Trilobites dominant Explosive evolution of marine life	

Time	Eon*
Precambrian	Proterozoic
	Archean
	Hadean

* No further subdivisions into eras or periods are in common use.

	PLANT EVOLUTION	ANIMAL EVOLUTION	MAJOR GEOLOGICAL EVENTS
1,000			
1,500			Formation of early super-continent
2,000			
2,500			
3,000	Early algae	Early bacteria	
3,500			
4,000			Oldest Earth rocks
4,500			Oldest Moon rocks / Heavy meteorite bombardment
4,600			Formation of the Earth

Millions of years ago

Appendix C
Selected Physical Constants and Astronomical Data

Avogadro's number
6.02×10^{23}/mol

Charge on electron
1.6×10^{-19} C

Electron mass
$m_e = 9.10939 \times 10^{-31}$ kg

Gravitational constant
$G = 6.673 \times 10^{-11}$ N · m^2/kg^2

Planck's constamt
$h = 6.62608 \times 10^{-34}$ J · s

Proton mass
$m_p = 1.6726 \times 10^{-27}$ kg $= 1836.1\ m_e$

Speed of light in a vacuum
$c = 2.9979 \times 10^8$ m/s

Astronomical unit
AU $= 1.4959789 \times 10^{11}$ meters

Hubble's constant
$H \sim 20$ km/s/Mly

Light-year
ly $= 9.46053 \times 10^{15}$ meters
$= 6.324 \times 10^4$ AU

Mass of Sun
$M_{sun} = 1.989 \times 10^{30}$ kg

Radius of Sun
$R_{sun} = 6.96 \times 10^5$ km

Mass of Moon
$M_{moon} = 7.348 \times 10^{22}$ kg

Radius of Moon
$R_{moon} = 1.738 \times 10^3$ km

Appendix D
Properties of the Planets

Planet	Length of Day	Distance from Sun (millions of km)	Length of Year (Earth year)	Average Radius (km)	Radius (Earth radii)	Mass (kg)	Mass (Earth masses)
Mercury	58.65 days	57.9	0.24	2,439	0.38	3.30×10^{23}	0.0562
Venus	243.01 days (retrograde)	108.2	0.615	6,052	0.95	4.87×10^{24}	0.815
Earth	23 h 56 min 4.1 s	149.6	1.000	6,378	1.00	5.974×10^{24}	1.000
Mars	24 h 37 min 22.6 s	227.9	1.881	3,397	0.53	6.42×10^{23}	0.1074
Jupiter	9 h 50.5 min	778.4	11.86	71,492	11.19	1.899×10^{27}	317.9
Saturn	10 h 14 min	1424	29.46	60,268	9.45	5.68×10^{26}	95.1
Uranus	17 h 14 min (retrograde)	2872	84.01	25,559	4.01	8.66×10^{26}	14.56
Neptune	16 h 3 min	4499	164.8	25,269	3.96	1.03×10^{26}	17.24
Pluto	6.39 days (retrograde)	5943	248.6	1,140	0.18	1.1×10^{22}	0.0018

Glossary

AAAS (pronounced "triple-AS") See *American Association for the Advancement of Science.*

absolute magnitude The brightness a star appears to have when it is viewed from a standard distance. (Ch. 17)

absolute zero The temperature, 0 Kelvins, at which no energy can be expected from atoms; the coldest attainable temperature, which is equal to $-273.16°C$ or $-459.67°F$. (Ch. 4)

absorption One of three possible responses of an electromagnetic wave encountering matter, in which light energy is converted into some other form, usually heat energy. See also *transmission* and *refraction.* (Ch. 6)

absorption line A dark line in the absorption spectrum that corresponds to the absorbed wavelength of light. (Ch. 7)

absorption spectrum The characteristic set of dark lines used to identify a chemical element or molecule from the photons absorbed by those atoms or molecules. (Ch. 7)

AC See *alternating current.*

acceleration The amount of change in velocity divided by the time it takes the change to occur. Acceleration can involve changes of speed, changes in direction, or both. (Ch. 2)

acid Any material that when put into water produces positively charged hydrogen ions (i.e., protons) in the solution; for example, lemon juice and hydrochloric acid. (Ch. 9)

acid rain A phenomenon that occurs when nitrogen and sulfur compounds in the air interact with water to form tiny droplets of nitric and sulfuric acid, which makes raindrops more acidic than normal. (Ch. 25)

adaptation The structure, process, or behavior that helps an organism survive and pass its genes on to the next generation. (Ch. 23)

addition polymerization The formation of a polymer in which the basic building blocks are simply joined end to end; for example, polyethylene. (Ch. 9)

adenosine triphosphate (ATP) An energy-carrying molecule, found in a cell, that contains three phosphate groups, the sugar ribose, and the base adenine. (Ch. 21)

aerobic A process that requires the presence of oxygen; for example, respiration. (Ch. 21)

air pollution A serious environmental problem, with immediate consequences for urban residents, from the emission of NO_x compounds, sulfur dioxide, and hydrocarbons into the atmosphere. (Ch. 25)

algae Single-celled organisms (or simple multicelled ones) that carry out between 50 and 90% of the Earth's photosynthesis. (Ch. 24)

alkali metals Elements that are highly reactive, such as lithium, sodium, and potassium; listed in far left-hand column of the periodic table of elements. These elements possess one valence electron. (Ch. 7)

alkaline earth metals Elements that combine with oxygen in a one-to-one ratio and form colorless solid compounds with high melting temperatures. Listed in the second column in the periodic table of elements: beryllium, magnesium, calcium, and others. These elements possess two valence electrons. (Ch. 7)

alkane A family of molecules based on the methane molecule, which burn readily and are used as fuels. (Ch. 9)

alloy The combination of two or more chemical elements in the metallic state; for example, brass (a mixture of copper and zinc) or bronze (an alloy of copper and tin). (Ch. 9)

alpha decay The loss by an atom's nucleus of a large and massive particle composed of two protons and two neutrons. (Ch. 11)

alpha particle A subatomic radioactive particle, made of two protons and two neutrons, used by Ernest Rutherford in a well-known experiment in which the nucleus was discovered. (Ch. 7)

alternating current (AC) A type of electric current, commonly used in household appliances and cars, in which charges alternate their direction of motion. (Ch. 5)

alveoli Tiny thin-walled sacs in the lungs which come in close contact with the blood vessels for the exchange of oxygen and carbon dioxide. (Ch. 24)

AM See *amplitude modulation.*

American Association for the Advancement of Science (AAAS) One of the largest scientific societies, representing all branches of the physical, biological, and social sciences. AAAS is a strong force in establishing science policy and promoting science education. (Ch. 1)

amino acid The building block of protein, incorporating a carboxyl group (COOH) at one end, an amino group (NH_2) at the other end, and a side group (which varies from one amino acid to the next). (Ch. 20)

amino group A group of atoms of nitrogen and hydrogen (NH_2) that forms one end of an amino acid. See *carboxyl group.* (Ch. 20)

amp See *ampere.*

ampere A unit of measurement for the amount of current (number of charges) flowing in a wire or elsewhere. (Ch. 5)

amphibians The first vertebrates adapted to live part of their lives on land; modern descendants include frogs, toads, and salamanders. (Ch. 24)

amplifier A device that takes a small current and converts it into a large one to do work. (Ch. 10)

amplitude modulation (AM) A process by which information is transmitted by varying the amplitude of a radio-wave signal being transmitted. After transmission, the signal is converted to sound by the radio receiver. (Ch. 6)

amplitude The height of a wave crest above the undisturbed level of the medium. (Ch. 6)

anaerobic A process that can occur in the absence of oxygen; for example, fermentation. (Ch. 21)

angiosperm The class of vascular plants that are flowering. (Ch. 24)

animals Multicelled organisms that get their energy by eating other organisms. One of five kingdoms in the modern Linnaean classification. (Ch. 19)

annihilation A process that occurs when a particle collides with its antiparticle, completely converting both masses to energy. (Ch. 12)

antibodies A set of molecules that are recognized and bind to a specific foreign molecule, triggering the immune response. (Ch. 21)

antigen Any substance that elicits an immune response. (Ch. 21)

antimatter Particles that have the same mass as their matter twins, but with an opposite charge, magnetic characteristics, and other properties. (Ch. 12)

apparent magnitude The brightness a star appears to have when it is viewed from the Earth. (Ch. 17)

applied research The type of research performed by scientists with specific and practical goals in mind. This research is often translated into practical systems by large-scale research and development projects. (Ch. 1)

aquifer An underground body of stored water, often a layer of water-saturated rock bounded by impermeable materials. (Ch. 16)

arthropods All invertebrate animals with segmented bodies and jointed limbs. The most successful phylum in the animal kingdom, in terms of numbers of species and total mass; includes insects, spiders, and crustacea. (Ch. 24)

artificial intelligence A field of research based on the idea that computers eventually will be able to perform all functions of thought that we normally think of as being distinctly human. (Ch. 10)

artificial selection The process of conscious breeding for specific characteristics in plants and animals. (Ch. 23)

asteroid belt A collection of small rocky planetesimals, located in a circular orbit between Mars and Jupiter, debris of material that never managed to collect into a single planet. (Ch. 14)

asteroids Small rocky objects, concentrated mostly in an asteroid belt between Mars and Jupiter, that circle the Sun like miniature planets. (Ch. 14)

astronomy The study of objects in the heavens. (Ch. 17)

atmospheric cycle The circulation of gases near the Earth's surface, which includes the short-term variations of weather and the long-term variations of climate. (Ch. 16)

atom Fundamental building blocks for all matter; the smallest representative sample of an element. It consists of a positively charged nucleus and electron in orbit. (Ch. 7)

atomic number The number of protons in the nucleus, which determines the nuclear charge, and therefore the chemical identity of the atom. (Ch. 11)

atomism The hypothesis that for each chemical element there is a corresponding species of indivisible objects called atoms. (Ch. 7)

ATP See *adenosine triphosphate.*

Australopithecus The first hominid, a primate closer to humans than any other; lived approximately 4.5 to 1.5 million years ago, walked erect, and had a brain about the size of that of a modern chimpanzee. (Ch. 23)

autonomic nervous system Chains of nerves that control actions such as the beating of the heart and the contractions of the gut. (Ch. 24)

axon The longest filament connecting one nerve cell to another, along which nerve signals move. (Ch. 5)

basalt A dense, dark, even-textured volcanic rock forming the oceanic plates; rich in oxides of silicon, magnesium, iron, calcium, and aluminum. (Ch. 15)

base A class of corrosive materials that when put into water produce negatively charged hydroxide ions; usually tastes bitter and feels slippery. (Ch. 9)

base pair One of four possible bonding combinations of the bases adenine, thymine, guanine, and cytosine on the DNA molecule: AT, TA, GC, and CG. (Ch. 20)

basic research The type of research performed by scientists who are interested simply in finding out how the world works, in knowledge for its own sake. (Ch. 1)

battery A device that converts stored chemical energy into kinetic energy of charged particles (usually electrons) running through an outside wire. (Ch. 5)

beta decay A kind of radioactive decay in which a particle such as the neutron spontaneously transforms into a collection of particles that includes an electron. (Ch. 11)

big bang theory The idea that the universe began at a specific point in time and has been cooling and expanding ever since. (Ch. 18)

biology The branch of science devoted to the study of living systems. (Ch. 19)

bird Modern descendant of reptiles with an anatomical adaptation to flight and feathers evolved from scales. (Ch. 24)

bit Binary digit: a unit of measurement for information equal to "yes-no" or "on-off." (Ch. 10)

black hole Formed at the death of a very large star, an object so dense, with a mass so concentrated, that nothing—not even light—can escape from its surface. (Ch. 17)

blueshifted The result of the Doppler effect on light waves, when the source of light moves toward the observer: light-wave crests bunch up and have a higher frequency. (Ch. 6)

Bohr atom A model of the atom, developed by Niels Bohr in 1913, in which electrons exist only in allowed orbits located at fixed distances from the center of an atom. In these orbits the electrons maintain fixed energy for long periods of time, without giving off radiation. (Ch. 7)

boiling A change of state from liquid to gas caused by an increase in temperature or pressure of the liquid, which speeds up the vibration of individual molecules of the liquid, allowing them to break free and form a gas. (Ch. 9)

bone The rigid part of the skeleto-muscular system. (Ch. 24)

brain The central organ of the human nervous system, which receives signals from sense organs, as well as signals that keep it apprised of the status of internal organs; sends out signals to keep the body functioning; and serves as the seat for all higher functions, such as thought and speech. (Ch. 24)

Brownian motion A phenomenon that describes the rapid, random movements caused by atomic collisions of very small objects suspended in a liquid. (Ch. 7)

bryophyte The phylum of primitive terrestrial plants, including mosses, which can use photosynthesis, are anchored to the ground by rhizoids, and absorb water directly through above-ground structures. (Ch. 24)

byte In a computer, a group of eight switches storing eight bits of information; the basic information unit of most modern computers. (Ch. 10)

c The speed of light and other electromagnetic radiation; a constant whose value is 300,000 kilometers per second (about 186,000 miles per second) equal to the product of the wavelength and frequency of an electromagnetic wave. (Ch. 6)

calorie A common unit of energy defined as the amount of heat required to raise 1 gram of room-temperature water 1 degree Celsius in temperature. (Ch. 4)

Cambrian explosion The sudden change in life on Earth, well-documented in the fossil record, when hard-bodied organisms first appeared 570 million years ago. (Ch. 23)

Cambrian period The geological period, beginning about 570 million years ago, during which animals first began to develop shells and skeletons. (Ch. 23)

carbohydrate A class of modular molecules made from carbon, hydrogen, and oxygen that form the solid structure of living things and play a central role in how living things acquire oxygen. (Ch. 20)

carboxyl group A group of atoms of carbon, hydrogen, and oxygen (COOH) that forms one end of an amino acid string. See *amino group*. (Ch. 20)

cell A complex chemical system with the ability to duplicate itself; the fundamental unit of life. (Ch. 21)

cell membrane A structure, formed from bilayers of lipids, that separates the inside of the cell from the outside, or separates one part of a cell from another. (Ch. 21)

cell wall A solid framework made from cellulose molecules and other strong polymers, by which plant cells are separated from one another. (Ch. 21)

central processing unit (CPU) The part of a computer in which transistors store and manipulate relatively small amounts of information at any one time. (Ch. 10)

Cepheid variable A type of star with a regular behavior of steady brightening and dimming, which is related to the star's luminosity. Cepheid variables are used to calculate distances to many millions of light-years. (Ch. 17)

ceramic A crystalline solid that includes a broad class of hard, durable solids, including bricks, concrete, pottery, porcelain, and numerous synthetic abrasives. (Ch. 9)

cerebral cortex The place in the brain where all activities normally associated with higher human faculties are carried out; for example, speech, rational thought, and memory. (Ch. 24)

CFCs See *chlorofluorocarbons*.

chain reaction The process in a nuclear reactor in which nuclei undergoing fission produce neutrons that will cause more splitting, resulting in the release of large amounts of energy. (Ch. 11)

change of state Transition between the solid, liquid, and gas states caused by changes in temperature and pressure. The processes involved are freezing and melting (for solids and liquids), boiling and condensation (for liquids and gases), and sublimation (for solids and gases). (Ch. 9)

chaos A field of study modeling systems in nature that can be described in Newtonian terms but whose futures are, for all practical purposes, unpredictable; for example, the turbulent flow of water or the beating of a human heart. (Ch. 2)

chemical bond The attraction that results from the redistribution of electrons between two or more atoms, leading to a more stable configuration particularly by filling the outer electron shells, and that holds the two atoms together. The principal kinds of chemical bonds are ionic, covalent, and metallic. (Ch. 9)

chemical evolution An area of research concerned with the process by which simple chemical compounds present in Earth's early atmosphere became an organized, reproducing cell. (Ch. 23)

chemical potential energy The type of energy that is stored in the chemical bonds between atoms, such as the energy in flashlight batteries. (Ch. 3)

chemical reaction The process by which atoms or smaller molecules come together to form large molecules, or by which larger molecules are broken down into smaller ones; involves the rearrangement of atoms in elements and compounds, as well as the rearrangement of electrons to form chemical bonds. (Ch. 9)

chlorofluorocarbons A class of stable and generally nonreactive chemicals widely used in refrigerators and air-conditioners until the late 1980s. (See *ozone hole*.) (Ch. 25)

chlorophyll A molecule, found in the chloroplasts of plant cells, that absorbs energy from sunlight and uses that energy to transform atmospheric carbon dioxide and water into energy-rich sugar molecules such as glucose and oxygen (as a by-product). (Ch. 21)

chloroplasts The main energy transformation organelles in plant cells; places where the molecules of chlorophyll are found and photosynthesis occurs. (Ch. 21)

cholesterol An essential component of the cell membrane synthesized by the body from saturated fats in the diet; in high levels, can cause fatty deposits that clog arteries. (Ch. 20)

chordate A phylum of animals with a thickened set of nerves down their backs; includes the subphylum vertebrate. (Ch. 19)

chromosomes A long strand of the DNA double helix, with the strand wrapped around a series of protein cores. (Ch. 21)

chromosphere One of the Sun's outer layers, visible for a few minutes as a spectacular halo during a total eclipse of the Sun. (Ch. 17)

circulatory system The system that distributes blood through the body; includes the blood vessels and the heart. (Ch. 24)

cladistic analysis A classification scheme based on the idea that descent and ancestry rather than present structure are what should be used to classify living things. The scheme records branchings in family trees without reference to when they occurred. (Ch. 19)

class The third broadest classification in the Linnaean classification system; humans are in the class of mammals. (Ch. 19)

classical genetics The laws developed from the observations of Gregor Mendel: (1) traits are passed from parent to offspring by genes; (2) each parent contributes one gene for each trait; and (3) genes are either dominant or recessive. (Ch. 22)

climate The average weather conditions of a place or area over a period of years. (Ch. 16)

closed ecosystem An ecosystem through which energy will flow but materials will cycle. (Ch. 25)

closed system A type of system in which matter and energy are not freely exchanged with the surroundings; an isolated system. (Ch. 3)

closed universe A universe in which the expansion will someday reverse because the universe holds enough matter to exert a strong enough gravitational force to reverse the motion of receding galaxies. (Ch. 18)

cluster A collection of galaxies. (Ch. 18)

COBE See *Cosmic Background Explorer*.

codon The set of three bases on mRNA that determines which of the possible tRNA molecules will attach at that point, and which amino acid will appear in a protein. (Ch. 22)

cold-blooded Animals, such as amphibians and reptiles, that must absorb heat from their environment to maintain body temperature. (Ch. 24)

combustion A rapid combination with oxygen, producing heat and flame. (Ch. 9)

comet   An object, usually found outside the orbit of Pluto, composed of chunks of materials such as water ice and methane ice embedded with dirt. A comet may fall toward the Sun, if its distant orbit is disturbed, and create a spectacular display in the night sky. (Ch. 14)

compass   A needle-shaped magnet designed to point at the poles of the Earth's magnetic field. (Ch. 5)

composite material   A combination of two or more substances in which the strength of one of the constituents is used to offset the weakness of another, resulting in a new material whose strengths are greater than any of its components; for example, plywood and reinforced concrete. (Ch. 10)

compressional wave   One of two principal types of seismic waves, in which the molecules in rock move back and forth in the same direction as the wave; a longitudinal wave. (Ch. 15)

compressive strength   A material's ability to withstand crushing. (Ch. 10)

computer   A machine that stores and manipulates information. (Ch. 10)

condensation   A change of state from gas to liquid caused by a decrease in temperature or pressure of the gas, which slows down the vibration of individual molecules of the gas, allowing them to form a liquid. (Ch. 9)

condensation polymerization   A chemical reaction often used to manufacture plastics and other polymers, and which, in the body, occurs during the formation of a peptide bond. (Ch. 9)

condensation reaction   The formation of a polymer in which each new polymer bond releases a water molecule as the ends of the original polymer molecules link up. (Ch. 9)

conduction   The movement of heat by collisions between vibrating atoms or molecules; one of three mechanisms by which heat moves. (Ch. 4)

conduction electron   An electron in a material that is able to move in an electric field. (Ch. 10)

conductor   A material capable of carrying an electric current; any material through which electrons can flow freely. (Ch. 10)

cone   A light-absorbing cell in the eye, which is sensitive to red, blue, or green light, enabling color vision. Compare to *rod*. (Ch. 6)

conservation law   Any statement that says that a quantity in nature does not change. (Ch. 3)

constructive interference   A situation in which two waves act together to reinforce or maximize the wave height at the point of intersection. (Ch. 6)

convection   The transfer of heat by the physical motion of masses of fluid. Dense, cooler fluids (liquids and gases) descend in bulk and displace rising warmer fluids, which are less dense. One of three mechanisms by which heat moves. (Ch. 4)

convection cell   A region in a fluid in which heat is continuously being transferred by a bulk motion of heated fluid from a heat source to the surface of the fluid, where heat is released. The cooled fluid then sinks and the cycle repeats. (Ch. 4)

convection zone   The outer region of the Sun, comprising the upper 200,000 km (about 125,000 mi) where the dominant energy transfer mechanism changes from collision to convection. (Ch. 17)

convergent plate boundary   A place where two plates are coming together. (Ch. 15)

core   (*a*) In geology, the heaviest elements of the Earth's mass, primarily iron and nickel, concentrated at the center with a radius of about 3400 km (2000 mi). (*b*) In astronomy, a small region in the center of a star where hydrogen burning is generally confined. (Ch. 14, Ch. 17)

corona   One of the Sun's outer layers, visible for a few minutes as a spectacular halo during a total eclipse of the Sun. (Ch. 17)

Cosmic Background Explorer (COBE)   An orbiting observatory that measured the presence of microwave radiation present as background noise in every direction of the sky. (Ch. 17)

cosmic microwave background   Microwave radiation, characteristic of a body at about 3K, falling to Earth from all directions. This radiation is evidence for the big bang. (Ch. 18)

cosmic rays   Particles (mostly protons) that rain down continuously on the atmosphere of the Earth after being emitted by stars in our galaxy and in others. (Ch. 12)

cosmology   The branch of science that is devoted to the study of the structure and history of the entire universe. (Ch. 18)

coulomb   (pronounced "*koo*-loam") The unit for measuring the magnitude of an electric charge. (Ch. 5)

Coulomb's law   An empirically derived rule that states that the magnitude of the electrostatic force between any two objects is proportional to the charges of the two objects, and inversely proportional to the square of the distance between them. (Ch. 5)

covalent bond   A chemical bond in which neighboring molecules share electrons in a strongly bonded group of at least two atoms. (Ch. 9)

cpu   See *central processing unit*.

critical mass   The minimum number of uranium-235 atoms needed to sustain a nuclear chain reaction to the point where large amounts of energy can be released. (Ch. 11)

crust   A thin layer at the surface of the Earth formed from the lightest elements, ranging in thickness from 10 km (6 mi) in parts of the ocean to 70 km (45 mi) beneath parts of the continents. (Ch. 14)

crystal   A group of atoms that occur in a regularly repeating sequence. Crystal structure is described by first determining the size and shape of the repeating boxlike group of atoms and then recording the exact type and position of every atom that appears in the box. (Ch. 9)

current   A river of moving water on an ocean's surface, found in each of the ocean basins. (Ch. 16)

cyclone   A great rotational pattern in the atmosphere, hundreds of kilometers in diameter, that can draw energy from warm oceanic waters and create low-pressure tropical storms. (Ch. 16)

cyclotron   The first of the particle accelerators, for which Ernest Lawrence won the 1939 Nobel prize in physics. (Ch. 12)

cytoplasm   The fluid that takes up the spaces between the organelles of a cell. (Ch. 21)

cytoskeleton   A series of protein filaments that extend throughout the cell, giving the cell a shape and, in some cases, allowing it to move. (Ch. 21)

dark matter   Material that exists in forms that does not interact with electromagnetic radiation and that may constitute 90% of the matter of the universe. (Ch. 18)

DC   See *direct current.*

decay chain   A series of decays, or radioactive events, ending with a stable isotope. (Ch. 11)

deep ocean trench   A surface feature associated with convergent plate boundaries in which no continents are on the leading edge of either of the two converging plates and one plate penetrates deep into the Earth. (Ch. 15)

degeneracy pressure   A permanent outward force, associated with the Pauli exclusion principle, exerted by electrons or neutrons that cancels the inward force of gravity in some stars. (Ch. 17)

dendrite   One of a thousand projections on each nerve cell in the brain through which nerve signals move; each is connected to different neighboring nerve cells in the brain. (Ch. 10)

deoxyribonucleic acid (DNA)   A strand of nucleotides with alternating phosphate and sugar molecules in a long chain, and with base molecules adenine, guanine, cytosine, and thiamine at the side. The nucleotide strand bonds with a second nucleotide strand to make a molecule with a ladder-like double-helix shape. DNA stores the genetic information in a cell. (Ch. 20)

depolymerization   The breakdown of a polymer into short segments. (Ch. 9)

destructive interference   A situation in which two waves intersect in a way that decreases or cancels out the wave height at the point of intersection. (Ch. 6)

diatomic   The simplest molecules, containing two atoms of the same element, such as the diatomic gases hydrogen (H_2), nitrogen (N_2), and oxygen (O_2). (Ch. 9)

differentiation   The process by which heavy, dense materials (such as iron and nickel) sank under the force of gravity toward the molten center of the planet, while lighter, less dense materials floated to the top, resulting in the layered structure of the present-day Earth. (Ch. 14)

diffraction   A special kind of scattering of electromagnetic waves in which the waves are strongly reinforced in certain directions by the process of interference; the bending of a wave around an obstacle or the edges of an opening. (Ch. 6)

diffuse scattering   A process by which light waves are absorbed and re-emitted in all directions by a medium such as clouds or snow. (Ch. 6)

diffusion   The transfer of molecules from regions of high concentration by ordinary random thermal motion. (Ch. 21)

digestive system   The parts of the body that are responsible for breaking down food so that its stored energy can be used by cells; includes the stomach, small intestine, liver, pancreas, and gall bladder. (Ch. 24)

diode   An electronic device that allows electric current to flow in only one direction. (Ch. 10)

dipole field   The magnetic field that arises from the two poles of a magnet. (Ch. 5)

direct current (DC)   A type of electric current in which the electrons flow in one direction only; for example, in the chemical reaction of a battery. (Ch. 5)

distillation   A process by which engineers separate the complex mixture of petroleum's organic chemicals into much purer fractions. (Ch. 9)

divergent plate boundary   A spreading zone of crustal formation; a place where neighboring plates move away from each other. (Ch. 15)

Divine Calculator   An eighteenth-century idea, proposed by Pierre Simon Laplace, which stated that if the position and velocity of every atom in the universe is known, with infinite computational power, the future position and velocity of every atom in the universe could be predicted. (Ch. 2)

DNA   See *deoxyribonucleic acid.*

DNA mapping   The process of finding the location of genes on chromosomes. (Ch. 22)

DNA sequencing   The process of determining, base pair by base pair, the exact order of bases along a specific stretch of a DNA molecule. (Ch. 22)

domain   Region in magnetic material where neighboring atoms line up with each other to give a strong magnetic field. (Ch. 10)

dominant trait   A genetic characteristic that always appears, or is expressed. (Ch. 22)

doping The addition of a minor impurity to a semiconductor. (Ch. 10)

Doppler effect The change in frequency or wavelength of a wave detected by an observer because the source of the wave is moving. (Ch. 6)

double bond The type of covalent bond formed when two electrons are shared by two atoms. (Ch. 9)

double helix The twisted double strand of nucleotides that forms the structure of the DNA molecule. (Ch. 20)

ear A sense organ that includes a membrane that vibrates at the arrival of sound waves. (Ch. 24)

earthquake Disturbance caused when stressed rock on Earth suddenly snaps, converting potential energy into released kinetic energy. (Ch. 15)

ecological niche The habitat, functional role(s), requirements for environmental resources, and tolerance ranges within an ecosystem. (Ch. 25)

ecology The branch of science that studies interactions among organisms as well as the interactions of organisms and their environment. (Ch. 25)

ecosystem Interdependent collections of living things; includes the plants and animals that live in a given area together with their physical surroundings. (See *open ecosystem* and *closed ecosystem*.) (Ch. 25)

efficiency The amount of work you get from an engine, divided by the amount of energy you put in; a quantification of the loss of useful energy. (Ch. 4)

El Niño A weather cycle in the Pacific basin that recurs every four to seven years and can cause severe storms and flooding all along the western coast of the Americas, as well as drought from Australia to India. (Ch. 16)

elastic limit The point at which a material stops resisting external forces and begins to deform permanently. (Ch. 10)

elastic potential energy The type of energy that is stored in a flexed muscle, a coiled spring, and a stretched rubber band. (Ch. 3)

electric circuit An unbroken path of material that carries electricity and consists of three parts: a source of energy, a closed path, and a device to use the energy. (Ch. 5)

electric current A flow of charged particles, measured in amperes. (Ch. 5)

electric field The force that would be exerted on a positive charge at a position near a charged object. Every charged object is surrounded by an electric field. (Ch. 5)

electric generator A source of energy producing an alternating current in an electric circuit through the use of electromagnetic induction. (Ch. 5)

electric motor A device that operates by supplying current to an electromagnet to make the magnet move and generate mechanical power. The motor employs permanent magnets and rotating loops of wire inside the poles of this magnet. (Ch. 5)

electrical charge An excess or deficit of electrons on an object. (Ch. 5)

electrical conductivity The ease with which a material allows electrons to flow. The inverse of electrical resistance. (Ch. 10)

electrical potential energy The type of energy that is found in a battery or between two wires at different voltages; energy associated with the position of a charge in an electric field. (Ch. 3)

electrical resistance The quantity, measured in ohms, that represents how hard it is to push electrons through a material. High-resistance wires are used when electron energy is to be converted into heat energy. Low-resistance wires are used when energy is to be transmitted from one place to another with minimum loss. (Ch. 5)

electricity A force, more powerful than gravity, that moves objects both toward and away from each other, depending upon the charge. (Ch. 5)

electroweak force A force resulting from the unification of the electromagnetic and weak forces. (Ch. 12)

electromagnet A device that produces a magnetic field from a moving electrical charge. (Ch. 5)

electromagnetic induction A process by which a changing magnetic field produces an electric current in a conductor, even though there is no other source of power available. (Ch. 5)

electromagnetic radiation See *electromagnetic wave*.

electromagnetic spectrum The entire array of waves, varying in frequency and wavelength, but all resulting from an accelerating electrical charge; includes radio waves, microwaves, infrared, visible light, ultraviolet, X-rays, gamma rays, and others. (Ch. 6)

electromagnetic wave A form of radiant energy that reacts with matter by being transmitted, absorbed, or scattered. A self-propagating wave made up of electric and magnetic fields fluctuating together. A wave created when electrical charges accelerate, but requiring no medium for transfer. Electromagnetic radiation. (Ch. 6)

electromagnetism A term used to refer to the unified nature of electricity and magnetism. (Ch. 5)

electron Tiny, negatively charged particles that circle in orbits around a positively charged nucleus of an atom. (Ch. 5)

electron microscope An instrument, introduced in the 1930s, that was a major new advance in microscopes because it used electrons instead of light to illuminate objects and had resolving power up to

100,000 times that of the optical microscope. (Ch. 21)

electron shell   A specific orbit in an atom that can be filled with a predetermined number of electrons. (Ch. 7)

electrostatic   The type of electrical charge that doesn't move once it has been placed on an object, and the forces exerted by such a charge. (Ch. 5)

element   A material made from a single type of atom, which cannot be broken down any further. (Ch. 7)

elementary-particle physics   The study of particles that are believed to comprise the basic building blocks of the universe; for example, the particles that make up the nucleus, and particles such as electrons. Also known as high-energy physics. (Ch. 12)

elementary particles   Particles that make up the nucleus, together with particles such as the electron; the basic building blocks of the universe. (Ch. 12)

ELF radiation   Extremely low-frequency waves associated with the movement of electrons to produce the alternating current in household wires. (Ch. 6)

emission spectrum   The characteristic set of lines used to identify a chemical element or molecule from the total collection of photons emitted during quantum leap. (Ch. 7)

emitter   The region in a transistor that first receives an electrical current. (Ch. 10)

endocrine system   A group of glands that secrete hormones that are taken by the bloodstream to produce specific chemical effects; part of the body's control system. (Ch. 24)

endoplasmic reticulum   One of the cellular organelles that contribute to protein and lipid synthesis. (Ch. 21)

endothermic   A chemical reaction in which the final energy of the electrons in the reaction is greater than the initial energy; energy must be supplied to make the reaction proceed. (Ch. 9)

energy   The ability to do work; the capacity to exert a force over a distance. A system's energy can be measured in joules or foot-pounds. (Ch. 3)

entropy   The thermodynamic quantity that describes the degree of randomness of a system. The greater the disorder or randomness, the higher the statistical probability of the state, and the higher the entropy. (Ch. 4)

enzyme   A molecule that facilitates reactions between two other molecules, but which is not itself altered or taken up in the overall reaction. (Ch. 20)

eukaryote   An advanced single-celled organism and all multicelled organisms that are made from cells containing a nucleus. (Ch. 21)

excited state   All energy levels of an atom above the ground state. (Ch. 7)

exothermic   A chemical reaction in which the final energy of the electrons is less than the initial energy,

and therefore energy is given off in some form. (Ch. 9)

experiment   The manipulation of some aspect of nature to observe an outcome. (Ch. 1)

extinction   The disappearance of a species on Earth. (Ch. 23)

eye   The most important of the five sense organs, through which human beings become aware of their environment. (Ch. 24)

family   A grouping of similar genera in the Linnaean classification system; humans are in the family of hominids. (Ch. 19)

fault   A fracture in a rock along which movement occurs. (Ch. 15)

fermentation   An anaerobic cellular process in which pyruvic acids are broken down and the energy is used by the cell to keep glycolysis going. (Ch. 21)

ferromagnetism   The property of a few materials in nature, such as iron, cobalt, and nickel metals, in which the individual atomic magnets are arranged in a nonrandom manner, lined up with each other into small magnetic domains. (Ch. 10)

field   The force—magnetic, gravitational, or electric that would be felt at a particular point. For example, forces exerted by one object that would be felt by another object in the same region. (Ch. 5)

first law of thermodynamics   The law of the conservation of energy. In an isolated system, the total amount of energy, including heat energy, is conserved. (Ch. 3)

first trophic level   All plants that produce energy from photosynthesis. (Ch. 25)

fission   A reaction that produces energy when heavy radioactive nuclei split apart into fragments that together have less mass than the original isotopes. (Ch. 11)

flat universe   A model of the future of the universe in which the expansion slows and comes to a halt after infinite time has passed. (Ch. 18)

fluorescence   A phenomenon in which energy contained in ultraviolet wavelengths is absorbed by the atoms in some materials and partially emitted as visible light, or "black light." (Ch. 6)

FM   See *frequency modulation.*

foot-pound   The amount of work done by a force of one pound acting through one foot; the unit of energy in the English system. (Ch. 3)

force   A push or pull that, acting alone, causes a change in acceleration of the object on which it acts. (Ch. 2)

fossil   A replica in stone of an organism, created when calcium and other atoms in the hard parts of the buried organism are replaced by minerals in the water flowing through the surrounding area. (Ch. 23)

fossil fuel Carbon-rich deposits of ancient life that burn with a hot flame and have been the most important energy source for 150 years. Examples include coal, oil, and natural gas. (Ch. 3)

fossil record A term that refers to all of the fossils that have been found, catalogued, and studied since human beings first began to study them in a systematic way. (Ch. 23)

frames of reference The physical surroundings from which a person observes and measures the world. (Ch. 13)

freezing A change of state from liquid to solid caused by a decrease in temperature or pressure of the liquid, which slows the vibration of individual molecules and forms the solid structure. (Ch. 9)

frequency The number of wave crests that go by a given point every second. A wave completing one cycle (sending one crest by a point every second) has a frequency of one hertz, 1 Hz. (Ch. 6)

frequency modulation (FM) A process by which information is transmitted by varying the frequency of a signal. After being transmitted, the signal may be converted to sound by the circuits in the receiver. (Ch. 6)

fungi One of the five kingdoms in the Linnaean classification. Organisms that obtain energy by absorbing materials through filaments and reproduce by production of spores. May be single-celled (e.g., yeasts) or multicellular (e.g., mushrooms). (Chs. 19, 24)

fusion A process in which two nuclei come together to form a third, larger nucleus. When this reaction combines light elements to make heavier ones, the mass of the final nucleus may be less than the mass of its constituent parts. The "missing" nuclear mass can be converted into energy. (Ch. 11)

G The constant of proportionality in Newton's law of universal gravitation. (Ch. 2)

g See *gravitational constant.*

galaxy A large assembly of stars (between millions and hundreds of billions), together with gas, dust, and other materials, that is held together by the forces of mutual gravitational attraction. (Ch. 18)

gall bladder Part of the digestive system that secretes specialized substances that break down starches, carbohydrates, proteins, and fats. (Ch. 24)

gamma radiation A kind of radioactivity involving the emission of energetic electromagnetic radiation from the nucleus of an atom, with no change in the number of protons or neutrons in the atom. (Ch. 11)

gamma ray The highest-energy wave of the electromagnetic spectrum with wavelengths less than the size of an atom, less than one trillionth of a meter; normally emitted in very high-energy nuclear particle reactions. (Ch. 6)

Gamma Ray Observatory (GRO) An orbiting observatory that detects the highest-energy end of the electromagnetic spectrum, gamma rays. It was one of the first permanent orbiting observatories launched by NASA's Great Observatories Program for monitoring all parts of the electromagnetic spectrum. (Ch. 17)

gas Any collection of atoms or molecules that expands to take the shape and fill the volume available in its container. (Ch. 9)

gauge particles Particles whose exchange produces the fundamental forces that hold everything together; corresponds to every force between two particles. (Ch. 12)

GCM See *global circulation models.*

gene A unit of biological inheritance, or a section of a long molecule of DNA. One gene carries the information needed to assemble one protein. (Ch. 22)

general relativity The second and more complex of two parts of Einstein's theory of relativity, which applies to any reference frames whether or not those frames are accelerating relative to each other. (Ch. 13)

genetic code The correspondence between base–pair sequences and amino acids. The connection, in all living things, between the codons and the amino acid for which they code. (Ch. 22)

genetic disease A hereditary mutation that can cause sickness or death; for example, hemophilia. (Ch. 22)

genetic engineering A technology in which foreign genes are inserted into an organism, or existing genes altered, to modify the function of living things. (Ch. 22)

genetics The study of ways in which biological information is passed from one generation to the next. (Ch. 22)

genome The sum of all information contained in the DNA for any living thing; the sequence of all the bases in all the chromosomes. (Ch. 22)

genus A grouping of similar species in the Linnaean classification system; humans are in the genus Homo. (Ch. 19)

glacier Large body of ice that slowly flows down a slope or valley under the influence of gravity; found primarily in Greenland and Antarctica. (Ch. 16)

glass A solid with predictable local environments for most atoms, but no long-range order to the atomic structure. Compared to crystal, glass lacks the repeating unit of atoms. (Ch. 9)

global circulation models (GCM) Complex computer models of the atmosphere that are the best attempts to date to predict long-term climate and to discuss various types of ecological changes such as global warming. (Ch. 16)

glucose An important sugar ($C_6H_{12}O_6$) in the energy cycle of living things, figuring prominently in the energy metabolism of every living cell. (Ch. 20)

gluon A massless particle, confined to the interior of particles, that mediates the force holding the quarks together. (Ch. 12)

glycolysis The first step in the extraction of energy from glucose, which takes place in nine separate steps, each of which is governed by a specific enzyme, and which splits each glucose molecule into two smaller molecules called pyruvic acids. (Ch. 21)

gneiss The metamorphic rock formed from slate under extreme temperature and pressure. (Ch. 16)

Golgi apparatus One of the cellular organelles that takes part in the synthesis of molecules. (Ch. 21)

gradualism A hypothesis which holds that most evolutionary change occurs as a result of the accumulation of small adaptations. (Ch. 23)

granite A rock that is lower density than the mantle rock it caps, and which forms much of the continents. (Ch. 15)

gravitational constant (g) A constant numerical value for the specific acceleration that all objects experience at the Earth's surface, determined by measuring the actual fall rate of objects in a laboratory. It is equal to $9.8 \text{ m/s}^2 = 32 \text{ feet/s}^2$. (Ch. 2)

gravitational escape One way that a planet's atmosphere can evolve and change. Molecules in the atmosphere heated by the Sun may move sufficiently fast so that appreciable fractions of them can actually escape the gravitational pull of their planet. (Ch. 14)

gravitational potential energy Energy associated with the position of a mass in a gravitational field. The gravitational potential energy of an object on the surface of the earth equals its weight (the force of gravity exerted by the object) times its height above the ground. (Ch. 3)

graviton The gauge particle of gravity. (Ch. 12)

gravity An attractive force that acts on every object in the universe. (Ch. 2)

great bombardment An event following the initial period of planetary formation in which meteorites showered down on planets, adding matter and heat energy. (Ch. 14)

greenhouse effect A global temperature increase caused by the fact that the Earth's atmospheric gases trap some of the Sun's infrared (heat) energy before it radiates out into space. (Ch. 25)

GRO See *Gamma Ray Observatory*.

ground state The lowest energy level of an atom. (Ch. 7)

groundwater Fresh water from the surface, which typically percolates into the ground and fills the tiny spaces between grains of sandstone and other porous rock layers. (Ch. 16)

gymnosperm The class of vascular plants that produce seeds without flowers, such as fir trees. (Ch. 24)

gyre A circulation of water at the surface of the ocean, transporting warm water from the equator toward the cooler poles, and cold water from the poles back to the equator to be heated and cycled again. (Ch. 16)

hadron Particles, including the proton and neutron, that are made from quarks and are subject to the strong force. (Ch. 12)

half-life The rate of radioactive decay measured by the time it takes for half of a collection of isotopes to decay into another element. (Ch. 11)

heart Part of an animal's circulatory system which, in birds and mammals, has four separate chambers. (Ch. 24)

heat (thermal energy) A measure of the quantity of atomic kinetic energy contained in every object. (Ch. 3)

heat capacity A measure of the ability of a material to absorb heat energy, defined as the quantity of heat required to raise the temperature of one gram of that material by 1°C. Water displays the largest heat capacity of any common substance. (Ch. 4)

heat transfer The process by which heat moves from one place to another, through three different mechanisms: conduction, convection, or radiation. (Ch. 4)

helium burning The final energy-producing stage of a star in which the temperature in the interior becomes so hot that the helium itself begins to undergo nuclear fusion reactions to make carbon. The net reaction is: $^4\text{He} + {}^4\text{He} + {}^4\text{He} \rightarrow {}^{12}\text{C}$. (Ch. 17)

hertz (Hz) The unit of measurement for the frequency of waves; one wave cycle per second. (Ch. 6)

Hertzsprung-Russell (H-R) diagram A simple graphical technique widely used in astronomy to plot a star's temperature (determined by its spectrum) versus the star's energy output (measured by its energy and brightness). (Ch. 17)

high-energy physics (See *elementary-particle physics*.)

high-grade energy Sources of energy that can be used to produce very high-temperature reservoirs; for example, petroleum and coal. (Ch. 4)

high-quality protein Foods that supply amino acids in roughly the same proportion as those in human proteins, such as meat and dairy products. (Ch. 20)

high-temperature reservoir Any hot object from which energy is extracted to do work. Within the cylinder of a gasoline engine is a high-temperature reservoir. (Ch. 4)

hole   The absence of an electron; in a silicon crystal, for example, a hole is left behind after the conduction electron is shaken loose. (Ch. 10)

homeostasis   The process the body uses to maintain equilibrium in its internal environment. (Ch. 24)

hominid   The family of the order primate whose members walk erect; includes humans, which are the only hominids that are not extinct. (Ch. 19)

Homo erectus   ("man the erect") The species of modern human's genus who first walked erect and learned to use fire; disappeared about 500,000 years ago. (Ch. 23)

Homo habilis   The first member of the genus of modern humans who appeared about 2 million years ago in East Africa; distinguished by a larger brain and stone tools. (Ch. 23)

Homo sapiens   The single species that includes all branches of the human race; recognized in fossils as old as 200,000 years. (Ch. 19)

hormones   Specific molecules that are secreted by the glands of the endocrine system to regulate body chemistry. (Ch. 24)

horsepower   Unit of power equal to 550 foot-pounds per second in the English system of measurement; commonly used to assess the power of engines and motors. (Ch. 3)

hot spot   A dramatic type of volcanism indirectly associated with plate tectonics. Large isolated chimney-like columns of hot rock, or mantle plumes, rising to the surface in certain places of the Earth; for example, Yellowstone National Park, Iceland, and Hawaii. (Ch. 15)

HST   See *Hubble Space Telescope.*

Hubble Space Telescope (HST)   A reflecting telescope, launched in 1990, with a 2.4-meter mirror designed to give unparalleled resolution in the visible and ultraviolet wavelengths. Manufacturing flaws in the main mirror were corrected by astronauts in late 1993. (Ch. 17)

Hubble's law   The law relating the distance to a galaxy, D, and the rate at which it recedes from Earth as measured by the redshift: $V = HD$. (Ch. 18)

Human Genome Project   A large-scale scientific project that will result in a complete knowledge of the entire human genome, which includes 46 chromosomes and 3 billion base pairs. (Ch. 22)

hurricane   Tropical storms having winds in excess of 120 km/h (75 mi/h) that begin in the Atlantic Ocean off the coast of Africa and affect North America. (Ch. 16)

hybrid   An individual whose parents possess different genetic traits. (Ch. 22)

hydrocarbon   A chainlike molecule from a chemical compound of carbon and hydrogen, which provides the most efficient fuels for combustion, with only carbon dioxide and water as products. (Ch. 9)

hydrogen bond   A bond that may form when polarized hydrogen atoms link to other atoms by a covalent or ionic bond. (Ch. 9)

hydrogen burning   A three-step process, generally confined to a small region in the center of a star, in which four protons are converted into a ^{4}He nucleus, two protons, and a photon. (Ch. 17)

hydrogenation   The addition of hydrogen atoms into the carbon chains of polyunsaturated products; a process that eliminates the carbon-carbon double bonds (Ch. 20)

hydrological cycle   The combination of processes by which water moves from repository to repository near the Earth's surface. (Ch. 16)

hydrophilic   Attracted to water. (Ch. 20)

hydrophobic   Repelled by water. (Ch. 20)

hypothesis   A tentative guess about how the world works, based on a summary of experimental or observational results and phrased so that it can be tested by experimentation. (Ch. 1)

Hz   See *hertz.*

ice age   A period of several million years during which glaciers have repeatedly advanced and retreated, causing radical changes in climate and influencing human evolution. (Ch. 16)

ice cap   Layers of ice that form at the north and south polar regions of the Earth. (Ch. 16)

igneous rock   The first rock to form on a cooling planet, solidified from hot, molten material; intrusive or extrusive (volcanic). (Ch. 16)

immune system   The system that defends an organism against harmful microorganisms by recognizing the geometric shape of molecules on foreign invaders and destroying them without harming the body cells themselves. (Ch. 21)

inertia   The tendency of a body to remain in uniform motion; the resistance to change. (Ch. 2)

inflation   A short period of rapid expansion of the universe, which, according to the grand unified theories, accompanied the "freezing" at 10^{-35} seconds. (Ch. 18)

inflationary theories   Those theories that incorporate the phenomenon of inflation. (Ch. 18)

Infrared Astronomical Satellite (IRAS)   An orbiting observatory launched in 1983 by the United States, United Kingdom, and Netherlands to view infrared radiation in the universe. IRAS is no longer functioning. (Ch. 17)

infrared energy   A form of electromagnetic radiation that travels from a source to an object, where it can be absorbed and converted into the kinetic energy of molecules. (Ch. 4)

infrared radiation Wavelengths of electromagnetic radiation that extend from a millimeter to micron; felt as heat radiation. (Ch. 6)

insulator A material that will not conduct electricity. (Ch. 10)

integrated circuit A microchip made of hundreds or thousands of transistors specially designed to perform a specific function. (Ch. 10)

interference When waves from two different sources come together at a single point, they interfere with each other. The observed wave amplitude is the sum of the amplitudes of the interfering waves. (Ch. 6)

intrusive Igneous rock that cools and hardens underground. (Ch. 16)

inversely proportional The relationship between two variables such that if the value of one variable increases, the other variable decreases, and vice versa, by a constant proportion. (Ch. 2)

invertebrates Organisms without backbones. (Ch. 24)

ion An atom that has an electrical charge, from either the loss or gain of an electron. (Ch. 7)

ionic bond A chemical bond in which the electrostatic force between two oppositely charged ions holds the atoms in place, often formed as one atom gives up an electron while another receives it, lowering chemical potential energy when atom shells are filled. (Ch. 9)

ionization Stripping away one or more of an atom's electrons to produce an ion. (Ch. 11)

IRAS See *Infrared Astronomical Satellite*.

isolated system A type of system in which matter and energy are not exchanged with the surroundings; a closed system. (Ch. 3)

isomer A molecule that contains the same atoms as another molecule, but has a different structural arrangement. (Ch. 9)

isotopes Atoms whose nuclei have the same number of protons but a different number of neutrons. (Ch. 11)

jet stream A high-altitude stream of fast-moving winds that marks the boundary between the northern polar cold air mass and the warmer air of the temperate zone. (Ch. 16)

joule The amount of work done when you exert a force of one newton through a distance of one meter. (Ch. 3)

Jovian planets Huge worlds also known as "gas giants" located in the outer solar system and made up primarily of frozen liquids and gases such as hydrogen, helium, ammonia, and water, with atmospheres of nitrogen, methane, and other compounds: Jupiter, Saturn, Uranus, Neptune. (Ch. 14)

kidneys The pair of organs in the body to which waste products from the metabolism of the cells are carried by the blood and through which the blood is filtered, with selective materials being reabsorbed. (Ch. 24)

kilowatt A commonly used measurement of electrical power equal to 1000 watts and corresponding to the expenditure of 1000 joules per second. (Ch. 3)

kinetic energy The type of energy associated with moving objects; the energy of motion. Kinetic energy is equal to the mass of the moving object times the square of that object's velocity, multiplied by $1/2$. (Ch. 3)

kingdom The broadest classification in the Linnaean classification system, corresponding to the coarsest division of living things. (Ch. 19)

Krebs cycle A complex series of chemical cellular reactions in which the the products of glycolysis are broken down completely into carbon dioxide and water, releasing some energy to the ATP molecules and storing some in other energy-carrying molecules. (Ch. 21)

large intestine An organ in the body where water is removed and feces are formed and, ultimately, voided. (Ch. 24)

laser An instrument that uses a collection of atoms, energy, and mirrors to emit photons that have wave crests in exact alignment. The instrument name is the acronym for *l*ight *a*mplification by *s*timulated *e*mission of *r*adiation. (Ch. 7)

law of nature An overarching statement of how the universe works, following repeated and rigorous observation and testing of a theory or group of related theories. (Ch. 1)

law of unintended consequences A phenomenon demonstrating the interdependent nature of ecosystems: It is virtually impossible to change one aspect of an ecosystem without affecting something else, often inadvertently. (Ch. 25)

length contraction The phenomenon in relativity in which moving objects appear to be shorter than stationary ones in the direction of motion. (Ch. 13)

lepton A particle (such as the electron, muon, and neutrino) that participates in the weak and electromagnetic, but not the strong, interaction (Ch. 12)

lichen (pronounced "*lie* kin") A combination of a fungus and a single-celled organism that can use the Sun's energy through photosynthesis; important to the processes of weathering rock and creating soil (Ch. 24)

ligaments The tough tissues that hold bones together at joints. (Ch. 24)

light   A form of electromagnetic wave to which the human eye is sensitive. Light travels at a constant speed and needs no medium for transfer. (Ch. 6)

light-year   The distance light travels in one year, 10 trillion kilometers (about 6.2 trillion miles). (Ch. 17)

limestone   A sedimentary rock formed from the calcium carbonate ($CaCO_3$) skeletons of sea animals, shells, and coral. (Ch. 16)

linear accelerator   A device for making high-velocity particles, which relies on a long, straight vacuum tube into which particles are injected to ride an electromagnetic wave down the tube. (Ch. 12)

Linnaean classification   A systematic attempt by Swedish naturalist Carolus Linnaeus to catalogue the diversity of all living things according to their shared characteristics so that each organism is as close as possible to those things it resembles, and as far apart as possible from those it does not. (Ch. 19)

lipid   An organic molecule that is insoluble in water. At the molecular level, lipids form the cell membranes that separate living material from its environment. Lipids are also an extremely efficient storage medium for energy; for example, fat in foods, wax in candles, and grease for lubrication. (Ch. 20)

liquid   Any collection of atoms or molecules that has no fixed shape but maintains a fixed volume. (Ch. 9)

liquid crystal   A recently synthesized substance, used in digital displays, which is formed from very long molecules that may adopt a very ordered arrangement even in the liquid form. (Ch. 9)

liver   Part of the digestive system that secretes specialized substances that break down starches, carbohydrates, proteins, and fats. (Ch. 24)

load   The location in an electric circuit where the useful work is done, such as the filament in a light bulb or the heating element of a dryer. (Ch. 5)

longitudinal wave   A kind of wave in which the motion of the medium is in the same direction as the wave movement; pressure wave or sound wave. (Ch. 6) Also, one of two principal types of seismic waves in which molecules in the rock move back and forth in the same direction as the wave; a compressional wave. (Ch. 15)

Lorentz factor   A number, equal to the square root of $[1-(v/c)^2]$, that appears in relativistic calculations and is an indication of the magnitude of change in time and scale. (Ch. 13)

low-quality protein   Foods, such as that from plants, which lack one or more of the amino acids found in human proteins. (Ch. 20)

low-temperature reservoir   The ambient atmosphere into which the waste heat generated by an engine is dumped; for example, from a cylinder in a gasoline engine to the atmosphere. (Ch. 4)

Lucy   The name given to a nearly complete skeleton of a female *Australopithecus*, found in Ethiopia in 1974. (Ch. 23)

luminosity   The total energy produced by a star. (Ch. 17)

lymph nodes   Part of the lymphatic system. (Ch. 24)

lymphatic system   An extensive network of capillaries and veins, parallel to the blood system and linked to about 500 lymph nodes in the human body. (Ch. 24)

lymphocyte   The main type of working cells of the immune system, designated B and T. (Ch. 21) Also, a white blood cell important in protecting against infections. (Ch. 24)

lysosome   One of the cellular organelles that has digestion and breakdown of wastes as its primary function. (Ch. 21)

magma   Subsurface molten rock, concentrated in the upper mantle or lower crust, which can breach the surface and harden into new rock. (Ch. 15)

magnetic field   A collection of lines that map out the direction that compass needles would point in the vicinity of a magnet. (Ch. 5)

magnetic force   The force exerted by magnets on each other. (Ch. 5)

magnetic monopole   A hypothetical single isolated north or south magnetic pole, existing in theory but not yet located through experimentation. (Ch. 5)

magnetic potential energy   The type of energy stored in a magnetic field. (Ch. 3)

magnetism   A fundamental force in the universe. (Ch. 5)

main-sequence star   A star that derives energy from the fusion reactions of hydrogen burning; found on the Hertzsprung-Russell (H-R) diagram within a bandlike pattern. (Ch. 17)

mammal   One of a group of vertebrates made up of individuals that are warm-blooded, have hair, and whose females nurse their young. Human beings are mammals. (Ch. 19)

mantle   The thick layer rich in oxygen, silicon, magnesium, and iron that contains most of the Earth's mass; it overlies the metal core of the Earth. (Ch. 14)

mantle convection   A force deep within the Earth, driven by internal heat energy, that moves continents and the plates of which they are a part. (Ch. 15)

marble   A metamorphic rock that begins as limestone that is subjected to intense pressure and high temperatures. (Ch. 16)

mass   The amount of matter contained in an object, independent of where that object is found. (Ch. 2)

mass extinctions   Rare and catastrophic events in the past that have caused large numbers of species to become extinct suddenly. (Ch. 23)

mass number The number of neutrons plus the number of protons, which determines the mass of an isotope. (Ch. 11)

Maxwell's equations Four fundamental laws of electricity and magnetism: (1) Coulomb's law, like charges repel and unlike charges attract; (2) magnetic monopoles do not exist in nature; (3) magnetic phenomena can be produced by electrical effects; and (4) electrical phenomena can be produced by magnetic effects. (Ch. 5)

mechanics The branch of science that deals with the motions of material objects and the forces that act on them; for example, a rolling rock or a thrown ball. (Ch. 2)

meiosis The division process that produces cells with one-half the number of chromosomes in each somatic cell. Each resulting daughter cell has half the normal complement of DNA. (See *mitosis*.) (Ch. 21)

meltdown The most serious accident that can occur at a nuclear reactor, in which the flow of water to the fuel rods is interrupted and the enormous heat stored in the central part of the reactor causes the fuel rods to melt. (Ch. 11)

melting A change of state from solid to liquid caused by an increase in temperature or pressure of the solid, which increases the vibration of individual molecules and breaks down the structure of the solid. (Ch. 9)

messenger RNA (mRNA) The single-stranded molecule that copies the sequence for one gene and carries that DNA information to the region of the cell where proteins are made. (Ch. 22)

metabolism The process by which a cell derives energy from its surroundings. (Ch. 21)

metal An element or combination of elements in which the sharing of a few electrons among all atoms results in a more stable electron arrangement; characterized by a shiny luster and ability to conduct electricity. (Ch. 9)

metallic bond A chemical bond in which electrons are redistributed so that they are shared by all the atoms as a whole. (Ch. 9)

metamorphic rock Igneous or sedimentary rock that is buried and transformed by the Earth's intense internal temperature and pressure. (Ch. 16)

meteor A piece of interplanetary debris that hits the Earth's atmosphere and forms a bright streak of light from friction with atmospheric particles: a "shooting star." (Ch. 14)

meteor showers A set of spectacular, regularly occurring events in the night sky, caused by the collision of the Earth with clouds of small debris that travel around the orbits of comets. (Ch. 14)

meteorite The fragment of a meteor that hits the Earth. (Ch. 14)

microchip A complex array of *p*- and *n*-type semiconductors, which may incorporate hundreds or thousands of transistors in one integrated circuit. (Ch. 10)

microwave Electromagnetic waves, with wavelengths ranging from approximately 1 meter to 1 millimeter, which are used extensively for line-of-sight communications and cooking (Ch. 6)

Milky Way A collection of about 100 billion stars that forms the galaxy of which the Sun is a part. (Ch. 18)

Miller-Urey experiment A demonstration of chemical evolution, performed in 1953 by Stanley Miller and Harold Urey, which showed that a combination of gases, believed to be present in the early atmosphere, and a series of electric sparks, simulating the lightning on the early Earth, will produce amino acids, a basic building block of life. (Ch. 23)

mineral In a nutritional context, all chemical elements in food other than carbon, hydrogen, nitrogen, and oxygen. (Ch. 20)

mitochondria Sausage-shaped organelles that are places where molecules derived from glucose react with oxygen to produce the cell's energy. (Ch. 21)

mitosis The process of cell division producing daughter cells with exactly the same number of chromosomes as in the mother cell. (Ch. 21)

moderator In a nuclear reactor, the fluid whose function is to slow down neutrons that leave the fuel rods. (Ch. 11)

mole The SI unit for the amount of a substance; the mass in grams equal to the atomic weight of an element or compound; a quantity of chemical substance that contains 6×10^{23} atoms or molecules. (Ch. 9)

molecular genetics The study of how the mechanism that passes genetic information from parents to offspring functions on the basis of molecular chemistry. (Ch. 22)

molecule A cluster of atoms that bound together; the basic constituent of many different kinds of material. (Ch. 7)

monera Single-celled organisms without cell nuclei; the most primitive living things in the Linnaean classification of kingdoms. (Ch. 19)

monosaccharide An individual sugar molecule. (Ch. 20)

monsoon Any wind system on a continental scale that seasonally reverses its direction because of seasonal variations in relative temperatures over land and sea. (Ch. 16)

Moon The Earth's only satellite, which may have formed when a planet-sized body hit the Earth early in its history. (Ch. 14)

mRNA See *messenger RNA*.

mudstone A sedimentary rock formed from sediments that are much finer-grained than sand. (Ch. 16)

muscles   The part of the skeleto-muscular system that produces movement; made of long cells that contract when stimulated. (Ch. 24)

mutation   A change in the genetic material of the parent that is inherited by the offspring. (Ch. 22)

N   See *newton*.

***n*-type semiconductor**   A type of conductor, formed from doping, that has a slight excess of mobile negatively-charged electrons. (Ch. 10)

National Academy of Sciences   A nationally recognized association of scientists, elected to membership by their peers to provide professional advice for the government on policy issues ranging from environmental risks and natural resource management, to education and funding for science research. (Ch. 1)

National Institutes of Health   A federal agency that provides funding for basic and applied research in medicine and biology. (Ch. 1)

National Science Foundation   A federal agency that funds American scientific research and education in all areas of science. (Ch. 1)

natural selection   The mechanism by which nature can introduce wide-ranging changes in living things over long periods of time by modifying the gene pool of a species. (Ch. 23)

Neanderthal Man   A type of human with a large brain who lived until 35,000 years ago in groups with a complex social structure; either a separate species of the genus *Homo* or a subspecies of *Homo sapiens*. (Ch. 23)

nebula   Dust and gas clouds, common throughout the Milky Way galaxy, rich in hydrogen and helium. (Ch. 14)

nebular hypothesis   A model that explains the formation of the solar system from a large cloud of gas and dust floating in space 4.5 billion years ago. This cloud collapsed upon itself under the influence of gravity and began to spin faster and faster, eventually forming the planets and the rest of the solar system along a flattened disk of matter surrounding a central star. (Ch. 14)

negative charge   An excess of electrons on an object. (Ch. 5)

nervous system   One of two control systems in the body that mediate responses to the environment. (Ch. 24)

neurotransmitter   A group of molecules, produced in nerve cells, that transfer a nerve signal from one nerve cell to another. (Ch. 5)

neutrino   A subatomic particle, emitted in the decay of the neutron, which has no electric charge, travels at the speed of light, and has no rest mass. (Ch. 11)

neutron   A type of subatomic particle, located in the nucleus of the atom, which carries no electrical charge but has approximately the same mass as the proton; one of two primary building blocks of the nucleus. (Ch. 7)

neutron star   A very dense, very small star, usually with a high rate of rotation and a strong magnetic field; the core remains of a supernova, held up by the degeneracy pressure of neutrons. (Ch. 17)

newton (N)   A unit of force defined as the force needed to accelerate a mass of 1 kg by 1 m/s^2, or in kilogram-meter-per-second-squared. (Ch. 2)

Newton's law of universal gravitation   Between any two objects in the universe there is an attractive force (gravity) that is proportional to the masses of the objects and inversely proportional to the square of the distance between them. In other words, the more massive two objects are, the greater the force between them will be, and the farther apart they are, the less the force will be. (Ch. 2)

Newton's laws of motion   Three basic principles, expressed as laws, that govern the motion of everything in the universe, from stars and planets to cannonballs and muscles. *The first law* states that a moving object will continue moving in a straight line at a constant speed, and a stationary object will remain at rest, unless acted on by an unbalanced force. *The second law* states that the acceleration produced on a body by a force is proportional to the magnitude of the force and inversely proportional to the mass of the object. *The third law* states that for every action there is an equal and opposite reaction. (Ch. 2)

noble gases   Elements listed in the far right-hand column of the periodic table of elements, including helium, argon, and neon, which are odorless, colorless, and slow to react. (Ch. 7)

nonrenewable energy   Sources of energy that, once used, are not quickly replaced; for example, petroleum and coal. (Ch. 4)

nonrenewable resources   Resources such as coal and petroleum, which are forming at a much slower rate than they are being consumed. (Ch. 3)

nuclear reactor   A device that controls fission reactions to produce energy when heavy radioactive nuclei split apart. (Ch. 11)

nucleic acid   Molecule originally found in the nucleus of cells that carries and interprets the genetic code; includes DNA and RNA. (Ch. 20)

nucleotide   Molecule that is the basic element from which all DNA and RNA are built; formed from a sugar, a phosphate group, and one of four bases (adenine, guanine, cytosine, and thymine or uracil). (Ch. 20)

nucleus   (1) The very small, compact object at the center of an atom; made up primarily of protons and neutrons. (2) A prominent structure in the interior of a cell that contains the cell's genetic material—the DNA—and controls the cell's chemistry. (Ch. 7)

observation The act of observing nature without manipulating it. (Ch. 1)

ohm A unit of measurement for the electrical resistance of a wire. (Ch. 5)

oil shale A form of fossil fuel in which petroleum is dispersed through solid rock. (Ch. 3)

Oort cloud A region beyond the orbit of Pluto that contains billions of comets circling the Sun; the reservoir for new comets. (Ch. 14)

open ecosystem An ecosystem in which materials are free to move in and out. (Ch. 25)

open system A type of system within which an object can exchange matter and energy with its surroundings. (Ch. 3)

open universe A model of the future of the universe in which the expansion will continue forever because the universe lacks the matter to exert a gravitational force to slow receding galaxies. (Ch. 18)

order The fourth broadest classification in the Linnaean classification system; humans are in the order of primates. (Ch. 19)

organelle Any specialized structure in the cell, including the nucleus. (Ch. 21)

organic chemistry The branch of science devoted to the study of carbon-based molecules and their reactions. (Ch. 9)

osmosis A special case of molecular movement in which materials such as water are transferred across a membrane while at the same time molecules dissolved in the water are blocked. (Ch. 21)

outgassing Release of gases from nongaseous materials; extrusion of gases from the body of a planet after its formation. (Ch. 14)

oxidation A chemical reaction in which an atom such as oxygen accepts electrons while combining with other elements; for example, rusting of iron metal into iron oxide, or animal respiration. (Ch. 9)

oxides Chemical compounds that contain oxygen, such as most common minerals and ceramics. (Ch. 10)

ozone A molecule made up of three oxygen atoms, instead of the usual two, which absorbs ultraviolet radiation. (Ch. 25)

ozone hole A volume of atmosphere above Antarctica during September through November in which the concentration of the trace gas ozone has declined significantly. (Ch. 25)

ozone layer A region of enhanced ozone (O_3) 20 to 30 miles above the Earth's surface where most of the absorption of the Sun's ultraviolet radiation occurs. (Ch. 25)

***p*-type semiconductor** A type of conductor, formed from doping, that has a slight deficiency of electrons, resulting in mobile positively charged holes. (Ch. 10)

paleomagnetism The field devoted to the study of remnant magnetism in ancient rock, recording the direction of the magnetic poles at some time in the past. (Ch. 15)

pancreas Part of the digestive system that secretes specialized substances that break down starches, carbohydrates, proteins, and fats. (Ch. 24)

parsec A unit of measurement equal to about 3.3 light-years, which roughly corresponds to the average distance between nearest neighbor stars in our galaxy. (Ch. 17)

particle accelerator A machine such as a synchrotron or linear accelerator that produces particles at near light speeds for use in the study of the fundamental structure of matter. (Ch. 12)

Pauli exclusion principle A statement that says no two electrons can occupy the same state at the same time. (Ch. 7)

peer review A system by which the editor of a scientific journal submits manuscripts considered for publication to a panel of knowledgeable scientists who, in confidence, evaluate the manuscript for mistakes, misstatements, or shoddy procedures. Following the review, if the manuscript is to be published, it is returned to the author with a list of modifications and corrections to be completed. (Ch. 1)

peptide bond A connection between two atoms that remains after hydrogen (H) at one end of an amino acid and the hydroxyl (OH) from the end of another amino acid combine, releasing a water molecule (H_2O). The process is identical to the condensation polymerization reaction. (Ch. 20)

periodic table of the elements An organizational system, first developed by Dmitri Mendeleev in 1869, now listing 109 elements by atomic weight (in rows) and chemical properties (in columns). The pattern of elements in the periodic table reflects the arrangement of electrons in their orbits. (Ch. 7)

petroleum Thick black liquid found deep underground, derived from many kinds of transformed molecules of former life forms. (Ch. 9, 20)

phosphate group One phosphorous atom surrounded by four oxygen atoms. (Ch. 20)

phospholipid The class of molecules that form membranes in cells. Lipids have a long, thin structure with a carbon backbone, and a phosphate group at one end of the molecule. One end of these molecules is hydrophilic, one hydrophobic. (Ch. 20)

photoelectric effect A phenomenon that occurs when photons strike one side of a material and cause electrons of that material to be emitted from the opposite side. This effect is observed in modern cameras, CAT scans, and fiber optics. (Ch. 8)

photon A particle-like unit of light, emitted or absorbed by an atom when an electrically charged electron changes state. The form of a single packet of electromagnetic radiation. (Ch. 7)

photosphere The gaseous layers of the Sun's outer part, which emit most of the light we see. (Ch. 17)

photosynthesis The mechanism by which plants convert the energy of sunlight into energy stored in carbohydrates, the chemical energy for virtually all life on Earth: energy + CO_2 + H_2O → carbohydrate + oxygen. (Ch. 21)

phylogeny A classification scheme based on the idea that descent and ancestry, rather than present structure, are what should be used to classify living things. Presents family trees along with estimates of when specific splitting took place. Evolutionary history of a species. (Ch. 19)

phylum The second broadest classification in the Linnaean classification system; humans are in the phylum chordata, subphylum vertebrate. (Ch. 19)

pituitary gland An organ in the lowest part of the forebrain that controls the body's system of hormones. (Ch. 24)

Planck's constant (h) A constant named after German physicist Max Planck that is the central constant of quantum physics, equal to 6.63×10^{-34} joule-seconds in SI units. (Ch. 8)

planetesimal Small objects, which range in size from boulders to several miles across, formed from the accretion of solid material during the formation of the planets. (Ch. 14)

plants Multicelled organisms that get their energy directly from the Sun through photosynthesis. One of five kingdoms in the Linnaean classification. (Ch. 19)

plasma A state of matter existing under extremely high temperatures in which electrons are stripped from their atoms during high-energy collisions, forming an electron sea surrounding positive nuclei. (Ch. 9)

plastic A solid composed of intertwined chains of molecules, with an ability to be molded or formed into virtually any desired shape: clear film, dense casting, strong fiber, colorful molding. (Ch. 9)

plate A rigid moving sheet of rock up to 100 km (60 mi) thick, composed of the crust and part of the upper mantle. (See *plate tectonics*.) (Ch. 15)

plate tectonics The model of the dynamic Earth that has emerged from studies of paleomagnetism, rock dating, and much other data. A theory that explains how a few thin, rigid tectonic plates of crustal and upper mantle materials are moved across the Earth's surface by mantle convection. (Ch. 15)

Pluto A rocky body that is located beyond the Jovian planets and is the smallest of all planets in the solar system. (Ch. 14)

polar molecule Atom clusters with a positive and negative end; exerts electrical force on neighboring atoms. Water is a polar molecule. (Ch. 9)

polarization The subtle electron shift from negative to positive that takes place when the electrons of an atom or a molecule are brought near a polar molecule such as water, resulting in a bond caused by the electrical attraction between the negative end of the polar molecule and the positive side of the other molecule. (Ch. 9)

poles The two opposite ends of a magnet, named north and south, which repel a like magnetic pole and attract an unlike magnetic pole. (Ch. 5)

polymer Extremely long and large molecules that are formed from numerous smaller molecules, like links in a chain, with predictable repeating sequences of atoms along the chain. (Ch. 9)

polymerization A reaction that includes all chemical reactions that form long strands of polymer fibers by linking small molecules. (Ch. 9)

polypeptide A bonded chain of amino acids. (See *peptide bond*.) (Ch. 20)

polysaccharide A molecule that is the result of many sugar molecules strung together in a chain; for example, starch and cellulose. (Ch. 20)

polyunsaturated A type of lipid that forms when two or more kinked "double bonds" between carbon atoms are in the molecule. (See *unsaturated*.) (Ch. 20)

positive charge A deficiency of electrons on an object. (Ch. 5)

positron The positively charged antiparticle of the electron. (Ch. 12)

potential energy The energy a system possesses if it is capable of doing work, but is not doing work now. Types of potential energy include magnetic, elastic, electrical, and chemical. (Ch. 3)

potential energy Any type of energy waiting to be released; stored energy. (Ch. 3)

power stroke The downward motion of a piston in a gasoline engine, in which the actual work is done, and the energy released by combustion is translated into the motion of the car. (Ch. 4)

power The rate at which work is done or energy is expended. The amount of work done, divided by the time it takes to do it. Power is measured in watts in the metric system, horsepower in the English. (Ch. 3)

precession The circular motion of the spinning axis of the Earth in space, which causes the tilt of the Northern Hemisphere to change on a 26,000-year cycle. (Ch. 16)

precipitation A chemical reaction that is the reverse of a solution reaction, producing a solid that separates from very concentrated solutions. (Ch. 9)

prediction A guess about how a particular system will behave, followed by observations to see if the system did behave as expected within a specified range of situations. (Ch. 1)

prevailing westerly   Wind in the temperate zones that blows primarily from west to east, causing weather patterns to move in the same direction. (Ch. 16)

primary structure   The simplest of the four stages of the organization of amino acids in a protein molecule. The exact order of amino acids along the protein string. (Ch. 20)

primates   An order of mammals that have grasping fingers and toes, eyes at the front of their heads, large brains, and fingernails instead of claws; includes monkeys, apes, and humans. (Ch. 19)

primordial soup   A rich broth of amino acids and other molecules thought to have been produced in the early oceans over a period of several hundred million years, re-created by the Miller-Urey process. (Ch. 23)

probability   The likelihood that an event will occur or that an object will be in one state or another; how nature is described in the subatomic world. (Ch. 8)

prokaryote   A type of primitive cell in which the DNA is coiled together, but not separated in the nucleus. Prokaryotes constitute the kingdom monera, including all cells that do not have a nucleus. (Ch. 21)

protein   An extremely complex molecule, which can consist of thousands of amino acids and millions of atoms formed in a chain structure. Proteins function as enzymes and direct the cell's chemistry. (Ch. 20)

protista   Single-celled organisms with nuclei, and a few multicelled organisms that have a particularly simple structure. One of five kingdoms in the Linnaean classification. (Ch. 19)

proton   One of two primary building blocks of the nucleus; with a positive electrical charge of +1 and a mass ($1.6726430 \times 10^{-24}$ g) approximately equal to that of the neutron. (Ch. 11)

pseudo-science   A kind of inquiry falling in the realm of belief or dogma, which includes subjects that cannot be proved or disproved with a reproducible test. The subjects include creationism, extrasensory perception (ESP), unidentified flying objects, astrology, crystal power, and reincarnation. (Ch. 1)

publication   A peer-reviewed paper written by a scientist or a group of scientists to communicate the results of their research to a larger audience. A publication will include the technical details of the methodology, so that the research can be reproduced, and a concise statement of the results and conclusions. (Ch. 1)

pulsar   A neutron star in which fast-moving particles speed out along the intense magnetic field lines of the rotating star, giving off electromagnetic radiation that we detect as a series of pulses of radio waves. (Ch. 17)

pumice   Frothy volcanic rock rich in silicon from magmas that mix with a significant amount of water or other volatile substance. (Ch. 16)

pumping   The process in a laser that adds energy to the system from the outside to return atoms continuously to their excited states so that coherent photons can be produced. (Ch. 7)

punctuated equilibrium   A hypothesis which holds that evolutionary changes usually occur in short bursts separated by long periods of stability. (Ch. 23)

pyruvic acid   Three-carbon molecules in a cell that were formed by glycolysis, splitting a glucose molecule into two smaller molecules. (Ch. 21)

quantized   Whenever energy or another property of a system can have only certain definite values, and nothing in between those values, it is said to be quantized. (Ch. 8)

quantum jump   See *quantum leap.*

quantum leap   A process by which an electron changes orbit without ever traversing any of the positions between the original and the final orbits; also known as quantum jump. (Ch. 7)

quantum mechanics   The branch of science that is devoted to the study of the motion of objects that come in small bundles, or quanta, which applies to the subatomic world. (Ch. 8)

quarks   *(pronounced "quorks")* The truly fundamental building blocks of the hadrons. Particles that have fractional electrical charge and cannot exist alone in nature. (Ch. 12)

quartzite   A durable rock in which the original sand grains of sandstone, under high temperature and pressure, recrystallize and fuse into a solid mass. (Ch. 16)

quasar   Quasi-stellar radio source. Objects in the universe, where as-yet unknown processes pour vast amounts of energy into space each second from an active center no larger than the solar system; the most distant objects known. (Ch. 18)

quaternary structure   The joining of separate protein chains, each with its own secondary and tertiary structures. (Ch. 20)

R&D   See *research and development.*

radiation   The transfer of heat by electromagnetic radiation. The only one of the three mechanisms of heat transfer that does not require atoms or molecules to facilitate the transfer process. (Ch. 4) Also, the particles emitted during the spontaneous decay of nuclei. (Ch. 11)

radio wave   Part of the electromagnetic spectrum that ranges from the longest waves—wavelengths longer than the Earth's diameter—to waves a few meters long. (Ch. 6)

radioactive decay The process of spontaneous change of unstable isotopes. (Ch. 11)

radioactivity The spontaneous release of energy by certain atoms, such as uranium, as these atoms disintegrate. The emission of one or more kinds of radiation from an isotope with unstable nuclei. (Ch. 3)

radiometric dating A technique based on the radioactive half-lives of carbon-14 and other isotopes that is used to determine the age of materials. (Ch. 11)

radon A colorless, odorless inert gas that can cause an indoor pollution problem when it undergoes radioactive decay. (Ch. 11)

receptor A large structure found in the cell membrane and made of proteins folded into a geometrical shape that will bond chemically only to a specific type of molecule. (Ch. 21)

recessive A gene that is present in offspring and can be passed along to subsequent generations, but does not determine the offspring's physical characteristics. (Ch. 22)

recessive trait A characteristic that will appear only if no dominant gene is present. (Ch. 22)

red giant An extremely large star that emits a lot of energy but whose surface is very cool and therefore appears somewhat reddish in the sky; found in the upper right-hand corner of the H-R diagram. (Ch. 17)

redshift An increase in the wavelength of radiation received from a receding celestial body as a consequence of the Doppler effect. A shift toward the long-wavelength (red) end of the spectrum. (Ch. 17)

redshifted The result of the Doppler effect on light waves, when the source of light moves away from the observer: light-wave crests are farther apart and have a lower frequency. (Ch. 6)

reduction A chemical reaction in which electrons are transferred from an atom to other elements, resulting in a gain in electrons for the material being reduced; for example, smelting of metal ores, and photosynthesis. (Ch. 9)

reductionism The quest for the ultimate building blocks of the universe. An attempt to reduce the seeming complexity of nature by first looking for an underlying simplicity and then trying to understand how that simplicity gives rise to the observed complexity. (Ch. 12)

reflection A process by which light waves are scattered at the same angle as the original wave; for example, from the surface of a mirror. (Ch. 6)

refraction One of three responses of an electromagnetic wave encountering matter, in which electromagnetic waves slow down and change direction in the matter. (See also *absorption* and *transmission*. (Ch. 6)

relativity An idea that the laws of nature are the same in all frames of reference, and that every observer must experience the same natural laws. (Ch. 13)

reproducible A criterion for the results of an experiment. In the scientific method, observations and experiments must be reported in such a way that anyone with the proper equipment can verify the results. (Ch. 1)

reptiles The first animals fully adapted to life on land; includes lizards, turtles, and snakes. (Ch. 24)

research and development (R & D) A kind of research aimed at specific problems, usually performed in government and industry laboratories. (Ch. 1)

residence time The average length of time that any given atom will stay in ocean water before it is removed by some chemical reaction. (Ch. 16)

respiration The process by which animals retrieve energy stored in glucose, in a complex series of cellular chemical reactions, which include breathing in oxygen produced by plants, burning carbohydrates ingested for food, and breathing out carbon dioxide. (Ch. 21)

respiratory system The system that takes oxygen from the air and transfers it to the circulatory system; includes the lungs and the alveoli. (Ch. 24)

ribonucleic acid (RNA) A molecule that consists of one string of nucleotides put together around the sugar ribose, and with the bases adenine, guanine, cytosine, and uracil. RNA plays a crucial role in the synthesis of proteins in the cell. (Ch. 20)

ribosomal RNA (rRNA) A constituent of ribosomes; involved in the synthesis of protein. (Ch. 22)

ribosome One of the cellular organelles that is the site of protein synthesis. (Ch. 21)

RNA See *Ribonucleic acid.*

rock cycle An ongoing cycle of internal and external Earth processes by which rock is created, destroyed, and altered. (Ch. 16)

rod One type of light-absorbing cell in the eye providing night vision; sensitive to light and dark. Compare to *cone.* (Ch. 6)

Roentgen Satellite (ROSAT) An X-ray satellite launched in 1990 by the United States, United Kingdom, and Germany as the latest in a series of satellites equipped to detect X-rays. (Ch. 17)

ROSAT See *Roentgen Satellite.*

rRNA See *ribosomal RNA.*

sandstone A sedimentary rock formed mostly from sand-sized grains of quartz (silicon dioxide) and other hard mineral and rock fragments. (Ch. 16)

saturated A fully bonded carbon atom in a lipid. In a straight lipid chain, every carbon atom bonds to two adjacent carbons along the chain and two hydrogen atoms on the sides. (Ch. 20)

scattering A process by which electromagnetic waves may be absorbed and rapidly re-emitted; can be diffuse scattering or reflection. (Ch. 6)

schist Metamorphic rock formed from slate under extreme temperature and pressure. (Ch. 16)

scientific method A continuous process used to collect observations, form and test hypotheses, make predictions, and identify patterns in the physical world. (Ch. 1)

second trophic level All herbivores, including cows, rabbits, and many different kinds of insects, that get their energy by eating plants. (Ch. 25)

secondary structure Shapes taken by the string of amino acids that makes up the primary structure of a protein. (Ch. 20)

sedimentary rock A type of rock that is formed from layers of sediment produced by the weathering of other rock or by chemical precipitation. (Ch. 16)

seismic tomography A branch of earth science that is enabling geophysicists to obtain three-dimensional pictures of the Earth's interior. (Ch. 15)

seismic wave The form through which an earthquake's energy is transmitted, causing the Earth to rise and fall like the surface of the ocean. (Ch. 15)

seismology The study and measurement of vibrations within the Earth, dedicated to deducing our planet's inner structure. (Ch. 15)

semiconductor Materials that conduct electricity but do not conduct it very well. Neither a good conductor nor a perfect insulator; for example, silicon. (Ch. 10)

shale A sedimentary rock formed from sediments that are much finer-grained than sand. (Ch. 16)

shear strength A material's ability to withstand twisting. (Ch. 10)

shear wave One of two principal types of seismic waves, in which molecules move perpendicular to the direction of the wave motion. A transverse wave. (Ch. 15)

single bond The type of covalent bond formed when only one electron is shared. (Ch. 9)

skeleto-muscular system An internal structure that supports the weight and produces movement of the human body. (Ch. 24)

slate A brittle and hard metamorphic rock, formed from shale or mudstone. (Ch. 16)

small intestine The part of the digestive system that completes the process of breaking down food into small molecules able to pass through intestine walls into the bloodstream. (Ch. 24)

smog The brownish stuff of modern urban air pollution that you often see over major cities during the summer, caused by a photochemical reaction. (Ch. 25)

solar system The Sun, the planets and their moons, and all other objects gravitationally bound to the Sun. (Ch. 14)

solar wind A stream of charged particles—mainly ions of hydrogen and electrons—emitted constantly by the Sun into the space around it. (Ch. 17)

solid Any material that possesses a fixed shape and volume, with chemical bonds that are both sufficiently strong and directional to preserve a large-scale external form. (Ch. 9)

solution reaction A chemical reaction in which a solid such as salt or sugar is dissolved in a liquid. (Ch. 9)

somatic nervous system Nerves in the body that control ordinary volitional motion, such as muscle contraction. (Ch. 24)

sound wave A longitudinal wave created by a vibrating object and transmitted only through the motion of molecules in a solid, gas, or liquid. The energy of the sound wave is associated with the kinetic energy of those molecules. (Ch. 3)

special relativity The first of two parts of Einstein's theory of relativity that deals with reference frames that do not accelerate. (Ch. 13)

species The basic unit of the Linnaean classification; an interbreeding population of individual organisms. (Ch. 19)

spectroscopy The study of emission and absorption spectra of materials in order to discover the chemical makeup of that material; a standard tool used in almost every branch of science. (Ch. 7)

spectrum The characteristic signal from the total collection of photons emitted by a given atom that can be used to identify the chemical elements in a material; the atomic fingerprint. (Ch. 7)

speed of light (c) The velocity at which all electromagnetic waves travel, regardless of their wavelength or frequency; equal to 300,000 kilometers per second (about 186,000 miles per second). (Ch. 6)

speed The distance an object travels divided by the time that it takes to travel that distance. (Ch. 2)

spreading The widening of the seafloor, as magma comes from deep within the Earth and erupts through fissures on the seafloor. (Ch. 15)

standard model Theories, supported by experimental evidence, that predict the unification of the strong force with the electroweak force. (Ch. 12)

star Objects such as our Sun that form from giant clouds of interstellar dust and generate energy by fusion. (Ch. 17)

starch A polysaccharide, with glucose constituents linked together at certain points along the ring. A large family of molecules found in many plants, such as potatoes and corn. (Ch. 20)

states of matter   Different modes of organization of atoms or molecules, which result in properties of gases, plasmas, liquids, or solids. (Ch. 9)

static electricity   A phenomenon caused by the transfer of electrical charge between objects. Often observed as lightning or as sparks produced when walking across a wool rug on a cold day. (Ch. 5)

steady-state universe   A model, no longer believed to be valid, that describes a universe that is constantly expanding and forming new galaxies, but with no trace of a beginning. (Ch. 18)

stomach   The organ in the digestive system where food is broken down by strong acids and other chemicals into molecules that can be used by individual cells. (Ch. 24)

strength   The ability of a solid to resist changes in shape; directly related to chemical bonding. (Ch. 10)

strong force   The force responsible for holding the nucleus together; one of the four fundamental forces in nature. This force operates over extremely short distances and between quarks to hold elementary particles together. (Ch. 11)

subduction zone   The regions of the Earth where plates converge and old crust returns to the mantle. (Ch. 15)

sublimation   The direct transformation of a solid to a gaseous state, without passing through the liquid state. (Ch. 9)

sugar   The simplest of the carbohydrates. Common sugars contain five, six, or seven carbon atoms arranged in a ringlike structure. (Ch. 20)

superclusters   Large collections of clusters and groups of thousands of galaxies. (Ch. 18)

superconductivity   The ability of some materials to exhibit the complete absence of any electrical resistance, usually when cooled to within a few degrees of absolute zero. (Ch. 10)

supernova   A stupendous explosion of a star, which increases its brightness hundreds of millions of times in a few days; results from the implosion of the core of a massive star at the end of its life. (Ch. 17)

synchrotron   A particle accelerator in which magnetic fields are increased as particles become more energetic, keeping them moving on the same track. (Ch. 12)

system   A part of the universe under study and separated from its surroundings by a real or imaginary boundary. (Ch. 3)

taxonomy   The science of cataloging living things, describing them, and giving them names. (Ch. 19)

technology   The application of the results of science to specific commercial or industrial goals. (Ch. 1)

tectonic plate   One of a dozen large, and some smaller, sheets of moving rock forming the surface of the Earth. (Ch. 15)

telescope   A device that focuses and concentrates radiation from distant objects; used by astronomers to collect and analyze radio waves, microwaves, light, and other radiation. (Ch. 17)

temperature   A quantity that reflects how vigorously atoms are moving and colliding in a material. (Ch. 4)

tendon   The part of the skeleto-muscular system that attaches the muscles to the bones. (Ch. 24)

tensile strength   A material's ability to withstand pulling apart. (Ch. 10)

terminal electron transport   The final stage of respiration in which the energy is used to produce more ATP molecules. (Ch. 21)

terrane   A mass of rock as much as several hundred kilometers across, found in most of the western part of the United States, which was once a large island in the Pacific Ocean and carried toward the North American continent by plate activity. (Ch. 15)

terrestrial planets   Those relatively small, rocky, and high-density planets located in the inner solar system nearest the Sun: Mercury, Venus, Earth, the Earth's moon, and Mars. (Ch. 14)

tertiary structure   The complex folding of a protein, caused by the cross-linking of chemical bonds from side groups in the amino acid chain. (Ch. 20)

the second law of thermodynamics   Any one of three equivalent statements: (1) heat will not flow spontaneously from a colder to a hotter body; (2) it is impossible to construct a machine that does nothing but convert heat into useful work; and (3) the entropy of an isolated system always increases. (Ch. 4)

theory   A description of the world that covers a relatively large number of phenomena and has met many observational and experimental tests. A conclusion based upon observations of nature. (Ch. 1)

thermal conductivity   The ability of a material to transfer heat energy from one molecule to the next by conduction. When thermal conductivity is low, as in wood or fiberglass insulation, the transfer of heat is slowed down. (Ch. 4)

thermodynamics   The study of the movement of heat; the science of heat, energy, and work. (Ch. 3)

time dilation   A phenomenon in special relativity in which moving clocks appear to tick more slowly than stationary ones. (Ch. 13)

tornado   The most violent weather phenomenon known. A rotating air funnel some tens to hundreds of meters across, descending from storm clouds to the ground, causing intense damage along the path where the funnel touches the ground. (Ch. 16)

trace element   A chemical, such as iodine in the thyroid gland and iron in the blood, which is needed in minor amounts by the body. (Ch. 20)

trace gas   A gas that constitutes less than one molecule in a million in the Earth's atmosphere; for example, ozone. (Ch. 25)

transcription   A process by which a cell transfers information in DNA to molecules of mRNA. (Ch. 22)

transfer RNA (tRNA)   The molecule with special configuration that attracts amino acids at one end, and at the other end, attaches to a specific codon of mRNA. (Ch. 22)

transform plate boundary   The type of boundary between plates that occurs when one plate scrapes past the other, with no new plate material being produced; for example, California's San Andreas Fault. (Ch. 15)

transistor   A device that sandwiches p- and n-type semiconductors in an arrangement that can amplify or redirect an electrical current running through it; a device that played an essential role in the development of modern electronics. (Ch. 10)

transmission   One of three responses of an electromagnetic wave encountering matter, in which light energy passes through the matter unaffected. See also *absorption* and *refraction*. (Ch. 6)

transverse wave   A kind of wave in which the motion of the wave is perpendicular to the motion of the medium on which the wave moves. (Ch. 6)

triangulation   A geometrical method used to measure the distances to the nearest stars up to a few hundred light-years away. The angle of sight to the star is measured at opposite ends of the Earth's orbit and the distances are worked out. (Ch. 17)

tRNA   See *transfer RNA*.

trophic level   All organisms that get their energy from the same source. (Ch. 25)

tropical storm   A severe storm that starts as a low-pressure area over warm ocean water and, while drawing energy from the warm water, grows and rotates in great cyclonic patterns hundreds of kilometers in diameter. (Ch. 16)

tsunami   A great wave, which can devastate low-lying coastal areas, occurring when the energy of an earthquake under or near a large body of water is transferred through the water. (Ch. 15)

typhoon   A tropical storm that begins in the North Pacific Ocean. (Ch. 16)

ultraviolet   radiation   High-frequency wavelengths, shorter than visible light, ranging from 400 nanometers to 100 nanometers. (Ch. 6)

uncertainty principle   The idea quantified by Werner Heisenberg in 1927 that at a quantum scale the location and velocity of an object can never be known at the same time, because quantum-scale measurement affects the object being measured. Specifically, "the error or uncertainty in the measurement of an object's position, times the error or uncertainty in that object's velocity, must be greater than a constant, h, divided by the object's mass. (Ch. 8)

unified field theory   The general name for any theory in which fundamental forces are seen as different aspects of the same force. (Ch. 12)

uniform motion   The motion of an object if it travels in a straight line at a constant speed. All other motions involve acceleration. (Ch. 2)

unsaturated   In a lipid chain, a carbon atom bonded with two carbons and one hydrogen, and with one kinked "double bond" with one of the carbons. (See *saturated*.) (Ch. 20)

vacuole   In plants, a cellular organelle that is responsible for waste storage. (Ch. 21)

valence electron   An outer electron of an atom that can be exchanged or shared during chemical bonding. (Ch. 9)

valence   The combining power of a given atom, determined by the number of electrons in an atom's outermost orbit. (Ch. 9)

van der Waals force   A net attractive bond and weak force resulting from the polarization of electrically neutral atoms or molecules. (Ch. 9)

vascular plant   The phylum of plants that have internal "plumbing" capable of carrying fluids from one part of the plant to another. (Ch. 24)

velocity   The distance an object travels divided by the time it takes to travel that distance, including the direction of travel. The velocity of a falling object is proportional to the length of time that it has been falling. (Ch. 2)

vertebrate   A subphylum of chordates in which the nerves along the back are encased in bone. (Ch. 19)

vesicle   The vehicle by which a particle moves around inside a cell; a tiny container formed from the cell membrane, and a particle with its receptor. (Ch. 21)

vestigial organ   An internal bodily feature that serves no useful function at present and is compelling evidence for evolution. (Ch. 23)

virus   A short length of RNA or DNA wrapped in a protein coating that fits cell receptors and replicates itself using the cell's machinery. (Ch. 22)

visible light   Electromagnetic waves with a wavelength that can be interpreted by nerve receptors in the brain; wavelengths range from 700 nanometers for red light to 400 nanometers for violet light. (Ch. 6)

vitamin   One of a host of complex organic molecules that, in small quantities, play an essential role in good health; for example, by mediating the body's chemical reactions. May be fat soluble and stored, or water soluble and not retained by the body. (Ch. 20)

volcanic rock   Extrusive igneous rock that solidifies on the surface of the Earth. (Ch. 16)

volcano   Places where subsurface molten rock breaks through to the surface of the Earth to form dramatic short-term changes in the landscape. (Ch. 15)

voltage   The pressure produced by the energy source in a circuit, measured in volts. (Ch. 5)

warm blooded   Animals, such as birds and mammals, that have a four-chambered heart and can maintain a constant body temperature in any environment. (Ch. 24)

watt   A unit of measurement that is the expenditure of 1 joule of energy in 1 second. (Ch. 3)

wave   A traveling disturbance that carries energy from one place to another without requiring matter to travel across the intervening distance. (Ch. 6)

wave mechanics   Another term for quantum mechanics, indicating the dual (wave and particle) nature of quantum objects. (Ch. 8)

wavelength   The distance between crests, the highest points of adjacent waves. (Ch. 6)

weather   Daily changes in rainfall, temperature, amount of sunshine, and other variables resulting partly from the general circulation in the atmosphere, and partly from local disturbances and variations. (Ch. 16)

weight   The force of gravity on an object. (Ch. 2)

white dwarf   A star that has a very low emission of energy but very high surface temperature; plots on the lower left-hand corner of the H-R diagram. (Ch. 17)

wind shear   Violent air turbulence created from sudden downdrafts, which can cause an extremely dangerous condition near airports. (Ch. 16)

work   The force that is exerted times the distance over which it is exerted; measured in joules in the metric system, in foot-pounds in the English. (Ch. 3)

X-rays   High-frequency and high-energy electromagnetic waves that range in wavelength from 100 nanometers to 0.1 nanometer, used in medicine and industry. (Ch. 6)

Credits

tom): Courtesy College Physicians of America. Figure 11–3: Philippe Plailly/Science Photo Library/Photo Researchers. Page 283: Ed Braveman/FPG International. Page 284 (*top*): Patrick Mesner/Gamma Liaison. Page 284 (*bottom*): Steve Krasemann/Photo Researchers. Page 288: Philip Bailey/The Stock Market. Page 289: Courtesy Princeton University Physics Laboratory. Page 290: Thornton/Archive Photos.

Chapter 12

Opener: Terry Vine/Tony Stone Images. Page 302 (*top*): Courtesy of Lawrence Berkeley Laboratory. Page 302 (*bottom*): Courtesy Fermi National Accelerator Laboratory. Page 303 (*left*): Courtesy Fermi National Accelerator Laboratory. Page 303 (*right*): Bill W. Marsh/Photo Researchers. Page 305 (*left*): Courtesy Carl Anderson, California Institute of Technology. Page 305 (*right*): Courtesy Lawrence Berkeley Laboratory, University of California. Figure 12–1*a*): Courtesy Brookhaven National Laboratory. Figure 12–1*b*) & *c*): Dan McCoy/Rainbow. Page 311: Courtesy National Archives.

Chapter 13

Opener: Eric Hayman/Tony Stone Images. Page 323: UPI/Bettmann. Pages 330 and 331: © 1989 Hsiung. Page 335: Imtek Imagineering/Masterfile

Chapter 14

Opener: Courtesy NASA. Page 342: Trans. #1307, Photo by Helmut Wimmer, Dept. of Library Services, American Museum of Natural History. Page 344: © Finley Holiday Films. Page 345: Lowell Observatory Photograph. Page 346: Bob Martin/Tony Stone Images. Pages 347 and 348: Courtesy NASA. Page 349: Jim Riffle/Astroworks, Corp. Page 352: Van Valte. Page 354: By Chesley Bonestell/Space Art International/© Life Picture Service. Page 358: JPL/NASA. Page 359: Dennis DiCicco/Sky and Telescope Magazine. Page 360: Breck Kent/Earth Scenes.

Chapter 15

Opener: Sagara/AllStock, Inc. Page 367: David A. Rosenberg/AllStock, Inc. Page 368 (*left*): J. A. Kraulis/Masterfile. Page 368 (*right*): Carr Clifton/AllStock, Inc. Page 370 (*left*): Ralph Perry/AllStock, Inc. Page 370 (*center*): Richard A. Cooke, III/AllStock, Inc. Page 370 (*right*): Gamma Liaison. Figure 15–3: Marie Tharp. Page 381: Courtesy NASA. Page 382: Dewitt Jones Prod./AllStock, Inc. Page 387: Andrew Rafkind/Tony Stone Images. Figure 15–14: Photo by Paul Morin/data by Zhang & Tanimoto.

Chapter 16

Opener: Daryl Benson/Masterfile. Page 397: Courtesy NASA. Figure 16–4: NOAA/Photo Researchers. Page 411 (*left*): Michael J. Howell/ProFiles West, Inc. Page 411 (*right*): Robert Frerck/Odyssey Productions. Figure 16–8: Tom Bean/DRK Photo. Page 413: David Ball/AllStock, Inc. Page 414: David Muench/AllStock, Inc. Page 415:

Bill Ross/H. Armstrong Roberts. Page 417: Dr. K. Roy Gill.

Chapter 17

Opener: David Malin/Anglo Australian Telescope Board. Figure 17–2: National Center for Atmospheric Research. Page 428 (*bottom*): Johnny Johnson/Earth Scenes. Page 431 (*left*): Richard Winscoat/Douglas Peebles Photography. Page 431 (*right*): Courtesy National Radio Astronomy Observatory. Pages 432, 441, & 442: Courtesy NASA.

Chapter 18

Opener: Roy Morsch/The Stock Market. Page 449 (*top right*): Courtesy AIP Emilio Segre Visual Archives/Hale Observatories. Figure 18–1: Dennis DiCicco/Sky and Telescope Magazine. Figure 18–2: David Malin/Anglo-Australian Telescope Board. Figure 18–3: Courtesy of Palomar Observatory, California Institute of Technology. Figure 18–6: Courtesy of C. Park and J. R. Gott, Princeton University. Page 460: Courtesy of NASA/Goddard Space Flight Center.

Chapter 19

Opener: Charles Krebs/AllStock,Inc. Page 474 (*top*): John Visser/Bruce Coleman, Inc. Page 474 (*bottom*): Frank Rossotto/The Stock Market. Page 478 (*left*): David M. Phillips/Visuals Unlimited. Page 478 (*center*): David M. Phillips/Visuals Unlimited. Page 478 (*right*): David M. Phillips/Visuals Unlimited. Page 479 (*left*): Manfred Kage/Peter Arnold, Inc. Page 479 (*right*): Eric Grave/Science Source/Photo Researchers.

Chapter 20

Opener: Rick Altman/The Stock Market. Figure 20–12: Joel Gordon. Figure 20–16*a*): Courtesy USDA. Figure 20–16*b*): Robert Capece.

Chapter 21

Opener: Charles Thatcher/Tony Stone Images. Page 517 (*top*): Bob Thomason/Tony Stone Images. Figure 21–2: David Scharf/Peter Arnold, Inc. Figure 21.5*b*): Courtesy Cell Research Institute, University of Texas, Austin. Page 530: CNRL/Science Photo Library/Photo Researchers.

Chapter 22

Opener: Stephanie Maze/Woodfin Camp & Associates. Page 541: Frank P. Rossotti/The Stock Market. Page 553 (*top*): Art Wolfe/AllStock, Inc. Page 553 (*bottom left*): Jim Balog/Black Star. Page 553 (*bottom right*): Courtesy R. L. Brinster, Laboratory for Reproductive Physiology, University of Pennsylvania.

Chapter 23

Opener: David Barnes/The Stock Market. Figure 23–2: BPS/Terraphotographics. Page 569: Gerard Lacz/Natural History Photographic Agency. Figure 23–3: Courtesy

Professor Lawrence Cook, University of Manchester. Page 572: from Stephen Jay Gould, *Wonderful Life,* NY, 1989, W. W. Norton & Company. Page 574 (*top*): Courtesy Professor Seilacher. Page 574 (*center*): William E. Ferguson. Page 574 (*bottom*): Courtesy Professor George Poinar, University of California, Berkeley. Page 580: Courtesy Institute of Human Origins. Page 583: Photofest.

Chapter 24

Opener: BIOS/Visage/Peter Arnold, Inc. Figure 24–1: Carr Clifton. Page 597 (*left*): John Cancalosi/Tom Stack & Associates. Page 597 (*center*): John Shaw/Tom Stack & Associates. Page 597 (*right*): Tom McHugh/ Photo Researchers. Figure 24–12: Courtesy Lennart Nilsson, From *A Child Is Born.*

Chapter 25

Opener: Walter Iooss Jr./The Image Bank. Page 614: Robert Frerck/The Stock Market. Figure 25–3: Courtesy NASA. Page 625: Ray Pfortner/Peter Arnold, Inc. Page 626: Jim Mendenhall. Page 627 (*bottom*): Courtesy Environmental Elements Corporation. Figure 25–4: Courtesy Westfälisches Amt für Denkmalpflege.

LINE ART

Figure 1–4
Courtesy New York Public Library.

Figure 2–2
Data from monograph of Dr. John Snow, 1854, as published in M. Goldstinge and I. Goldstinge, *How We Know* (New York: Plenum Press, 1978).

Figure 3–5
Adapted from Fulkerson, Judkins, and Sanghvi, "Energy from Fossil Fuels," *Scientific American,* September 1990, p. 131.

Figure 15–4
Courtesy P. J. Wyllie, *The Dynamic Earth: Textbook in Geosciences,* 1st edition (New York: John Wiley & Sons, 1971), Figure 13–2, p. 310.

Figure 19–3
Adapted from E. O. Wilson, *The Diversity of Life* (New York: W. W. Norton & Co., 1993).

Figure 19–4
Adapted from E. O. Wilson, *The Diversity of Life* (New York: W. W. Norton & Co., 1993).

Figure 23–1
S. L. Miller and L. E. Orgel, *The Origins of Life on Earth* (New Jersey: Prentice-Hall, 1974), p. 84.

Figure 24–2
Adapted from Curtis & Barnes, *Biology,* 5th edition (New York: Worth, 1989), p. 500.

Figure 25–6
C. D. Keeling et al. A three-dimensional model of atmospheric CO_2 transport based on observed winds: 1. Analysis of observational data (AGU Monograph 55, 1989), Figure 16.

Figure 25–7
Adapted from *Energy and the Industrial Society,* p. 82.

Index

A

I

J

K

Q